Philosophy and the Neurosciences

To Roger Gibson

A true friend, whose vision and leadership guided
the PNP Program into being

Philosophy and the Neurosciences

A Reader

Edited by

William Bechtel, Pete Mandik,
Jennifer Mundale, and
Robert S. Stufflebeam

Copyright © Blackwell Publishers Ltd 2001
Editorial arrangement and
introductions copyright © William Bechtel, Pete Mandik, Jennifer Mundale,
and Robert S. Stufflebeam 2001

First published 2001

2 4 6 8 10 9 7 5 3 1

Blackwell Publishers Inc.
350 Main Street
Malden, Massachusetts 02148
USA

Blackwell Publishers Ltd
108 Cowley Road
Oxford OX4 1JF
UK

Library of Congress Cataloging-in-Publication Data has been applied for

ISBN   0-631-21044-X (hardback); ISBN   0-631-21045-8 (paperback)

British Library Cataloguing in Publication Data

A CIP catalogue record for this book is available from the British Library.

Typeset in 10 ½ on 12 ½ pt Ehrhardt
by Best-set Typesetter Ltd., Hong Kong

This book is printed on acid-free paper.

Contents

Preface and Acknowledgments

The interplay between philosophical and neuroscientific ideas has been present from the earliest inception of the disciplines comprising the neurosciences (neuroanatomy, neurophysiology, parts of molecular biology, biophysics, pharmacology, genetics, etc.). Indeed, the roots of their conceptual entanglement can be found in the prior engagement of philosophy with the medical, biological, and psychological sciences. Subtle philosophical questions and controversies lay just beneath the surface of much early neuroscientific research. Occasionally, these latent philosophical currents would erupt within the context of a larger controversy, such as those surrounding Darwinism, the localization of function, structure versus function relationships, psychological and neurological taxonomy, intelligence and language research. Not surprisingly, many of these nascent neurophilosophical domains have themselves matured and deepened in tandem with the growth and development of neuroscience.

From a more contemporary point a view, today's research at the intersection of philosophy and neurosciences was catalyzed in the 1950s with the articulation of the mind–brain identity theory by Place, Smart, and Armstrong. In the tradition of those philosophers who have looked at the neurosciences with a focus on how the brain supports mental or cognitive capacities, we have focused primarily on those parts of neuroscience that constitute cognitive neuroscience. The recent explosion of philosophical interest in the neurosciences owes much to the work of philosophers Paul Churchland and Patricia Churchland, and particularly to Patricia Churchland's *Neurophilosophy*, published in 1986. One of the critical features of the Churchlands' work is that it showed philosophers the benefits of incorporating neuroscientific data into their philosophical research. We seek to continue and extend this interdisciplinary program; whereas previous work has mainly focused on the benefits of neuroscience for philosophers, our work also addresses the other side of the interdisciplinary dialogue and makes some preliminary headway toward a philosophy of neuroscience.

This volume is intended to serve a variety of purposes. As one of the first few books in the philosophy of neuroscience, it provides an introduction for both philosophers and neuroscientists to the philosophical issues in neuroscience. It is also suitable as a textbook for advanced undergraduates or graduate students. It is partly for this reason that we have included several classic texts on neuroscience, for example, papers by Broca, Wernicke, and Hubel and Wiesel. Other readings, such as those by Petersen and Fiez and by Van Essen and Gallant, provide clear introductions to state-of-the-art work in neuroscience that is of great philosophical relevance.

This volume is a product of an innovative program linking philosophy with neuroscience and psychology, the Philosophy–Neuroscience–Psychology (PNP) Program in the Department of Philosophy at Washington University in St. Louis, Missouri. The program was the brainchild of the then chair of the Philosophy Department, Roger Gibson, who conceptualized the program in 1989. With the financial support of the James S. McDonnell Foundation and the strong encouragement of the President of the McDonnell Foundation, John Bruer, the program was inaugurated in 1993. Gibson and Bruer foresaw the opportunity of drawing upon the enormous resources in psychology and the neurosciences to build a distinctive program in which graduate students (and subsequently undergraduates) would relate their interests in philosophy to empirical investigations in psychology and neuroscience and draw upon findings in those disciplines in developing their philosophical ideas. One of the editors of this volume (Bechtel) is a professor and acting director of the program, and the other three are all graduates of the PNP doctoral program.

The PNP program has been home to an incredibly stimulating collection of individuals, professors, post-doctoral researchers, and students alike. Past and present PNP colleagues to whom we are especially grateful include Adele Abrahamsen, Irene Applebaum, Dave Chalmers, Morten Christiansen, Andy Clark, Chris Eliasmith, Rick Grush, Brian Keeley, Jesse Prinz, Whit Schonbein, Jon Waskan, and Tad Zawidzki.

In addition to the above-mentioned people, the editors would also like to thank the following people: Beth Remmes and Steve Smith of Blackwell Publishers for their support and encouragement of this project, Valery Rose for guiding the book through desk editing, and Tamara Cassanova for her help in preparing the electronic manuscripts.

William Bechtel thanks his students in his "Philosophy of Neuroscience" course who provided an initial testbed for teaching some of this material. In addition, he thanks Patricia and Paul Churchland, Huib Looren de Jong, George Graham, W. F. G. Haselager, Robert McCauley, Robert Richardson, Cees van Leeuwen, and William Wimsatt for many years of fruitful discussion of issues in philosophy of science relevant to neuroscience.

For intense and continued discussion, Pete Mandik thanks his associates in the McDonnell Project for Philosophy and the Neurosciences at Simon Fraser University. He also extends special thanks to project director Kathleen Akins, who, as his undergraduate professor of philosophy at the University of Illinois at Urbana–Champaign, ignited his initial interest in neurophilosophy. Thanks also to fellow project members Anthony Atkinson, Jose Luis Bermudez, Chris Eliasmith, Phillip Gerrans, Ian Gold, Rick Grush, Guven Guzeldere, Valerie Hardcastle, Brian Keeley, Sean Kelly, Victoria McGeer, Thomas Metzinger, Jonathan Opie, Steven Quartz, Adina Roskies, and Evan Thompson.

The following selections are reprinted with permission:

Akins, Kathleen (1996) Of Sensory Systems and the "Aboutness" of Mental States, *Journal of Philosophy*, 93 (7): 337–72.

Bates, Elizabeth (1994) Modularity, Domain Specificity and the Development of Language, *Discussions in Neuroscience*, 10: 136–55.

Broca, Paul (1861) Remarks on the Seat of the Faculty of Articulate Language, Followed by an Observation of Aphemia. Originally published as "Remarques sur le siège de la faculté du langage articulé; survies d'une observation d'aphemie," *Bulletin de la Société Anatomique*, 6: 330–57; translation by G. von Bonin (ed.), *Some Papers on the Cerebral Cortex* (Springfield, IL: Charles C. Thomas, 1960) pp. 49–72.

Churchland, Paul M. and Churchland, Patricia S. (1991) Intertheoretic Reduction: A Neuroscientist's field guide, *Seminars in the Neurosciences*, 2: 249–56.

Churchland, Paul M. and Churchland, Patricia S. (1996) McCauley's Demand for a Co-level Competitor. In R. McCauley (ed.), *The Churchlands and their Critics* (Oxford: Blackwell, 1996) pp. 222–31.

Crick, Franas and Koch, Christof (1998) Consciousness and Neuroscience, *Cerebral Cortex*, 8: 97–107.

Daugman, John G. (1990) Brain Metaphor and Brain Theory. In E. L. Schwartz (ed.), *Computational Neuroscience* (Cambridge, MA: MIT Press, 1990) pp. 9–18.

Grush, Rick (1997) The Architecture of Representation, *Philosophical Psychology*, 10 (1): 5–23.

Hubel, David H. and Wiesel, Torsten N. (1979) Brain Mechanisms of Vision, *Scientific American*, 241 (3): 150–62.

McCauley, Robert N. (1996) Explanatory Pluralism and the Co-evolution of Theories of Science. In R. McCauley (ed.), *The Churchlands and their Critics* (Oxford: Blackwell, 1996) pp. 17–47.

Mishkin, Mortimer, Ungerleider, Lēslie G., and Macko, Kathleen A. (1983) Object Vision and Spatial Vision: Two Cortical Pathways, *Trends in Neurosciences*, 6: 414–17.

Petersen, Steven E. and Fiez, Julie A. (1993) The Processing of Single Words Studied with Positron Emission Tomography, *Annual Review of Neuroscience*, 16: 509–30.

Van Essen, David C. and Gallant, Jack L. (1984) Neural Mechanisms of Form and Motion Processing in the Primate Visual System, *Neuron*, 13: 1–10.

Wernicke, Carl (1885) Recent Works on Aphasia. Originally published as *Einige neure Arbeiten über Aphasie, Fortschritte der Medizin*, volumes 3 and 4, trans. G. H. Eggert in *Wernicke's Works on Aphasia: A Sourcebook and Review* (The Hague: Mouton, 1977) pp. 173–205.

List of Authors

Kathleen Akins, Department of Philosophy and Coordinator of the McDonnell Project in Philosophy and the Neurosciences, Simon Fraser University, Vancouver, BC, Canada V5A 1S6.

Elizabeth Bates, Department of Cognitive Science, University of California, San Diego, La Jolla, California 92093-0119, USA.

William Bechtel, Department of Philosophy and Philosophy–Neuroscience–Psychology Program, Washington University in St. Louis, St. Louis, MO 63124, USA.

Paul Broca, nineteenth-century surgeon and pioneer in physical anthropology who spent his career at the University of Paris.

Patricia S. Churchland, Department of Philosophy, University of California at San Diego, La Jolla, CA 92093-0119, USA.

Paul M. Churchland, Department of Philosophy, University of California at San Diego, La Jolla, CA 92093-0119, USA.

Francis Crick, The Salk Institute, La Jolla, California 92037-1099, USA.

John G. Daugman, Computer Laboratory, Cambridge University, Cambridge CB2 3QG, England.

Julie A. Fiez, Department of Psychology and Center for the Neural Basis of Cognition, University of Pittsburgh, Pittsburgh, PA 15260, USA.

Jack L. Gallant, Department of Psychology, University of California at Berkeley, Berkeley, CA 94720-2020, USA.

Rick Grush, Department of Philosophy, University of California at San Diego, La Jolla, CA 92093-0119, USA.

Valerie Hardcastle, Center for Technology Studies and Department of Philosophy, Virginia Polytechnic Institute and State University, Blacksburg, VA 24061-0227, USA.

David H. Hubel, Department of Neurobiology, Harvard Medical School, 220 Longwood Avenue, Boston, MA 02115, USA.

Christof Koch, Division of Biology, California Institute of Technology, Pasadena, CA 91125, USA.

Robert N. McCauley, Department of Philosophy, Emory University, Atlanta, GA 30322, USA.

Pete Mandik, Assistant Professor and Associate Director, Cognitive Science Laboratory, Department of Philosophy, William Paterson University of New Jersey, 265 Atrium Building, 300 Pompton Road, Wayne, NJ 07470, USA.

Mortimer Mishkin, Section on Cognitive Neuroscience, National Institutes of Health, Building 49, Suite 1B-80, 49 Convent Drive, Bethesda, MD 20892, USA.

Jennifer Mundale, Department of Philosophy, University of Central Florida, Orlando, FL 32816, USA.

Steven E. Petersen, Neuropsychology Division, Department of Neurology and Radiology, Washington University School of Medicine, St. Louis, MO 63110, USA.

Jesse Prinz, Department of Philosophy and Philosophy–Neuroscience–Psychology, Washington University, St. Louis, MO 63124, USA.

Robert S. Stufflebeam, Assistant Professor, Department of Philosophy, Illinois State University, Normal, IL 61790-2200, USA.

Leslie G. Ungerleider, Laboratory of Brain and Cognition, National Institute of Mental Health, 10 Center Drive, MSC 1366, Bethesda, MD 20892-1366, USA.

David C. Van Essen, Department of Anatomy and Neurobiology, Washington University School of Medicine, St. Louis, MO 63110, USA.

Carl Wernicke, nineteenth-century German psychiatrist and neurologist who initially practiced in Berlin before taking a professorship at the University of Breslau.

Torsten N. Wiesel, President Emeritus, Rockefeller University, New York, and Secretary-General of the Human Frontier Science Program.

Part I

Neurophilosophical Foundations

Introduction

Pete Mandik

A combination of neuroscience and philosophy may raise the eyebrows of more than a few neuroscientists and philosophers, not to mention the lay reader. But upon a bit of reflection, the connection becomes quite clear. Philosophers have long been concerned to think about thinking, and the mind in general. Among their concerns is the question of the relation of mental phenomena to physical reality. Is the human soul the sort of thing that can survive the destruction of the body? In a world of causes and effects, what room can there be for free will? Such questions are the natural province of philosophers, but are not theirs alone. Recent centuries have witnessed an explosion of scientific approaches to the topic of the mind, among them cognitive neuroscience. In addition to addressing the many versions of the age-old question of the relation of mind to brain, the cognitive neurosciences raise many novel questions of interest to philosophers, questions introduced in these initial chapters.

In chapter 1, Bechtel, Mandik, and Mundale provide a historical sketch of developments in philosophy and neuroscience that eventuated in their contemporary interaction. That chapter also provides broad backgrounds on two major areas in contemporary philosophy that are especially pertinent to neurophilosophical investigation: the philosophy of science and the philosophy of mind. Chapter 2 continues the historical overview of philosophical and scientific approaches to the mind and brain. In this chapter, Daugman explores this history through the organizing theme of the multiple metaphors that have guided understanding of the mind and brain over the years. Chapters 3 and 4 view foundational neuroscientific issues through the eyes of philosophers of science. In chapter 3 Mundale traces the compelling history of how the brain's micro-level structure came to be understood, including the controversy concerning its neuronal composition and organization. For those with little previous exposure to the brain, this chapter will provide an introduction to some of its major features. In chapter 4 Bechtel and Stufflebeam examine techniques of data acquisition in the neurosciences: nature does not hand data to the scientist on a silver platter. Securing data, and evaluating theories, are a hard and tricky business, and this is no less the case when the nature of the brain is at stake.

Philosophy Meets the Neurosciences

William Bechtel, Pete Mandik, and Jennifer Mundale

1 Cognitive Science, Neuroscience, and Cognitive Neuroscience: New Disciplines of the Twentieth Century

The Greek oracle admonished "Know thyself." But for more than two millennia, the only avenues to self-knowledge were to examine one's own thoughts or to review one's behavior. The idea of knowing oneself by knowing how one's brain worked was at best a philosopher's thought experiment. When, in the middle of the twentieth century, the philosopher Herbert Feigl (1958/1967) proposed the idea of an auto-cerebroscope through which people could examine the activities of their own brains, no one imagined that by century's end we would be close to realizing this fantasy. New tools for studying the brain, especially positron emission tomography (PET) and functional magnetic resonance imaging (fMRI), provide avenues for revealing which brain areas are unusually active when individuals perform specific tasks (see chapter 4, this volume). Knowing which brain areas are activated when a subject performs a given task helps us to better understand the mental processes involved in performing that task. While the false-color images produced by these techniques are captivating, they are only part of what has opened up the study of mental processes in the brain in the last half century. Over the last 150 years careful analyses of behavioral deficits resulting from brain damage have offered further clues about what different brain regions do. In addition, scalp recordings of electrical or magnetic activity (through the electroencephalograph or EEG machine, or the magnetoencephalograph or MEG machine) evoked by particular stimulus events have provided detailed information about the time course of brain processing.

The second half of the twentieth century has witnessed not just an explosion of research at the behavioral end of neuroscience, but also extraordinary advances in understanding the basic cellular, synaptic, and molecular processes in the brain. The idea that neurons constitute the basic functional, cellular units of the brain was not widely accepted until the beginning of the twentieth century, but this provided the foundation for subsequent micro-level research (see chapter 3, this volume). This research, in turn, has increasingly been integrated with research into higher brain functions. For example, the introduction of electrodes that allow recording from awake, behaving primates has provided a means for linking local neural behavior to particular cognitive tasks. More recently new tools for manipulating genetic material have extended the inquiry even further. These advances have required the collaboration of scientists trained in a number of specialties, including neuro-anatomy, neurochemistry, and neurophysiology. In the 1960s, the term *neuroscience* was introduced for this collaborative inquiry, which has since expanded rapidly; the Society for Neuroscience, founded in 1970, now has 25,000 members.

Simultaneous with the introduction of new techniques for studying the brain has been the development of sophisticated ways of analyzing behavior so as to determine the information-processing mechanisms that generate it. Until recently, psychologists did not have access to the tools for examining brain activity, and so had to rely on indirect measures. One was to measure the time it took for a subject to respond to a particular stimulus (known as *reaction time* or RT); another was to note the error patterns that could be induced by manipulating the conditions under which the stimulus was presented, from which researchers could hypothesize about what operations the brain must be performing. With these tools cognitive psychologists succeeded in developing detailed and well-supported models of the operations occurring as people carry out a variety of cognitive tasks including categorization, problem solving, planning, recognition, and recall. In these endeavors cognitive psychologists often collaborated with researchers in other professions, especially linguistics and computer science. The metaphor of the mind as an information-processing system united researchers from these disciplines in the 1950s into a common enterprise which has come to be called *cognitive science* (Bechtel et al., 1998). In the late 1970s these efforts became institutionalized with the creation of the journal *Cognitive Science* and the establishment of the Cognitive Science Society.

Although both neuroscience and cognitive science were robust interdisciplinary enterprises through the 1960s, 1970s, and 1980s, until the late 1980s there was little intellectual interaction between them. Investigations in the two fields were pursued independently from each other and there were prominent philosophical arguments on behalf of maintaining such autonomy. Hilary Putnam (1967), for example, argued that mental states were multiply realizable – they could be realized by different neural processes in different species, by different patterns of electrical activity in different computers, and by potentially very different kinds of processes in whatever extra-terrestrial life forms might exist. Similarly, Jerry Fodor (1975) argued that the tax-onomies of cognitive science and neuroscience would cross-cut each other, spelling failure for any reductionist program. Fodor used as an example of such cross-cutting the way chemistry and economics may cross-cut each other. Different materials can constitute units of money (e.g. a silver dollar and a paper dollar bill) but even very similar objects made out of the same material (e.g. paper) would not count as genuine (but instead counterfeit) money. What makes a chunk of matter a genuine monetary unit and not a mere counterfeit is the role it plays in an economy of minters, bankers, and spenders. Analogously, according to Fodor, what makes something a psychological state is its role in an economy of psychological states, not its intrinsic material (in this case, neural) properties. Consequently, laws in psychology would be independent of any laws characterizing brain processes.

Towards the end of the 1980s, however, spurred in part by exciting new results stemming from the analysis of various forms of brain deficits and the development of PET imaging, neuroscientists and cognitive scientists began to collaborate and integrate their methodologies in a sustained examination of how brain processes underlie cognitive processes. Psychologist George Miller and neurobiologist Michael Gazzaniga coined the term *cognitive neuroscience* to designate the collaborative inquiry that integrates the behavioral tools of the psychologist with the techniques for revealing brain function to determine how the brain carries out the information processing that generates behavior. Today, cognitive neuroscientists routinely study both psychological processes and neural activity, and since the 1990s cognitive neuroscience has taken off as one of the fastest-developing and most exciting areas of scientific study.

2   Why a Philosophy of Neuroscience?

Two hundred years ago, the world of scientific inquiry and academic scholarship had not yet been divided into specialized disciplines. Law, medicine, theology, literature, and history each had separate faculties, but most of the other scholarly pursuits were the province of philosophy. Those individuals now recognized as major figures in the development of philosophy, such as Descartes, Locke, and Kant, directed many of their inquiries at the natural sciences (e.g. by providing epistemological foundations for the new sciences) and drew upon the results of those sciences. Gradually, however, the various natural and social sciences developed their own techniques, modes of inquiry, and bodies of knowledge, and split off from philosophy into separate disciplines. Psychology, for example, is one of the most recent defectors from the philosophic fold, and was established as a distinct discipline by the end of the nineteenth century, largely through the efforts of the philosopher William James.

As a result of the diminished scientific content of the field, philosophy became identified primarily with inquiries into values (ethics) and attempts to address foundational and general questions about ways of knowing (epistemology) and conceptions of reality (metaphysics). Thus, epistemology became preoccupied with whether *justified true belief* sufficed for knowledge, and metaphysics addressed such questions as whether events or objects and properties are the basic constituents of reality. In the hands of some of its practitioners, philosophy became *purified*, relying only on what it took to be its own tools, such as logic, conceptual analysis, or analysis of ordinary language, to address its own specialized questions.

Not all philosophers accepted the divorce of philosophy from other disciplines. They attempted to maintain philosophy's links to the inquiries that separated from it while nonetheless addressing foundational epistemic, metaphysical, or value questions. These philosophers have tried to relate their investigations to the ongoing inquiries in other fields, often focusing their philosophical analyses on foundational issues that arise within these fields. Thus, one finds subspecialties in philosophy for philosophy of art, philosophy of economics, philosophy of physics, and philosophy of biology. In particular, the emergence of psychology as an experimental discipline, and more recently of cognitive science has resulted in the increased popularity of philosophy of psychology and philosophy of cognitive science as focal areas within philosophy. Philosophy of neuroscience is a natural continuation of these efforts, comprising an inquiry into foundational questions (especially epistemic and meta-physical ones) that apply to neuroscience (a first step in developing this inquiry was Churchland, 1986).

Philosophy of neuroscience, like philosophy of psychology and philosophy of cognitive science, however, represents more than an attempt to address foundational issues in neuroscience. Insofar as psychology, cognitive science, and neuroscience all address the cognitive and intellectual capacities of humans (and other intelligent animals), their results can inform philosophical thinking about epistemology and metaphysics (especially metaphysical questions about human beings such as the relation of mind to body and the conditions for personal identity). Philosophers who think that results in the sciences themselves may provide material addressing philosophical questions often refer to themselves as *naturalized philosophers* (Callebaut, 1993). Naturalized philosophy involves a dialogue with the sciences, not just an analysis of the science. More generally, the *naturalized* approach to understanding the mind and brain involves seeing them as part of the natural world (rather than as miraculous or supernatural anomalies) and recognizing the biological, evolutionary, and environmental pressures which have helped to shape them. The approach to philosophy of neuroscience represented in this volume is naturalistic. The contributions represent either work in the neurosciences which is especially philosophically relevant or work by philosophers drawing upon or analyzing the scientific research. In part because the areas of neuroscience where discoveries and theories are of most consequence for philosophical issues have been primarily those that focus on neural systems and their relation to cognitive processes, rather than more basic processes such as the chemical events involved in neural transmission (but for an example of philosophy directed toward lower-level neuroscience, see Machamer et al., 2000), most of the focus in this volume will be on developments in systems and cognitive neuroscience.

The philosophical issues concerning neuroscience that are addressed in this book are characteristic of those that arise in philosophy of science and philosophy of mind. To appreciate these issues, some understanding of both areas is helpful. Although we cannot offer a detailed introduction to either in a short chapter (for such introductions, see Bechtel, 1988a, 1988b), the next two sections do provide a synopsis of the central issues in philosophy of science and philosophy of mind. We then focus briefly on four aspects of neuroscience that are especially interesting from a philosophical perspective and which will be examined further in other chapters.

3　Philosophy of Science

One of the main objectives of science is to provide explanation; accordingly, a major goal of philosophy of science is to specify what constitutes an explanation. We briefly review several of the most influential approaches to explanation that have been advanced in recent philosophy of science and point to how these approaches would apply to research in the neurosciences.

One of the most common views of explanation, which traces back to Aristotle and was developed in great detail earlier in this century (Hempel, 1965, 1966), holds that explanation of a phenomenon requires the logical deduction of the occurrence of the phenomenon from laws. In this approach, known as the *covering law model* or the *deductive-nomological* (D-N) model of explanation, laws specify relations between events. Laws are taken to specify general relations (as in Newton's law that force equals mass times acceleration ($f = ma$); to apply these general relations to particular events, one must specify conditions holding at a previous time, which are usually called *initial conditions*. Recognizing that multiple laws and initial conditions may be involved in a given explanation, such explanations can then be represented in the following canonical form (where L designates a law, C an initial condition, and E the event to be explained):

$$L1, L2, L3, \ldots$$
$$\underline{C1, C2, C3, \ldots}$$
$$\therefore E$$

Advocates of the D-N perspective generally assumed that the Cs and Es were sentences whose truth or falsity could be determined directly through observation. These *observation* sentences, accordingly, provided a grounding for the meaning of terms figuring in the laws. In addition to providing a basis of meaning, these observation sentences also provided the empirical support for the laws. In particular, just as one could derive a statement about an event already known to have happened so as to explain it, one could derive a statement about an event not yet known. In this way, the framework allowed for predictions, and the success of these predictions provided a basis for accepting or rejecting proposed laws. The D-N model was extremely influential in some areas of psychology earlier in the twentieth century. Many behaviorists, for example, sought to discover general laws of learning to characterize how various kinds of experiences (e.g. reinforcement) would change the behavior of organisms.

Recognizing that one might want to explain why laws held, the proponents also generalized this framework, allowing for the derivation of one or more laws from other laws. These other laws might be more general ones from which, under specific boundary conditions, the first set of laws might be derived. (Thus, the boundary conditions replace the initial conditions in the above formalism.) Proponents also suggested that this approach might be extended to relations between laws in one science and those of a more basic science by providing bridge laws relating the vocabularies of the two sciences, giving rise to the following schema:

Laws of the lower-level science
Bridge laws
Boundary conditions
∴ Laws of the higher-level science

These derivations are known as *reductions*; they figure prominently in discussions about the relation between psychology and neuroscience in which some theorists propose that the laws of psychology ought to reduce to those of neuroscience (see the papers in Part VI of this volume).

In contexts of explanation, one already knows that the E events have occurred and one is trying to explain why they occurred. But the same formalism provided by the D–N model can be employed in cases where the E events are not yet known to have occurred; the formalism then provides for predictions. This is an extremely important aspect of the D–N framework. Finding a law-like statement under which one could subsume an event known to have occurred is extremely easy, but one has no check on whether the purported laws are true. By making predictions which turn out to be true, the logical positivists thought we could justify laws.

For this claim, however, they were criticized by Karl Popper (1935/1959), who noted that such arguments had the invalid form of affirming the consequent:

If L were true, then prediction P would be true
P is true
∴ L is true

This formalism is invalid since it is possible for both premises to be true, but the conclusion false. (To see this is so, consider the following case: if Lincoln were beheaded, then Lincoln would be dead. Lincoln is dead. Both of these statements are true. But the conclusion that Lincoln was beheaded is false.) Because it is invalid, the truth of the premises does not *guarantee* the truth of the conclusion. However, neither is this particular formalism completely without value, since a number of confirming instances do lend inductive (though inconclusive) support to the initial law or hypothesis. Popper argued that the only way evidence could bear with certainty on laws was through the use of *modus tollens* arguments in which failed predictions could be used to falsify a purported law:

If L were true, then prediction P would be true:
P is false
∴ L is false

Accordingly, Popper emphasized that the method of science was a method of conjectures and refutations in which one proposed explanatory laws and then sought evidence showing that the hypothesized law was false. If a proposed law resisted all attempts at falsification, Popper would speak of it as *corroborated*, but not as shown true or confirmed, *recognizing that future evidence could always reveal it to be false.* Later, Hempel (1966) pointed out that even the logic of falsification is problematic, since the law or hypothesis that is ostensibly falsified may itself be a complex conjunction of several auxiliary hypotheses. In this case, a falsifying instance may be the result of a single, false auxiliary hypothesis, which, of course, will give the logical result of falsifying the larger hypothesis.

The D-N model of explanation and accompanying account of reduction coheres well with the general textbook account of the scientific method wherein a scientist is presented as first observing a range of phenomena, hypothesizing a law, testing it by deriving new predictions, and revising the law if the predictions are not borne out. Starting in the late 1950s, however, a number of philosophers and historians of science objected that this picture does not describe the usual practice of science. Thomas Kuhn (1962/1970), for example, argued that in the normal practice of science, researchers are not so much testing theoretical ideas as trying to make them fit nature. Much of the ongoing work of science involves developing and modifying experimental protocols to develop evidence that more and more phenomena fit already-accepted theoretical ideas, which he termed *paradigms*. Rather than testing whether or not $f = ma$ applied to a new range of phenomena, a scientist would be trying to devise ways of showing that it did apply. Only when these normal practices of science began to encounter repeated failures, would scientists explore alternative paradigms. Once a seemingly adequate alternative was found, they would abandon the pursuit of "normal science" and try to extend the range of application of a new, revolutionary framework. Although Kuhn's ideas of how science develops through normal science and revolutionary changes of paradigms have been adopted by many scientists and historians of science to characterize the development of science, they have also proven extremely controversial.

While the logical positivists were themselves very interested in the science of their time, their account was grounded primarily in logic, not in the details of scientific practice. (A consequence of this is that they viewed it as a normative model characterizing any possible science.) Kuhn's work drew philosophers' attention (as well as that of historians and sociologists of science) to the specific details of the process of scientific research. One consequence of this has been the recognition that there may be fundamental differences between scientific disciplines. Philosophers focusing on biology, for example, found that there are few laws associated with biology and that laws do not play a central role in biological explanations. While this might be evidence that biology is not a real science, another interpretation is that another explanatory framework is at play in biology. In particular, biological explanation typically makes references to goals, purposes, and functions. For example, an evolutionary account of an organism's features explains them in terms of what they are "for," in the sense of how they contributed to its predecessors' reproductive success. This kind of explanation is inherently *teleological*, and unlike explanations in physics (atoms are not "for" anything, they just are). Though the legitimacy of teleological

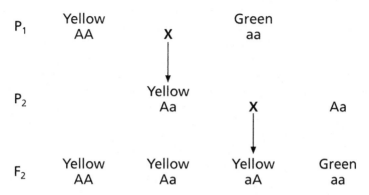

Figure 1.1 A typical diagrammatic representation of the recombination of Mendelian factors in inheritance. In the P_1 generation, a pure-breeding yellow pea is crossed with a pure-breeding green pea. The offspring in the F_1 generation will all be yellow hybrids. If two of them mate, the offspring in the F_2 generation will consist of one pure-breeding yellow pea, two hybrids, and one pure-breeding green pea.

explanation has been frequently challenged, most philosophers of science now accept it (see Mundale and Bechtel, 1996, for an example of how a teleological approach can serve to integrate neuroscience, psychology, and biology).

To further illustrate the importance of functional analysis in biological explanation, note the frequent use of figures and diagrams in biological texts, which often bear an uncanny resemblance to the flowcharts devised by computer scientists. Consider, for example, the familiar Mendelian diagrams of heredity in which the distribution of traits in successive generations are explained in terms of recombination of Mendelian factors (see figure 1.1). What these diagrams do is identify the operations that are being performed in a biological system as it carries out a given task. As such, they propose a *functional decomposition* of the system. At this stage, one does not have to specify what the components are that carry out the different tasks. While we now construe the units postulated as genes and link them with bits of DNA, Mendel merely referred to them as factors and offered no proposals as to how they were realized in the organism. But of course if the account is correct, then there ought to be components of the system that perform these tasks, and a major part of providing evidence that the proposed functional decomposition is correct is the *localization* of the functions in different components of the system.

Diagrams decomposing a system into its functional components (and generally identifying the physical components that perform the functions) provide an account of the mechanism operative in the system. Accordingly, to contrast this account with the D-N model, we will speak of such explanations as *mechanistic explanations*. The following are some of the major differences between mechanistic explanations and D-N explanations. First, as we note by the discussion of diagrams above, the explanations are not necessarily framed linguistically. While we can use language to present such explanations, or use words in diagrams to identify what the features of the diagram represent, what is crucial to such explanations is the decomposition of the actual system into component functions and component parts. It is by identifying the parts of the system, what they do, and how they are organized to work together, that one explains how a mechanical system performs its operations. Second, as a consequence of the first point, logical deduction is not the *glue* that holds an explanation together. Rather, a diagram portrays a relation between operations (or components), and it is by *envisaging* what will happen when the component functions are performed in the manner portrayed that we appreciate the connection between the explanandum and the explanans. Third, laws do not have a central place in mechanistic explanations. Rather, it is the details of the particular organization of functions (or parts) that do the explanatory work. (This is not to deny any role for laws, or for the D–N framework. Sometimes one does appeal to laws to specify how a component will behave. But what is critical to a mechanistic explanation is the putting together of component functions.)

Here we cannot develop the mechanistic alternative to the D–N approach in detail (but see Bechtel and Richardson, 1993). However, we will see numerous examples through this volume of attempts to develop mechanistic explanations. The neuroscientific study of language, for example, has sought to identify the contributions of various brain areas, such as those identified by Broca (chapter 5, this volume) and Wernicke (chapter 6, this volume), to language processing. Likewise, the neuroscientific study of vision (Part III) has sought to identify the contributions of different brain areas to visual processing. But we will conclude this section by noting one difference in emphasis in developing mechanistic explanations in neuroscience that distinguishes it from some other efforts to develop mechanistic explanations. As we noted above, Mendel carried out his decomposition of the mechanism of heredity without knowledge of the physical components that were involved. Likewise, biochemists have often worked out models of chemical reactions underlying vital phenomena without having discovered the responsible enzymes. And psychologists and other cognitive scientists often developed functional decompositions of cognitive tasks without knowing the brain mechanism involved. But neuroscience inquiries often begin with information relating one or more brain areas with a given cognitive performance. The challenge is often then to develop a functional decomposition that identifies the particular functions performed by the different brain areas, thus insuring, in these cases, that information about implementation figures in the explanation of the higher function to be understood. This difference in approach may make neuroscience a useful place for discovering an important variation in the way scientists develop mechanistic explanations.

4   Philosophy of Mind

Philosophers since antiquity have been enticed by the distinctive character of the mental processes of which we are aware – our thoughts and feelings; our reasoning processes; and our affects and emotions. As they present themselves to our phenomenal consciousness, these events and states seem very different from the physical events and states in the world. When we are aware of these processes within us, we are not (at least, not obviously) aware of any physical processes. It seems as if we could have the thoughts and feelings we do even if we lacked a physical body. Consequently, the fundamental question in philosophy of mind has been to explain the relation mental states bear to physical states. This is known as the *mind–body problem*.

One venerable position on the mind–body problem is *dualism*, the view that minds are indeed distinct from physical bodies. Plato advanced one form of dualism as a result of his attempt to understand knowledge. He construed knowledge as involving contact between a mind and what he termed *Forms* or *Ideas*: eternal, non-physical entities which provided the patterns of which all physical entities are imperfect instantiations. Accordingly, in various myths he describes the mind as engaging in direct interaction with the Forms prior to incorporation within a physical body. The experience of being inserted in a body results in a profound loss of memory, and the epistemic challenge of life is to regain the pure knowledge of the Forms unimpeded by physical bodies.

In contrast to Plato, Aristotle rejected the idea of a separate, non-physical realm of Forms; he retained the notion of Forms, but construed them as realized in physical objects. They were what defined an object as of a particular kind. But they were not identical to the matter that comprised the object, and knowledge for him consisted of internalizing the Form of objects dissociated from the matter they possessed when they were realized in physical objects. Thus, even though Aristotle identified a central role for sense perception in learning about Forms, cognitive activity for him entailed a dissociation from the physical domain and he entertained the idea of pure intelligence as dissociated from anything physical.

Modern philosophy of mind originates with the seventeenth-century philosopher René Descartes, who maintained a division between mind and body as great as Plato had proposed. He identified mind as a thinking, unextended substance (occupying no space), and the physical body as non-thinking extended substance. While Descartes attributed many mental activities to non-physical minds, he did allow the brain a role in more basic activities that we now construe as cognitive, including perception and memory. For Descartes, thinking proper was reasoning, which is manifest more clearly in the use of language. Descartes was fascinated with the abilities of complex mechanisms to produce patterns of behavior much like those produced by non-human animals, but linguistic processes, he thought, exhibited a creativity that could not be produced by any machine (or by any non-human animals, which he viewed as mere machines). The creative use of language, exemplified by the ability to construct a sentence never uttered before, required, for him, the non-material mind. One of the challenges for Descartes and others who have maintained that the mind is totally different from physical objects was to explain how brains and minds could interact. The interaction could not be like ordinary physical interactions, since minds were not even located in space. His own proposal was that by effecting small perturbations in the location of the pineal gland, the mind could alter the course of the "animal spirits" (fine fluids) flowing through the nerves and thus affect behavior. Even though such a proposal might minimize the physical work that the mind was required to do, the challenge remains to explain how the mind might perform any physical work (and be affected itself by physical work). Such difficulties have made dualism a rather unattractive framework for thinking about mind and brain, although even in neuroscience there have been some prominent dualists such as Sir John Eccles. The attraction of dualism is that it seems to enable the mind to be a creative agent outside the ordinary causal nexus and perhaps account for various religious convictions about the spiritual element of humans.

Philosophers of mind have advanced a variety of non-dualistic accounts. In the first half of the twentieth century theorists such as Gilbert Ryle (1949) attempted to diffuse the mind–body problem by arguing that separating mind from body involved a category mistake (comparable to the mistake involved in supposing that, after one has met each of the players on an athletic team, there is still some sense in which one has yet to meet the team). Mental states, Ryle proposed, are exhibited in the behavior of organisms such as humans. Thus, a belief might consist in the propensities to behave in ways we associate with such a belief. Because of its attempt to link mental states with behaviors, this view came to be known as *philosophical behaviorism*. While overcoming the need to account for mind–body interactions, behaviorism encountered its own problems, such as specifying the set of behaviors that were to be identified with particular mental states. Another worry was how it could account for affective states, such as pains or emotions, of which we have direct phenomenal awareness.

To overcome the problem posed by affective states, philosophers such as U. T. Place (1956) and J. C. Smart (1959) proposed in the 1950s that mental states were identical with brain states. For example, feeling pain consisted in being in a particular kind of brain state. Generalizing beyond sensations, what came to be known as the *identity theory* held that mental states in general were identical with brain states. A common objection to the identity theory was that we might characterize our mental states without knowing anything about the underlying brain states. But identity theorists argued that this failed to undercut the identity theory for the same reason that the FBI's initial ignorance that Kaczynski was the Unabomber failed to undercut their ability to characterize the Unabomber.

An objection that many philosophers took to be far more telling against the identity theory was the claim (discussed above) by Putnam (1967) and Fodor (1974) that the same mental states might be realized in very different kinds of systems, including alien life forms and computer systems (science fiction has long contemplated encounters with aliens and computers who have beliefs, desires, etc.). If the same the mental state could be realized in different ways, they contended, then the mental state could not be *identical* with the brain state. Rather, they proposed that the relation of mental states to brain states was comparable to that of software to hardware: a single piece of software characterizes processes that can be performed by many different kinds of hardware. Their view, which came to be known as *functionalism*, held that mental states are defined by their relation to each other and to sensory inputs and motor outputs. As we noted above, functionalism has been taken as supporting the autonomy of cognitive inquiry from neuroscience.

Although the multiple realizability claim has been widely accepted on faith, its empirical warrant has been questioned (Bechtel and Mundale, 1999), especially with respect to terrestrial animals. The main thrust of the objection is that if we attend to the practice of neuroscientists, we discover that they use a coarse-grained criterion for identifying brain states and treat brain states in different species as the same (despite their more fine-grained differences) just as cognitivists use a coarse-grained criterion for identifying psychological states across species. When grain sizes are matched, researchers are able to advance identity claims between psychological and brain states. Moreover, even though it was a source of inspiration for functionalism, the main functionalist claim that mental states are defined in terms of their functional relations with other mental states can be maintained independently of the claim of multiple realizability. If one does dissociate the claims, then it is possible to adopt both functionalism and the identity theory, holding that mental processes are defined by their interactions, but allow that they are identical with neural processes. Such a stance on the mind–body problem seems most congenial to recent work in cognitive neuroscience and to the account of mechanistic explanation in terms of decomposition and localization we presented in the previous section. The emphasis on functional decomposition adopts functionalism's emphasis on characterizing mental processes in terms of their relations to other mental processes, while the concern for localization adopts the identity theory's claim that mental states are brain states.

5   Special Philosophical Issues for Understanding Neuroscience

The previous two sections provide an overview of some of the main philosophical issues that frame philosophers' consideration of the neurosciences, especially systems and cognitive neuroscience. But there are a number of more specific questions, which we briefly introduce here.

The indirectness of studies of mind and brain

One of the epistemic challenges confronting studies of both mind and brain is the indirectness of the inquiry. The concern with the inaccessible character of mental processes was one of the factors leading the behaviorist B. F. Skinner to attempt to explain behavior totally in terms of observable stimuli and responses. We do seem to be aware of our mental processes (e.g. we know the sequence of our internal thoughts), but each one of us is limited to such awareness of our own thoughts. Moreover, we have no awareness of many of the mental processes psychologists are interested in (e.g. *how* we remember things or recognize objects we see as opposed to *what* we remember or see), and so any knowledge of these processes must be arrived at indirectly. Accordingly, over the last 150 years psychologists have developed a variety of indirect measures of the processes occurring in us. One measure is the patterns of errors we make on a variety of cognitive tasks (e.g. failing to remember some of the words on a list of words we are asked to remember, or falsely remembering that a word was on the list which was not). Perhaps the most powerful indirect measure psychologists have employed has been reaction time – the time it takes for subjects to perform particular tasks.

Compared to the indirect strategies on which psychologists must rely, it might seem that studying brains is rather direct. However, brains present their own set of problems. Merely looking at a brain is not informative; one must determine what processes occurring in brains are related to cognition. The brains we have been most preoccupied with are human brains, but ethical considerations prevent us from carrying out many invasive studies. Accordingly, until very recently the main source of information about human brains was derived from lesions manifested in patients with various forms of brain damage. Two problems with this source of evidence are that naturally occurring deficits are often quite diffuse in nature and there are challenges in inferring normal function from damaged systems. Brain imaging techniques such as PET and fMRI do provide a window into processes occurring in human brains, but these measures are themselves indirect. Thus, although difficult ethical issues are involved, much of what we know about brains stems from studies of other species in which more invasive approaches are used. From these studies researchers have been able to examine details of neuroanatomy, conduct electrophysiological studies recording from individual neurons, and surgically induce lesions to determine the effects of removing specific brain parts. But even these methods of studying the brain are indirect. For example, the details of neuroanatomy require the use of stains whose own mechanism of operation is often poorly understood.

Thus, rather than relying on some direct forms of observation for their data, psychologists and neuroscientists must rely on techniques and instruments, which may themselves generate artifacts and mislead scientific inquiry. On this topic, chapter 4 below examines the epistemic challenges faced, especially in neuroscience inquiries. While these challenges are not unique to studies of mind and brain (they arise in many other biological disciplines such as biochemistry and cell biology, as well as in the more basic sciences of physics and chemistry), they are not often attended to. They raise, however, significant questions for philosophers of science.

Relations between psychological and neuroscientific inquiries

The logical positivists put forward the model of theory reduction introduced above as a framework for relating different sciences, and this still guides much of the discussion about the relation between psychology and neuroscience. In chapter 22 below, for example, Paul and Patricia Churchland argue for the utility of the reduction model in developing the relationship between psychology and neuroscience. They acknowledge, however, that not all aspects of psychology will be successfully reduced to neuroscience. For those aspects of psychology that resist reduction they propose elimination in much the same manner as theories of phlogiston chemistry were eliminated at the beginning of the nineteenth century. The Churchlands' claim that some areas of psychology, especially those invoking folk concepts such as belief and desire, should be eliminated has become the focus of much philosophical controversy. McCauley, for example, in chapter 23, this volume, argues that reductions usually emerge when a theory in one discipline is replaced by an improved theory in its own discipline, not as a result of theories in other disciplines.

There are other serious issues raised by the use of the reduction model as a way of relating disciplines. Reduction, as it is understood in most philosophical accounts, involves deriving one *theory* from another, where theories are construed as sets of laws. Within this framework the focal questions have been whether or not psychological theories can be derived from neuroscientific ones. If so, psychological theories seem to lose their autonomy. Accordingly, those arguing for the special status of psychology or other higher-level sciences have argued that such derivations are not possible. (It is, of course, precisely this failure of derivation that the Churchlands cite as the basis for elimination.) However, as we have argued above, most neuroscience explanations do not take the form of D-N explanations in which phenomena are derived from laws, but rather are models of mechanisms. This casts a different light on the issue of reduction. Models of mechanisms are inherently reductionist: each proposed mechanism is designed to show how a phenomenon ascribed to a system is due to its constituent parts and their interaction. On the other hand, reduction no longer threatens the autonomy of the higher-level science: the higher level characterizes the interaction of processes, the lower level accounts for the performance of individual processes. For example, the higher level may account for language processing in terms of the interactive performance of several functions, while the lower level explains how a particular brain part performs one of those functions. Both employ decomposition and localization to offer explanations, but each is explaining a different phenomenon in terms of a system located at one level in the natural hierarchy, its components and their functions, and the organization of the several components into a functioning system.

The perspective presented in the previous paragraph is one that incorporates both reduction and a form of autonomy of higher levels (they are concerned with the integration of components, something not addressed at the lower level) and one that provides a framework for understanding much of the research in contemporary cognitive neuroscience which tries to link explanatory frameworks in psychology with information about neural mechanisms. But there are times when both disciplines are focused on essentially the same phenomena and are working at the same level of organization (the level of integrated neural systems). The reason this happens is that disciplines are not distinguished just by the particular levels of organization they consider. As Abrahamsen (1987) argues, they are also differentiated by the manner in which they approach the phenomenon. Specifically, the behavioral sciences, including psychology, focus on the mental and behavioral aspects of the functioning of organisms, whereas the biological sciences, including most of the neurosciences, focus on the organic features of the physical world. Each discipline has developed special tools for conducting its inquiry, tools such as reaction time measures for diagnosing cognitive processes through their behavioral products, and tools such as electrophysiological recording for detecting the physiological processes occurring in organisms. Thus, even when the phenomena being examined overlap, the approach of practitioners from different disciplines is still different. But problems often require integrating the approach of different disciplines. The tools of cognitive neuroscience, such as neuroimaging and single-cell recording, are examples of this, since they require both a focus on the physiological processes and a focus on the behavioral activities these processes are subserving. In instances such as this, collaboration between disciplines does not involve reduction at all, but the integration of perspectives and experimental skills.

Modularity

Decomposition and localization inherently involve fractionating a system into components. In cognitive and neuroscience inquiries, these components are often referred to as *modules*; the most prominent example of a proposed module in the cognitive domain is that of a module for language. A critical issue is just what is intended in segregating a module. Is the module assumed to be totally responsible for the process? If so, what is the nature of its inputs and outputs? And how does it come to acquire such a dedicated capacity? And if there is no such dedicated capacity, what is the significance of assigning the process to a particular brain region?

In a 1983 book *Modularity of Mind*, the philosopher Jerry Fodor advanced a strong statement of what the commitments of such an explanation were. Fodor contended that the following properties were conjointly satisfied by what he termed a *module*: (1) domain specificity, (2) mandatory operation, (3) limited output to central processing, (4) rapidity, (5) information encapsulation, (6) shallow outputs, (7) fixed neural architecture, (8) characteristic and specific breakdown patterns, and (9) characteristic pace and sequencing of development. Of these, Fodor has placed the greatest emphasis on information encapsulation, which is the claim that processing within modules only has access to the limited information represented within the module, not to information stored elsewhere in the system. For Fodor, it is the fact that modules rely only on encapsulated information that allows them to be extremely fast in their processing, but limits them to specific domains of information, reduces their flexibility, and results in their outputs being shallow.

An important feature of Fodor's account is that he does not treat the whole cognitive system as modular; rather, he distinguishes central cognition from modules for sensory systems and language. Central cognition performs the general reasoning of which humans are capable. Fodor characterizes such reasoning as *isotropic* in that any information a person knows can be invoked in reasoning about any subject and *Quinean* in that one's judgment about any proposition may depend upon its relation to all other propositions one believes. The modules provide input into this central system, but are not affected by its isotropy and Quinean character. One epistemic benefit of such an arrangement, according to Fodor, is that perception can provide an objective account of the world one is sensing, uncontaminated by one's beliefs and feelings; Fodor hopes thereby to avoid the epistemic relativism Kuhn and others have proposed.

While Fodor's account of modules has been very influential, few theorists who appeal to modules have actually adopted all of the characteristics Fodor associates with them. Neuroscientists emphasize that a characteristic feature of brain organization is backward projection. Areas in the temporal cortex involved in higher visual processing send backward projections to primary visual areas in the occipital cortex, and they in turn send projections back to the lateral geniculate nucleus of the thalamus and ultimately the retina. Although the function of these backward projections is not fully understood, they appear to allow downstream processing to modulate processing earlier in the system. If so, the earlier processing is not encapsulated. Basing arguments on behavioral data rather than neural data, Appelbaum (1998) argues that evidence from speech perception shows the effects of higher-level processing (e.g. lexical processing) on lower-level processing (e.g. phonetic processing) and that, as a result of the fact that these influences vary with context, Fodor's attempts to answer this objection by letting some apparently higher-level processing into the speech perception module cannot work.

What is left of modularity if one gives up informational encapsulation? At a minimum, modules would cease to be units which operate independently of the rest of the system except for inputs and outputs. But the consequences might be even more dire. Abandoning the requirement of information encapsulation might jeopardize decomposition and localization, especially if the alternative was to assume that the whole cognitive system was one integrated system. But these are not the only options – one can have a differentiated system in which different components specialize in performing particular tasks without encapsulation. Moreover, information flow between components can be limited without being encapsulated so that not all information in the system is available to every component. Information may reach a component, such as early visual processing, only through specific pathways and must be mediated by activity along that pathway in accordance with the functions of the intermediate processing areas. This provides for sufficient modularity for the strategy of decomposition and localization to be successful without the extreme consequences of Fodor-style modularity.

Computational or representational analysis of brain processing

Perhaps the feature that most clearly distinguished the cognitivism of cognitive psychology, cognitive science, and cognitive neuroscience from the behaviorism that dominated psychology and even neuroscience in the first half of the twentieth century was the information-processing metaphor. The crucial idea was that various states within a system (computer or brain) would represent information about other things (e.g. objects in the world, events in the world, or even states of the system itself) and that these representations could be manipulated in fruitful ways such as those articulated in formal logic. What is crucial about representations in this account (see Newell, 1980), is that they are states within a system that stand in for that which they represent and enable the system that employs them to deal with that which is represented; in the language of Franz Brentano (1874), they exhibited *intentionality*. In our social world, representations take many forms, including pictures and diagrams, but many theorists found language-like representations to be especially suggestive for cognitive modeling. Fodor (1975), for example, defended the claim that cognition requires a *language of thought*. If cognitive representations were language-like, then the computations would consist of syntactical operations specified by formal rules (i.e. rules making reference only to syntactic structure, not the meaning or reference of the representations).

Although this conception of computation and representation was quite popular in artificial intelligence research of the 1980s, the language-like character of the representations made it seem very unpromising for characterizing brain-based cognitive processing (Churchland, 1986; Churchland, 1989). An alternative approach to computational modeling, known as *neural network modeling* or *connectionism*, on the other hand, has been seen as far more promising. In this approach, the computational system is construed as a network of very simple units partially analogous to neurons. Whereas neurons discharge or spike, these units become activated or deactivated and, depending on their activation, excite and inhibit other units to which they are connected. To model cognitive processing, some of these units are designated as inputs and others as outputs; cognitive tasks are supplied to a network by activating some of its input units and allowing activation to spread through the network until the network stabilizes or a pattern is produced on the output units (Bechtel and Abrahamsen, in press; Clark, 1993). Although patterns of activations in networks are very different from language-like representations in traditional artificial intelligence programs, many theorists construe them also as representations (often referring to them as *distributed* representations – see van Gelder, 1990). In particular, researchers often try to analyze the activation patterns on hidden units (units that are neither input nor output units) as constituting intermediate representations which the network employs in the course of trying to perform a cognitive task (Elman, 1991).

Neuroscientists often speak of representations in the brain when, for example, they are able to show that particular neurons fire most actively in response to a specific stimulus (see chapter 18, this volume), and their usage appears to be rather similar to that of neural network modelers. Moreover, when they speak of computation, they tend to focus on changes in neural processes that can be modeled mathematically, not on formal operations (as in traditional artificial intelligence models). But there are increasingly vocal critics of the whole concept of representation. Some object to the rather minimal notion of representation invoked, arguing that the standing-in-for relation is insufficient to render something a representation; these critics offer proposals as to what more is needed to turn something (including a brain state) into a representation (see chapter 19, this volume). Others question the utility of analyzing neural systems in terms of representations altogether (chapters 20 and 21, this volume). Some of those raising questions are advocates of dynamical systems theory who emphasize the interdependent relationship of elements in the brain and the interactive relations of these with parts of the body and features of the world. They propose that it is often holistic, emergent features of such systems (such as the system itself settling into different attractor states in response to different environmental circumstances) that provide the key to understanding the behavior of these systems, and they advocate the tools of dynamics as the means of developing such explanations (Port and van Gelder, 1995).

As we noted above, it was the idea of information processing in which internal states were construed as representations that characterized the cognitivists' challenge to behaviorism. It is also the appeal to representations and to computational analyses of the processing of such representations in the brain that has helped spawn the collaboration of cognitive scientists and neuroscientists in the enterprise of cognitive neuroscience. If the appeals to representation and computation in analyses of the brain turn out to be viable, then this integration may have a secure foundation. If not, then alternative bases may need to be sought if the integration is to be successful. In any case, the analysis of representations, or any replacement notion, is a key issue in the foundation of neuroscience and cognitive neuroscience.

6 Summary

Our goal in this chapter has been to identify some of the key issues that arise as philosophy confronts the neurosciences. In particular, we have introduced some of the key issues in philosophy of science and philosophy of mind that are pertinent to the neurosciences, and identified four specific philosophical issues of particular relevance to the neurosciences.

References

Abrahamsen, A. A. 1987: Bridging boundaries versus breaking boundaries: Psycholinguistics in perspective. *Synthese*, 72, 355–88.

Appelbaum, I. 1998: Fodor, modularity, and speech perception. *Philosophical Psychology*, 11, 317–30.

Bechtel, W. 1988a: *Philosophy of Mind: An Overview for Cognitive Science*. Hillsdale, NJ: Erlbaum.

Bechtel, W. 1988b: *Philosophy of Science: An Overview for Cognitive Science*. Hillsdale, NJ: Erlbaum.

Bechtel, W., Abrahamsen, A., and Graham, G. 1998: The life of cognitive science. In W. Bechtel and G. Graham (eds), *A Companion to Cognitive Science*, Oxford: Blackwell, 1–104.

Bechtel, W., and Abrahamsen, A. A. in press: *Connectionism and the Mind* (2nd edn). Oxford: Blackwell.

Bechtel, W., and Mundale, J. 1999: Multiple realizability revisited: Linking cognitive and neural states. *Philosophy of Science*, 66, 175–207.

Bechtel, W., and Richardson, R. C. 1993: *Discovering Complexity: Decomposition and Localization as Scientific Research Strategies*. Princeton, NJ: Princeton University Press.

Brentano, F. 1874: *Psychology from an Empirical Standpoint* (A. C. Pancurello, D. B. Terrell, L. L. McAlister, trans.). New York: Humanities.

Callebaut, W. 1993: *Taking the Naturalistic Turn, or, How Real Philosophy of Science is Done*. Chicago: University of Chicago Press.

Churchland, P. S. 1986: *Neurophilosophy*. Cambridge, MA: MIT Press.

Churchland, P. M. 1989: *A Neurocomputational Perspective: The Nature of Mind and the Structure of Science*. Cambridge, MA: MIT Press.

Clark, A. 1993: *Associative Engines*. Cambridge, MA: MIT Press.

Elman, J. L. 1991: Finding structure in time. *Cognitive Science*, 14, 179–211.

Feigl, H. 1958/1967: *The "Mental" and the "Physical": The Essay and a Postscript*. Minneapolis: University of Minnesota Press.

Fodor, J. A. 1974: Special sciences (or: the disunity of science as a working hypothesis). *Synthese*, 28, 97–115.

Fodor, J. A. 1975: *The Language of Thought*. New York: Crowell.

Fodor, J. A. 1983: *The Modularity of Mind*. Cambridge, MA: MIT Press/Bradford Books.

Hempel, C. G. 1965: Aspects of scientific explanation. In C. G. Hempel (ed.), *Aspects of Scientific Explanation and Other Essays in the Philosophy of Science*. New York: Macmillan.

Hempel, C. G. 1966: *Philosophy of Natural Science*. Englewood Cliffs, NJ: Prentice-Hall.

Kuhn, T. S. 1962/1970: *The Structure of Scientific Revolutions* (2nd edn). Chicago: University of Chicago Press.

Machamer, P., Darden, L., and Craver, C. 2000: Thinking about mechanisms. *Philosophy of Science*, 67, 1–25.

Mundale, J., and Bechtel, W. 1996: Integrating neuroscience, psychology, and evolutionary biology through a teleological conception of function. *Minds and Machines*, 6, 481–505.

Newell, A. 1980: Physical symbol systems. *Cognitive Science*, 4, 135–83.

Place, U. T. 1956: Is consciousness a brain process? *British Journal of Psychology*, 47, 44–50.

Popper, K. 1935/1959: *The Logic of Discovery*. London: Hutchinson.

Port, R., and van Gelder, T. 1995: *It's About Time*. Cambridge, MA: MIT Press.

Putnam, H. 1967: Psychological predicates. In W. H. Capitan and D. D. Merrill (eds), *Art, Mind and Religion*. Pittsburgh: University of Pittsburgh Press.

Ryle, G. 1949: *The Concept of Mind*. New York: Barnes and Noble.

Smart, J. J. C. 1959: Sensations and brain processes. *Philosophical Review*, 68, 141–56.

van Gelder, T. 1990: What is the "D" in "PDP"? An overview of the concept of distribution. In S. Stich, D. Rumelhart, and W. Ramsey (eds), *Philosophy and Connectionist Theory*. Hillsdale, NJ: Lawrence Erlbaum Associates.

2

Brain Metaphor and Brain Theory

John G. Daugman

Then do you now model the form of a multitudinous, many-headed monster, having a ring of heads of all manner of beasts, tame and wild, which he is able to generate and metamorphose at will?

Plato, The Republic, Book IX

In its ideas about itself . . . as in all of its other endeavors, the mind goes from mastery to enslavement. By an irresistible movement, which imitates the attraction death exercises over life, thought again and again uses the instruments of its own freedom to bind itself in chains.

Roberto Mangabeira Unger, Knowledge and Politics

1 The Rise and Fall of Metaphors

Contemporary philosophers of science have emphasized the centrality of metaphor both in normal scientific discourse and in the transformative processes of scientific revolution.[1] The life cycle of the dominant metaphor in a scientific theory demarcates the life cycle of the scientific paradigm itself, and the adoption of new metaphors is said to be the signature of transitions between paradigms. More generally, it is believed that reasoning through metaphor is a pervasive and inescapable habit of thought itself (see, for example, MacCormac, 1985), both in the ordinary language of everyday life as well as in scientific and philosophical reasoning. In *Philosophy and the Mirror of Nature* (1979), Rorty claims that in general it is pictures rather than propositions, metaphors rather than statements, that determine most of our philosophical convictions. And perhaps nowhere else in the history of ideas has there been a more striking pattern of reliance on metaphors than in the history of reflection about the brain and the causes of behavior, and about the enigmatic relationship among brain, mental life, and personhood.

Invariably the explanatory metaphors of a given era incorporate the devices and the spectacles of the day, and, in perhaps subtler ways, they may reflect the prevalent social forms and daily texture of life. Theorizing about brain and mind has been especially susceptible to sporadic reformulation in terms of the technological experience of the day. For example, the water technology of antiquity (fountains, pumps, water clocks) underlies (see Vartanian 1973) the Greek pneumatic concept of the soul (*pneuma*) and the Roman physician Galen's theory of the four humours; the clockwork mechanisms proliferating during the Enlightenment are ticking with seminal influence inside La Mettrie's *L'Homme machine* (1748); Victorian pressurized steam engines and hydraulic machines are churning underneath Freud's hydraulic construction of the unconscious and its libidinal economy; the arrival of the telegraph network provided Helmholtz with his basic neural metaphor, as did reverberating relay circuits and solenoids for Hebb's theory of memory; and so on. In this historical framework, surely it would be folly for us to regard the recent computer bewitchment of theoretical work in psychology and neuroscience as an *entirely different* kind of breakthrough in the history of ideas. More than folly, it would constitute a "Whiggism": the tendency to view the past history of our subject as the stumbling progression toward its inevitable culmination in today's understanding that the brain turns out to be a computer.

Yet there are many among the forward ranks of theorists today who ask precisely that we not think of computation as just the contemporary metaphor, but instead that we adopt it as the literal description of brain function. Thus, for example, Zenon Pylyshyn complains that "there has been a reluctance to take computation as a *literal* description of mental activity, as opposed to being a mere heuristic metaphor . . . there is no reason why computation ought to be treated merely as a metaphor for cognition, as opposed to a hypothesis about the literal nature of cognition" (1980, p. 111). But, however meritorious the currently fashionable metaphor may be, it behooves us to regard it as the latest in a long sequence of compelling metaphors that, like the technologies from which *they* sprang, also had their day. Not that through such awareness we could somehow transcend metaphor in favor of some higher or more direct language. Rather, the salutary effects of considering the history of brain metaphors may be twofold. (1) It makes us aware of the origins, genealogy, possible hidden agendas, and latent content of our present metaphors. (Or, as King Lear observes, "'Tis a wise man who can say who his father is." This might be called the Problem of Progenitive Epistemology.) And (2), it alerts us to the markings of paradigm shifts when the old metaphors prove inadequate or transparent and novel metaphors take their place. Historians of science such as Thomas Kuhn identify these changes in the metaphorical language as being characteristic of radical and revolutionary science, the brief punctuated episodes of real intellectual vitality, separating the long intervals of stable and sluggish "normal" science. The enlivening effect of a new metaphor and the deadening effect of embracing one too literally or too ideologically or too long are eloquently (and again, metaphorically) described by the Brazilian philosopher Roberto Mangabeira Unger (1975): "in its ideas about itself and about society, as in all its other endeavors, the mind goes from mastery to enslavement. By an irresistible movement, which imitates the attraction death exercises over life, thought again and again uses the instruments of its own freedom to bind itself in chains. But whenever the mind breaks its chains, the liberty it wins is greater than the one it had lost." In this spirit, we might review the history of metaphors for brain and mind not with the goal of exorcizing metaphor from scientific discourse, which may be neither desirable nor possible, but rather to recognize our enclosure within metaphors and to identify their risings and settings as transitions that are the harbingers of insight and intellectual vitality.

2   A Taxonomy of Mind/Brain Metaphors

One of the edifying if uncomfortable lessons to be learned from the history of brain metaphors presents itself whenever we feel astonished by how inappropriate the categories and metaphors of past eras seem to be: when we marvel at the possibility that anyone could ever have thought *that*. At such times we must try to imagine the amused reception that our own are likely to receive from the intellectual historians of the future.

Something like this edifying lesson was presumably intended by Michel Foucault when he began *Les Mots et les choses* (trans. *The Order of Things: An Archaeology of the Human Sciences*) by citing Jorge Luis Borges's discovery of a tenth-century Chinese Encyclopedia,[2] in which the following classification of the world's animals was given: (1) Those Belonging to the Emperor; (2) Embalmed; (3) Tame; (4) Suckling Pigs; (5) Sirens; (6) Fabulous; (7) Stray Dogs; (8) Included in the Present Classification; (9) Frenzied; (10) Innumerable; (11) Drawn with a Very Fine Camelhair Brush; (12) *Et Cetera*; (13) Having Just Broken the Water Pitcher; and (14) That From a Long Way Off Look Like Flies (originally from Borges, *El Libro de Los Seres Imaginarios*).

Foucault proceeds: "In the wonderment of this taxonomy, the thing we apprehend in one great leap, the thing that, by means of the fable, is demonstrated as the exotic charm of another system of thought, is the limitation of our own. . . ."

3   Embodied Spirits and Helmsmen

It might be said that a cornerstone of Western thought and social institutions is the notion that persons are embodied spirits. Many of the antinomies and dichotomies of Judeo-Christian culture spring from this idea that we have on the one hand, bodies, and on the other hand, minds. These antinomies include determinism and accountability, cause and volition, the temporal and the eternal, the descriptive and the normative, immanence and transcendence, and the profane and the sacred. As Descartes asked, what then is the relationship, both moral and mechanical, between the body and the spirit that is embodied by it? Our art and literature are filled with images of spirit or consciousness being infused into an earthen body: recall Michelangelo's Sistine fresco of Adam, made from clay, being imbued with spirit, or the Greek legend of the sculptor Pygmalion whose ivory statue is infused with life, or the variations in transmutation myths in which consciousness is conferred through the eyes or by touch (as in Pinocchio), by breath (as in Genesis), by heat (as for the alchemists), or by electricity (as for Frankenstein). In this same tradition, today, near my house in Cambridge there is a company named "Thinking Machines Corporation," which has apparently managed to confer thought through parallelism.

Michael Arbib points out in his book *The Metaphorical Brain* (1972) that much of modern thought about brain and behavior can be distilled to two fundamental metaphors: (1) the cybernetic metaphor ("Humans are machines"), and (2) the evolutionary metaphor ("Humans are animals"). The very term *cybernetics*, coined by Norbert Wiener to invoke the concepts of control and communication, reflects the tradition of embodied spirits, since the *cyber* (Gk. *kybernētēs*) is the helmsman controlling the ship. This concept of *control* is absolutely central to our reflection about the relationship between the mental and the physical. It is expressed as not only a mechanical problem but also a moral problem: *self*-control is the great Judeo-Christian problem, with ever-recurrent incarnations throughout our culture ranging from the moral allegories of the brothers Grimm, to Freud's psychodynamic conflict, to the anti-utopian solutions of Skinner's *Walden Two* or Huxley's *Brave New World*. Our social sciences of psychology and economics similarly conceive of personhood largely through expressions of control and choice: person as puppet, role-player, poker-player, script-follower, bundle of desires, optimizer. Very broadly, whether within the Judeo-Christian framework or the cybernetic one or the social scientific one, the recurrent and fundamental metaphor for personhood and consciousness in relation to brain, body, and action has been that of a ship's helmsman.

4 Hydraulic and Mechanical Metaphors

Hydraulic and mechanical concepts of the psyche began in pre-Socratic thought[3] (figure 2.1). In the fifth century BC, the School of Hippocrates advocated an early hydraulic model of mind based on the four humors, whose preponderance or imbalance could be related to mental dispositions. An excess of phlegm resulted in apathy; an excess of black bile led to melancholy; an excess of yellow bile produced biliousness; while an excess of blood yielded sanguinity. This basic humoral model evolved in the second century BC into the Roman physician Galen's theory of "animal spirits," a highly rarefied fluid that flowed inside the nerves. Through these animal spirits, the brain received sensory messages and sent back commands. Thus began the pneumatic/hydraulic concept of the mind, a concept whose many descendants include Victorian "hydraulic" theories of the psychodynamic libidinal forces of desire and repression (the libidinal economy).

Probably the first modern exposition of the mechanical metaphor for the brain was Descartes's contention that *animals*, as distinct from men, were pure automata

Figure 2.1   An early triumph of the mechanical metaphor for cognition. Baron Wolfgang von Kempelen, widely celebrated for his mechanical genius and many notable inventions, while holding the post of Aulic Counselor on Mechanics to the Royal Chamber of Austrian-Hungarian Empress Maria Theresa, in 1770 unveiled before the Habsburg Royal Palace in Vienna a chessplaying automaton. Operating only with brass gears and springs, the "Terrible Turk" had the upper body of an Ottoman sultan whose eyes scanned the chessboard and glowered, whose head nodded from side to side, and who puffed on a Turkish pipe held in his right arm while he moved the chess pieces with his left arm. His chessplay was legendary for its brilliance, even though the cabinet could be opened from all four sides and peered through to reveal nothing but whirring clockwork and springs. This early touring machine made the rounds of all the royal courts of Europe, drawing throngs in Paris, Dresden, Leipzig, Brussels, and London, defeating nearly all opponents (including Joseph II and Napoleon Bonaparte) and making a fortune for von Kempelen. It finally reached the New World via Cuba, and eventually perished in a museum fire in Philadelphia.

– the "bête-machine" doctrine. This difference between animals and men reflected both Descartes's metaphysical dualism (that the mental and the corporeal are two different sorts of substance) and his Christian theology (which precluded awarding animals souls). Hence animals were unconscious bodies, and, for example, he argued that we should feel no qualms about inflicting pain on them. We cannot do so.

Just as Aquinas had earlier used the metaphor of clockwork mechanisms to explain animals' instincts, Descartes (1664, from Vartanian, 1953, p. 136) attributed our own mental experiences to the function of the body's organs as clockwork mechanisms within an automaton:

> . . . the reception of light, sounds, odors, tastes, warmth, and other like qualities into the exterior organs of sensation; the impression of the corresponding ideas upon a common sensorium and on the imagination; the retention or imprint of these ideas in the Memory; the internal movements of the Appetites and Passions; and finally, the external motions of all the members of the body . . . I wish that you would consider all of these as following altogether naturally in this Machine from the disposition of its organs alone, neither more nor less than do the movements of a clock or other automaton from that of its counterweight and wheels. . . .

Among other major seventeenth–century and Enlightenment invocations of mechanical and clockwork metaphors, Hobbes (1658), *De Homine* (from Vartanian,

1953, p. 138), sought to trace the physical basis for ideas and associations to minute mechanical motions in the head, and thus to turn epistemology into a branch of the new physics of Newtonian mechanics. In the same spirit, Hartley (1749), *Observations on Man* (from Vartanian, 1953, p. 141), proposed that all mental phenomena result from *vibratory motions* in the brain, by virtue of which arise the laws of association, causing ideas to cohere. And perhaps the most seminal of all Enlightenment appearances of this metaphor was La Mettrie's *L'Homme machine* (1748, from Vartanian, 1953, p. 139) which, based on profuse evidence of correlation between mental and physical states, described the human brain and body as "a machine that winds its own springs – the living image of perpetual motion . . . man is an assemblage of springs that are activated reciprocally by one another."

Among contemporary incarnations of the hydraulic metaphor, probably the most pervasive one is not so much a literal theory of brain function as a theory of the drives of the psyche. Psychodynamic theory holds that both individual behavior and social history are the outgrowth of invisible forces of desire and repression. For structuralists such as Freud and Marx, the visible is to be interpreted in terms of the hidden: economic and social forms are the superstructure thrown up by an invisible "deep structure" of class struggle in the Marxist perspective, just as in the Freudian framework, the individual's conscious experience and behavior are the manifestation of a surging unconscious libidinal struggle between desire and repression. From the psychodynamic viewpoint, the present is pregnant with the past ("the child is father to the man"), and from the social-historic viewpoint, the present is pregnant with the future (class struggle being the midwife), but in both cases, what is visible in the behavior of the individual or of society must be explained in terms of underlying hydraulic forces that eventually will have their way. Similar to water or steam pressure, which cannot build up indefinitely without release, the internal psychic or social pressures must inevitably express themselves in one form or another. According to these modern incarnations of the hydraulic metaphor, war, artistic movements, work, passion, religion, revolution, and new economic forms are the result.

In contemporary brain theory, sometimes the hydraulic metaphor appears in reverse form, in which something analogous to pressure (or voltage) *drops*, rather than rises. One phenomenon that this *complementary* hydraulic metaphor is meant to illuminate is the hemispheric lateralization of the brain. Kosslyn (1987) has recently proposed that the "activation" of either hemisphere consumes a depletable resource, like energy; the more "activation" is used up by one hemisphere, the less there is available for the other, rather like water pressure or electric power. Kosslyn uses the metaphor of an electric toaster (not because it resembles the brain's bilateral morphology), with an intuition that some sort of depletable resource is consumed when part of the brain is activated: "If so, then the hemisphere that receives input directly may draw more activation (rather like the way in which using a toaster will draw more current from the power grid), leaving less for the other hemisphere" (1987: 160). Although Kosslyn refers to an electric toaster on a power grid, this is in essence a reincarnation of the hydraulic metaphor.

Finally, perhaps one of the more beautiful twentieth-century incarnations of the mechanical metaphor, albeit in an expressly poetic form, was that of Sherrington (1906), who described the brain as an "enchanted loom," in which "millions of flashing shuttles weave a dissolving pattern."

5 Electronic and Optical Metaphors

The notion of the neuron as a switch, gate, or relay probably reflects earlier electro-mechanical devices for communication and computation. Long before Howard Aiken, John von Neumann, and ENIAC, mechanical devices had been proposed for adding, subtracting, and multiplying by Wilhelm Schickard (1623), Blaise Pascal (1642), and Charles Babbage (1822); even Leibniz had imagined[4] a universal logical machine in which by a general method, "all the truths of reason would be reduced to a kind of calculation." Echoes of electromechanical relays appear in Hebb's (1949) theory of reverberating circuits underlying memory. Electronic communications systems such as the telegraph provided a basic nerve metaphor for Helmholtz, and indeed in the 1950s, the differential equation that describes coaxial cable transmission (the spatiotemporal "Telegrapher's Equation," which had been developed to model signal propagation for the design of the transatlantic undersea cable), was adopted directly by Hodgkin and Huxley (1952) in their Nobel Prize-winning studies of nerve action potential generation and propagation.

Optical metaphors for cognition have a long history, reaching back to antiquity, and appear even in such vernacular expressions for understanding as, "I see." Greek epistemology held that our knowledge of the external world was mediated through optically transmitted copies (*eidola*) that objects emitted of themselves, entering our brains through the *pneuma*. In contemporary brain theory, one optical metaphor that recently became widely popularized was the holographic metaphor. Inspired by the obviously distributed character of neural information storage, and its relative insensitivity to brain injury, Pribram (1969) applied van Heerden's (1963) research on holographic information storage properties to propose a general holographic metaphor for perception and memory.

6 Networks and Societies of Simple Automata

Earlier we examined seventeenth- and eighteenth-century visions of simple mechanical automata underlying brain and mental function. Important recent theoretical movements have been based on whole armies of simple automata, each of which obeys very simple rules, unintelligently or "automatically." A key motivation for such models is the notion of *synergy*, that the whole is significantly greater than the sum of its parts, so that remarkably intelligent behavior might emerge from surprisingly elementary subunits if they are collectively configured properly, and with sufficiently rich interconnections. This idea lies at the core of many contemporary brain theories based on "cellular automata," "neural network," "connectionism," and "parallel distributed processing."

Dennett (1978) has pointed out the philosophical merit of this approach to implementing the reductionist program. He argues that whereas visions of a *single* internal homunculus (who "watches" the internal movie screen of the world, and so on) are of course philosophical anathema, a vision of an *army* of homunculi, provided they are reductively stupid, may be philosophically felicitous and fruitful. The reductionist project might be achieved by positing hierarchies of successively more stupid homunculi, each of which behaves in a way that can be understood by its reduction to still less intelligent homunculi, until at the base of this hierarchy, the homunculi are true automata.

The metaphors based on networks or societies of simple automata may be coarsely divided into deterministic and stochastic versions.

Deterministic

Most of the seminal notions of networks of automata have been deterministic, with each formal neuron or daemon obeying specific and simple rules. Thus the ancestral prototype of such models, by its combination of microscopic simplicity with unpredictable emerging macroscopic pattern formation, is the *cellular automaton* first investigated systematically by John von Neumann. An idea with far-reaching influence, its descendants include the classic McCullough and Pitts (1943) treatment of formal neurons whose behavior is governed by a simple logical calculus, and the negatively influential study of simple perceptual automata by Minsky and Papert (1969), *Perceptrons*. Much current connectionist and neural network modeling makes the argument that well-orchestrated combinations of simple and stupid elements (thresholded summators, adaptive linear combiners, Boolean gates) can collectively achieve marvelous things.

Other variants of this idea include the general approach to human cognition as a society of simpler daemons, envisioned in Minsky's (1988) *Society of Mind*, which portrays the mind as emerging from its multitudinous cognitive components in much the same way as society emerges from its collectively interacting individuals. (This socio-political brain metaphor is clearly indebted to Hobbes's *Leviathan*, and offers the prospect of diverse forms of neuro-political organization such as social democratic, autocratic or winner-take-all, anarchic, and fascistic.) Finally, in the psychodynamic dimension, an earlier vision of the mind as a deterministic collection of unconscious, automatic mechanisms is Freud's (1904) *The Psychopathology of Everyday Life*, which, in a similar manner, decomposes the life of the psyche into ensembles of numerous unconscious automata.

Stochastic

Various of the "daemon" network models recently proposed have possessed stochastic spirits, either specified by probabilistic transfer functions, Markov transition rules, or more explicitly thermodynamic characteristics governed by stochastic differential equations. Those models which are serious about their thermodynamics, such as the Boltzmann machine (Ackley et al., 1985), Markov random field models, and relaxation labeling schemes, exploit certain analogies with statistical physics in which the evolution toward equilibrium or asymptotic configurations bears some resemblance to such cognitive phenomena as categorization, learning, recognition, memory, and decision-under-uncertainty.

Solid-state physics has provided salient metaphors for such theories. Perhaps the best-known of these is the spin–glass lattice popularized by Hopfield (1982) as a neural network formalism, which inspires Ising models of local domain formation by nearest-neighbor interactions. The dynamics of such networks are represented by trajectories (or flows) through a state space of Ising spins (Gutfreund et al., 1988), with these flows corresponding to the minimization of a global energy function. An essential element of the spin–glass metaphor is that in these networks every neuron is connected to every other neuron, so that distance has no meaning. Indeed, all "neurons" in such a lattice are nearest neighbors and, usually, all their interactions are symmetric; these two properties lead to a formal representation in an infinite-dimensional Hilbert space, in which all pairs of points are separated only by a unit distance. Although such globality is clearly a significant departure from brain connectedness, there have been several compelling demonstrations in which the chaotic dynamics that emerge can solve certain optimization and classification problems (Amari, 1988; Shaw, 1981). Recently, based on experimental observations in rabbit olfactory bulb, Skarda and Freeman (1987) have explicitly proposed that chaotic dynamics and their attractors may have great functional significance in neural mechanisms of olfactory recognition.

7 The Computational Metaphor

At its core the computational metaphor of brain function invokes the notion of formal rules for the manipulation of symbols, as well as certain ideas about data structures for representing information. Surprisingly, given the pervasive popularity of this metaphor, there remains today no well-established evidence of symbolic manipulation or formal logical rules at the neurobiological level in animal physiology. Perhaps the closest supporting neural evidence for symbol manipulation comes from human neuropsychological studies of language use (Geschwind, 1974), which demonstrate that specific brain lesions can result in specific *syntactic* deficits. But in general, while the computational metaphor often seems to have the status of an established fact, it should be regarded as a hypothetical, and historical, conjecture about the brain.

The computational or "Boolean brain" notion was originally expressed in the seminal proposal by McCullough and Pitts (1943) that nervous activity embeds a logical calculus, and was explored further in very general terms by John von Neumann (1958) in *The Computer and the Brain*. Alan Turing had proposed in 1950 the famous "Turing test" for justifying the assertion that under certain circumstances machines can be said to think, based on a positivistic criterion for distinguishing between artificial and human intelligence. A consequence of the Turing test is that if, in a teletype conversation with a computer no basis can arise for proving that there is not a real person producing the responses, then it is not valid to deny to the machine, in principle, any of the potential attributes or accoutrements of personhood, such as beliefs and intentionality.

Through formal work on the foundations of computation, augmented by results in mathematical logic by Alonso Church and Kurt Gödel on decidable propositions and computable functions, and second-order logic containing predication over predicates, the notion of the Universal Turing Machine was born together with the suggestion in some quarters that human brains might be instances of such machines. A central aspect of the Turing Machine concept is its focus on abstract and formal properties, independent of their hardware realization. The principle of the irrelevance of implementation has allowed cognitive scientists to adopt this metaphor in fairly direct form without requiring brains in any way to resemble computational devices physically. With the emergence of the terms "artificial intelligence" and "cognitive science" after the famous 1956 Dartmouth conference, which led to the creation of AI laboratories at major universities, the computational metaphor for brain function had become a slogan and a banner without par. Among the myriad recent expressions of adherence to the computational metaphor, perhaps the strongest statement advocating its adoption as the paramount theoretical framework for the cognitive and brain sciences is Pylyshyn's *Computation and Cognition* (1986).

Today's embrace of the computational metaphor in the cognitive and neural sciences is so widespread and automatic that it begins to appear less like an innovative leap than like a bandwagon phenomenon, of the sort often observed in the sociology and history of science. There is a tendency to rephrase every assertion about mind or brains in computational terms, even if it strains the vocabulary or requires the suspension of disbelief. In cognitive usage the meaning of the phrase "computational theory" has changed, so that now the term can be invoked to describe any theory that includes a *task analysis*. This appears to be an error of overextension (i.e. computational models include task analysis, so a theory with task analysis is a computational theory); and the descriptor "computational" then implies that the theory possesses properties and formal notions that in fact it does not. The irresistible temptation to announce "a computational theory of X," where X could be anything from the gill-withdrawal reflex of *Aplysia* (Gluck and Thompson, 1987) to the hemispheric lateralization of the entire brain and its variability among individuals (Kosslyn, 1987) typically reveals more about the sociological diffusion of slogans than about any proven or natural connection between X and computation. What yesterday would have been called a theory must today be called a computational theory, whether or not any deep relationship with computability (beyond some general notion of process or transformation) is established or even explicitly articulated.

As Earl MacCormac (1985) put it: "Theories require metaphors to be both hypothetical and intelligible." But perhaps like the goslings of Konrad Lorenz (to adopt another metaphor), we are too easily imprinted with the spectacle of the day, the "duckiest" device in sight. That we should have chosen to model the brain as a computing machine, and even to believe that we have found at last the essence of our personhood in computation, is perhaps in historical perspective no more surprising than the fact that Freud's pre-eminent metaphor for personhood involved hydraulic forces or that La Mettrie's was based on a clockwork mechanism.

What seems remarkable about today's pervasive computational metaphor in neuroscience and psychology is how great is the distance between everything that is actually known about neurobiology and cognition, on the one hand, and the basic logical structure, formal sequence, tokens, powers, and limitations of computation on the other hand. With the exception of human linguistic competence, there is very scarce evidence indeed in the nervous systems of animals for any biological mechanisms of formal symbol manipulation. In some ways, the properties of pattern recognition and representation found in the immune and genetic systems, as Edelman (1987) has recently suggested, may provide a more germane model for information processing (and intentionality) in biology. Even in very mundane terms, it almost belabors the obvious to point out that where syllogisms and symbols are concerned, people are terrible at computation, and that conversely, any current implementation of computation is terrible at doing what people do or simulating what they are like. Although conscious experience obviously may not reveal anything about underlying mechanisms, who among us finds any recognizable strand of their personhood, or of their experience of others and of the world and its passions, to be significantly illuminated by, or distilled in, the metaphor of computation? Perhaps the ascendancy of this metaphor within the discourse of neuroscience and psychology, like the images of its many technological predecessors in a great hall of mirrors, is more a reflection of the sociology of science, its bandwagons and its bewitchments, than any of us may care to admit. We should remember that the enthusiastically embraced metaphors of each "new era" can become, like their predecessors, as much the prisonhouse of thought as they at first appeared to represent its liberation.

Acknowledgments

I am grateful to Cathryn Downing, Gary Hatfield, and Scott Weinstein for their helpful criticisms and suggestions.

Notes

1   Expressions of this view have taken many forms, from Michel Foucault's perspective (in *The Order of Things: An Archaeology of the Human Sciences*) that scientific thought and belief should be regarded in the same way as a literary text, to Earl MacCormac's view (in *Metaphor and Myth in Science and Religion*) that scientific explanations would simply not work as explanations if not for their metaphors, to Thomas Kuhn's view (in *The*

Structure of Scientific Revolutions) that the preferred or permissible metaphors within a scientific paradigm determine what will be accepted as an explanation and what problems remain as unsolved puzzles.

2   The *T'ai P'ing Kuang Chi* Encyclopedia, known as the "Extensive Records Made in the Period of Peace and Prosperity." Completed in the year 978 and published in 981.

3   For the review of historical material in this section, I am indebted to Vartanian (1953, 1973).

4   For this material I am indebted to Resnikoff (1988).

References

Ackley, D. H., Hinton, G. E., and Sejnowski, T. J. 1985: A learning algorithm for Boltzmann machines. *Cognitive Science*, 9, 147–69.

Amari, S. 1988: Statistical neurodynamics of associative memory. *Neural Networks*, 1 (1), 63–74.

Arbib, M. A. 1972: *The Metaphorical Brain*. New York: Wiley Inter-science.

Barthes, R. 1964: *Elements of Semiology* (Translated from French.) New York: Hill and Wang.

Borges, C. 1967: *El Libro de Los Seres Imaginarios* (trans. *The Book of Imaginary Beings*). Buenos Aires: Editorial Kier, S.A.

Churchland, P. 1986: *Neurophilosophy: Towards a Unified Science of the Mind-Brain*. Cambridge, MA: MIT Press.

Dennett, D. C. 1969: *Content and Consciousness*. London: Routledge and Kegan Paul.

Dennett, D. C. 1978: Artificial intelligence as philosophy and as psychology. In M. Ringle (ed.), *Philosophical Perspectives on Artificial Intelligence*, New York: Humanities Press.

Descartes, R. 1664: *Traité de l'homme*. Trans. by T. Hall (1972), *Treatise on Man*. Cambridge, MA: Harvard University Press.

Dreyfus, H. L. 1972: *What Computers Can't Do: A Critique of Artificial Reason*. New York: Harper and Row. See also H. L. Dreyfus, *Alchemy and Artificial Intelligence*, RAND Publication P-3244. Santa Monica, December 1965.

Dreyfus H., and Dreyfus, S. 1986: *Mind over Machine*. New York: Free Press.

Edelman, G. 1987: *Neural Darwinism: The Theory of Neuronal Group Selection*. New York: Basic Books.

Egecioglu, O., Smith T., and Moody, J. 1987: Computable functions and complexity in neural networks. In J. Casti and A. Karlqvist (eds), *Real Brains, Artificial Minds*, New York: Elsevier.

Foucault, M. 1973: *Les Mots et les choses* (trans. *The Order of Things: An Archaeology of the Human Sciences*). New York: reprinted by Random House.

Geschwind, N. 1974: *Selected Papers on Language and the Brain*. Dordrecht: Reidel.

Gluck, M., and Thompson, R. 1987: Modelling the neural substrates of associative learning and memory: A computational approach. *Psychological Review*, 94 (2), 176–91.

Gunderson, K. 1985: *Mentality and Machines*, 2nd edn. Minneapolis: University of Minnesota Press.

Gutfreund, H., Reger, J. D., and Young, A. P. 1988: The nature of attractors in an asymmetric spin glass with deterministic dynamics. *Journal of Physics A: Mathematical and General*, 21, 2775–97.

Haugeland, J. 1985: *Artificial Intelligence: The Very Idea*. Cambridge, MA: MIT Press.

Haugeland, J. 1987: *Mind Design: Philosophy, Psychology, Artificial Intelligence*. Cambridge, MA: MIT Press.

Hebb, D. 1949: *The Organization of Behavior*. New York: Wiley.

Hinton, G. E., Sejnowski, T. J., and Ackley, D. H. 1984: Boltzmann machines: Constraint satisfaction networks that learn. *Technical Report CMU-CS-84-119*, Carnegie-Mellon University.

Hodgkin, A. L., and Huxley, A. F. 1952: A quantitative description of membrane current and its application to conduction and excitation in nerve, *Journal of Physiology*, 117, 500–44.

Hopfield, J. J. 1982: Neural networks and physical systems with emergent collective computational abilities. *Proceedings of the National Academy of Sciences (USA)*, 79, 2554–8.

Jay, M. 1973: *The Dialectical Imagination: A History of the Frankfurt School and the Institute of Social Research, 1923–1950*. Boston: Little, Brown.

Kosslyn, S. 1987: Seeing and imagining in the cerebral hemispheres: A computational approach. *Psychological Review*, 94 (2), 148–75.

Kuhn, T. 1962: *The Structure of Scientific Revolutions*. Chicago: University of Chicago Press.

La Mettrie 1748: *L'Homme machine*. Holland: Jean-Jacques Pauvert (ed.), Nr 40, 1966.

MacCormac, E. R. 1976: *Metaphor and Myth in Science and Religion*. Durham, NC: Duke University Press.

MacCormac, E. R. 1985: *A Cognitive Theory of Metaphor*. Cambridge, MA: MIT Press.

McCulloch, W. S., and Pitts, W. 1943: A logical calculus of the ideas immanent in nervous activity. *Bulletin of Mathematical Biophysics*, 5, 115–33.

Medawar, P. 1968: *Induction and Intuition in Scientific Thought*. Philadelphia: American Philosophical Society.

Minsky, M. 1988: *Society of Mind*. New York: Simon and Schuster.

Minsky, M., and Papert, S. 1969: *Perceptrons*. Cambridge, MA: MIT Press.

Pribram, K. 1969: The neurophysiology of remembering. *Scientific American*, 200, 73–86.

Putnam, H. 1960: Minds and machines. In A. Anderson (ed.), *Mind and Machines*, Englewood Cliffs, NJ: Prentice-Hall.

Pylyshyn, Z. W. 1980: Cognition and computation: Issues in the foundations of cognitive science. *Behavioral and Brain Science*, 3 (l), 111–32.

Pylyshyn, Z. W. 1986: *Computation and Cognition: Toward a Foundation for Cognitive Science*. Cambridge, MA: MIT Press.

Resnikoff, H. L. 1988: *The Illusion of Reality: Topics in Information Science*. Heidelberg: Springer-Verlag.

Rorty, R. 1979: *Philosophy and the Mirror of Nature*. Princeton, NJ: Princeton University Press.

Rosen, R. 1987: On the scope of syntactics in mathematics and science: The machine metaphor. In J. Casti and A. Karlqvist (eds), *Real Brains, Artificial Minds*, New York: Elsevier.

Scheffler, I. 1967: *Science and Subjectivity*. Indianapolis: Bobbs-Merrill.

Shaw, R. 1981: Strange attractors, chaotic behavior, and information flow. *Zeitschrift für Naturforschung*, 36 (a), 80–112.

Sherrington, C. 1906: *The Integrative Action of the Nervous System*. New Haven, CT: Yale University Press.

Simon, H. A. 1981: *The Sciences of the Artificial*, 2nd edn. Cambridge, MA: MIT Press.

Skarda, A., and Freeman, W. J. 1987: How brains make chaos in order to make sense of the world. *Behavioral and Brain Sciences*, 10 (2), 161–95.

Unger, R. M. 1975: *Knowledge and Politics*. New York: Free Press.

van Heerden, P. 1963: Theory of optical information in solids. *Applied Optics*, 2, 393–400.

Vartanian, A. 1953: *Diderot and Descartes: A Study of Scientific Naturalism in the Enlightenment*. Princeton, NJ: Princeton University Press.

Vartanian, A. 1973: *Dictionary of the History of Ideas: Studies of Selected Pivotal Ideas*, ed. P. P. Wiener. New York: Scribners.

von Neumann, J. 1958: *The Computer and the Brain*. New Haven, CT: Yale University Press.

Wiener, N. 1948: *Cybernetics, of Control and Communication in the Animal and the Machine*. Cambridge, MA: MIT Press.

3

Neuroanatomical Foundations of Cognition: Connecting the Neuronal Level with the Study of Higher Brain Areas

Jennifer Mundale

1 Introduction

In the philosophy of any science, it is important to consider how that science orga-
nizes itself with respect to its characteristic areas of inquiry. In the case of neuroscience, it is important to recognize that many neuroscientists approach the brain as
a stratified system. In other words, it is generally acknowledged that there is a wide
range of levels at which to investigate the brain, ranging from the micro to the macro
level with respect to both anatomical scope and functional complexity. The following list of neuroanatomical kinds, for example, is ordered from the micro to the macro
level: neurotransmitters, synapses, neurons, pathways, brain areas, systems, the brain,
and central nervous system.

For philosophers of neuroscience, it is heuristically useful to understand that
many neuroscientists view the brain in this way, because this conception of the brain
helps shape the disciplinary structure of the field and helps define levels of neuroscientific research. Language, for example, is usually studied at the macro level,
whereas synaptic transmission is investigated at the micro level. Although there is
overlap and integration among the different levels, each level has its own methods
and problem domains.

Furthermore, some levels lend themselves more easily than others to interdisciplinary involvement. Connections with biochemistry, for example, are most easily
made at the micro level, whereas connections with cognitive psychology are typically
made at more macro levels. Although philosophers can in principle make contact
with neuroscience at any level, they have most commonly intersected at the
macro levels of research. This is not surprising: it was, after all, the philosophers of
mind, more so than the philosophers of science, who lead us into this particular

empirical engagement. Except for the most rabid functionalists, all cognitivists find the relevance of higher brain systems to our theories of mental function intuitively obvious.

In its central organization, this book retains that emphasis on higher-level brain research, but we also want to stress the importance and philosophical appeal of integrating work at the micro level. Unfortunately, although the philosophical gain to be had is appreciable, it is also much less apparent. Why bother with *neurons* when you really want to understand the higher operations? One of the goals of this chapter is to show that, even if one's philosophical interests in neuroscience are confined mainly to its role in cognitive explanations, understanding some important issues at the micro level can enhance one's ability to draw from this science. Another goal is to encourage the philosophical appreciation of neuroscience as a science, apart from its contributions to other fields. So, though the lowly neuron generally receives less attention than, say, the enigmatic frontal lobe, the controversy over its structural and functional elucidation is the stuff of neuroscientific legend and is of great philosophical interest.

In what follows, while I discuss both cellular and systems-level perspectives on the brain, I attempt to do more justice to the former by emphasizing the neuron doctrine (briefly, the view that neurons are physically discrete, cellular units) and the relation of this doctrine to higher-level theories. After a brief introduction to the anatomy of a neuron, I begin my account with some of the formative events in nineteenth-century biology which led to the development of the neuron doctrine. The nineteenth century is a convenient point to begin the discussion since it was at this time that the microscope (a seventeenth-century invention) really came to the forefront of scientific research. In part, this was because the microscope had undergone significant refinement in the early nineteenth century, resulting in improved clarity and range. It was at this time that microscopy at the cellular level first became sophisticated enough to be useful to biological theory. In tracing some of the most important advances from this point into the late twentieth century, I hope to show the fundamental importance of addressing micro-level research in the philosophy of neuroscience. My goal, however, is not to deny the importance of higher-level views of the brain. Rather, it is hoped that such redress will only facilitate a deeper, more informed perspective of the grosser level, the level from which philosophers most commonly draw in relating neuroscience and philosophy.

2 Basic Neuron Anatomy

In order to appreciate the issues discussed below, it is necessary to understand a few rudiments of neuron anatomy. Neurons, or nerve cells, consist of three major parts: cell body (soma), axon, and dendrites. Most neurons have only one axon, but several dendrites. Axons and dendrites are known as processes. Dendrites receive incoming signals and carry them to the cell body. Axons carry signals (action potentials) away from the cell body and toward the synapse (see figure 3.1).

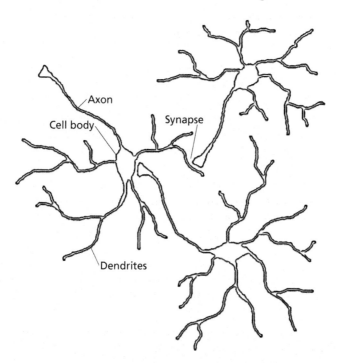

Figure 3.1   A neuron synapsing with another neuron. Note the soma, dendrites, axon, and synapse.

A synapse is a junction between two neurons, that is, between the axon terminal of the presynaptic cell and the dendrite (or sometimes, the cell body itself) of the postsynaptic cell. Synapses can be either chemical or electrical, though most are chemical. At the chemical synapse, there is no direct contact between the processes of the two cells. Instead, the signal transmission between them is mediated by chemicals known as neurotransmitters. The open space between the presynaptic cell and the postsynaptic cell is known as the synaptic cleft; this is the space into which neurotransmitters are released by the presynaptic cell.

Electrical synapses are mediated by the flow of electrical current from one cell to the next. At these synapses, the presynaptic and postsynaptic cells are linked by a very special kind of connection known as a *gap junction*, which appears to involve some cytoplasmic continuity between the connecting processes. For reasons which will be discussed below, gap junctions present a particular difficulty for the neuron doctrine.

There are several different kinds of neurons, differing according to features such as shape, number of processes, size, targets, location, neurotransmitter selectivity, and function. Estimates of the total number of neurons in the human brain vary widely, but 10^{11} appears to be a conservative estimate. Each neuron, in turn, forms an average of 1,000 synaptic connections with other neurons, resulting in at least

10^{14} synaptic connections within the brain (Kandel et al., 1991, p. 121). Altogether, we are confronted with a system of staggering complexity when looked at from the neuronal level.

For the purposes of this chapter, key concepts about neuronal structure can be summarized as follows: (1) in chemical synapses (the most common type), there is no physical contact between presynaptic and postsynaptic neurons; (2) neurons are discretely bounded, physically separate and distinct cells; (3) axonal and dendritic processes are physically integral to the nerve cell and continuous with the cell body; (4) functionally relevant differences among neuronal types can be microscopically observed (assuming the tissue has first been stained and subjected to other special treatments in order to make microscopic viewing possible). This latter feature makes it possible to use such cellular variation, otherwise known as cytoarchitectonics, as one technique for the localization and mapping of brain functions. The philosophical significance of each of these four features will become clearer in what follows.

3   The Cell Theory

That living organisms are made up of *cells*, and that neurons *are* cells, hardly seems like earthshaking news today, but at one time it was neither obvious nor uncontroversial. In fact, the establishment of the cell theory, or the theory that animal and plant tissues are composed of and generated from cells, was one of the chief advances of nineteenth-century biology. Although most historians acknowledge the key contributions of Schleiden and Purkinje, the cell theory is generally credited to Schwann, who proposed it in 1839 (Finger, 1994, pp. 43–4). Schwann's view was attractive in part because it allowed for powerful, theoretical unification. One can see him as a sort of Democritus with a biological twist: just as different material objects are composed, ultimately, of the same atomic constituents, so all living tissue is composed of the same, basic, cellular elements. What also made Schwann's theory attractive is that he proposed a general mechanism for cellular generation whereby cells are formed from the inside out, by the gradual accretion of new material, in a manner analogous to crystallization. This part of his theory proved to be erroneous, however, and by the 1850s, first Remak, and then Virchow proposed that cells are generated through cellular division. In spite of Schwann's false lead, however, the mechanism he proposed helped establish the larger idea that cells are the basic building blocks of life.

It is easy to suppose that the cell theory and the emerging structural picture of the neuron would have dovetailed so neatly as to have escaped notice. After all, given the variability of cell types found elsewhere in the body, it would seem to require very little effort to regard the neuron as just another kind of cell, and the neuron doctrine as nothing more than an extension of the cell theory. But now envision a neuron's shape and structure, and you will see an immediate problem researchers faced in classifying them as cells: neurons have long, thready processes sticking out of them, features not found in other cells of the body. To regard the threads as part of the neuron was to be faced with two difficult alternatives. Either the concept of a cell had to be revised in order to include the irregular neuron, or the neuron had to be stricken from the cellular category.

On the other hand, it was not clear that these threads were physically integral to the neuron. Perhaps they merely played a supporting role, and, though always seen with neurons, were not themselves part of the neuron. Valentin and Purkinje, for example, both prominent, mid-nineteenth-century investigators, examined brain tissue under a microscope, and concluded that the processes were not physically continuous with the cell body. They did hold, however, that the processes played some role in the transmission of neural signals. In this case, neurons could be regarded as just "normal" cells after all, and the brain would appear to be a "normal" tissue, composed of cells just like all other tissues. The connective relation between the neuron's cell body and its processes was not clearly understood by early neuromicroscopists, and only began to take shape in the late 1830s through the 1850s, with the work of such figures as Remak and Schwann (Shepherd, 1991, pp. 19–23).

This seemingly minor issue itself makes up a philosophically rich chapter in the history of neuroscience. It ties in with such large-scale issues as materialist vs. dualist (Cartesian) views of the mind, and mechanistic vs. vitalistic explanations of natural phenomena (vitalism, or the appeal to vital forces to explain living phenomena, was still common in the nineteenth century). An extended discussion of these topics is not possible here, but they are worth a passing mention. Jacobson (1995) suggests that depicting the nerve cell body as continuous with its input and output processes conflicts with Descartes's concept of the mind, because, in this view, "There was no room for the soul between input and output in that model, and although not explicitly stated, it was implied that the soul was not required for routing the nervous activity in the brain and spinal cord. The nucleated part of the nerve cell then occupied the place between input and output formerly occupied by the Cartesian soul" (p. 181). Furthermore, if brain *tissue* were to be relegated to the lowly status of all other living tissues, not fundamentally different in kind, this demotion can be seen as yet another slap at the Rylean ghost caught within the cellular machine, and as a hopeful nod toward a reductive, mechanistic explanation.

In any event, seeing the nerve fibers as physically continuous with the cell body not only stretched the structural concept of a cell, but generated another point of tension with the cell theory. This tension centered around the nature of inter-cellular connection. For the cell theorist, a cell is regarded as an individual, physically discrete unit. Neural tissue, though, when prepared (stained, sectioned, fixed, etc.) and examined under a microscope, looked like a tangled net, dotted with small black blobs. So, if the long, thready processes of the nerve cell are actually part of the cell, where does one nerve cell begin and another end? *Do* they end, or is neural tissue made up of a vast, physically continuous network? (See figure 3.2.)

These seem to be questions that could easily be settled by mere observation, but there were serious limitations on what could be observed at that time.

Histological methods of staining and preparation were not nearly up to the task of settling these questions, and they would not prove to be dispositive, in fact, until the advent of electron microscopy in the mid-twentieth century. Thus we have arrived at one of the most important neuroscientific debates of the twentieth century.

4 The Neuron Doctrine

In simple terms, the neuron doctrine consists of the view that neurons are discrete, individual cells, one physically discontinuous from another; whereas the opposing, reticular theory holds that the brain consists of a continuous web, or reticulum of nerve fibers. Camillo Golgi (1843–1926) was a committed reticularist, but he invented a method of staining neurons (discussed below) which helped Santiago Ramón y Cajal (1852–1934) make a convincing case for the neuron doctrine. The debate between Golgi and Ramón y Cajal (usually referred to just as "Cajal") is now a famous bedtime story for students of neuroscience. In the fabled version, it amounts to a heroic battle in which the great but doddering Golgi stubbornly clings to the old view 'til death, but nonetheless invents the very weapon which the brave young Spanish blade, Santiago Ramón y Cajal, uses to defeat him. Though Cajal was right and Golgi was wrong, they both won the Nobel Prize anyway, and the neuron doctrine lived happily ever after . . . or something like that.

The less condensed version of the story often flounders on just what the neuron doctrine consists of, for here things get a bit more complicated, and not all commentators agree (compare, for example, the contemporary accounts of Jacobson, 1995, and Shepherd, 1991, both with each other and with such classic sources as Clarke and O'Malley, 1968, 1996). Differences and difficulties in the precise formulation of this doctrine can be attributed to at least four major factors. First, the doctrine evolved over time and there are differences from one historical period to another. Second, its relation to the cell theory has been variously construed, and the earliest roots of the neuron doctrine were in place even before neurons themselves were widely accepted as just another kind of cell. Third, even those most closely involved with this research did not always agree on just what constituted the neuron doctrine. Golgi and Cajal, for example, articulated very different views about what constituted the neuron doctrine. For that matter, even the reticularist viewpoint is not easily encapsulated. Cajal, for example, complained in his Nobel address that the form of the reticular viewpoint "changes every five or six years" (1906, p. 241). Fourth, it was and is a very difficult matter to decide just what is to count as a part of the main doctrine itself, as opposed to the many closely related issues about neuronal structure and function.

The essence of the controversy, however, turns on whether or not the neuronal processes in the brain all run together into one vast, physically continuous network or reticulum (as Golgi and other reticularists saw it), or whether the neurons are physically distinct cells connected merely contiguously (as Cajal and other neuronists would have it). Certainly the neuronists had an easier time squaring their views with the cell theory, as discussed above, but the reticularists, as we will see more clearly below, had an easier time squaring their views with certain other theories.

On the foundation of the cell theory, the neuron doctrine had been building for some time before its star combatants, Golgi and Cajal, took center stage. As it happens, it was Golgi's development of a revolutionary, silver-based stain which laid the basis for their eventual confrontation. As noted above, it is one of the most famous of scientific ironies that Golgi himself invented the very method which Cajal exploited so successfully against him. The great advantage of Golgi's procedure is that it stains only a small percentage of the nerve cells in a given sample (those that are stained, however, are stained completely). Previous methods, on the other hand, tended to stain samples indiscriminately, thereby making it very difficult to discern much of anything, let alone the finer structure of a single neuron (see figure 3.2). It is important to keep in mind, however, that Golgi himself did not know why the reaction worked this way, and we still do not have a clear understanding of the underlying mechanism.

Golgi developed his technique in the early 1870s and published his findings in 1873, though his work did not attract immediate attention. In 1887, Cajal first became acquainted with the complicated Golgi method and was instantly captivated by it (Ramón y Cajal, 1988, p. 3). He set to work on it immediately, attempting both to improve the unpredictable nature of the reaction itself, as well as to capitalize on the histological advantages it offered. By 1888 he had already begun to publish the results of his early work with the Golgi technique, and by 1889 was making definite assertions in print which directly opposed Golgi's reticular theory.

From Cajal's perspective, he simply *saw* when he looked through the microscope that his Golgi preparations supported the neuron doctrine. Yet, if it were this obvious, why didn't Golgi come to the same conclusions? Consider figure 3.2. and it is easier to see why the debate wore on for some time even after the use of the Golgi stain. Visual observation of neural tissue did not provide conclusive warrant to decide between Cajal's neuronism and Golgi's reticularism. Although this kind of evidence alone would not make for decisive interpretation, the state of microscopy at that time could provide nothing better. What else was there?

Commonly, the larger theoretical structures with which the evidence coheres provides further grounds of appeal. But even when lodged within larger theoretical frameworks, neuronism and reticularism were still fairly well-matched rivals. Golgi's reticularism harmonized well with a holistic, or non-localizationist theory of brain function, to which he was also deeply committed. Here, he explicitly notes the tension between localizationist ideas and his network theory:

> Another observation occurs to me: The concept of the so-called location of the cerebral functions, should it be insisted on accepting it in a rigorous sense, would not be in perfect harmony with the anatomical data, or at the least, it should now be admitted only in a somewhat limited . . . sense. It being demonstrated, for example, that a nervous fibre is in relation with extensive groups of gangliar cells, and that the

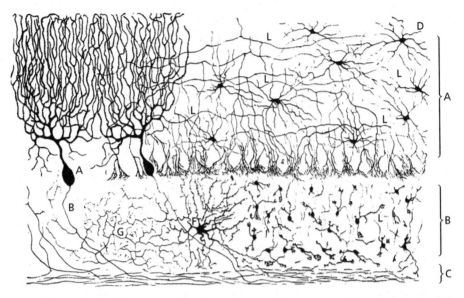

Figure 3.2 Cajal's illustration of brain tissue sample, taken from cerebellum, and prepared for microscopic observation according to Golgi method. From Observations sur la texture des fibres musculaires des pattes et des ailes des insectes. *International Journal of Anatomy and Physiology*, 5 (1888), 205–32; 253–76.

> gangliar elements of entire provinces, and also of various neighboring provinces, are conjoined by means of a diffuse network . . . it is naturally difficult to understand a rigorous functional localization, as many would have it.
>
> (Golgi, 1883, in Shepherd, 1991, p. 96)

Placing the neuron debate within the context of the localization debate enlarges the discussion but confers no obvious advantage for either side. Just as investigators were split over the neuron question, they were split on the question of cerebral localization. More will be said about the localization controversy below, but for now it is important to note that it was also a contentious issue with an even longer history than that of the neuron doctrine. In a more subdued tone, the debate continues even to the present day.

On other grounds of appeal, when it came to explaining the transmission of neural signals from one part of the brain to the next, the reticularists clearly win the prize for parsimony. If neurons form a continuous, uninterrupted network, understanding signal transmission becomes a simple matter. Even Cajal acknowledges this point, noting that

> [i]t is necessary to realize that for certain minds the reticular theory offers a most attractive and convenient explanation. Among other physiological advantages it would offer

the inestimable one of explaining in a simple manner the propagation of the nerve impulse from one neuron to another and its diffusion throughout the gray substance in a number of directions.

The important thing here is not to ponder the theoretical simplicity and facility (more apparent than real) of a theory but rather to evaluate to what extent it conforms with well-known, demonstrable facts.

(Ramón y Cajal, 1954, trans. Purkiss and Fox, 1954, p. 1)

Neuronists, by contrast, had to explain how the signal is propagated across an open space. Today, of course, we recognize this as the process of synaptic transmission, and have detailed, well-corroborated theories about its mechanism. Cajal did not.

Interestingly, though, at approximately the same time that Cajal was establishing the neuron doctrine, Charles Sherrington was developing the theory of synaptic transmission. Sherrington coined the term *synapse* in 1897, and his earliest work on the subject can be seen as providing mutual support with the neuron doctrine, as we can see from his statement below:

So far as our present knowledge goes, we are led to think that the tip of a twig of the arborescence is not continuous with but merely in contact with the substance of the dendrite or cell body on which it impinges. Such a special connection of one nerve cell with another might be called a synapse.

(Sherrington, 1897, in Shepherd, 1991, p. 228)

So, the reticularists had an easier way to account for neural signaling than the neuronists; yet even though neuronists had a more difficult story to tell, that story was already in progress, and being developed independently of Cajal's work.

Golgi never did capitulate to the neuron doctrine, though it was clear by the time they won the Nobel Prize (1906) that support had begun to sway in favor of the neuron doctrine. Even so, reticularism could not be ruled out as conclusively as one might expect. Until the development of electron microscopy the technology did not support a decisive verdict.

Even if the micrographic images themselves had presented an unambiguous case, this likely would not have settled the matter, for, as philosophers of science are wont to point out, it is not always clear what you see when you look through a microscope. Several issues arise here having to do with realism vs. anti-realism, the theory-ladenness of perception, and others, but most immediately, there is the problem of separating data from artifact.

It is a long, tortuous road from brain tissue to a thinly sliced, chemically stained and fixed slide preparation such as those on which Golgi, Cajal, and other late-nineteenth-century neuroanatomists were basing their conclusions. The transformation that takes place may not result in a veridical image of the object of study; artifacts of insertion, deletion, and distortion are all possible, and can lead to a misinterpretation of the underlying phenomena. Bechtel (1995) identifies several distinct factors underlying the difficulties separating data from artifact:

First, there are often a large number of intervening steps between the original phenomenon and the results that are construed as data. Each of these steps is potentially
a point that could give rise to an artifact. Second, many procedures are extremely brutal
since one often has to transform radically the phenomenon to achieve interpretable
results. Third, there is often very little knowledge about how exactly the procedures
work. . . . Last, it is often the case that procedures are extremely sensitive to the details
of the way in which they are carried out such that slight variation in procedures may
alter the results. . . .

(p. 167)

In the case of the Golgi preparation, each of these four factors is particularly striking. With respect to the first two, I have already noted some of the harsh, complicated procedures to which neural tissue is subjected in order to render images visible
under the microscope. With respect to the third, for example, we *still* do not have a
clear understanding of how and why the process works, and even less was understood in Golgi's time. With respect to the fourth, Golgi and Cajal were both frustrated by the capricious nature of the reaction and made constant adjustments in the
process in order to attain better results. In this sense, there was no single Golgi
method, but several variations on a broad technique.

Brief excerpts from Golgi's own 1875 description of his method underscore the
presence of all the complicating factors identified above. Consider, for example,
that *after* the samples have completed the long hardening and staining processes,
they still have to undergo the following procedure in order to be made ready for the
microscope:

For microscopic examination the sections are placed in damar varnish . . . or in Canada
balsam after they have been dehydrated through the use of absolute alcohol and have
been rendered transparent with creosote.

Time and light continually spoil the microscopic preparations obtained with my
method. . . .

(in Clarke and O'Malley, 1996, p. 842)

Golgi is also indefinite about the lengths of time required for each phase of the
reaction to take place and loosely remarks, for example, that the length of time he
specifies can be decreased in hot weather and increased in cold weather. Finally,
interspersed throughout his description are such caveats as: "I must equally declare
that I have not yet succeeded in determining with certainty why under the same
conditions . . . I have obtained very different results," and "Permit me to advise,
however, that I do not find myself as yet in a position to explain with precision all
the necessary procedures for the best results. They are still partly fortuitous" (in
1996, p. 845).

In time, of course, several converging factors led to the acceptance of the neuron
doctrine in spite of the complications described above. Improved technology,
multiple and independent confirmation, entrenchment within larger theories, all
of these contributed to establishing it as one of the cornerstones of neuroscience.
But many of the concerns raised in this debate, including the problem of artifacts,

retain their importance for this science; some of these will be discussed further below.

As a wry footnote to this section, recent evidence shows that we cannot quite yet consider this case closed. Shepherd's (1991) remarkable history of the neuron doctrine closes with an absorbing analysis of how some aspects of it may need to be gently revised in light of new research. The categories under which he considers challenges to the neuron doctrine are: the neuron as anatomical unit, physiological unit, genetic unit, and metabolic unit. One of the more significant developments he addresses is how the dramatic discovery of the gap junction (mentioned above) calls into question the status of the neuron as an anatomical unit. At gap junctions, there is some physical continuity, or a direct coupling between connecting neurons. Signals are carried electrically, by ionic current, rather than chemically, by neurotransmitters. Across gap junctions, aligned portals form in both presynaptic and postsynaptic neurons, creating a conduit through which ions and other molecules flow freely from one neuron to another. This direct flow of ionic current eliminates the synaptic delay found in chemically mediated transmission. For further comparison, distances across a chemical synapse (30–50 nm) are approximately 10 times greater than distances across gap junctions (3.5 nm). Both structurally and functionally there is continuity between the two neurons, which is exactly what the essence of the neuron doctrine denies. It should be borne in mind, however, that chemical synapses are in greater abundance than electrical ones. Nevertheless, if Golgi had lived to see this discovery, he would probably find some consolation in it.

5   Cytoarchitecture and Localization

We have seen how the cell theory provided support for neuron doctrine. Now I will connect the neuron doctrine with localization theory and another important progression of micro-level research: cytoarchitectonics. I will begin with localization theory.

Before addressing specific localization theories, it is useful to consider the concept of localization itself. What does it mean for a function to be localized? Early attempts to localize function were pretty straightforward, carving out a spatially defined region of the brain and correlating it with a distinct function. This is one means of analysis, and provides a convenient continuum along which to evaluate degrees of localization. Weak localizationists might be willing to attribute some functional differentiation between the two hemispheres, or perhaps among the different lobes, for example, but would not agree to specialization on a smaller scale. Strong localizationists would see functional differences among small, narrowly circumscribed regions, on the order of Brodmann's areas (discussed below), for example, or even smaller regions.

In this quantitative construal, functions are associated with a spatially defined area of the brain, and the magnitude at which functional differences are thought to arise differs from one theory to another. This also provides a convenient way of understanding the historical progression of localization theory. As Finger explains, "In the long history of the brain sciences, it is possible to conceive of the theory of localization as being applied to the whole and then to increasingly smaller parts" (1994, p. 3).

Another way to think of localization, compatible with the sense above, is qualitatively. In other words, one might begin to associate a function with a specific, neurological correlate, but that correlate might be of a given neurological kind, rather than a spatially well-defined one. In this way, we can speak about the functional specificity of different cortical layers, rods vs. cones, the purkinje cells of the cerebellum, dopaminergic pathways, etc.

It is also important to consider the function itself when analyzing the concept of localization. Some functions are considered more localizable than others, and this also helps define different degrees of localization. Many of the functions associated with the senses are considered to be more obviously localizable than consciousness, for example.

Finally, contemporary brain-mapping techniques stretch the notion of localization beyond these relatively simple analyses. In this case, a given function may correlate with repeatable activation patterns of several disparate regions of the brain. These activation patterns are neither quantitatively nor qualitatively well defined, but they are specific and regular. In many cases, it has also been possible to analyze these complex activations into the subfunctions associated with the smaller, better-defined regions of activation.

Historically, almost since the brain itself came to be seen as the seat of cognition, there have been attempts to correlate a given region of the brain with a specific psychological function. Some of the earliest localizationist theories arose in the fourth and fifth centuries, and involved metaphysically obscure, functional differentiation of the ventricles, or hollow cavities of the brain (then thought to be permeated with ethereal, animal spirits). Renaissance theories showed appreciable gain in terms of anatomical sophistication and detail, but the first modern theories of cortical localization did not appear until the nineteenth century.

One of the more influential localization schemes of this period was phrenology, developed by Gall and Spurzheim in the early 1800s (this movement is well known, so will only be summarized here). Although Gall's map delineated boundaries on the cranial surface, it was assumed that the cranial surface conformed perfectly to the underlying cortex, and it was the cortex Gall was actually mapping. He claims to have derived the map through years of human observation, correlating the most pronounced psychological characteristics of thousands of subjects with localized enlargements in the skull. Although his central assumptions were erroneous and his work widely discredited, many modern commentators agree that the essence of his project, to map the functional regions of the cortex, represented an important leap forward. Even Brodmann, a central figure of brain-mapping research, credits him with this important conceptual contribution (Brodmann, 1909, Garey trans. 1994, p. 250). As I will explain in more detail shortly, Brodmann's most important tool in mapping the cortex was cytoarchitectonics.

Cytoarchitectonics provides a means of identifying and delineating the functional areas of the brain according to neuronal population patterns, or cellular "demographics." Of course, the overarching assumption of cytoarchitectonics is that function *can* be localized to specific brain regions; it is the raison d'être of this research. Another assumption behind the research has to do with the relation between structure and function, particularly at the histological level. Neurons vary according to number of processes, length of processes, degree of arborization, cell body size, cell body shape, and other structural features. Some structural features provide clues to the functional significance of a given neuron. The degree of dendritic arborization, for example, is an indication of how many input connections a cell can accommodate, and the degree of axonal arborization tells us about the number of different output sites. Not surprisingly, regions which differ cytoarchitectonically are likely to differ functionally, and this is why cytoarchitectonic variation is one of the techniques used to map the functional regions of the brain. The operative principle is that there is a close connection between structure and function, such that, when one varies, the other also tends to vary. This assumption has been instrumental in twentieth-century neuroscience, particularly in brain-mapping research.

These two assumptions behind cytoarchitectonic research, that function can be localized, and that functional boundaries correspond to cytoarchitectonic boundaries, are difficult for a reticularist to accommodate. With respect to the first assumption, it is difficult for a reticularist to explain how or why one part of a continuous, uniform nerve net should behave any differently than another. Holism is much more compatible with reticularism, and, as discussed above, Golgi subscribed to both. A localizationist, on the other hand, sees structural variation, and couples that with functional variation. With respect to the second assumption, cytoarchitectonic variation is not terribly meaningful for reticularists since, in their view, the nerve cell lacks both structural and functional independence. As late as his Nobel lecture, Golgi expressly denied the physiological "individuality and independence of each nerve element," insisting, instead, that:

> nerve cells, instead of working individually, act together, so that we must think that several groups of elements exercise a cumulative effect. . . . However opposed it may seem to the popular tendency to individualize the elements, I cannot abandon the idea of a unitary action of the nervous system.
>
> (1906, p. 216)

Golgi's remarks make it even easier to see why reticularism and holism are complementary views. Similarly, it should also be easier to see the conceptual and historical connections among the localization of function, the neuron doctrine, and cytoarchitectonic research.

The holistic view of the brain (minus reticularism), presently endures; as, of course, does the localizationist view. These two, broad research traditions, holistic and localizationist, have both made important contributions to neuroscience. Before pursuing the latter topic, below, the holistic tradition deserves some mention. At approximately the same time that cytoarchitectonic research was beginning to achieve worldwide prominence in neuroscientific research, World War I produced the practical necessity of finding effective treatments of soldiers who had sustained severe head injuries. In this sphere, holistic principles dominated therapeutic assumptions and provided more optimistic prognoses than a strictly localizationist framework. It will come as no surprise that Golgi worked at a military hospital during the war, where he created a special center for the treatment, study, and rehabilitation of soldiers with neurological injuries. The numerous cases of rehabilitated soldiers who experienced full or partial recovery of function provided some vindication for holistically oriented thinking. Kurt Goldstein was another eminent holist to emerge from this crucible, and he later helped found the Gestalt movement in psychology and neurology. Though not adamantly opposed to the concept of some regional specialization in the brain, Goldstein was holistic both in the sense discussed above and in the following sense: he enlarged the scope of information which was brought to bear in guiding his treatment and understanding of his patients. He saw the recovering patient as an organism with altered abilities attempting to cope with an environment of constantly shifting demands and challenges. The spirit of this comprehensive approach, taking into account both the patient and the world in which the patient lives, also continues to influence neuropsychiatry, and is reflected in such notable figures as Oliver Sacks. In philosophy and cognitive science, one sees a similar approach in the current shift toward "situated cognition."

6 Brain Areas and Modern Neuroscience

Although the 1990s were considered the decade of the brain, for some areas of research, it might better be seen as the grande finale of an entire century of the brain. In the case of brain mapping, the endpoints of the twentieth century mark a particularly important period of progress. In the early 1900s, newly developed cytoarchitectonic techniques (see above) made possible the first scientifically significant maps of the cortical surface. By the century's close, sophisticated radiographic techniques were imaging the areas where increased activation occurred as live, human subjects performed specific tasks. In this section I discuss how the acceptance of both the neuron doctrine and the localizability of function was key to these developments.

To embark on a project of mapping the functional regions of the brain, of course, requires some commitment to the localizability of function, since that is the goal of such research. As I discussed in the previous section, *cytoarchitectonic* research involves the further assumption that different populations of cells, as distinguished at the histochemical level, perform different functions. This micro-level case is an instance of a general biological principle that physical differentiation tracks functional differentiation (and vice versa). Toward the end of the nineteenth century, Betz, and later Flechsig, published some of the earliest cytoarchitectonic work, but the method did not really come into its own until early in the twentieth century. From roughly 1905 to 1925 several researchers employed cytoarchitectonic techniques in deriving functional maps of the human cortex. Campbell produced a series of maps in 1905, Vogt and Vogt in 1919, and Economo in 1925, but the most famous and influential map (see figure 3.3) was produced in 1907 (and revised in 1909), by Korbinian Brodmann (1868–1918).

Brodmann's work and commentary provide us with a clear example of how ideas concerning neuronal structure and function became an integral part of subsequent brain cartography. Unlike many of his contemporaries, Brodmann often comments on the deeper, theoretical issues related to his research, and attempts to justify his overall approach to the problem of mapping the cortex. Possibly, this helps to explain why he had such a formative effect on future cartographic research. He makes his assumptions quite clear, for example, when he writes: "There is an undisputed axiom: physiologically dissimilar elements have dissimilar structures. Reversing this statement one may equally justifiably conclude: parts of organs that are structurally different must serve different purposes" (Brodmann, 1909, Garey trans. 1994, p. 253). He also clearly identifies his commitment to the specific case of this "axiom," which is that the level of dissimilarity which is functionally significant extends down to the cellular, or histological level: "It is a basic biological principle that the function of an organ is correlated with its elementary histological structure" (ibid., p. 243). Additionally, scattered throughout his work are several attempts to support this principle. Clearly, Brodmann's work depended on a view of the neuron as a largely, if not entirely, structurally and functionally independent unit. That, of course, is exactly what the neuron doctrine was all about.

Now, to connect Brodmann's work with later brain-mapping research, it is important to realize that in addition to cytoarchitectonics (which is still in use), many other methods have been employed to chart the functional regions of the brain, and Brodmann's famous map itself has undergone some modest revisions by other researchers (see, for example, Mundale, 1998). Yet Brodmann's map has served as a common reference point since it was first published, and continues to do so for contemporary neuroscientists. It also helped to support explanations of human behavior in terms of areas of functional activation. Though Brodmann's methods and results were seriously challenged by critics from the holistic and Gestalt schools (see especially Lashley and Clark, 1946, for example), Brodmann struck a lasting blow for localizationist thinking which continues to motivate cartographic research.

Other chapters in this volume will elaborate more fully on contemporary brain-mapping methods, particularly PET scanning and other radiographic techniques. And those, in turn, will be tied to a greater understanding of such high-level

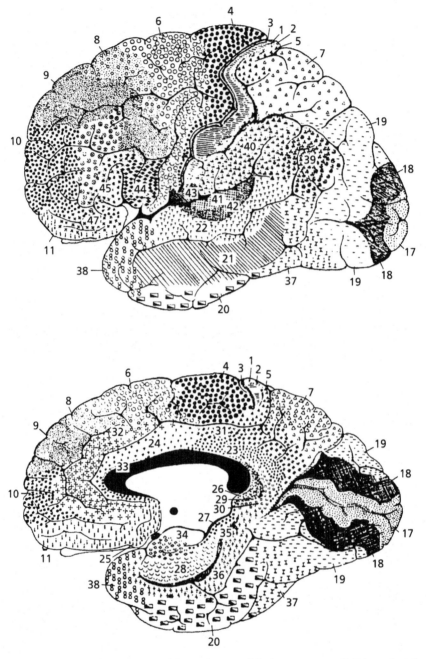

Figure 3.3 Brodmann's (1909) cytoarchitectonic map of the human cortex. Relying primarily on regional differences in cell type, density, and distribution – a method broadly referred to as *cyto-architectonics* – Brodmann identified over 40 distinct areas of human cortex.

functions as perception and language. But we can also work back down to the lowly neuron, with a more connected perspective about its place in the foundations of cognition.

7 Conclusion

I began with an explanation of how researchers approach the brain as a stratified, yet integrated system in both physical and functional respects. Working within this framework, my major concern has been to show how an appreciation of several critical developments at the micro level, though often overlooked by philosophers, can greatly enhance our understanding of the brain and its functions at higher levels. To show this, I traced one particular thread from the cell theory, the neuron doctrine, the localization of function, cytoarchitectonics, and modern brain-mapping research. Although other antecedent and subsequent developments in this thread, as well as several important collaterals, remain unmentioned, it is hoped that the main example itself will provide some sense of how to enlarge the picture and draw further, conceptually useful connections between micro-level and macro-level subjects.

References

Bechtel, W. 1995: Deciding on the data: Epistemological problems surrounding instruments and research techniques in cell biology. *PSA 1994*, 2, 167–78.

Brodmann, K. 1909: *Vergleichende Lokalisationslehre der Grosshirnrinde in ihren Prinzipien dargestellt auf Grund des Zellenbaues*. Leipzig: Barth, 1909. Trans. and ed. by Laurence J. Garey, *Brodmann's "Localisation in the Cerebral Cortex."* London: Smith-Gordon, 1994.

Clarke, Edwin, and O'Malley, C. D. 1968: *The Human Brain and Spinal Cord: A Historical Study Illustrated by Writings from Antiquity to the Twentieth Century*. Berkeley: University of California Press.

Clarke, E., and O'Malley, C. D. 1996: *The Human Brain and Spinal Cord: A Historical Study Illustrated by Writings from Antiquity to the Twentieth Century*, 2nd edn, revised and enlarged, with a new preface by Edwin Clarke San Francisco: Norman Publishing.

Finger, S. 1994: *The Origins of Neuroscience*. New York: Oxford University Press.

Golgi, C. 1906: The Neuron Doctrine – Theory and Facts. In *Nobel Lectures, Including Presentation Speeches and Laureates' Biographies: Physiology or Medicine, 1901–1921*, Amsterdam: Elsevier, 1967.

Jacobson, M. 1995: *Foundations of Neuroscience*. New York: Plenum Press.

Kandel, E. R., Scwartz, J., and Jessell, T. (eds) 1991: *Principles of Neural Science*, 3rd edn. New York: Elsevier.

Lashley, K. S., and Clark, G. 1946: The cytoarchitecture of the cerebral cortex of ateles: A critical examination of architectonic studies. *Journal of Comparative Neurology*, 85, 223–305.

Mundale, J. 1998: Brain mapping. In W. Bechtel and G. Graham (eds), *A Companion to Cognitive Science*, Cambridge, MA: Blackwell.

Ramón y Cajal, S. 1906: The Structure and Connexions of Neurons. In *Nobel Lectures, Including Presentation Speeches and Laureates' Biographies: Physiology or Medicine, 1901–1921*, Amsterdam: Elsevier, 1967.

Ramón y Cajal, S. 1954: *Neuron Theory or Reticular Theory?* Trans. M. U. Purkiss and C. Fox. Madrid: Consejo Superior De Investigaciones Científicas.

Ramón y Cajal, S. 1988: *Cajal on the Cerebral Cortex: An Annotated Translation of the Complete Writings.* J. DeFelipe and E. G. Jones (eds). New York: Oxford University Press.

Shepherd, G. M. 1991: *Foundations of the Neuron Doctrine*. New York: Oxford University Press.

4

Epistemic Issues in Procuring Evidence about the Brain: The Importance of Research Instruments and Techniques

William Bechtel and Robert S. Stufflebeam

1 The Epistemic Challenge Posed by Research Instruments and Techniques

According to traditional philosophical accounts of scientific methodology, the evidence for scientific theories stems from observation, especially observation with the naked eye. These accounts portray the testing of scientific theories as a matter of comparing the predictions of the theory with the data generated by these observations, which are taken to be an objective portrayal of reality. One lesson learned by philosophers of science in the last 40 years is that even observation with the naked eye is not as epistemically straightforward as is sometimes assumed. What one is able to see depends upon one's training: a novice looking through a microscope may fail to recognize the neuron and its processes (Hanson, 1958; Kuhn, 1962/1970).[1] But a second lesson is only beginning to be learned: the evidence in science is often not procured through simple observations with the naked eye, but through observations mediated by complex instruments and sophisticated research techniques. In order to acquire evidence about the phenomena under investigation, these instruments must alter it. (For a simple, prosaic example, consider the ordinary thermometer. It requires the transformation of the temperature of the surrounding air into the expansion of a liquid or metal so as to produce a display that we can see [Hacking, 1983].) The fact that evidence consists of altered phenomena then raises a serious question: to what degree is what is taken as evidence just the product of the alteration or in what respects does it reflect the original phenomena for which it is taken to be evidence?

Since most scientific evidence is procured through instruments and research techniques, the question of whether purported evidence is really an artifact is a frequent one, especially when a new instrument or technique is being introduced. One reason for this is that the purported evidence is often extremely variable. Even when different researchers try to use the same instrument and follow the same procedure, subtle variations in the instrument or technique can produce significantly different results. Another is that often when they are introduced, the ways in which the new instruments and techniques alter the phenomena are not understood, at least in sufficient detail to vindicate directly the purported evidence. A potent example of how purported evidence is produced through techniques not well understood is Golgi's silver nitrate stain which only stains selective neurons and played a critical role in establishing the claim that neurons were discrete individual cells (see chapter 3, this volume); yet 100 years after its introduction, its method of action and the reason why it only stains selective cells are not understood.

If purported evidence is only a reflection of the means of altering the phenomena, then it is only an *artifact*; if it is, in the relevant respects, reflective of the underlying phenomena, then it is genuine evidence. The challenge for scientists is to determine which it is. Scientists themselves are usually very much aware of the problem. Especially at a time when new sources of evidence are being introduced, they often engage in bitter and extended arguments as to whether the purported evidence should be trusted. Typically, these controversies are resolved after a short period of time. Protocols for the use of instruments and performance of the techniques that are accepted as informative are established and routinized; students learn them and treat the resulting evidence as authoritative. As a result, the epistemic issues disappear from sight. As Bruno Latour (1987) graphically describes it, the procedures become *black boxes*.

For those concerned with the epistemic status of science, the means by which new instruments and techniques are evaluated and, if accepted, become black boxes, is of central significance. Initially it is a bit of a mystery how scientists could provide evidence that new instruments and techniques are producing genuine evidence and not artifacts. The approach used to evaluate theoretical knowledge – comparing predictions of theories or models against evidence – is of no use since there is no independent body of evidence against which to evaluate the instruments and techniques.[2] As we will demonstrate later in this chapter in the context of discussing techniques for neuroimaging, scientists generally rely on a variety of indirect measures in their evaluations. These include (1) whether the instrument or technique is producing well-defined or determinate results, (2) the degree to which the results from one instrument or technique agree with results generated in other ways, and (3) the degree to which the purported evidence coheres with theories that are taken to be plausible. Before turning to that case study, we will first describe some of the instruments and techniques on which neuroscience, especially neuroscientific accounts of mental function, have relied and attempt to cultivate an appreciation for how much these instruments and techniques alter the underlying phenomena and raise the prospect of artifact. Our goal is not to promote skepticism about scientific inquiry, but rather an awareness of an epistemic challenge that is central to scientific practice.[3]

2   Instruments and Research Techniques Employed in Discovering How the Brain Performs Mental Processes

To study the brain, researchers must make its structure and processes accessible. The functioning brain is usually shrouded in a skull out of sight. Even when the skull is removed, little information about its operation is available through simple observation. One can identify some of its larger landmarks, such as the cerebellum and the major gyri and sulci in convoluted brains such as our own, but these observations do not reveal the relevant structural units in the brain (which tend to be much smaller and not indicated by surface structures; see chapter 3, this volume), let alone the occurrent chemical and electrical processes most significant for maintaining mental life. Even determining that chemical and electrical processes are occurring in the brain requires sophisticated tools of intervention.

In this section we describe a number of the major tools which have been employed in the neuroscience investigations recounted elsewhere in this volume. (We will focus on techniques that are thought to be most informative about the primary neural correlates of cognitive activity, the electrical activities of neurons. One should note, though, that other activities, especially chemical activities, are also highly relevant to cognitive performance and to other mental phenomena such as emotions.) In providing a brief introduction to the instruments and techniques, our emphasis will be on revealing the sort of intervention each requires, the kind of information about the brain each provides, and some of the limitations each encounters.

One general issue that we should note at the outset is that the various techniques for studying brains are employed with members of different species. Most of what we know of the neuroanatomy of the brain has been learned from studies of non-human primates such as the macaque and most of the single-cell electrophysiology has been done on cats and various species of monkeys. Results from these studies are then extrapolated to the human brain. Electroencephalography, PET, and fMRI studies, on the other hand, are generally done on humans, and although some comparative work on other species has begun, such investigations pose serious challenges. Yet, from what is known, it is clear that there are significant differences in the organization of brains across species, rendering the task of making inferences across species challenging.

Neuroanatomical methods

Neuroanatomy, the characterization of the structure of the brain, both at a macro and a micro level, has provided the foundation for a great deal of understanding of how the brain performs mental functions. The previous chapter described some of the classic discoveries of neuroanatomy – the discovery that neurons are the functional units of the nervous system, and the identification of areas of the brain with different neural composition, patterns of connectivity, etc. – and the procedures by

which these were discovered. To determine the physical composition and organization of the brain, neuroanatomists have had to do such things as cut it apart, slice it
into thin preparations, treat these with chemical baths of various sorts, and examine
the products through an (optical, electron, etc.) microscope. These processes are all
disruptive of the normal brain but are pursued because the resulting images are
thought to reveal important structures within the non-disrupted brain.

Critical to the functioning of the brain is the manner in which neurons are
connected to each other. Consider what is involved in discovering the connectivity
pattern in a portion of cortex. You might try viewing a stained preparation of neural
tissue, but all you will see is a tangled web of connections. To discover the relevant
organization within these complex mazes, neuroscientists have done such things as
cut axons from their neurons, causing the axons to die. This results in accumulation
of dense granular material along the axons' path which can be seen in histological
preparations after the animal dies. Another approach is to use retrograde and
anterograde tracers, chemical substances which are taken up by neurons at a given
point and transported backwards or forwards along their axons and dendrites.
Horseradish peroxidase, for example, is taken up by the axons of cells and transported back to the cell bodies where, over a few days, it oxidizes and takes on vivid
colors. This allows the cell bodies connected to particular axons to be readily identified in slice preparation after the animal is killed.

Insofar as neuroanatomical studies depend not just on microscopes to magnify
images, but on dissection of the brain and applications of various stains and tracers,
these studies clearly involve interventions. While many of these interventions are
reasonably well understood (e.g. the process and rate of radioactive decay in radioactive tracers), others are often less well understood (e.g. the process by which various
stains bind with substances within the cell). When successful, these procedures can
provide rich detail about the structures in the nervous system. But there are also
some clear limitations on neuroanatomical approaches. For example, while they may
reveal that neurons in one area project on to another, they do not show what kinds
of connections are involved (inhibitory or excitatory) and what information is conveyed. A critical example of how such limitations are restricting the emerging understanding of the nervous system is provided by the discovery of enormous numbers
of neurons projecting backwards from areas further within the system to areas closer
to the sensory periphery. Most neuroscientists think these recurrent connections are
extremely important for brain function, but at present it is not clear what information is carried by these connections and hence what information-processing
function they perform. Thus, as critical as neuroanatomy is for understanding the
brain, understanding function requires techniques that directly intervene in the
functioning of the brain and render the functional processes salient.

Deficits and lesions

One of the oldest approaches to identifying the function of brain components is
analysis of the deficits resulting from lesions (localized damage) to those compo-

nents. Lesions can originate either from illness or injury or from neuroscientists actually destroying neural tissue. Whatever the source of the lesion, the goal of this approach is to identify a psychological deficit associated with it and to infer from that what contribution the damaged area made to normal psychological function. Although the deficits to language (e.g. aphasia, dyslexia, agraphemia) are perhaps the best known (and the use of lesions to study language will be illustrated in the contributions to Part II), there is a wide range of mental abilities that can be selectively impaired by lesions to the brain (e.g. face recognition, ability to orient towards the locations of objects, encoding new events into episodic memory).

One challenge in lesion research is determining precisely what areas of the brain are injured. Until the recent introduction of imaging technology, one could only determine what areas in the human brain were damaged after the person died and an autopsy was performed. By then, though, the range of damage may have extended. This required Broca (see chapter 5, this volume), for example, to engage in protracted argument as to what the extent of damage was when the deficit appeared. Perhaps the greatest challenge in using lesions and deficits to understand brain operation is to infer precisely what the damaged component had contributed to normal function. The most general inference is that the damaged area was in some way necessary to the normal performance (e.g. inferring from the fact that lesioning the hippocampus or hippocampal region results in anterograde amnesia, to the conclusion that the region is necessary for encoding new episodic memories). Even this inference is problematic as sometimes an organism can recover or develop an alternative way of performing a function over time after brain injury. The challenge is even greater when one tries to specify just what aspect of the task (e.g. encoding new episodic memories) the damaged part played. It may have been responsible for the whole task, or it may have performed only one contributing function, perhaps even an ancillary one. One can gain an appreciation of the challenge involved by considering how one might go about trying to understand how a radio (or a computer) operates by selectively removing parts and examining the resulting performance. As Richard Gregory (1968) notes, removal of a transistor from a radio may cause it to hum, yet it would be a bad inference to assume that the removed transistor was the hum suppressor.

One strategy that is widely invoked in lesion research is to attempt to dissociate two mental functions by showing that damage to a given brain part may interfere with one but not another. Single dissociations, however, do not show that the damaged brain part is only involved in the impaired function, since it could be that the two functions differ in the demands they make on a component and that with increased damage, the same brain part might interfere with both functions. As a result, researchers often seek double dissociations, where damage to one area causes disruption in one function (while leaving the other largely unaffected), and damage to another area disrupts the other function (while leaving the first largely unimpaired). Double dissociations are often taken as compelling evidence that the two functions are performed separately in the brain (Shallice, 1988). Recent investigations with neural networks and other dynamical systems, however, have shown that double dissociations can result from differential damaging of a single system where it is known that there are not different subsystems carrying out separate tasks (e.g. applying rules for pronouncing words versus looking up pronunciations in a lexicon – see Hinton and Shallice, 1991; Van Orden et al., in preparation). Thus, double dissociations are not foolproof indicators that there are separate systems responsible for separately impaired functions.

We noted above the problem of determining the precise extent of naturally occurring lesions. Although there remains uncertainty with surgically induced lesions,[4] in general these offer much more control. However, for obvious ethical reasons, permanent lesions are only made in human brains in neurosurgical patients when it is anticipated that removal of a brain area is likely to have a beneficial effect such as reducing epileptic seizures, and in such cases the pre-existing medical problem is likely to complicate any functional interpretation of the consequences of the lesion. This has meant a restriction of experimental lesion studies to non-human animals. Recently, however, new techniques have been pioneered in which researchers can induce temporary lesions in humans. One of these techniques that currently affords great promise is transmagnetic stimulation; it involves application of a strong but localized magnetic field so as to disrupt the activity in the affected brain area. Early reports (Walsh and Cowey, 1998) indicate that one can disrupt very specific functions, but the critical question of what the affected area contributed positively to the function remains a challenge, one which typically requires complementing lesion studies with electrical stimulation or recording studies.

Electrophysiological studies

The discovery of the nature of electricity and that the brain in part operates on electrical principles enabled neuroscientists to study the brain as one would study other electrical systems, probing it with electrical stimuli or recording its electrical activity. It is important to note, though, that the idea that the brain is an electrical system was only formulated in recent times. One of the eighteenth-century discoverers of electricity, Luigi Galvani, proposed that nerve transmission was electrical on the basis of experiments stimulating peripheral nerves and muscles with an electrostatic device. His proposal was only definitively established in the mid-nineteenth century by Emil du Bois-Reymond, who, by using non-polarizable electrodes and a multiplier for nerve current, developed a galvanometer that was sufficiently sensitive to detect electrical currents in nerves. Only after this demonstration did researchers attempt to analyze the electrical processing within the brain either by stimulating it or by recording the electrical currents generated as the brain carried out various tasks.

Stimulation studies   The strategy in stimulation studies is to inject electrical current into the brain in the attempt to elicit responses, with the assumption that if one can elicit a response from a given area with an exogenous source of electricity, then normal electrical activity in the affected area would also generate the same

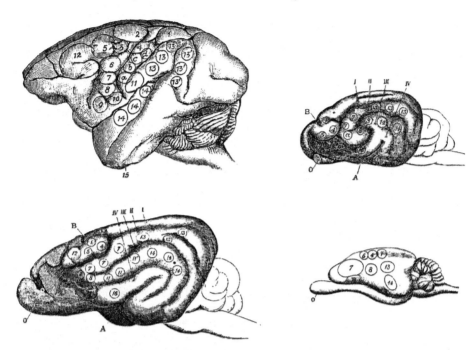

Figure 4.1   Ferrier's (1886) maps of sites on the left hemispheres of monkey (upper left), dog (lower left), cat (upper right) and rabbit (lower right) from which motor responses were elicited with mild electrical stimulation. The same numbering pattern is used for each species and designated a specific motor response.

response. Several nineteenth-century researchers pursued this strategy, but could not discover the right dosage of electricity to elicit a response. By using very mild stimulation, in 1870 Gustav Fritsch and Eduard Hitzig succeeded in eliciting muscle movement in dogs after electrical stimulation of their brains. The approach was generalized by David Ferrier (1876), who elicited responses to electrical stimulations in a variety of areas in the brains of many different species including macaque monkeys, dogs, jackals, cats, rabbits, guinea pigs, rats, pigeons, frogs, and fishes (see figure 4.1). Ferrier argued that many of these loci were not specifically motor, but reflected sensory or other psychological processes leading to motor responses. Ferrier construed stimulation as the natural complement of lesion studies – where deficits resulting from lesions would show what areas were necessary for a function, response after stimulation would show what brain activity was sufficient for a particular response.[5]

Since the pioneering work of Fritsch and Hitzig and Ferrier, electrical stimulation has been widely used in attempts to map brain areas from which particular mental or motor responses could be elicited. Among the major contributors to this work were Walter Rudolf Hess (1949), who applied the technique to subcortical

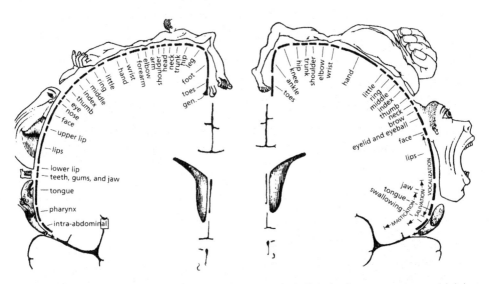

Figure 4.2   Representations of the primary sensory cortex (left) and primary motor cortex (right), using homunculi to show the relative amounts of brain tissue devoted to different portions of the body. Redrawn from Penfield and Rasmussen (1950).

areas, where stimulation seemed to alter emotional behavior in cats, and Wilder Penfield (Penfield and Rasmussen, 1950), who employed it on humans who were candidates for neurosurgery to reduce epileptic symptoms. Penfield's immediate objective was avoiding damage to critical areas such as those responsible for speech, but his research played a major role in mapping cortical areas, including sensory and motor cortices, in humans (see figure 4.2).

Although electrical stimulation has provided a great deal of useful information about the functioning of different brain regions, it relies on some major assumptions. The first is that the electrical stimulus supplied by the electrode induces electrical activity comparable to that which arises within the brain itself. One reason to be cautious is that generally only the joint activity of multiple neurons, transmitted chemically across numerous synapses, is sufficient to elicit a response in another neuron or muscle; to stimulate a neuron artificially, a relatively large burst of electricity must be applied at one location. Such an electrical discharge could easily spread a considerable distance through the cortex and activate areas beyond that which is directly stimulated. Second, just as with lesion studies, there is a serious question of how one should interpret the cognitive contribution of the stimulated area. In the case of the sensory and motor cortices, the idea that these were the projection and major motor command areas in the cortex is supported on other grounds, including neuroanatomy and lesion studies, and what electrical stimulation facilitated was the detailed mapping of, for example, different locations on the primary motor strip which generated motions in different parts of the body. Numerous researchers in the 1960s applied the technique of electrical stimulation more widely and drew conclusions that have proven extremely controversial. For example, James Olds claimed to have discovered a pleasure center upon finding that electrical stimulation probably to an area in or near the hypothalamus would cause rats to press a bar repeatedly when an apparatus was configured so as to produce further stimulation (Olds, 1965). Other researchers have contested his conclusion, advancing alternative interpretations of his findings. For example, William Uttal (1978) concluded: "it now appears that many of the hypothalamic effects on feeding and drinking are mediated not by hypothalamic nuclei but rather by interruption of the sensory-motor signals conveyed by fiber tracts that pass in close proximity to these nuclei" (p. 340). Likewise, José Delgado's (1969) interpretation of his finding, that stimulation in the brain of a bull would stop its aggressive charge, as indicating a center for inhibiting aggression was challenged by Elliot Valenstein, who notes that the stimulated bull continually circles in one direction and concludes, "any scientist with knowledge in this field could conclude only that the stimulation had been activating a neural pathway controlling movement" (Valenstein, 1973, p. 98). Just as with lesion studies, the challenge in electrical stimulation studies is to constrain the interpretation of the contribution of the stimulated site to normal mental function.

Electroencephalogram and evoked response potentials Electroencephalography involves recording aggregate electrical signals from electrodes placed either on the skull or directly on the cortex. In the late nineteenth century such recordings were made from animals by Richard Caton and Adolph Beck, but it was Hans Berger, a German psychiatrist, who applied the technique to humans by adapting methods designed to record the much stronger electrical signal from the heart muscle. He first recorded electroencephalograms in humans with skull defects, then in 1924, from the skull of his then 15-year-old son. Berger distinguished several different patterns of waves, including what he called alpha waves (large-amplitude, low-frequency waves which appeared when subjects closed their eyes) and beta waves (smaller, higher-frequency waves, which appeared as soon as the subject received sensory stimuli or were asked to solve a problem) (Berger, 1929).

Berger's research provides a clear example of the development of an instrument and technique which seemed to be revealing evidence about the operation of the brain, but which was difficult to interpret. On the one hand, as researchers recorded EEGs from a wide variety of individuals in various different states (e.g. different stages of sleep), a complex classification system for EEG patterns began to develop. But, on the other hand, it was not at all clear either what was the source of the signal or what it was telling researchers about brain processes. Initially researchers assumed it reflected some sort of summation of nerve firing. However, early attempts to record from individual neurons (see below) revealed no correlation between individual neural activity and the EEG (Li et al., 1952). It is now accepted that the EEG signal originates principally with pyramidal cells which are aligned in columns in the cortex; when these cells are stimulated, ion flows into and out of the cell are created, resulting in a dipole. When the cells are aligned spatially and activated synchro-

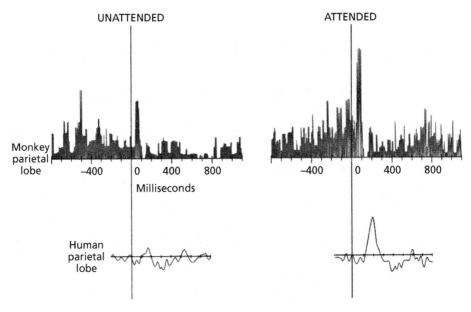

Figure 4.3 Examples of both recordings from a single cell and event-related potentials. The upper traces are the electrical activity shown by a single parietal neuron following stimulus, delivered at a time indicated by the long vertical line. The left side was recorded while the monkey was not attending to the location of the stimulus, and the right side was recorded while it was. The lower traces are the average event-related potentials recorded from a scalp electrode over the parietal lobe of a human subject under the same stimulus conditions. Reprinted from M. I. Posner and M. E. Raichle, *Images of Mind*, New York: Scientific American Library, 1994, p. 21.

nously, these ion flows create an electric field strong enough to be recorded at the scalp (Kutas and Dale, 1997).

EEG recordings have turned out to be useful for studying sleep-wake cycles, identifying brain damage, detecting the origin of epileptic tumors, and monitoring the depth of anesthesia. However, the basic EEG has not been useful for studying information processing in the brain, since recorded EEG activity is a combined measure of many different neural processes. To employ EEG for studying information processing, it was necessary to identify a response in the EEG that was specific to a particular stimulus presented to the subject. This was accomplished when G. D. Dawson (1951) adapted a procedure originally designed for detecting lunar tides in the atmosphere by averaging over a large number of cases. Over many cases in which the same stimulus was presented, background noise would likely be randomly distributed whereas any specific response to the stimulus would stand out. Through use of this averaging procedure, researchers have been able to develop evoked response potentials (ERPs) in which a specific electrical response pattern generated in response to a particular stimulus provides a temporal record of neural activity (figure 4.3).

ERPs have turned out to be extremely valuable in relating neural processing and mental activity because of their temporal precision. For example, by studying the changes in the ERP signal under various attention-directing conditions, researchers are able to resolve in part a long-enduring controversy over whether attention operates early or late in the course of processing a stimulus. Attending to the location of a stimulus resulted in increased P100 and N100 waves (P and N stand for positive and negative currents respectively and the numbers refer to the time in milliseconds after stimulus presentation that the effect is produced) (Luck and Ford, 1998). ERP studies have also been informative about many aspects of language processing. For example, sentences with semantic anomalies generate a negative wave beginning around 200 milliseconds after the presentation of the anomalous word. This wave peaks at approximately 400 milliseconds and so is referred to as an N400 wave. By identifying what word in a semantically anomalous sentence elicited the N400, Garnsey and her colleagues provided evidence that people commit to a particular parsing of ambiguous sentences early rather than waiting for disambiguating information (Garnsey et al., 1989).

One of the chief advantages of ERPs is that they can provide very fine-grained information about the timing of neural processes. On the other hand, it is extremely difficult to determine the spatial origin in the brain of the ERP signal recorded on the skull. The problem, known as the *inverse problem*, is to determine what set of generators within the skull could produce the observed pattern of activation. The difficulty is that there are usually multiple solutions to this problem. A further limitation of ERP studies is that while they can show that there is a distinctive electrical activity related to a particular stimulus, they neither reveal the precise nature of the response at the neuronal level nor specifically what information-processing role that underlying brain activity is playing.

Recording from single cells One of the most powerful ways of relating brain activity to function is to record electrical activity from individual neurons, either by inserting an electrode into the neuron or by placing it next to the neuron, and relate the activity there to ongoing mental processes (see top half of figure 4.3). The first challenge neurophysiologists faced in recording from single cells was to amplify the electrical activity sufficiently to record it. Success in this effort came from the work of Edgar D. Adrian (1926), who connected an intact sensory nerve in the frog to the input of an amplifier and recorded the resulting spike pattern on a oscillograph with a moving strip of photographic film. The recording technique was further advanced by two Washington University investigators, Joseph Erlanger and Herbert Gasser, who in 1922 introduced a cathode ray oscillograph, which they employed to discover that nerve impulses traveled at different velocities depending on the diameter of the fiber, resulting in the classification of different types of nerve fiber (including the philosophers' favorite example, C-fibres; see chapter 16, this volume).

Once the technology for recording from individual neurons was worked out, the challenge was to interpret the activity by relating it to mental processes. To do so, researchers require independent access to mental processes. This is done most easily with neurons involved in sensory processing or controlling motor activity. In the course of his research in which he removed the optic nerve from the frog and identified cells in it that responded to stimuli presented at particular locations in the visual field, Haldan Keffer Hartline (1938) introduced the idea that individual neurons had specific *receptive fields*. The first successful attempt to record from cells in living organisms was Stephen Kuffler's recording from retinal ganglion cells; Kuffler discovered that these cells responded most when a stimulus was presented at the center of the receptive field and not in the surround (on center, off surround) or vice versa (Kuffler, 1953).

Many of the pioneering neuroscience discoveries made in the middle of the twentieth century employed cell recording. In a extremely influential paper entitled "What the frog's eye tells the frog's brain," Jerome Lettvin and his colleagues identified retinal ganglion cells which responded to specific stimuli, including some that responded to small moving spots that the researchers characterized as bug detectors (Lettvin et al., 1959). Hubel and Wiesel's pioneering work in the 1950s, which they summarize in chapter 10, below, used cell recording to map out the response characteristics of cells in primary visual cortex. Working in auditory cortex of the cat during the same period, Clinton Woolsey (1960) identified cells at different locations that responded to stimuli at different frequencies. Cell recording has also been used to identify cells engaged in tasks further removed from the sensory and motor periphery; for example, Goldman-Rakic (1987) has identified cells that continue to fire after a stimulus has been removed when the animal must retain that information for a short interval before performing an action.

As useful a technique as cell recording is, it does have limitations. First, since the technique is primarily correlational, it requires identifying a sensory stimulus, motor response, or ongoing cognitive activity that can be correlated with the neural activity. Second, although when successful it allows researchers to identify what stimulus drives the cell, it does not reveal what contribution the cell is making to processing that information. As Marr (1982) argued, this requires an analysis of the task the cognitive system is performing and accounts (pitched at the psychological, information-processing level) of how it is carrying out that task. Third, it assumes that electrical responses of individual cells are the proper correlate of psychological function. Increasingly, researchers are exploring the possibility that the proper correlate may be a pattern distributed over many cells. Procedures for recording from many, possibly hundreds, of cells simultaneously are now being developed, but these pose serious challenges in terms of analyzing the resulting information.

Neuroimaging

As we noted above, until very recently researchers had to wait until they could perform an autopsy to identify structural features of an individual's brain. One of the major recent developments has been the introduction of procedures that allow researchers to image the brain while the person or organism is still alive. Some of these are principally useful for examining neural structure whereas others have been employed to investigate functions performed by brain structures. Since we will focus on the epistemic issues concerning neuroimaging below, we will defer introducing those issues and simply describe the various neuroimaging techniques.

Conventional radiographs   X-rays, like other forms of radiation, are absorbed by the material through which they pass, with the density of the matter determining the degree of absorption. Those X-rays passing clear through the material irradiate the photographic plate, producing a white image, whereas a black image results when X-rays have been absorbed by the substrate. Dense materials, such bone and tumors absorb many X-rays and show up as black in images. The gray and white matter of the cortex absorbs very few X-rays and so appears as white and cannot be distinguished from other soft tissue. Examination of the brain with traditional X-rays therefore produces a two-dimensional image that primarily reveals the cranial structure and tumors; by adding compounds to the blood that absorb X-rays, it is also possible to create X-ray images of the arteries and veins of the brain. For neuroscientists interested in the organization of gray matter, however, X-ray images are of little use.

Computerized tomography (CT)   In a two-dimensional image such as that produced by X-rays, one cannot determine where in the third dimension an absorbent material might be. By rotating the radiation source (initially, the source was X-rays) and detectors around the object and summing the readings from all of the beams that pass through a given location within the object, researchers developed a means of locating where in the third dimension a structure was located. The Greek word *tomos* means "cut" and this technique is referred to as *tomography* since locations are specified in a single plane cut through the object. Not only do X-ray CTs provide detailed depth information, they also are able to differentiate gray and white matter, blood, and cerebrospinal fluid, thereby providing a much more detailed account of the anatomy of the brain. But the CT technique has also been extended for use with numerous other radiation sources and is used in the last two imaging techniques described below.

Autoradiography   To image function, one needs to identify a correlate of neural firing which provides a recordable signal. Whereas EEG and ERP uses the electrical fields generated by ion flows in and out of cells, functional neuroimaging relies on signals associated with basic cell metabolism or blood flow. One of the first techniques, autoradiography, relies on labeling metabolically active cells by tagging metabolites such as deoxyglucose with a radioactive element (often C^{14} or F^{18}). After injection, an animal performs a task, then is sacrificed. Slices of its brain are laid on photographic plates to develop images which reveal the areas of the brain most metabolically active. While in some instances this approach can provide stunning images of brain activity (Tootell et al., 1982, for example, demonstrated the topographical layout of visual cortex by having a monkey view a figure resulting in the reproduction of the pattern on the monkey's visual cortex upon autopsy), it has the obvious disadvantages of requiring the sacrifice of the subject and being able to gather data on only one mental activity per animal.

Positron emission tomography (PET) By combining the use of a emittive radioactive element with computerized tomography, PET generates images of metabolic activities in functioning brains. One strategy for PET imaging adapted a technique developed by Louis Sokoloff and his colleagues for labeling glucose metabolism. It employs radioactively labeled 2-deoxyglucose, a close analog of the glucose that figures in basic cell metabolism but which builds up rather than being metabolized in the cell. As the cell requires more energy, more radioactively labeled deoxyglucose-6-phosphate builds up in the cell; as it decays, positrons are ejected from the radioactive atom, travel a short distance until they collide with an electron, whereupon they annihilate in the emission of two gamma rays directed at 180° to each other. The PET scanner contains detectors surrounding the head which record an event only when two gamma rays arrive at different locations simultaneously; sophisticated computational techniques are then employed to determine the site of the annihilation. Although the 2-deoxyglucose strategy is often employed in diagnostic uses of PET, the more common approach in functional neuroimaging is to use labeled H_2O which has a short half-life (permitting several sequential scans of the same subject); the labeled H_2O is carried in the bloodstream, so what is being measured is the increased bloodflow that accompanies neural firing. Since PET was developed before fMRI, much of the pioneering research in functional neuroimaging employed it. However, it encounters a number of serious limitations. Since it uses radioactive tracers, for health reasons subjects can only be scanned while performing a limited number of tasks on one occasion in their lifetime. Since the signal is relatively weak, subjects must perform the task repeatedly during the 20-second scan, and data must be averaged over multiple subjects. Increasingly, therefore, fMRI is replacing PET as the neuroimaging technique of choice.

Magnetic resonance imaging (MRI) and functional magnetic resonance imaging (fMRI) In a strong magnetic field, the nuclei of elements which have an odd atomic weight (e.g. hydrogen) are induced to align the axes of their spin. A brief pulse of radiowaves can perturb this alignment by tipping the orientation of spin, thereby increasing the energetic state of the nuclei. When the pulse ends, they precess back into their aligned state, releasing energy in the form of radio waves in which the frequency reflects the particular atom and its environment. Since hydrogen atoms in gray and white matter have different relaxation frequencies, MRI can clearly differentiate them and provide detailed structural images of the brain. The ability to find an MRI signal correlated with function rests on the fact that brain activity generates increased blood flow in excess of oxygen utilization, resulting in a task-dependent reduction in deoxyhemoglobin, a paramagnetic molecule (Fox et al., 1988). This gives rise to what is termed the blood oxygen level – dependent (BOLD) contrast between conditions of heightened and less heightened neural activity.

Ogawa et al. (1990) predicted that this would permit BOLD-based MRI, a prediction borne out in processing of simple sensory stimuli in Ogawa et al. (1992). In a very short time, techniques for using fMRI (functional MRI) to study cognition have been dramatically improved, making it now the neuroimaging technique of choice. One indication of the potential for fMRI is the recent development of techniques to relate changes in the MRI signal to individual events (still averaging over multiple such events), thereby avoiding the need to have subjects perform the same cognitive task over 20-second intervals (Rosen et al., 1998).

3 Epistemic Evaluation of New Techniques: The Case of Neuroimaging

In our survey of methods for gaining information about brain function related to mental activity, we have emphasized that each involves an indirect measure of the brain's activity, often generated by intervening in the normal activity of the brain. This raises the prospect that the results are artifacts, not data informative about the brain activity normally underlying mental activities. Thus, an extremely important aspect of scientific practice is the process by which scientists evaluate new instruments and research techniques to determine whether results from them should be accepted as data. What makes the issue of evaluating whether a technique is producing data or an artifact challenging is that, unlike the evaluation of theories against independently evaluated data, there is not a prior source of data which one can generally use to evaluate a research technique. In evaluating the reliability of the data, researchers must invoke very different kinds of criteria. We propose that three criteria figure prominently in such evaluations: (1) the definitiveness of the results themselves, (2) the consilience of the results with those generated by other procedures, and (3) the coherence of the results with plausible theoretical accounts.

To make the case that these criteria figure prominently in scientist's actual evaluation of new instruments and techniques, we will focus on the new functional neuroimaging techniques, PET and fMRI. In both cases, the basic physical processes used to produce the image are reasonably well understood, and not the source of any skepticism. Thus, in PET, the processes by which a radioactive isotope emits a positron which, once it collides with an electron, annihilates and produces two gamma rays directed 180° from each other is well understood and not contested. Similarly, the processes by which nuclei align themselves in a magnetic field, are induced to tilt by a pulsed radio wave, then precess back when the pulse ends in fMRI are not at issue. Further, the fact that both reliably measure increased blood flow in brain areas is not a concern for those concerned with artifacts. Rather, much of the concern focuses on the relation between cognitive processes and increased blood flow.

We can differentiate two components of the relation between cognition and blood flow. On the one hand, there is the mechanism responsible for changes in blood flow.

On the other, there is the cognitive interpretation of the activities. Raichle (1998) traces inquiry into the relation of brain activity and blood flow to nineteenth-century researchers such as Angelo Mosso, but shows that the mechanism responsible for increasing blood flow is still not known. The details of the relation between increased neural activity and increase in blood flow was not critical for evaluating PET. For fMRI the issue is more important since, as we noted, the BOLD technique relies on blood flow increasing more than is required by oxygen consumption by neurons. At present there are several proposals to explain the increased blood flow (for a review, see Raichle, 1998), which have differing implications for how the resulting images relate to neural activity.

While not denying the importance of this first issue, we will concentrate on the second – the process of bridging to cognition. We should note first that both PET and fMRI place considerable constraints on the cognitive activities that can be studied. The person has to perform the activities while lying motionless on his or her back within the confines of a scanner. Consequently, the cognitive activities studied in neuroimaging will not be the ordinary ongoing cognitive activities of life, but tasks specifically designed to be carried out under such circumstances (e.g. reading words from a list). Second, the goal of imaging is to localize specific cognitive processes or operations by identifying the increased blood flow associated with them (and, to the degree possible, determine the relation between different processes).[6] Imaging studies thus depend critically on construction of tasks for which researchers already have a plausible cognitive decomposition which they employ to guide interpretation of the imaging results. Information-processing psychology has produced cognitive decompositions of a number of activities into successive stages of processing and neuroimagers have frequently availed themselves of these. But these decompositions are themselves contested, especially by advocates of dynamical systems models who deny that overall behavior is not the result of successive stages of processing, but instead an emergent product of highly distributed dynamical processes (see MacKay, 1998; van Orden and Paap, 1997). Even if one accepts that there is a decomposition of cognitive processes in the brain, there remains the question of exactly what decomposition the brain employs. One point to thus stress is that any imaging study is only as good as the assumption of decomposition of processing components on which it relies.

One of the major ways in which imagers have tried to link the results of imaging with proposed decompositions of tasks into cognitive operations is to image a person while performing two different tasks thought to differ only in that one employs one or more cognitive operations additional to those employed by the other, and then subtracting the second image from the first, generating a difference image. One then identifies the area(s) revealed in the difference image as the locus of the additional operation. For example, in a landmark early imaging study that we will take as our main example in what follows, Petersen et al. (1988, 1989) subtracted the image produced when a subject reads a noun and pronounces it aloud from the image produced when a subject reads nouns, generates a related verb, and pronounces the verb.

They thereby hoped to identify those brain areas required to generate the verb (see chapter 7, this volume).

The procedure just described for relating areas of increased activation in images to cognitive operations is known as the *subtractive method*. It was initially developed by F. C. Donders (1868) for use in chronometric studies of cognitive process: reaction times for one task were subtracted from those for another task and the difference was thought to reflect the time required for the additional processes required in the longer task. In chronometric studies, the subtractive method was broadly criticized in the 1960s. Sternberg (1969) pointed out, for example, that the subtractive method assumed that the additional cognitive activity was a pure insertion into a sequential set of processes, and this assumption might well be false. As a result, he advocated replacing the subtractive method in studies of mental chronometry by techniques which measure whether different tasks interfere with each other (for detailed discussion, see Posner, 1978). Neuro-imagers have returned to the original simple subtraction approach of Donders, a move Raichle (1998) defends by arguing that because they will be able to observe any changes in activation in other brain areas that might arise in the more complex task, researchers will be able to detect failures of pure insertion. There are, however, reasons for skepticism: imaging procedures will only identify statistically significant changes in activation elsewhere in the brain; if there are resulting accommodations elsewhere in the brain, they may fall below this threshold and thus not be noted.

Amongst skeptics, a major source of doubt about the reliability of localization based on imaging studies stems from the variability in results that have been obtained. Through a meta-analysis of PET studies of rhyming, David Poeppel (1996) revealed considerable variability in the areas researchers identified. The five studies he reviewed identified 22 different brain areas, only eight of which were identified in more than one study, and only three appeared in as many as three of the studies. Such variability is not universal in neuroimaging studies (Corbetta, 1998, for example, emphasizes the agreement of different imaging studies in identifying areas with increased activation in tasks involving covert attentional or eye movements to target locations), but when it occurs, it raises questions for researchers and skepticism amongst critics. Poeppel is not himself a skeptic of imaging research, but uses the variability he identified to point to the critical importance of psycholinguistic theories in PET results – differences in the theories used to guide the decomposition can result in very different linkages between brain areas and cognition. An additional factor is the specific way in which the studies are carried out. Often differences in the way the tasks are administered that are not thought to be significant can produce major differences in resulting images. In fact, neuroimagers themselves uncovered one such difference in method that can generate different results – the degree of practice subjects have with the task prior to imaging. After Petersen et al.'s study of verb generation, the researchers decided to repeat the study with subjects who had practiced the task. Performance after even quite short practice resulted in different patterns of activation when the practiced lists were used (Raichle et al., 1994).

Given the grounds for skepticism, why have so many cognitive neuroscientists quickly adopted neuroimaging? We will focus on three factors that figure in the assessment of neuroimaging as well as other newly introduced techniques: the definiteness of the images themselves, the agreement of the imaging results with those arrived at through other techniques, and finally the ability of the results to support plausible theoretical models.

By *definiteness* of the images, we are referring to the fact that the techniques generate specific, reasonably well delineated images which change under different task conditions. For a lay person, one of the compelling features of the images produced by PET and fMRI is their coloration which makes it appear that particular areas of the brain "light up" on specific tasks. In fact, the results of raw scans or of subtractions are actually numbers, either indicating amount of blood flow or the degree of significance of the increase in blood blow. These numbers are translated into a coloring scheme which was specifically chosen for its suggestiveness – hot colors (reds and yellows) indicating increases in blood flow, cold colors (blues) indicating decrease. The lay impression is, thus, a consequence of the means of presentation researchers have adopted. But what is important is that the images reveal increases and decreases in activation in reasonably well delineated areas of the brain.[7] Moreover, these areas remain roughly constant across trials and subjects (when the tasks remain constant). If the PET or fMRI signal was an artifact not linked to brain processes associated with specific tasks, multiple individual performances would vary and cancel each other out, or yield a pattern in which the pixels whose activations are statistically significant would be distributed randomly over the cortex. The fact that continuous areas are all activated above threshold indicates that there is a brain activity related to the cognitive task that gives rise to the image.

The constancy is revealed most significantly in studies (especially PET studies) which rely on averaging across subjects. Since even the size and shape of the brain varies across individuals, such averaging relies on techniques for mapping individual brains on to a common atlas (such as the Talairach Atlas (Talairach and Tournoux, 1988). The differences between brains could easily have resulted in diluting the activation so that no areas would show up as having significantly increased activation. The fact that areas of localized increased activations appeared in spite of averaging suggests that the imaging results are robust and not likely to be artifacts.

A second strategy is to demonstrate that the results are consistent with the results of other ways of studying cognitive function in the brain such as lesion studies and single cell recording. Prior to the advent of PET, these tools had provided a modestly rich account of the tasks performed by different areas in primate cortex. These studies had provided good grounds for believing that most visual processing occurs in occipital cortex and surrounding areas of temporal and parietal cortex. Likewise, cortical stimulation studies had revealed the motor and somatosensory cortices in Brodmann's areas 1–4. As a result, the first studies using PET to study cognitive performance only sought to demonstrate that simple motor tasks such as finger tapping would produce activation in motor cortex and that looking at visual stimuli would produce activation in primary visual cortex (Fox et al., 1987; Fox et al., 1986).

As researchers have moved to more cognitive tasks, it has become routine to identify similarities between the areas identified in imaging with those identified in lesion and single cell recording research. For example, the stage in the Petersen et al. (1989) study in which they subtracted the activations generated when subjects simply listened to or read words, from those where they pronounced them, yielded bilateral activations in the motor and sensory face areas as well as the cerebellum. They found these results to be highly credible since they cohered with a long history of evidence that the motor and sensory face areas are involved in language production, especially articulatory processing (for example, that lesions in the most focal parts of Broca's area are known to result in speech production deficits). Another example that relates to work discussed elsewhere in this volume is provided by Haxby et al. (1991), who demonstrated that activations on tasks requiring determining the location of visual stimuli versus the identity of the stimulus produced activations in approximately the same areas as Mishkin, Ungerleider, and Macko (chapter 11, this volume) had shown resulted in deficits when lesioned in monkeys.

Insisting on consistency with findings from other techniques creates something of a paradox. The goal of introducing new techniques is to revise and extend our knowledge, but the consistency requirement would seemingly prevent that. One part of the resolution of this paradox is to require that the new technique maintain consistency with established techniques only in the domain of overlap, and on the basis of that overlap to assume that the new technique is also providing correct information when it provides results that extend beyond those techniques. Imaging studies often find activations both in areas which had been associated with a particular cognitive function through lesion or cell-recording studies as well as in new areas not previously expected. For example, Petersen et al. found activation during the verb-generating task in parts of the cerebellum. Traditionally the cerebellum has been viewed as principally engaged in motor activity, but this study and others has contributed to the growing recognition that the cerebellum plays a role in a variety of cognitive tasks (Thach, 1998).

The more challenging situation is when a new technique such as imaging produces results that are at odds with those produced by earlier techniques. Since we cannot be sure that the older techniques were themselves reliable, one would not want to simply dismiss new techniques where they generate conflicting evidence. But clearly there is a greater burden on those advancing the new technique when such discrepancies emerge. Again, the Petersen et al. study with the verb-generating task provides an instructive example since they conceived of this task as adding a semantic component to the pronunciation task. It was expected that the areas of increased activation after subtracting the areas active in the corresponding noun pronunciation task would be those involved in semantic processing. Following the modern reinterpretation of Wernicke's deficit studies by Bradley et al., (1980) and others (see chapters 7 and 8, this volume), Wernicke's area in the superior temporal lobe has been interpreted as the site of semantic processing. But in the verb-generating task, which requires semantic processing, Petersen et al. found increased activation in the left prefrontal cortex and not in Wernicke's area. They contend that increased activation in Wernicke's area is limited to situations requiring phonological encoding of the input (such as when words are presented auditorily). Their PET results are thus inconsistent with a long history of lesion studies; accordingly, the burden was on Petersen et al. to show that their results were not artifacts.

Our goal is not to endorse the Petersen et al. interpretation, but to examine the strategy of arguing for it. A starting point is to note that lesions only indicate weak points in a processing system (a point through which information is conducted, for example), not necessarily the location where a failed process is performed. Thus, lesion results are not definitive. But Petersen et al. required a more positive defense for their alternative proposal. One strategy they employed was to emphasize consilience with earlier, non-PET blood flow studies. However, since PET also relies on blood flow, that just raises the possibility that both sets of studies rest on artifacts involving blood flow. They also emphasized consilience across PET studies. For example, they report PET scans on five subjects while they carried out a semantic judgment task, in which subjects judged whether words referred to objects in the same category; this study showed increased activation in a very similar area of prefrontal cortex (although these increases did not reach the threshold for statistical significance).

The challenge to Petersen et al. was increased by the results of a PET study by another laboratory, one using related but different tasks that were also expected to tap semantic processing. Thus, Christopher Frith and his colleagues (Frith et al., 1991) found activation in a semantic task in both Wernicke's area and prefrontal areas. Frith et al. present their results as *disconfirmation* of Petersen's results since they did get activation in Wernicke's area, and they attribute the prefrontal activation to "intrinsic generation rather than semantics" (p. 1146). Petersen and Fiez (chapter 7, this volume) respond to this claimed disconfirmation by analyzing the tasks employed in the Frith et al. study. Specifically, they argue that since all of the input conditions in Frith et al.'s study involve auditory input, the activations Frith et al. found in Wernicke's area are due to auditory processing, not semantic processing and that only the activation in prefrontal cortex found in Frith et al.'s study is really a candidate for semantic processing.

While their critique of the Frith et al. study may suffice to neutralize its challenge to their results, the burden is still primarily on Petersen and his colleagues to show that the prefrontal activations reflect semantic processing since it is their result which is at odds with the history of lesion studies. To make their case, Petersen et al. invoke a third strategy for providing credibility for new techniques, showing that the new evidence fits a compelling theory. The theory claims that the type of processing required for semantics is characteristic of that performed in prefrontal cortex. Particularly important here is the idea that prefrontal cortex is involved in withholding responses to stimuli, a cognitive process that is likely to be important in semantic processing. Evidence for this proposal is offered by studies by Goldman-Rakic (1987) showing that lesions in anterior prefrontal cortex in monkeys leaves the animals unable to withhold responses to false cues. (This theoretical framework, and its possible relevance to semantic processing, has recently been developed more fully by Deacon, 1997, who argues that what is critical for a semantic system is the ability to establish contrasts between lexical items with different meanings, a capacity Deacon also links with processes in prefrontal cortex.)

In advancing the theoretical perspective in which they situate their PET results, Petersen et al. contrast two theoretical approaches, a serial, single-route framework and a dual-route framework. The serial, single-route framework they associate with Norman Geschwind (1979), who interprets the processing in reading aloud as following a pathway from visual cortex through Wernicke's area in the angular gyrus, to Broca's area, and finally to primary motor cortex. In contrast, the dual-route models distinguish a lexical route (required in order to read words with non-standard pronunciations such as *pint*) and a non-lexical route that utilizes phoneme-to-grapheme correspondence rules (required to read non-words such as *rint*). Dual-route models have received independent support from neuropsychological research that reveals a double-dissociation between patients who can read words with non-standard pronunciations but not non-words, and patients who read non-words, but not words with non-standard pronunciations (Coltheart, 1987); however, they have also been challenged by a variety of investigators (Plaut, 1995). Petersen et al. advocate the dual-route model as indicated in figure 4.4, thereby rejecting the single-route framework which has traditionally assigned a major role in semantic processing to Wernicke's area. It is important to note that what serves to explain away the traditional interpretation of the semantic function of Wernicke's area is not the two routes distinguished by Coltheart (since neither the lexical nor the non-lexical route involves Wernicke's area), but a separate distinction between the processing of visually presented words and of auditorily presented words. The latter necessarily entails phonological encoding of words, while such encoding can be by-passed with visually presented words. The dual-route models of word reading thus provide a framework which Petersen et al. can *expand* upon to incorporate their results indicating lack of semantic processing in Wernicke's area. (Petersen et al. also indicate a variety of pieces of supporting psycholinguistic evidence for this extension of the dual route framework.) With the expanded dual-route framework, they can advance a theoretical model that renders their results plausible, thereby countering the objection that their results are inconsistent with more traditional results indicating a semantic function for Wernicke's area.

From a strongly empiricist point of view, in which evidence or data is the foundation upon which theoretical frameworks are developed and evaluated, the suggestion that researchers use theoretical frameworks to support their experimental results seems seriously misguided. In practice, however, such an approach is rather common (see Bechtel, 2000). Failure to find a plausible theoretical framework in which to understand the results of a technique leads scientists to suspect that the technique is generating artifacts, whereas success in identifying such a framework reduces that

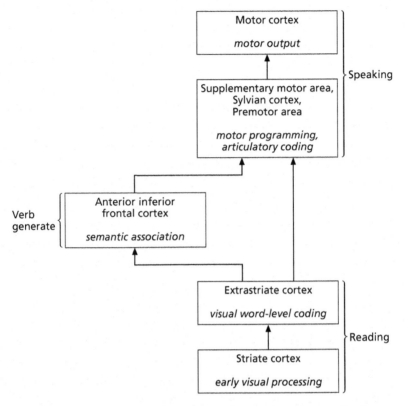

Figure 4.4 Two routes for processing visual and auditory language inputs (after Petersen et al., 1989). Note that neither the direct route from visual processing to motor output nor that through semantic association employs Wernicke's area.

suspicion and can even counter the threat posed by results generated by more traditional techniques.

4 Conclusions

The attempt to link studies of the brain to psychological processes is critically dependent on a variety of instruments and research techniques, many of which we described briefly in section 2 of this chapter. Because these techniques all rely on indirect measures and generally upon intervening in normal brain processes, each has confronted an epistemic challenge to show that they are generating reliable evidence about normal brains. In addition to identifying some of the reasons for these epistemic questions we have pointed to some of the limitations faced by the various instruments and techniques. Our objective, again, is not to promote skepticism, but an appreciation of the basis on which scientific claims about brain function are based. In the third section we examined these epistemic issues in greater depth by focusing on the new neuroimaging techniques of PET and fMRI. Particularly important in our analysis is the challenge of drawing psychological interpretations of these measures of blood flow. Using Petersen et al.'s study of single-word processing with PET, we argued that in fact, in evaluating the reliability of such a procedure, scientists rely on the definitiveness of their results, the consilience of their results with those of other techniques, and the plausibility of theoretical models in terms of which they interpret their results.

Notes

1   The idea that perception, the recognition of objects and events, is indirect and rests on "unconscious inferences," was clearly articulated in the nineteenth century by Helmholtz. A number of perceptual theorists (for example, James Gibson) have taken issue with the reference to "inferences" in perception, but the development of a neurobiological understanding of how the visual system operates (see Part III) clearly supports the idea that there are many processing steps between the registration of light on the retina and the recognition of the object or events seen. Accordingly, many of the same issues that arise in evaluating instruments arise in perception itself. Bechtel (2000) argues for parallels between the way we settle the occasional but rare disputes about what is visually perceived and the disputes between scientists over whether an instrument has generated an artifact.

2   Sometimes scientists offer theories of how instruments or techniques work, in which case the more standard account of evaluation by comparing these theories against their evidence can be invoked. But although some theoretical ideas often lie behind the development of instruments and techniques, they are often incomplete and incapable of settling controversies that arise over artifact versus evidence. Detailed theoretical knowledge of how the instrument or technique operates is generally not available until long after the instruments and techniques have been employed and the questions about artifacts resolved.

3   A second objective is to make philosophers who are inclined to draw upon the results of empirical inquiry more sensitive to the procedures on which such inquiry rests. Responsible philosophical utilization of scientific results requires cognizance of how those results were obtained.

4   For example, the focus on the role of the hippocampus in encoding memory resulted from lesions in a patient, H.M. that were made in the attempt to control incurable epilepsy. Although the surgeon thought he had removed the whole hippocampus, more recent MRI studies reveal that much of it remained (although over time it has atrophied).

5   Ferrier is rightly construed as one of the major figures in the development of neuroscience, but that does not imply that most of his fundamental claims turned out to be correct. As we shall see in chapter 13, drawing upon both lesion and stimulation studies, he contended that the angular gyrus in the parietal cortex, not the occipital lobe, was the primary site of visual processing in the brain.

6 If we focus on whole activities a person performs, the associated brain processes will not
   be restricted to just one brain area. Rather, performance is generally distributed over
   numerous areas, each of which does part of what is needed to perform the overall tasks.
   As noted by Petersen and Fiez (chapter 7, this volume), "a functional area of the brain is
   not a task area: there is no 'tennis forehand area' to be discovered. . . . Any task or 'function' utilizes a complex and distributed set of brain areas." Accordingly, what must be
   localized or mapped on to the brain through neuroimaging are "simple operations." The
   aim is to determine the distinctive contributions, or simple operations, performed in
   different regions of the brain.
7 By referring to specific areas, though, we are not suggesting that these areas are known
   to correspond to neuroanatomically delineated areas which are thought to be functionally
   significant. Imagers often try to link areas of activation with anatomically or physiologically identified areas (see Zeki et al., 1991), but this is in general a very difficult task due
   to the paucity of information about the locus of these neuroanatomical areas in humans.
   Most of the relevant neuroanatomy has in fact been done on monkeys or other non-human
   species.

References

Adrian, E. D. 1926: The impulses produced by sensory nerve endings, part I. *Journal of Physiology (London)*, 61, 49–72.

Bechtel, W. 2000: From imaging to believing: Epistemic issues in generating biological data. In R. Creath and J. Maienschein (eds), *Biology and Epistemology*, Cambridge, UK: Cambridge University Press, 138–63.

Berger, H. 1929: Über das elektroenkephalogramm des menschen. *Archiv für Psychiatrie und Nervenkrankheiten*, 87, 527–70.

Bradley, D., Garrett, M., and Zurif, E. 1980: Syntactic deficits in Broca's aphasia. In D. Caplan (ed.), *Biological Studies of Mental Processes*, Cambridge, MA: MIT Press, 269–86.

Coltheart, M. 1987: Cognitive neuropsychology, and the study of reading. In M. I. Posner and O. S. M. Marvin (eds), *Attention and Performance*, vol. 11, Hillsdale, NJ: Lawrence Erlbaum, 3–37.

Corbetta, M. 1998: Frontoparietal cortical networks for directing attention and the eye to visual locations: identical, independent, or overlapping neural systems? *Proceedings of the National Academy of Sciences, USA*, 95, 831–8.

Dawson, G. D. 1951: A summation technique for detecting small signals in a large irregular background. *Journal of Physiology*, 115, 2P.

Deacon, T. W. 1997: *The Symbolic Species*. New York: Norton.

Delgado, J. 1969: *Physical Control of the Mind*. New York: Harper.

Donders, F. C. 1868: Over de snelheid van psychische processen. Onderzoekingen gedaan in het Pysiolish Laboratorium der Utrechtsche Hoogeschool: 1868–1869. *Tweede Reeks*, 2, 92–120.

Ferrier, D. 1876: *The Functions of the Brain*. London: Smith, Elder, and Company.

Fox, P. T., Burton, H., and Raichle, M. E. 1987: Mapping human somatic sensory cortex with positron emission tomography. *Journal of Neurosurgery*, 67, 34–43.

Fox, P. T., Minton, M. A., Raichle, M. E., Miezin, F. M., Allman, J. M., and van Essen, D. C. 1986: Mapping human visual cortex with positron emission tomography. *Nature*, 323, 806–9.

Fox, P. T., Raichle, M. E., Mintun, M. A., and Dence, C. 1988: Nonoxidative glucose consumption during focal physiologic neural activity. *Science*, 241, 462–4.

Frith, C. D., Friston, K. J., Liddle, P. F., and Frackowiak, R. S. J. 1991: A PET study of word finding. *Neuropsychologia*, 29 (12), 1137–48.

Garnsey, S. M., Tanenhaus, M. K., and Chapman, R. M. 1989: Evoked potentials and the study of sentence comprehension. *Journal of Psycholinguistic Research*, 18, 51–60.

Geschwind, N. 1979: Specializations of the human brain. *Scientific American*, 238 (3), 158–68.

Goldman-Rakic, P. S. 1987: Circuitry of primate prefrontal cortex and regulation of behavior by representational memory. In J. M. Brookhart, V. B. Mountcastle, and S. R. Geiger (eds), *Handbook of Physiology: The Nervous System*, vol. 5, Bethesda, MD: American Physiological Society, 373–417.

Gregory, R. L. 1968: Models and the localization of functions in the central nervous system. In C. R. Evans and A. D. J. Robertson (eds), *Key Papers in Cybernetics*, London: Butterworth, 91–102.

Hacking, I. 1983: *Representing and Intervening*. Cambridge, UK: Cambridge University Press.

Hanson, N. R. 1958: *Patterns of Discovery*. Cambridge, UK: Cambridge University Press.

Hartline, H. K. 1938: The response of single optic nerve fibers of the vertebrate retina. *American Journal of Physiology*, 113, 59–60.

Haxby, J. V., Grady, C. L., Horwitz, B., Ungerleider, L. G., Mishkin, M., Carson, R. E., Herscovitch, P., Schapiro, M. B., and Rapoport, S. I. 1991: Dissociation of object and spatial visual processing pathways in human extrastriate cortex. *Proceedings of the National Academy of Sciences, USA*, 88, 1621–65.

Hess, W. R. 1949: The central control of the activity of internal organs, *Nobel Lectures, Physiology or Medicine 1942–1962*. New York: Elsevier.

Hinton, G. E., and Shallice, T. 1991: Lesioning a connectionist network: Investigations of acquired dyslexia. *Psychological Review*, 98, 74–95.

Kuffler, S. W. 1953: Discharge patterns and functional organization of mammalian retina. *Journal of Neurophysiology*, 16, 37–68.

Kuhn, T. S. 1962/1970: *The Structure of Scientific Revolutions*, 2nd edn. Chicago: University of Chicago Press.

Kutas, M., and Dale, A. 1997: Electrical and magnetic readings of mental function. In M. D. Rugg (ed.), *Cognitive Neuroscience*, Cambridge, MA: MIT Press, 197–242.

Latour, B. 1987: *Science in Action*. Cambridge, MA: Harvard University Press.

Lettvin, J. Y., Maturana, H. R., McCulloch, W. S., and Pitts, W. H. 1959: What the frog's eye tells the frog's brain. *Proceedings of the Institute of Radio Engineers*, 47, 1940–51.

Li, C.-H., McLennan, H., and Jasper, H. H. 1952: Brain waves and unit discharges in cerebral cortex. *Science*, 116, 656.

Luck, S. J., and Ford, M. A. 1998: On the role of selective attention in visual perception. *Proceedings of the National Academy of Sciences (USA)*, 95, 825–30.

MacKay, D. G. 1998: Stage theories refuted. In W. Bechtel and G. Graham (eds), *A Companion to Cognitive Science*, Oxford: Basil Blackwell, 671–8.

Marr, D. C. 1982: *Vision: A Computation Investigation into the Human Representational System and Processing of Visual Information*. San Francisco: Freeman.

Ogawa, S., Lee, T. M., Kay, A. R., and Tank, D. W. 1990: Brain magnetic resonance imaging with contrast dependent on blood oxygenation. *Proceedings of the National Academy of Science, USA*, 87, 9868–72.

Ogawa, S., Tank, D. W., Menon, R., Ellerman, J. M., Kim, S., Merkle, H., and Ugurbil, K. 1992: Intrinsic signal changes accompanying sensory stimulation: Functional brain mapping with magnetic resonance imaging. *Proceedings of the National Academy of Science, USA*, 89, 5951–5.

Olds, J. 1965: Pleasure centers in the brain. *Scientific American*, 195, 105–16.

Penfield, W., and Rasmussen, T. 1950: *The Cerebral Cortex in Man: A Clinical Study of Localization of Function*. New York: Macmillan.

Petersen, S. E., Fox, P. T., Posner, M. I., Mintun, M., and Raichle, M. E. 1988: Positron emission tomographic studies of the cortical anatomy of single-word processing. *Nature*, 331, 585–8.

Petersen, S. E., Fox, P. T., Posner, M. I., Mintun, M., and Raichle, M. E. 1989: Positron emission tomographic studies of the processing of single words. *Journal of Cognitive Neuroscience*, 1 (2), 153–70.

Plaut, D. C. 1995: Double dissociation without modularity: Evidence from connectionist neuropsychology. *Journal of Clinical and Experimental Neuropsychology*, 17, 291–321.

Poeppel, D. 1996: A critical review of PET studies of phonological processing. *Brain and Language*, 55 (3), 317–51.

Posner, M. I. 1978: *Chronometric Explorations of Mind*. Hillsdale, NJ: Lawrence Erlbaum Associates.

Raichle, M. E. 1998: Behind the scenes of functional brain imaging: A historical and physiological perspective. *Proceedings of the National Academy of Sciences*, 95, 765–72.

Raichle, M. E., Fiez, J. A., Videen, T. O., MacLeod, M.-A. K., Pardo, J. V., Fox, P. T., and Peterson, S. E. 1994: Practice-related changes in human brain functional anatomy during nonmotor learning. *Cerebral Cortex*, 4, 8–26.

Rosen, B. R., Buckner, R. L., and Dale, A. M. 1998: Event-related functional MRI: Past, present, and future. *Proceedings of the National Academy of Sciences*, 95, 773–80.

Shallice, T. 1988: *From Neuropsychology to Mental Structure*. New York: Cambridge University Press.

Sternberg, S. 1969: The discovery of processing stages: Extension of Donders' method. *Acta Psychologica*, 30, 276–315.

Talairach, J., and Tournoux, P. 1988: *Co-planar Stereotoxic Atlas of the Human Brain*. New York: Thieme Medical Publishers.

Thach, W. T. 1998: What is the role of the cerebellum in motor learning and cognition. *Trends in Cognitive Science*, 2 (9), 331–7.

Tootell, R. B., Silverman, M. S., Switkes, E., and DeValois, R. L. 1982: Deoxyglucose analysis of retinotopic organization in primate striate cortex. *Science*, 218, 902–4.

Uttal, W. R. 1978: *The Psychobiology of Mind*. Hillsdale, NJ: Lawrence Erlbaum.

Valenstein, E. 1973: *Brain Control: A Critical Examination of Brain Stimulation and Psychosurgery*. New York: John Wiley and Sons.

van Orden, G. C., and Paap, K. R. 1997: Functional neural images fail to discover the pieces of the mind in the parts of the brain. *Philosophy of Science*, 64 (4), S85.

Van Orden, G. C., Pennington, B. F., and Stone, G. O. in preparation: What do double dissociations prove? Inductive methods and isolable systems. Submitted for Publication.

Walsh, V., and Cowey, A. 1998: Magnetic stimulation studies of visual cognition. *Trends in Cognitive Sciences*, 2, 103–10.

Woolsey, C. 1960: Organization of the cortical auditory system: a review and a synthesis. In *Neural Mechanisms of the Auditory and Vestibular Systems*, eds. G. L. Rasmussen and W. F. Windle, pp. 165–80. Springfield, Ill.: Charles C. Thomas.

Zeki, S. M., Watson, J. D. G., Lueck, C. J., Friston, K. J., Kennard, C., and Frackowiak, R. S. J. 1991: A direct demonstration of functional specialization in human visual cortex. *Journal of Neuroscience*, 11, 641–9.

Questions for Further Study and Reflection

1   What is the difference between direct and indirect kinds of neuroscientific evidence such that indirect evidence is more problematic?
2   Would it be possible to collect scientific evidence that the mind was not the brain? If so, what would be examples of such evidence? If not, why not?
3   Why are metaphors useful for scientific understanding? Are there any examples of things you can understand with a metaphor that you couldn't understand without one? Are there risks in invoking metaphors in developing scientific accounts?
4   Daugman's article was written in 1990. What sorts of things have happened since then that might suggest new brain metaphors? In the far future, what brain metaphors might people use?
5   What implications would the truth or falsity of the neuron doctrine have for understanding the nature of high-level mental phenomena?
6   Why is localization significant in cognitive neuroscience? Are there analogous cases of localization in other sciences in which knowing where something is has contributed to the scientific understanding of what it is and how it works?
7   Give cognitive scientific examples of a datum, a theory, and an artifact. What are the most significant relations that each bears to the others?
8   In what ways are neuroimaging techniques sources of more direct (as opposed to indirect) evidence of neurocognitive functioning? In what ways are they less direct than other techniques?
9   What would count as a claim about the brain that is not a metaphor? How would you know whether it was or was not?

Part II

Language

Introduction

William Bechtel

The cognitive lives of humans revolve around language. For Descartes, language was a distinctively human capacity that indicated we were partially constituted by a non-material substance, a mind. His reasoning focused on the creative power of language – the fact that any human could constantly construct new sentences. Not seeing how a machine might do this, he opted for dualism. In the mid-twentieth century, Noam Chomsky showed how an automaton could in fact generate ever more well-formed sentences, obviating the need to appeal to a nonmaterial mind to explain linguistic capacities. But that has not diminished the special nature of language. Even though some other species seem capable of exhibiting some linguistic abilities, none has as developed a linguistic capacity as we have. Moreover, in many respects the advent of language has transformed our mental lives. Most of us report reasoning in language. And with language we are able to think thoughts that would not be possible otherwise (consider trying to reason about what constitutes justice in a particular situation or planning for complex contingencies in the future without language).

Thus, it is not surprising that when researchers first began to investigate how mental processes were supported by the brain, one of the first capacities they sought to localize in the brain was language. Moreover, some of the first successes in relating mental abilities to the brain involved language. Chapter 5 presents an excerpt from the classic 1861 paper by Paul Broca, the French anatomist and anthropologist, who presented his findings connecting his patient Leborgne's deficit in articulate speech with a lesion in his frontal cortex. The paper is often regarded as presenting the first successful attempt to establish a connection between a mental capacity and a brain region.

In his research, Broca employed a faculty psychology in which mental abilities (such as the ability to produce articulate speech) were identified with faculties responsible for them; it was these faculties that were then localized in a part of the brain. Carl Wernicke was also interested in the relations of language to the brain, but he worked out of a different, associationist, framework in which mental capacities were explained in terms of associations or connections between primary sensory and motor areas. In the 1870s and 1880s he studied patients who had deficits in understanding speech, which he attributed to an area in the temporal cortex in which sound images of words were stored. When patients suffered damage in this area they would be unable to connect sounds with meanings, and so could not understand speech. In chapter 6 Wernicke describes both his core ideas, and his extensions of them to account for deficits in reading and writing.

As chapter 9 discusses in more detail, the localizationist claims, as developed by either Broca or Wernicke, fell into disrepute in the earlier decades of the twentieth century, before being revived by Norman Geschwind in the 1960s and 1970s. Geschwind defended the view that Wernicke's area was a comprehension center while Broca's was a production center. Since then there have been multiple reinterpretations of the contributions of different areas involved in language processing; one prominent view has been that Broca's area is involved in syntactic analysis while Wernicke's is involved in semantic analysis. Most of this research has relied on analysis of deficits following lesions to Broca's or Wernicke's areas, but more recently researchers have started to employ neuroimaging of normal individuals performing language tasks. In chapter 7, Petersen and Fiez describe several of the seminal studies using positron emission tomography (PET) to study people performing linguistic tasks, typically involving processing of individual words. One of the major findings is that in tasks involving the meaningful processing of words (that is, semantic tasks) it is not Wernicke's area that is active, but areas in the left dorsolateral prefrontal cortex. In chapter 8 Bates takes up the issues of the innateness, localization, and domain specificity of language. She supports claims for the innateness and localization of language, but denies that language processing is domain-specific or modular. She discusses a variety of examples that indicate the plasticity of the neural structures subserving language and, in particular, examines research on adult aphasia and on normal and abnormal language development, including language acquisition by deaf children and individuals with Williams syndrome, which she presents as telling against the idea of a language module. Finally, chapter 9 discusses the distinctions between the ways different disciplines study language and how inquiry in each relates to neural investigations of language. It also discusses three major controversies in the study of language – whether it is really unique to humans, whether it relies on domain-specific processes, and whether linguistic capacities are innate.

5

Remarks on the Seat of the Faculty of Articulate Language, Followed by an Observation of Aphemia

Paul Broca

The observations which I present to the Anatomical Society support the ideas of M. Bouillaud on the seat of the faculty of language. This question which is both physiological and pathological deserves more attention than most doctors have given it so far and the matter is sufficiently complicated and the subject sufficiently obscure so that it seems useful to me to make a few remarks before relating the facts which I have observed.

1

[...]

There are cases where the general faculty of language persists unaltered, where the oratory apparatus is intact, where all muscles without exception, even those of voice and articulation follow volition and where nonetheless the cerebral lesion has abolished articulate language. This abolition of speech in individuals, who are neither paralyzed nor idiots, is a sufficiently important symptom so that it seems useful to designate it by a special name. I have given it the name Aphemia (α privativum + $\varphi\acute{\eta}\mu\eta$, voice, I speak), for what is missing in these patients is only the faculty to articulate the words; they hear and understand all that is said to them, they have all their intelligence and they emit easily vocal sounds. They execute with their tongue and their lips movements larger and more energetic than would be necessary to articulate sounds, and nonetheless, the well-sensed response which they would like to make, becomes reduced to a very small number of articulate sounds. They are always the same and always arranged in the same way. Their vocabulary, if one can use that word, is composed of a short series of syllables, sometimes of a monosyllable which expresses everything or rather nothing, for this unique word is most often a stranger to all vocabularies. Certain patients do not even have this vestige of articulate language. They make vain efforts without pronouncing one syllable. Others have in a certain way two degrees of articulation. Under ordinary circumstances they invariably pronounce their word of predilection. But when they experience wrath they become capable of articulating a second word, generally a gross swear word which was probably familiar to them before their disease. Then they stop after this last effort. M. Auburtin has observed a patient who is still alive and who does not need any excitation to pronounce the stereotype swear word. All his responses begin with a bizarre word of six syllables and end invariably with this supreme invocation "Sacré nom de Dieu."

Those who have studied for the first time these strange facts could believe that the faculty of language in such cases was abolished if they did not make a sufficient analysis. But it remains evidently intact for the patients understand completely articulate language and written language. Those who do not know or who cannot write but have sufficient intelligence (and they need a lot in such cases) to find the means to communicate their ideas and those finally who can write and have the free use of their hands, bring their ideas well on paper. Therefore, they know the sense and the value of words in auditive form as well as in graphic form. The articulate language which they used to speak is still familiar, but they cannot bring about the series of methodic movements which correspond to the syllable they want. What they lost is therefore not the faculty of language, is not the memory of the words nor is it the action of nerves and of muscles of phonation and articulation, but something else. It is a particular faculty considered by M. Bouillaud to be the faculty to coordinate the movements which belong to the articulate language, or simpler, it is the faculty of articulate language; for without it no articulation is possible.

[. . .]

One can . . . make at least two hypotheses on the nature of the special faculty of articulate language. In the first hypothesis this would be a superior faculty, and aphemia would be an intellectual disturbance. In the second hypothesis, this would be a faculty of much less elevated order and the aphemia would be only a disturbance of locomotion. Although this latter interpretation seems to be much less probable than the former, still I would not dare to be categorical, if I were restricted merely to clinical observations.

However that may be from the point of view of functional analysis, the existence of a special faculty of articulate language – as I have defined it – can no more be doubted, because a faculty which can perish isolated without those which are in its neighborhood is evidently a faculty independent of all others, i.e. a special faculty.

If all cerebral faculties were as distinct and as clearly circumscribed as this one, one would finally have a definite point from which to attack the controversial question of cerebral localization. Unfortunately, this is not the case, and the greatest obstacle in this part of physiology comes from the insufficiency and the uncertainty of the functional analysis which necessarily has to precede the search of the organs which are coordinated to each function.

In this respect science has so little advanced that it has not even found its base, and what is today in doubt is not this or that phrenological system but the principle of localization itself. The question which has to be answered first is whether all parts of the brains which contribute to the thought make identical or different contributions?

A communication of M. Gratiolet pertaining to the parallelism of cerebral and intellectual development of human races, some time ago, has led the Anthropological Society of Paris to examine this important problem and M. Auburtin, who believes in the principle of localization, has rightly thought that the localization of a single faculty would suffice to establish the truth of the principle. He has therefore tried to demonstrate in conformity with the doctrine of his master M. Bouillaud that the faculty of the articulate language resides in the anterior lobes of the brain.

In order to do this he first reviewed a series of cases where spontaneous cerebral affection abolished the faculty of articulate language without destroying other cerebral faculties, and on autopsy a deep lesion of the anterior convolutions of the brain was found. The special nature of the symptoms of aphemia did not depend upon the nature of the disease but only on its seat, because the lesion was sometimes a softening, sometimes an apoplexia, sometimes an abscess or a tumor. To complete this demonstration M. Auburtin has called in another series of cases in which the aphemia was a consequence of a traumatic lesion of the brain. According to him these facts are equivalent to vivisections, and he closed by saying that, to the best of his knowledge, the anterior lobes of the brain were never found to be completely intact (or even relatively intact) at the autopsies of individuals who had lost the faculty of articulate language but not the rest of their intelligence.

Against him were held several remarkable facts concerning individuals who had spoken to the last moment and in whom, nonetheless, the anterior lobes of the brain had profound traumatic lesions; however, his answer was that this proved nothing, that even a large lesion of the anterior lobes could fail to reach those parts where the faculty of articulate language resides; that the objection would only be valid if all frontal convolutions were destroyed on both sides, in other words, in their whole extent back to the sulcus of Rolando, but in the cases which one held against him the destruction of these convolutions was only partial. He has, therefore, recognized that a lesion of the anterior lobes does not necessarily lead to a loss of language, but he has maintained that it is a certain indication and that one can diagnose it; that this diagnosis has been made several times during life and was always confirmed at autopsy. Finally, after citing the observation of the individual who is still alive and who has presented for several years in the perfect manner the symptoms of aphemia who is actually at the Hospice for the Incurables, he declared that he would renounce forever the doctrine of M. Bouillaud if the autopsy of this patient did not confirm the diagnosis of a cerebral lesion occupying exclusively or principally the anterior lobes.

I thought that I should sum up in a few words this discussion to bring out the topical interest of the observation which I present today at the Anatomical Society.

Without doubt the value of the facts depends not on the circumstances in which one observes them, but our impression depends, to a large extent, on these circumstances, and when a few days after having heard the argument of M. Auburtin, I found one morning on my service a dying patient who 21 years ago had lost the faculty of articulate language, I gathered with the greatest care these observations which seem to serve as a touchstone for the theory of my colleague.

So far without repulsing this theory, without denying the importance of the facts which are in its favor, I have shown much caution in the presence of contradictory facts which exist in the science. Although I believe in the principle of localization, I have asked and still ask myself within what limits this principle can be applied. I think there is a point which is fairly well established by comparative anatomy, by the parallel of anatomy and physiology in the human races and finally by comparison of normal, abnormal or pathological individual variations of people of the same race; namely that the most elevated cerebral faculties such as judgment, reflection, the faculties of comparison and abstraction have their seat in the frontal convolutions, while the convolutions of the temporal, parietal and occipital lobes are affected by sentiments, predilections and passions. In other words, there are in the human mind a group of faculties and in the brain groups of convolutions, and the facts assembled by science so far allow to state, as I said before, that the great regions of the mind correspond to the great regions of the brain. It is in this sense that the principle of localization appears to be, if not rigorously demonstrated, so at least probable. But to know whether each particular faculty has its seat in a particular convolution, is a question which seems completely insoluble at the present state of science.

The study of the facts pertaining to loss of the faculty of articulate language is one of those which has the greatest chance to lead to either a positive or a negative conclusion. The independence of this faculty is made evident by pathological observations, and although some doubts could be raised about its nature, and although one could ask, as we have seen above, whether it is part of the intellectual functions or of those cerebral functions which have to do with motion, it is permissible to base us, at least provisionally, on the standpoint of the former hypothesis, which at first sight seems the more probable one and in whose favor the pathological anatomy of aphemia establishes strong evidence. In fact, in almost all cases in which an autopsy could be performed, it was found that the substance of the convolutions is profoundly altered to a notable extent. In some subjects the lesions were even confined to the convolutions; from this one can conclude that the faculty of articulate language is one of the functions of the convolutional mass. But it is generally admitted that all faculties, called intellectual, have their seat in this part of the brain, and it seems therefore very probable, that all faculties that reside in the cerebral convolutions are of the intellectual nature.

Assuming this point of view, we recognize easily that the pathological anatomy of aphemia can give something more than a solution of a particular question, that it can throw much light on the general question of cerebral localization, giving the physiology of the brain a point of departure or rather a point of comparison which is very valuable. If it were proven, for instance, that aphemia could be the result of lesions which affect indifferently any convolution, any lobe, then one would have the right to conclude not only that the faculty of articulate language is not localized but also that in all likelihood the other faculties of the same order are not localized either. If on the other hand it were demonstrated that the lesions which abolish the speech constantly occupy the same convolution, then one could hardly help the admission that this convolution is the seat of the faculty of articulate language. Thus the existence of the first localization once admitted, the principle of localization by convolutions would be established. Finally, between these extreme alternatives there is a third one which would lead to a mixed doctrine. Let us assume that the lesions of aphemia always occupy the same cerebral lobe, but that in this lobe they do not always occupy the same convolution, then it would follow that the faculty of articulate language had its seat in a certain region, in a certain group of convolutions, but not in a particular convolution. It would become then very probable that the cerebral faculties are localized by regions and not by convolutions.

It is important, therefore, to study with the greatest care the specific question which could have theoretical consequences of such a general and important nature. We have to investigate not only in what parts of the brain are situated the regions of aphemia, but we also have to designate by their name and by their rank the diseased convolutions and the degree of alterations of each of them. So far one has not proceeded in that way. Even in the most thorough observations, it was only said that the lesions began and ended so and so many cm from the anterior end of the hemisphere, so and so many cm from the median fissure or from the fissure of the Sylvius. But this is quite insufficient: because with these indications, however minute, the reader cannot guess which is the diseased convolution. Thus, there are cases where the disease is situated in the most anterior part of the hemisphere, others where it is situated five or even eight cm behind that point, and it would appear therefore that the seat of the lesion is very variable. But when one keeps in mind, that the three anteroposterior convolutions of the convexity of the frontal lobe begin at the level of the supra–orbital arcade and run together, side by side, from the front backwards to end all three in the transverse convolution (which forms the anterior border of the sulcus of Rolando); when it is kept further in mind that this sulcus is situated more than 4 cm behind the coronal suture, and that the three frontal convolutions occupy more than 2/5 of the whole length of the brain, one will understand that the same convolution can be attacked by lesions situated at very different points far from each other. It is, therefore, much less important to indicate the level of the disease than to say which are the diseased convolutions.

This description is without a doubt less comfortable than the other one because the classical treatises of anatomy have so far not disseminated widely the study of cerebral convolutions which even the phrenologists neglected far too much. One allowed oneself to be dominated by the old prejudice that the cerebral convolutions are in no way fixed, that they are simply pleats made by chance, comparable to the disorderly flexions of the intestinal loops. What has given credit to this idea was that the so-called secondary folds which depend on the degree of development of the fundamental convolutions, vary not only from individual to individual but frequently in the same individual from one side to the other. It is no less true that the fundamental convolutions are fixed and constant in all animals of the same species and that in the animal series they behave as so many completely distinct organs. The description and the enumeration of fundamental convolutions, of their connections and their relations are here out of place. One can find them in the special papers by MM. Gratiolet and Rudolph Wagner.

[. . .]

But I have to beg your forgiveness for having given so much space to these preliminary remarks. It is time to talk about my observation on aphemia.

2

Aphemia for twenty-one years produced by the chronic and progressive softening of the second and third convolution of the superior part of the left frontal lobe

On 11 April 1861, to the general infirmary of the Bicêtre, to the service of surgery, was brought a man 51-years-old called Leborgne who had a diffused gangrenous cellulitis of the whole right inferior extremity, from the foot to the buttocks. To the questions which one addressed to him on the next day as to the origin of his disease he responded only by the monosyllable "*tan*," repeated twice in succession and accompanied by a gesture of the left hand. I tried to find out more about the antecedents of this man, who had been at Bicêtre for 21 years. I asked his attendants, his comrades on the ward, and those of his relatives who used to see him and here is the result of this inquiry.

Since his youth he was subject to epileptic attacks, but he could become a last-maker at which he worked until he was 31 years old. At that time he lost the ability to speak, and that is why he was admitted at the Hospice of Bicêtre. One could not find out whether this loss of speech came on slowly or fast, nor whether some other symptom accompanied the beginning of this affection.

When he arrived at Bicêtre he could not speak for 2 or 3 months. He was then quite healthy and intelligent and differed from a normal person only by the loss of articulate language. He came and went in the Hospice where he was known under the name of "Tan." He understood all that was said to him. His hearing was actually very good. Whatever question one addressed to him, he always answered, "*tan, tan*," accompanied by varied gestures, by which he succeeded in expressing most of his ideas. If one did not understand his gestures, he usually got irate, and added to his vocabulary a gross swear word, exactly the same that I have indicated above when I talked about a patient of M. Auburtin. Tan was considered an egoist, vindictive and objectionable, and his friends who detested him even accused him of stealing. These defects could be due largely to his cerebral lesion. They were not pronounced enough to be considered pathological, and although the patient was at Bicêtre, one never thought of passing him into the division of alienated persons. On the contrary, he was considered as being completely responsible for his acts.

He had lost his speech for 10 years when a new symptom appeared. The muscles of the right arm began getting weak and ended by being completely paralyzed. Tan continued to walk without difficulty, but the paralysis gradually extended to the inferior right extremity, and after having trained the leg for some time the patient had resigned himself to stay in bed. About four years had elapsed, when the beginning of a paralysis of the arm set up. At this time the paralysis of the leg was sufficiently advanced to make standing absolutely impossible. Before he was brought to the infirmary Tan was in bed for almost seven years. This last period of his life is the one for which we have the least information. Since he was incapable of doing harm, his comrades had nothing to do with him any more, except to amuse themselves at his cost which caused him much wrath, and he lost the little celebrity which the singularity of his disease had once given him at the hospital. It was noticed that his vision had become notably weaker during the last two years. This was the only aggravation one could notice since he kept to his bed. Otherwise he never was in the way. One changed his linen only once a week, so that the diffused cellulitis for which he was brought to the infirmary on the 11th April, 1861, was not recognized by the attendants until it had made considerable progress and had involved the whole leg from foot to buttocks.

The study of this unhappy person who could not speak and who, being paralyzed in his right hand, could not write, offered some difficulty. Moreover, his general state was so grave that it would have been cruel to torment him by long interviews.

I found, in any case, that the general sensitivity was present everywhere although it was unequal. The right half of the body was less sensitive than the left, and this undoubtedly contributed to the decrease of pain on the site of the diffuse cellulitis. As long as one did not touch him, the patient did not suffer much but palpation was painful and incisions which I had to make provoked agitation and cries.

The two right extremities were completely paralyzed. The left one could be moved voluntarily, and although weak, could without hesitation execute all movements. Emission of urine and fecal matters was normal, but deglutition was difficult. Mastication, on the other hand was executed very well. The face did not deviate; however, in whistling, the left cheek appeared a little less inflated than the right, which indicated that the muscles of this side of the face were a little weak. There was no tendency to strabismus. The tongue was completely free, did not deviate, the patient could move it everywhere and stretch it out of the mouth. Both sides of the tongue were of the same thickness. The difficulty in the deglutition, which I indicated, were due to a beginning paralysis of the pharynx, and not to a paralysis of the tongue, for it was only the third stage of deglutition which appeared labored. The muscles of the larynx did not seem to be altered. The timber of the voice was natural, and the sounds which the patient uttered to pronounce his monosyllable were quite pure.

Hearing kept its acuity. Tan heard well the noise of the watch, but his vision was weak. When he wanted to see the time he had to take himself the watch in his left hand, and place it in a peculiar position about twenty cm from the right eye which seemed better than the left.

The state of his intelligence could not be exactly determined. Certainly, Tan understood almost all that was said to him, but since he could express his ideas or his desires only by movements of his left hand, the moribund patient could not make himself understood as well as he understood others. Numerical responses were his best: by opening or closing his fingers. I asked him several times, for how many days had he been ill? He sometimes answered 5, sometimes 6 days. How many years was he in Bicêtre? He opened the hand four times, and then added one finger; this made 21 years, and we saw that this was the correct answer. The next day, I repeated the same question and obtained the same answer, but when I tried to come back to it a third time, Tan understood that I wanted to make an exercise out of it; he became irate, and he uttered the above-mentioned swear word which I heard only this one time from him. I showed him my watch for two days in a row. The second hand did not move, he could therefore distinguish the three hands only by their shape and length. Nonetheless, after having looked at the watch for a few seconds, he could indicate correctly the hour each time. It cannot be doubted, therefore, that the man was intelligent, that he could think, that he had, to a certain extent, kept the memory of old habits. He could even understand quite complicated ideas: thus I asked him in which order his paralyses had developed. He first made with his left index finger a short horizontal gesture which meant that he had understood, then he showed successively his tongue, his right arm and his right leg. This was perfectly correct; he attributed the loss of language to the paralysis of the tongue, which was quite natural.

Nonetheless, several questions to which a man of ordinary intelligence would have found the means to respond by gesture, remained without response. At other times, one could grasp the sense of his answers which seemed to annoy the patient considerably. At still other times the answer was clear but wrong, for instance, although he had no children, he pretended to have some. Doubtless therefore, the intelligence of this man had deteriorated either under the influence of his cerebral lesion, or of his devouring fever, but he has obviously much more intelligence than it is necessary to talk.

From the anamnesis and from the state of the patient it was clear that he had a cerebral lesion which was progressive, which, at the beginning and for the first 10 years of the disease remained limited to a fairly well circumscribed region, and which during this first period, had attacked neither the organs of motility nor of sensitivity; that after 10 years the lesion had spread to one or more organs of motion, still respecting the organs of sensitivity, and that still more recently the sensitivity had suffered together with the vision, particularly the vision of the left eye. A complete paralysis of movement occupied the two extremities of the right side, and the sensitivity of these two extremities was slightly less than normal. The principal cerebral lesion should therefore be in the left hemisphere. This opinion was confirmed by the incomplete paralysis of the left cheek, and of the left retina, for, needless to say, paralyses of cerebral origin are crossed for the trunk and the extremities, but direct for the face.

One should determine more exactly, if possible, the seat of the first lesion although the last session of the Society of Anthropology had left some doubts about the doctrine of M. Bouillaud. While waiting for the autopsy, I want to reason as though this doctrine were correct. This seemed the best way to prove it. Since M. Auburtin had declared a few days previously that he would renounce it if we could show him a single case of well characterized aphemia without lesions in the frontal lobes, I invited him to see my patient, first of all to know his diagnosis. Was this one of the cases which he would admit as a conclusive evidence? In spite of the complications which had piled up for the last 11 years, my friend found the actual state and the anamnesis sufficiently clear to affirm, without hesitation, that the lesions should have begun in one of the frontal lobes.

Reasoning from here to complete the diagnosis, I considered that the corpus striatum was that motor organ which was nearest to the frontal lobes. Undoubtedly, by spreading to this organ, the original lesion had produced the hemiplegia. The probable diagnosis was then: original lesion of the left frontal lobe, spreading to the corpus striatum of the same side. As to the nature of the lesion, everything indicated that it was a chronic softening of a regressive character, but very slow, for the absence of any phenomenon of compression excluded the idea of an intracranial tumor.

The patient died on April 17, at 11 o'clock a.m. Autopsy was performed as soon as possible, that is to say, after 24 hours. The temperature was a little elevated. The cadaver showed no signs of putrefaction. The brain was shown a few hours later in the Society of Anthropology, then put immediately into alcohol. This organ was so altered that one had to be very careful to preserve it. Only after two months and after several changes of the fluid, the piece began to harden. Today it is in perfect condition, and it has been deposited in the Musée Depuytren under the number 55a of the nervous system.

I pass over in silence the details concerning the diffuse cellulitis. The muscles of the two right extremities were fatty and reduced to a small volume. All the viscera were healthy, with the exception of the brain.

The skull was opened with the saw with great care. All the sutures were closed, the bones were a little thickened, and the diploe was replaced by compact tissue. The inner surface of the skull presented in its whole extent a finely mottled surface, proof of a chronic osteitis (55b).

The outer surface of the dura mater was red and very vascular; the membrane itself was very thick, and very vascular, like meat, and covered on the inside with a pseudomembranous layer, infiltrated with serum, and appearing like lard. The dura mater and the pseudomembranous layer together had an average thickness of 5 mm (minimum 3 mm, maximum 8 mm) from which it follows that the brain must have lost a considerable part of its original volume.

After lifting up the dura mater, the pia mater appeared very injected in certain places, was thick everywhere, and opaque in places, infiltrated with a plastic, yellowish mass of the color of pus, but solid and, on microscopical examination, not containing purulent globules.

On the lateral side of the left hemisphere, at the level of the Sylvian fissure, the pia mater is lifted up by a collection of transparent serum which lies in a large and profound depression of the cerebral substance. When this fluid was evacuated by a puncture, the pia mater became profoundly depressed, and there resulted a long cavity about as large as a hen's egg, which corresponded to the Sylvian fissure, hence separated the frontal and the temporal lobes. It reached backwards as far as the sulcus of Rolando which separates, as one knows, the frontal from the parietal convolutions. The lesion is therefore completely in front of the sulcus, the parietal lobe is intact, at least relatively, since no part of the hemisphere is absolutely intact.

When incising and reflecting the pia mater from the cavity which I just indicated, one recognized at once that this cavity corresponded not to a depression but to a loss of substance in the cerebral mass; the liquid which filled it was later secreted to fill up the void which had been formed, as happens in chronic softenings of the cerebrum or cerebellum. A study of the convolutions which limit the cavity shows clearly that they were the seat of one of those chronic softenings which go so slowly that the cerebral molecules, in some way dissociated one after the other, can be absorbed and be replaced by the secreted serum. A considerable part of the left hemisphere had thus gradually become destroyed, but the softening extended well beyond the limits of the cavity; this cavity is not circumscribed, and can in no way be compared with a cyst. Its walls, almost everywhere irregular and indented, are constituted by the cerebral substance itself, which is here very soft, and the innermost layer of which, in direct contact with the secreted serum, was in the process of gradual and slow destruction when the patient succumbed. Only the inferior wall offers a fairly firm consistency.

It is clear therefore, that the first focus of softening occurred where there is today the loss of substance, that the disease then spread gradually by contact, and that the point where it started should not be looked for among the organs actually softened or in the process of softening, but among those which are more or less completely destroyed. We have therefore, after the inspection of the parts which limit the loss of substance, made a list of those which have disappeared.

The cavity which we have described is situated, as we saw, at the level of the Sylvian fissure, and is therefore between the frontal and the temporal lobes and if the organs which surround it were only rolled back without being destroyed, one should find on its inferior or temporal border the inferior marginal convolution, on its superior border the third frontal convolution, and finally, in its depth the lobe of the island. But this is not the case. (1) The inferior border is limited by the second temporal convolution, which is intact and firm. The marginal temporal convolution has therefore been destroyed in its whole thickness, that is to say, as far as the parallel sulcus. (2) The deep wall of the cavity shows no trace of the insula. This lobe is entirely destroyed, just as the inner half of the extra-ventricular nucleus of the striate body. Finally, the loss of substance is prolonged on this side up to the ventricular nucleus of the striate body, so that our cavity communicates, by an opening of about ½ cm and of irregular borders, with the lateral ventricle of the brain. (3) The superior border, finally, or rather the superior wall of the hemisphere, involves the frontal lobe which at this place shows a large and deep ulceration. The posterior part of the third frontal convolution is completely destroyed in its whole thickness, the second convolution is less affected. At least its outer two-thirds have disappeared, the inner third which is still present, is very soft. Further back, the inferior third of the transverse frontal convolution is destroyed in its whole thickness, up to the sulcus of Rolando.

To sum up, the destroyed organs are the following.

The small inferior marginal convolution of the temporal lobe, the small convolutions of the insula and the underlying part of the striate body, finally, in the frontal lobe, the inferior part of the transverse frontal, and the posterior part of those two great convolutions designated as the second and third frontal convolutions. Of the four convolutions which form the superior part of the frontal lobe, only one, the superior and most medial one, has been conserved, not in its integrity, for it is softened and atrophied, but in its continuity, and if one puts back in one's thought all that has been lost, one will find that at least three quarters of the cavity has been traversed by the frontal lobe.

Now we have to decide where the lesion started. An examination of the cavity caused by the loss of substance, shows at once that the center of the focus corresponds to the frontal lobe. Consequently, if the softening spread out uniformly in all directions, it would have been this lobe in which the disease began. But we should not only be guided by the study of the cavity, we should also keep an eye on the parts which surround it. These parts are very unequally softened, especially to a very variable extent. Thus the second temporal convolution, which limits the focus below, shows a smooth surface of firm consistency, it is without a doubt softened, but not much, and only in its superficial parts. On the opposite side, in the frontal lobe the softening is almost fluid near the focus, as one goes away from it, the substance of the brain becomes gradually firmer, but the softening extends, in reality, for a considerable distance and involves almost the whole frontal lobe. It is here, therefore, that the softening has mainly progressed and it is almost certain that the other parts have only later been invaded.

If one wanted to be more precise, one could remark that the third frontal convolution is the one which shows the greatest loss of substance, that it is not only cut transversely at the level of the anterior end of the Sylvian fissure, but is also completely destroyed in its posterior half, it alone has undergone a loss of substance, equal to about one-half of the total; that the second or middle frontal convolution, although deeply affected, still preserves its continuity in its innermost parts, and that, consequently, it is most likely in the third convolution that the disease began.

[. . .]

After having described the lesions and tried to determine their nature, their anatomical seat and their evolution, we have to compare now these results with those of the clinical observations, in order to establish, if possible, a relation between the symptoms and the material disorders.

The anatomical inspection shows us that the lesion was still progressing when the patient died. The lesion was therefore progressive but it progressed very slowly, because it took 21 years to destroy quite a limited part of the brain. It is therefore permitted to believe that at the beginning there was a considerable period during which it did not go past the limits of the organ where it started. But we saw that the original focus of the disease was situated in the frontal lobe and very likely in its third convolution. This makes us say, from the point of view of pathological anatomy, that there were two periods, one in which only one frontal convolution, probably the third one, was attacked, the other in which the disease gradually spread to other convolutions, to the island or to the extra-ventricular nucleus of the corpus striatum.

When we examine now the succession of the symptoms, we also find two periods, the first which lasted 10 years during which the faculty of speech was destroyed but all other functions of the brain were intact and a second period of 11 years during which paralysis of movement, first partial then complete, successively involved the superior and the inferior extremity of the right side.

With this in mind it is impossible not to see that there was a correspondence between the anatomical and symptomatological periods. Everybody knows that the cerebral convolutions are not motor organs. The corpus striatum of the left hemisphere is of all the attacked organs the only one where one could look for the cause of the paralysis of the two right extremities, and the second clinical period where the motility had changed, corresponded to the second anatomical period when the softening had gone beyond the limit of the frontal lobe and had invaded the insula and the corpus striatum.

Hence, the first period of 10 years clinically characterized only by the symptom of aphasia must correspond to the period during which the lesion was still limited to the frontal lobe.

[. . .]

Facts as these which have to do with the broader questions of doctrine can not be given in too great detail nor discussed with too much care. I need this excuse to forgive the aridity of the descriptions and the length of the discussion. I have now only to add a few words to point out the consequences of this observation.

(1)   The aphemia, i.e. the loss of speech, before any other intellectual trouble and before any paralysis, was a consequence of a lesion of one of the frontal lobes of the brain.

(2)   Our observation confirms thus the opinion of M. Bouillaud who places in these lobes the seat of the faculty of articulate speech.

(3)   The observations which have so far been made, at least those accompanied by clear and precise anatomical description, are not numerous enough to consider as

definitely demonstrated the localization of a particular faculty in a particular lobe, but one can consider it at least extremely probable.

(4) It is a much more doubtful question to decide whether the faculty of articulate language depends on the whole frontal lobe or specially on one of its convolutions; in other words to know whether the localization of cerebral faculties takes place by faculty and convolution or by groups of convolutions. In order to solve this problem further observations must be collected. For this purpose one must indicate exactly the name and the range of the diseased convolution and, if the lesion is very large, to try to determine as much as possible by anatomical methods the point or rather the convolution where the disease seems to have started.

(5) In our patient, the original seat of the lesion was in the second or third frontal convolution, more likely in the latter. It is therefore possible that the faculty of articulate speech is in one or the other of these two convolutions. However, it is difficult to know it at present, because the former observations do not describe the state of each convolution and one cannot even forecast it because the principle of localization by convolution is not yet firmly established.

(6) In any case, it is enough to compare our observation with the preceding ones, to dismiss the idea that the faculty of articulate language resides in a circumscribed fixed point situated under a certain elevation of the skull. The lesions of aphemia have been found most often in the most anterior part of the frontal lobe, not far from the eyebrow and above the orbital roof, whereas in my patient they were much further back, much nearer to the coronal suture than the superciliary arch. This difference in the localization is incompatible with the system of bumps. It could well be brought into line, however, with the system of localization by convolutions, because each of the three convolutions of the superior part of the frontal lobe runs successively in its antero-posterior extent through all these regions where, so far, lesions of aphemia were found.

6

Recent Works on Aphasia

Carl Wernicke

During the past few years a succession of new works on aphasia have appeared, and we shall attempt to review the most pertinent findings which have emerged from such studies. . . . Professional interest in this subject has yielded two major points of view. The first, which is essentially psychological in orientation – at least in its present state of development – tries to analyze the aphasia symptom-complex into its component parts. In so doing, it attempts for the first time to establish a precise theory of the existence of cortical centers in the human brain whose correlative connections and functions are achieved by means of open circuits, i.e. the so-called association pathways. The second concept proposes localization of these various circuits and centers to different areas of the brain. It can be dealt with only when the reader has been prepared by a review of the clinical analyses of individual case studies and has classified them according to the centers and conduction pathways which have failed.

1 The Symptomatology of Aphasia

1

. . . The cerebral centers under consideration are, it is true, anatomically preformed, but their functional content represents the unique acquisition of each individual, usually during childhood. The child learns to understand the speech of others and thereby develops a center for speech comprehension. He then learns to express himself aloud, thereby acquiring a center for complex motor speech patterns. Speech mimicry, which is of paramount importance in this training process, is dependent on the usage of a conduction pathway connecting these two centers in such a way that each acoustic image arouses the corresponding combination of movements. This leads to the following schema of the speech apparatus which was first submitted in my work of 1874 and which since that time has found almost universal acceptance among clinicians.

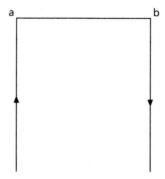

Figure 6.1 Schema of the psychic speech reflex. (a) Center of acoustic speech imagery. (b) Motor speech imagery.

A center (a), in figure 6.1 located in the central projection area of the acoustic nerve, contains the store of memory images of the speech sounds or the "acoustic images." The center (b), situated in the so-called motor zone of the cortex, is likewise a depository of memory images of motor speech patterns which we may refer to as "motor speech-representation" or "motor speech images." The latter cause activation of speech movements by means of a centrifugal pathway leading to the pertinent bulbar nuclei. A centripetal tract, that of the acoustic nerve, transmits to the first center. These centers are connected by means of an association pathway (a– –b) which is utilized in mimicry of speech sounds. The destruction of center (b) causes motor aphasia, that is, loss of speech with intact speech comprehension, while destruction of center (a) results in sensory aphasia, that is, loss of speech comprehension with preservation of basic speech capacity. The disruption of the pathway (a– –b) causes word-confusion in speech, a symptom to which Kussmaul has aptly applied the term "paraphasia," and which I have designated conduction aphasia, provided that speech comprehension in center (a) and speech expression in center (b) are undamaged. . . .

[. . .]

One might even go a step further, as has already occurred in all quarters, and try to explain the reading and writing disorders so frequently observed in aphasia on the basis of two other completely analogous centers. Analogous to the speech comprehension center in the acoustic area is a center serving comprehension of written material in the visual cortical area. Analogous to the speech motor center a center for movements involved in writing. The symptoms of alexia and agraphia (to retain the old, established terminology), are explained on the basis of the failure of these centers.

The following schema, illustrated in figure 6.2, may then be formulated. Let χ be the location of visual memory images of written symbols, and let β be that of graphic

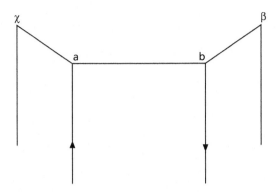

Figure 6.2   Schema of reading and writing. (a) Acoustic speech imagery. (χ) Visual letter imagery. (b) Motor speech imagery. (β) Graphic motor imagery.

motor imagery. Destruction of χ would then result in alexia as an analog of sensory aphasia, while destruction of β would produce agraphia as an analog of motor aphasia. However, the hypothesis of the existence of such specific centers may be justified only on the basis of appropriate clinical case studies, for even if memory images of written signs are undoubtedly located in the visual cortical region, just as those of the speech sounds are contained in the acoustic area, it is nevertheless questionable, and, I believe, as yet unproven, if they always lie together and are localized at a particular point of the visual cortical region. The same is true of the graphic motor images. Undoubtedly, they are also contained somewhere in the cortical area involved in innervation of the upper extremities. However, confirmation by an adequate number of clinical studies is necessary before their deposition in a specific area can be definitely established. . . .

In my first work on aphasia I took pains to show that in such an interpretation of the speech process as has been reviewed above, we had probably found the scheme of cortical function as a whole, that memory images were the psychic elements populating the cortex in a mosaic-like arrangement as a functional development which may very well be localized according to the regions of the nerve-endings, so that the acoustic images find their abode within the cortical terminals of the acoustic nerve; the visual images, within the cortical endings of the optic nerve; and the olfactory images in that of the olfactory nerve and so on. Likewise, the motor memory images or movement-representation could be located in the cortical sites of the motor nerve origins. For example, the images of speech movements would then be found in the Broca gyrus and those of writing within the cortical area serving arm movements, etc. Apart from that I assumed only that the discharge-sites of voluntary movements and the depository of motor images were identical. Any higher psychic process, exceeding these mere primary assumptions, could not, I reasoned, be localized, but rested on the mutual interaction of these fundamental psychic elements mediated by means of their manifold connections via the association fibers. Since that time I have become even more strongly convinced, particularly on the basis of clinical studies of aphasia, that we are not justified in going beyond this elementary hypothesis.

[. . .]

2

Now, before undertaking a discussion of the work of Lichtheim, that author who has been the most consistent in his support of our own theory, and who has, as I am bound to recognize, carried it further with great perspicacity, I must first explain to the reader my interpretation of the term "object-concept."

It can be readily seen that our interest in the speech mechanism, at least in the light of present knowledge, lies particularly in its role as an agent of consciousness. As was indicated at the conclusion of my last discussion, the cerebral hemispheres *in toto*, as the organ of consciousness, function as the executor of the motor speech center *b* in spontaneous speech production. Likewise, the organ of consciousness *in toto* receives the message which is first transmitted to the sensory speech center at *a*, functioning as a receiving station for acoustic messages. Therefore, it would seem that further localization within this single organ of consciousness is not the issue here. But as soon as we wish to point to a concrete example as a test of our schema, we can find other, more comforting results. For example, how might the process involved in comprehension and spontaneous expression of the word "bell" be explained? If we are to comprehend this word, the concept of a bell must be aroused within us by the acoustic message which has reached center *a*. The acoustic message must stimulate the memory images of a bell which are deposited in the cortex and located according to the sensory organs. These would then include the acoustic imagery aroused by the sound of the bell, visual imagery established by means of form and color, tactile imagery acquired by cutaneous sensation, and finally, motor imagery gained by exploratory movements of the fingers and eyes. Close association between these various memory images has been established by repeated experience of the essential features of bells. As a final result, arousal of each individual image is adequate for awakening the concept as a whole. In this way a functional unit is achieved. Such units form the concept of the object, in this case a bell. Thus when a spoken word is understood and provokes thought, these units are in a sense a second station, accessible to our own recognition, in the total activity of the hemispheres, a station which must be passed through if the spoken word is not to die away in our ears without having been understood. Moreover, our consciousness makes uses of this same station when the word "bell" is to be articulated spontaneously, i.e. as the result of what may be highly complex processes within our consciousness.

The first stage in this process then consists in the arousal of the concept of the object, "bell," and the second in the process of transmission to the pertinent motor memory images in *b*, the site involved in dispatch of the message. A schematic illustration of this process is indicated in the diagram in figure 6.3, in which (B)

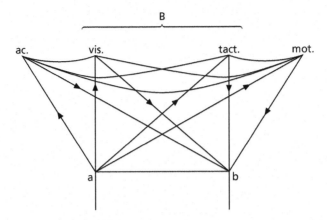

Figure 6.3 Composite schemata of language and conceptualization. (a) Acoustic speech center. (b) Motor speech center. (B) Conceptualization. (ac) Acoustic imagery. (vis) Visual imagery. (tact) Tactual imagery.

represents the concept of the object bell. The reader may find a very similar schema in Lichtheim's work. In the same way, if we attempt to construct the concept of the word or "symbol" as the name of an object is often called such as that of the word "bell" we find that, completely analogous to the object itself, it consists of the relevant firmly associated memory images in (a) and (b). These speculations then may suggest differentiation of speech comprehension into two stages, namely, (1) the arousal of the word and (2) arousal of the corresponding object-concept. The same process occurs in spontaneous speech, but in the reverse order, with the concept of the object emerging first, followed by that of the word.

This brief digression in regard to the word or symbol concept might seem unnecessary. However, we have an immediate need to use it. For we shall have to examine the extent to which such word-concepts are inseparable unities. Pathological research yields two lines of evidence relevant to this problem. On the one hand, if center *b* is destroyed, speech comprehension may remain completely intact; in other words, the acoustic imagery of the word is adequate for arousal of the concept of the object. If, however, center *a* is damaged, the independence of center *b* can be seen in the continued production of spontaneous speech. The latter, however, is characterized by inconsistent word-choice, with symptoms of word-transposition or paraphasia. Therefore, preservation of the word-concept is of greater significance in the active phase of the speech process than in the passive. Or, translated into terms of our schema, the association between the acoustic word image and the concrete object is firm and independent, but that between the object-concept and the pertinent motor word image is more fragile and not adequate to ensure accurate speech production. Presence of the word-concept in its entirety is necessary for production of spontaneous speech. This finding, gleaned from pathological studies, is comprehensible if

understood in the light of the mechanisms involved in speech acquisition. Undoubtedly the first knowledge of language which the child acquires consists of the comprehension of words, the association of acoustic images with concepts of concrete
objects, while in many cases a further period of years is necessary for the development of the faculty of active speech. A preliminary stage of this last is the ability,
by using the association pathway a–b, to imitate the speech sounds heard. On this
basis I hypothesize that centrifugal innervation of the word-concept from the area
(of sensory perception) of the concrete object follows a double path, namely the
simple path B–b and the more complicated route B–a–b. If a portion of the latter is
disrupted in any place incomplete activation of the word-concept will be reflected
in the transposition of words. . . .

[. . .]

If we now return to our original schema in figure 6.3 and consider it in relationship to a concept center, which for the sake of simplicity we shall reduce to a point
designated B, and if we also restrict the scope of aphasic symptoms to all such cases
of speech disturbance in which the concept of the object itself is preserved, damage
to the centers under question and their various conduction paths would yield seven
different forms of aphasia.
    Let us enumerate the causes of these, using Lichtheim's model but changing the
order.

1   Damage to center a.
2   Disruption of the acoustic path ending in a.
3   Disruption of the centripetal pathway between a and the concept center.

Forms 1–3 constitute the group of the sensory aphasias.

4   Damage to center b.
5   Disruption of the motor speech pathway.
6   Disruption of the centrifugal pathway B–b.

Forms 4–6 constitute the group of the motor aphasias.

7   Conduction aphasia (previously observed by us).

[. . .]

The following nomenclature of aphasia may then be formulated.

1   Cortical sensory aphasia
2   Subcortical sensory aphasia
3   Transcortical sensory aphasia

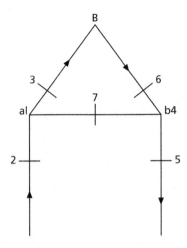

Figure 6.4 The Lichtheim aphasia schema using Wernicke's names. (B) Concept center, (1) Cortical sensory aphasia. (2) Subcortical sensory aphasia. (3) Transcortical sensory aphasia. (4) Cortical motor aphasia. (5) Subcortical motor aphasia. (6) Transcortical motor aphasia. (7) Conduction aphasia.

4   Cortical motor aphasia
5   Subcortical motor aphasia
6   Transcortical motor aphasia
7   Conduction aphasia

[. . .]

[Wernicke then elaborates on each of these forms of aphasia, using figure 6.4 to illustrate their origin.]

1   Cortical sensory aphasia is characterized by lack of comprehension of the spoken word and inability to mimic. However, the patient is able to speak spontaneously, although vocabulary is limited and characterized by frequent word-transposition, that is, paraphasia. See figure 6.4.
2   Subcortical sensory aphasia presents the same lack of comprehension of the spoken word and the same impairment in word mimicry. Spontaneous speech however is maintained, and the word-concept remains intact.

[. . .]

3   Transcortical sensory aphasia. Impairment in comprehension of the spoken word with preservation of mimicry. Symptoms of paraphasia are evident in sponta-

neous speech. See figure 6.4. It will be agreed that the features of forms 1–3 are complete without the necessity for consideration of disorders of written language. The assumption in all cases of sensory aphasia is that common deafness is not the basis of the lack of comprehension.

4   Cortical motor aphasia. Speech comprehension is intact, but the patient presents either muteness or a vocabulary limited to a few words. Spontaneous speech and mimicry as well as the voluntary mental sounding of the word are not possible. . . .

5   Subcortical motor aphasia. This form is differentiated from the preceding type by the complete integrity of the word-concept. See figure 6.4. The muteness is the same as that found in type 4. The patient is able to indicate the number of syllables contained in a word corresponding to an object presented to him.

6   Transcortical motor aphasia. . . . There is loss of spontaneous speech but no evidence of impairment in speech comprehension.

Of these three forms of motor aphasia, only the differentiation between forms 4 and 5, i.e. cortical motor and subcortical motor aphasia, presents problems which demand a critical discussion of written language. This function, writing, to anticipate a bit, is impaired in the first form of the aphasia mentioned, cortical motor aphasia, and is intact in the second, subcortical motor aphasia. Transcortical motor aphasia, however, may be identified without this feature. The cause of muteness in motor aphasia does not lie in paralysis of the speech musculature, just as common deafness is not the basis for impaired comprehension in sensory aphasia.

7   Conduction aphasia is primarily characterized by negative symptoms. If motor or sensory aphasia is not evident, but speech is paraphasic, presenting word-transposition, one may predict a disturbance in conduction between centers a and b.

I shall return later to the problem of so-called amnesic aphasia. This type is not related to the aphasia forms just reviewed, but is rather concerned with an actual memory disturbance.

3

The complex process involved in the acquisition of reading and writing may be more readily understood if one applies its obvious analogy to speech development.

One might then say that reading consists in the activation of the word-concept by the visual written image. Conversely, the process of writing involves the activation of the corresponding graphic motor image by the word-concept.

To produce a suitable schema (figure 6.5) let us use the same kind of device which we used earlier to denote the concept center B. In other words, let us write the word-

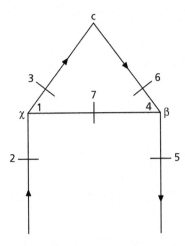

Figure 6.5 Schema of the alexias and agraphias. (c) Word-concept (acoustic and motor speech imagery). (χ) Visual written image. (β) Center of motor graphic imagery. (1) Cortical alexia. (2) Subcortical alexia. (3) Transcortical alexia. (4) Cortical agraphia. (5) Subcortical agraphia. (6) Transcortical agraphia. (7) Conduction agraphia.

concept as $a + b = c$, where the addition sign designates integrity of the association pathway $a-b$. This then leads to the schema formulated in the diagram in figure 6.5 in which χ denotes the visual written image and β the motor center of movements involved in writing. The pathway $\chi-\beta$, an analogue of pathway $a-b$ in the earlier schema diagrammed in figure 6.4, represents the crucial pathway by means of which writing is learned. Just as the critical process involved in speech-learning is repetition of the spoken word mediated by the pathway $a-b$, so the learning of writing is acquired under the controlling influence of the visual written image. The pathway $\chi-\beta$ therefore has the same significance for written language as the path $a-b$ for speech.

After these preliminary remarks, let us now turn to pathological case studies of written language disorders. The theoretical possibility of seven types of disorder therefore becomes immediately evident as was also true of disturbance in speech. Let us designate these disorders by the numbers 1 through 7, labeling disturbances of the centripetal pathway as alexia and those of the centrifugal path, agraphia. This provides us with the following summary.

1 Cortical alexia
2 Subcortical alexia
3 Transcortical alexia
4 Cortical agraphia
5 Subcortical agraphia
6 Transcortical agraphia
7 Conduction agraphia

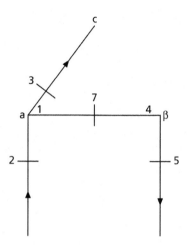

Figure 6.6 Schema of agraphias and alexias (with omission of tract *c–β*). (1) Cortical alexia. (2) Subcortical alexia. (3) Transcortical alexia. (4) Cortical agraphia. (5) Subcortical agraphia. [(6) Transcortical agraphia.] (7) Conduction agraphia.

A simplification of this schema is needed at this point, for reasons which shall be presented later, before we turn to a discussion of the symptom-pictures of the seven theoretically predictable forms, which is not written without practical interest. We can only briefly indicate here that there is much which contradicts the actual existence of a pathway *c–β* and therefore the schema illustrated in the accompanying diagram in figure 6.6 alone may be used as a basis for our discussion. In this way an analogy to disturbances in oral speech can be readily demonstrated, although with certain simplifications.

1 Cortical alexia. Loss of reading and writing. Disruption of the centrifugal pathway for writing at *χ*.
2 Subcortical alexia. Inability to read. No disturbance in writing with the exception of writing from a model.
3 Transcortical alexia. Loss of ability to read and write with preservation of the mechanical copying of printed and script material.
4 Cortical agraphia. Loss of fine motor patterns requisite to writing. Copying is not possible. No impairment in reading.
5 Subcortical agraphia. Essentially the same picture. (Further comment will be made later.)
6 Transcortical agraphia. This form is nonexistent.
7 Conduction agraphia. Reading unimpaired. Loss of writing.

If one assumes the existence of a direct pathway *c–β*, transcortical agraphia, form 6, would be characterized by the exclusive loss of spontaneous writing. In form 7, conduction agraphia, writing would still be possible, but paragraphia, analog of paraphasia observed in oral speech disorders, would be present.

Therefore, the control exerted by means of visual imagery may be considered much more indispensable to the writing process than that of acoustic imagery in respect of the speech process. Writing demands greater conscious activity than speech. In the latter, distracting factors may be predominant: the influence of the acoustic image *a* is always only slight, while the writing process, on the contrary, demands at least enough concentration to produce a series of letters.

Paragraphia, which is the chief issue here, therefore probably amounts to the same as complete agraphia; with impairment in formation of the letter itself. This can be seen when attempts to write produce only scribbles and lines which depart from the correct letter form. These samples of writing also occasionally contain correctly formed letters. Such, however, are infrequent, occurring only here and there among a group of incomprehensible scribbles. If correct and completely formed letters can be produced, and the disturbance is limited to transposition and interpretation of letters, one is dealing not with agraphia or paragraphia, as we interpret it, but rather with the symptoms of paraphasia, which may also be revealed in writing: in a word, written paraphasia.

A second point which must be briefly touched on here concerns the building of words by the combining of individual letters. The visual memory images deposited in χ and the motor memory images at β consist of single letters of the alphabet. The word-concept, on the contrary, always includes at least one syllable, and often several syllables, and is comprehended as a series of association of letters in a specific sequence. If we should take a very simple example of a one-syllable word such as *hand*, it can be seen that only the four letters in sequence, *h–a–n–d*, are able to call up the correct meaning. This is also true of multi-digit numbers. In alexia it will therefore be possible to determine whether whole words or only individual letters can be read, i.e. recognized. Likewise, in agraphia we must see whether only letters, but not whole words, can be written, or whether whole words, too, can be produced. For both the reading and the writing of individual letters it is necessary only for the concept of the letter to be intact, and this consists of χ, β, and the area between χ and β which I need not again describe in detail.

The fact that man, however, can learn to read letters without necessarily learning to write them – one need only point to the reading of printed characters – indicates the greater autonomy of χ in contrast to that of β.

We may now complete the classification of the individual forms of alexia presented earlier within the framework of this hypothesis. In cortical alexia, the letters presented to the patient apparently cannot be recognized but appear strange and foreign, a fact which may be confirmed by the use of leading questions. Furthermore, he is unable to copy, and in so attempting does not actually write, but merely draws the letters, and even then, later recognition is not possible.

Moreover, subcortical alexia spares spontaneous writing alone, and letters presented to the patient appear unfamiliar and must be laboriously copied from the model. Copying, in fact, is not possible. Occasionally letters may be recognized during such attempts to copy, and under certain conditions may even be read, a trick which has long been observed, since it often occurs in aphasia.

In transcortical alexia, on the contrary, the letters appear to be recognized and can be copied without difficulty. Reading aloud is not possible since disruption of the path χ–c results in inability to arouse the associated acoustic image of the letter. In none of the three forms of alexia is the reading of words intact. This is readily under-stood, since reading of individual letters is not possible. Nevertheless, one is impressed by the striking ability demonstrated by such patients in their immediate reading of the word itself, as soon as the letters of which it is composed can be read. Therefore, it would seem that impairment of the word-concept must be postulated. A similar situation exists in the writing of letters and words. Only in impairment of the word-concept itself does it happen that the letters alone, but not the word, can be written without aid.

[. . .]

7

The Processing of Single Words Studied with Positron Emission Tomography

Steven E. Petersen and Julie A. Fiez

1   Introduction

Language, the most profoundly human of all abilities, has been a focus of study in many disciplines. Cognitive psychological studies of normals have been used to describe language-related information processing. The correlation of language-related behavioral deficits with the site of nervous system injury has been used to relate language processing to brain function.

Over the past decade [to 1993], there has been an explosion in the use of imaging technology, including positron emission tomography (PET), to study the structure and function of the human brain. This article explores how functional imaging with PET has been applied to the study of language, particularly the study of the processing of single words (lexical processing).

To understand how PET is used in the study of lexical processing, one has to have some understanding of the technology of PET, a framework in which to explore the processing of words, and knowledge of the questions that can be addressed by the specific application of PET studies to issues in lexical processing. This paper follows that path with a brief general introduction to PET, an outline of some aspects of lexical processing, and a presentation of selected PET studies that have addressed issues in lexical processing.

2   PET Basics

Positron emission tomography creates pictures of the distribution of radiation within the central opening of a doughnut-shaped PET scanner. The rationale behind functional imaging is that those parts of the brain that "work harder" (i.e. have high levels of neuronal activity) have higher blood flow or metabolism (Lassen et al.,

1978), and damaged and/or disconnected areas may have aberrant levels of blood flow or metabolism (Baron et al., 1989). Radioactive substances, called tracers, are employed to "image" different physiological processes, such as brain blood flow or metabolism. By comparing images of activity, information relating to localization of certain functions can be obtained. For general reviews of PET technology and methodology, see Raichle (1989) and Stytz and Frieder (1990).

The methods and data analyses used are specific to the application, so those issues are addressed in the sections on specific studies below.

3 Lexical Processing Basics

In the design and interpretation of PET studies, it is necessary to utilize information from other types of studies on lexical processing. At an intuitive level, there are several different kinds of internal coding that can be done at the level of the single word, including how a word looks (orthographic codes) and sounds (phonological codes) and what it means (semantic codes). These internal codes can be experimentally isolated (for a review of cognitive psychological experimental methodology, see Posner, 1978). For instance, differences in the internal coding of phonology and articulation can be demonstrated by having subjects perform articulatory and phonological processing tasks concurrently (Shallice et al., 1985). Other investigations have focused on orthographic (e.g. Carr and Pollatsek, 1985; Glushko, 1979), phonological (e.g. Jackson and Morton, 1984; Tanenhaus et al., 1980), and semantic (e.g. Davidson, 1986; Neely, 1977) aspects of lexical processing.

Studies of behavioral deficits following damage to particular brain regions have provided the most information on the neural substrates underlying different kinds of lexical coding (see reviews by Caramazza, 1988, and Damasio, 1992), although nontomographic blood flow measurements (Lassen et al., 1978), event-related potential recording (Kutas and Van Petten, 1990), and other methods have also contributed. For instance, specific deficits in reading (e.g. Damasio and Damasio, 1983; Henderson, 1986), understanding auditory input (e.g. Yaqub et al., 1988), and programming speech output (e.g. Geschwind, 1979) have been documented following damage to particular brain regions.

4 Studies of Lexical Processing with PET

Functional studies of lexical processing with PET fall into two main categories. Although both utilize functional images, their general approaches are quite different.

The first type of study defines brain areas of functional abnormality in conjunction with behavioral studies of language deficits. This use of PET to study abnormal populations is a natural extension of work using classical lesion/behavior.

The second type of study uses activity-based measures to provide images of brain activation during performance of lexical processing tasks, most often in normal subjects. These studies represent a chance to study the function of the normal human brain *in vivo*.

Imaging of functional abnormality

Rationale   There is now strong evidence that structural damage is frequently associated with hypometabolism in structurally intact regions distant to the site of damage (Baron, 1989; Metter, Kempler, et al., 1989; Powers and Raichle, 1985). Based upon observations of the rapidity by which these distant effects can develop, the initial hypometabolism is most likely caused by the loss of afferent input to the secondary region. Behavioral deficits similar to those associated with structural damage have been associated with regions of secondary hypometabolism, which leads to the term "metabolic lesions." Despite uncertainty regarding mechanisms responsible for the production of and recovery from metabolic lesions, descriptions of the relationships between such lesions and behavioral deficits should be extremely useful.

Methods   Under normal conditions at rest, blood flow, oxygen metabolism, and glucose metabolism are tightly correlated (Lassen et al., 1978), and all three are useful for indirectly measuring neuronal activity. In such conditions as ischemia or transient activation, this coupling breaks down (Baron et al., 1989; Fox and Raichle, 1986). For this reason glucose metabolism, rather than blood flow, has been measured in nearly all PET studies of aphasic patients.

^{18}F-labeled fluorodeoxyglucose (FDG), a competitive substrate for glucose, is the tracer typically used to measure glucose metabolism. The accumulation of FDG in the brain is proportional to the amount of glucose utilized. The simplest and most commonly used method requires that the tracer reach a steady- or near steady-state before measuring the accumulation of FDG with the PET scanner (Reivich et al., 1979; Sokoloff et al., 1977). During this period (about 40 minutes), subjects must continuously perform the task of interest. The task condition for studies of abnormal populations is most often a "resting" condition, e.g. the patients are told to close their eyes, and their ears may be plugged.

To analyze metabolic images, each individual PET image is usually first divided into standard regions of interest (Herholz et al., 1985; Mazziotta et al., 1983). Two general approaches have been used to identify which regions are metabolically abnormal. For one approach, the average regional metabolic values of controls are determined for each region; based upon these values and the range of normal variation, a criterion is established to define abnormal regional metabolism (e.g. see Kushner et al., 1987; Metter, Kempler, et al., 1987). Because of the numerous comparisons to be made, and the small size of the abnormal group (in many cases, only an individual is compared), this criterion does not reflect a statistically rigorous value for significantly different regional metabolism, but rather a descriptive measure of regional metabolic normality. Another approach has been to avoid any comparison to normal metabolic regional values and instead determine, in a large patient group, which regional values correlate most closely with behavioral measures of impairment (e.g. see Karbe et al., 1989; Metter, Kempler, et al., 1989).

Areas of investigation   The use of PET to extend lesion-behavior analysis to include information about metabolic lesions is still relatively new, but has provided insight into mechanisms of aphasia. Several studies have provided evidence that the distribution of metabolic lesions can help account for inconsistencies in the relationship between structural lesions and behavioral deficits. For example, various types of aphasia have been reported following damage to thalamic and basal ganglia structures (Crosson, 1985; Naeser et al., 1982). However, aphasia is not always present following subcortical structural damage, which leads many investigators to question the importance of subcortical structures in language processing. Aphasia following subcortical structural damage is apparently correlated with cortical hypometabolism, particularly in left posterior temporoparietal regions (Karbe et al., 1989, 1990; Metter et al., 1986, 1988).

Other investigations have focused on identifying the structural and metabolic location of lesions associated with specific behavioral deficits. In one report, left temporoparietal hypometabolism, caused by either structural damage or a secondary metabolic effect, was found in all 44 aphasic patients examined (Metter et al., 1990). In another report, standard diagnostic aphasia exams were used to classify patients into a variety of subtypes, such as Broca's and Wernicke's aphasia. The location of structural lesions ranged from damage limited to subcortical structures to widespread damage of frontal, temporal, and parietal regions of the left hemisphere. The degree of temporoparietal hypometabolism correlated most strongly with impaired performance on tests of language comprehension. The presence of frontal hypometabolism, caused by either structural or metabolic lesions, reliably discriminated Broca's from Wernicke's aphasics (Metter et al., 1989). Other studies have also found correlations between temporal/temporoparietal hypometabolism and impaired language comprehension (Karbe et al., 1989, 1990; Tyrrell et al., 1990).

Finally, PET provides a unique means to monitor functional changes over time. Several different investigators have now reported the results of sequential PET scans of aphasic patients. Correlations between reductions in secondary hypometabolism and recovery of language functions have been reported. Conversely, patients who do not show significant improvement tend to show persistence of metabolic abnormalities (Kushner et al., 1987; Metter et al., 1986). (Also see a study by Vallar et al. (1988), who address this issue by using a related technique to measure metabolism.)

Mapping functional activation

Rationale   The basic rationale for using PET-activation studies is that the performance of any task places specific information-processing demands on the brain. These demands are met through changes in neural activity in various functional areas of the brain (Posner et al., 1988). Changes in neuronal activity produce changes in local blood flow (Frostig et al., 1990; Raichle, 1989), which can be measured with PET.

In the case of studies from our group, the design and interpretation of these experiments is based on a particular view of the localization of function in the brain (see Posner et al., 1988). Understanding from integrative neuroscience and cognitive science has encouraged a limited framework for understanding functional localization. The main idea of this framework is that elementary operations, defined on the basis of information-processing analyses of task performance, are localized in different regions of the brain. Because many such elementary operations are involved in any cognitive task, a set of distributed functional areas must be orchestrated in the performance of even simple cognitive tasks.

Although these seem like trivial statements, it is what is not said that is important. A functional area of the brain is not a task area; there is no "tennis forehand area" to be discovered. Likewise, no area of the brain is devoted to a very complex function; "attention" or "language" is not localized in a particular Brodmann area or lobe. Any task or "function" utilizes a complex and distributed set of brain areas.

The areas involved in performing a particular task are distributed in different locations in the brain, but the processing involved in task performance is not diffusely distributed among them. Each area makes a specific contribution to the performance of the task, and the contribution is determined by where the area resides within its richly connected parallel, distributed hierarchy [such as that described for the visual system by Felleman and Van Essen, 1991, or by Van Essen and Gallant in chapter 12, this volume]. The difficult but exciting job to which PET activation studies contribute is the identification of specific sets of computations with particular areas of the brain. Given the complexity of this job, no single experiment or image provides compelling evidence at this level of analysis. Information from multiple PET studies and the results from other methodologies must be considered. This section attempts not only to address specific issues in lexical processing, but also provide examples of how converging evidence can aid in the interpretation of specific results.

Methods   Positron emission tomography activation methods have been extensively described, and a brief description follows.

In designing or understanding PET activation studies, one should realize the performance characteristics of the methodologies for these studies. Currently, the spatial resolution of reconstructed images for activation studies is well above 1 cm, but the accuracy of localizing a single source is on the order of 2–5 mm (see, e.g. Fox, Miezin, et al., 1987; Mintun et al., 1989). These spatial capabilities allow for some topographic mapping within large primary and extra-primary sensory and motor areas (e.g. Grafton et al., 1991), but most of the study is at the level of the identification of functional areas involved in the performance of particular tasks (e.g. Zeki et al., 1991).

Although earlier studies utilized different tracers, including FDG, most activation studies in normals currently measure blood flow by using ^{15}O. ^{15}O has a short half-life (about 2 min) and acquisition time (about 40 s–4 min) compared with other tracers, which allows several scans to be performed in a single session (Herscovitch et al., 1983; Raichle et al., 1983).

Because several scans can be made in a scan session, different tasks can be performed by each subject. Because within-subject designs are used, comparison images between different task conditions can be created, with the idea of imaging activity changes related to specific elementary operations of a task. Different techniques have been developed to allow the creation of inter- or intrasubject averaged images, or data sets that increase the signal-to-noise ratio (Evans et al., 1988; Fox et al., 1985, 1988; Friston et al., 1991b; Mazziotta et al., 1991). Recent reports (Steinmetz and Seitz, 1991) have asserted that although such averaging might be appropriate for primary areas, association and higher-level cortical areas have much more anatomical variability and, thus, are inappropriately studied with averaging. Fox and Pardo (1991) have shown that although activation in frontal and visual association cortex shows slightly higher variability across subjects than in primary areas, this difference is small and the averaging techniques are still quite effective.

Once created, images are usually searched with computerized routines to localize areas of change in stereotactic coordinates (e.g. Mintun et al., 1989). Because these images have been cast into stereotactic coordinates, most often one of the Talairach stereotactic brain atlas spaces (Talairach et al., 1967; Talairach and Tournoux, 1988), the results of experiments from different groups can be compared with one another in a relatively direct way.

Different types of statistical analyses have been applied to identify areas of significant change. The most common approach is to survey images without a priori assumptions about the locations of significant changes. This presents more difficult statistical problems, because of the numerous spatial locations that are not statistically independent from one another and on which there are small numbers of observations (Fox et al., 1988; Friston et al., 1991a; Worsley et al., 1992). The issue of statistical analysis is an aspect of PET activation studies in which further advances should occur.

Functional mapping investigations A survey study of lexical processing (Petersen et al., 1988, 1989) was designed to examine, at several levels, the processing of single words (see table 7.1 and figures 7.1 to 7.3). This study identified several regions potentially involved in lexical processing and provided a framework for the development and interpretation of subsequent PET investigations of language processing. This study looked at three levels of change: simple presentation of the

Table 7.1 Paradigm design for Petersen et al. (1988) study

Control state	Active state	Added component
Fixation point only	Passive words (aud or vis)	Passive sensory processing, automatic word-level coding
Passive words (aud or vis)	Repeat words	Articulatory coding motor programming and output
Repeat words	Generate verbs (uses and actions)	Semantic/syntactic association, selection for action

Note: Each task (except fixation only) was performed in one set of scans with visual presentation of words and in another set of scans with auditory presentation. All subtractions are made within modality of presentation.

common English nouns, compared with no presentation; repetition aloud of the nouns, compared with simple presentation; and generation aloud of a verb appropriate to the presented noun, compared with repetition. The active state from one level acted as the control for the next level of analysis. Each of these levels was assessed in one set of scans with visual input and in another set of scans with auditory input.

Passive presentation of words appeared to activate modality-specific primary and extra-primary sensory-processing areas (figure 7.1). When words are presented visually (without any task demands), several areas of extrastriate visual cortex are activated in both hemispheres. The presentation of auditory words activated areas bilaterally along the superior temporal gyrus, as well as a left-lateralized area in the temporoparietal cortex.

When the repeat aloud tasks were compared with the passive tasks, similar areas of activation for auditory and visual presentations were found (as would be expected, because the sensory-specific activation would be subtracted away). For both auditory and visual cues, speech output produced activation in areas that have been implicated in some aspects of motor coding or programming, including primary sensorimotor mouth cortex, the supplementary motor area (SMA), and regions of the cerebellum. Several areas that might be considered lateral premotor regions in Sylvian-opercular cortex and a left-lateralized region on the lateral surface of the frontal cortex at or near inferior area 6 were also activated (figure 7.2).

The generate verb task made additional processing demands. Here, the subtraction condition was the repeat aloud condition, so the sensory input was identical in the active and control condition, and the motor output was very similar. Again, the comparison should have subtracted away differences between the auditory and visual input versions of the task, and this was most often the case. For the generate subtraction, two foci in anterior cingulate cortex were activated; several regions of

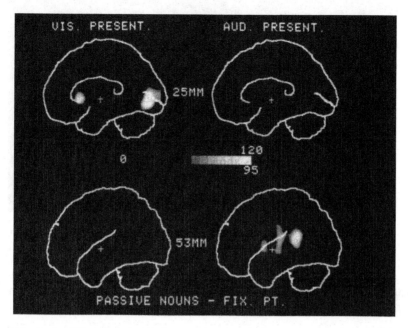

Figure 7.1 Sagittal slices (*anterior left, posterior right*) through averaged subtraction images during passive presentation of visual (*left slices*) and auditory words with a control condition of simple fixation. The upper slices are taken 25 mm left of midline, and the lower slices are taken 53 mm left of midline. The upper left image shows extrastriate visual cortex activation posteriorly for the visual presentation with no activation in the corresponding slice (*upper right*) for auditory activation. The opposite holds true for the more lateral slices. Several areas are active for auditory presentation (*lower right*) that are not present for visual presentation. The most posterior activation for auditory words is discussed in the text as the temporoparietal region.

the left anterior inferior prefrontal cortex and the right inferior lateral cerebellum (figure 7.3).

We now examine each of the areas discussed above.

Extrastriate visual cortex Recent studies (Marrett et al., 1990; Wise, Hadar, et al., 1991) have also shown extrastriate activation with the presentation of visual words. With this evidence alone, however, the activations in extrastriate cortex could be accounted for by any of several factors. The activations could be due to the simple visual features of the stimuli; processing at some higher level, such as the letter or word levels; or some intermediate between these stages. Subsequent experiments have addressed these questions.

In one experiment, different sets of words and word-like visual stimuli were presented in separate scans, without other task demands (Petersen et al., 1990). Again, the main control condition was simple fixation on a crosshair. One of the sets consisted of real common nouns to replicate the condition from the original study. In

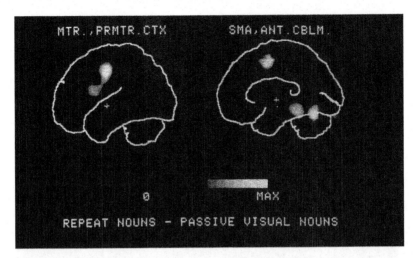

Figure 7.2 Sagittal slices through averaged subtraction images when subjects repeat aloud visually presented words with a control condition of visual presentation. The left image is taken about 40 mm left of midline and shows activation in primary motor (upper activation) and premotor cortex (lower activation). The right image is taken near the midline and shows activation in supplementary motor cortex (SMA, upper activation) and midline cerebellum (lower activation).

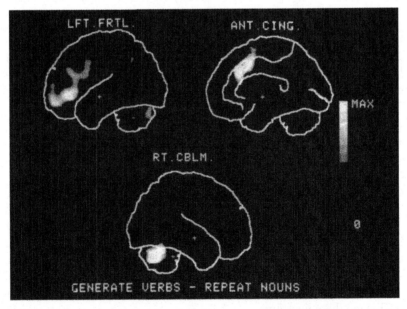

Figure 7.3 Sagittal slices through averaged subtraction images when subjects generate verbs appropriate to visually presented nouns with a control condition of simple repetition of visually presented nouns. The upper left image, taken about 40 mm left of midline, shows activation in several prefrontal regions. The upper right image is taken near midline and shows activation in the anterior cingulate. The lower image (*anterior right, posterior left*) is taken 25 mm right of midline and shows activation in lateral cerebellum.

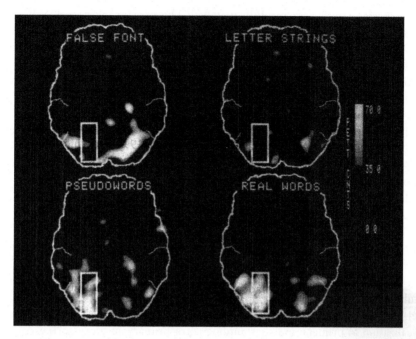

Figure 7.4   Horizontal slices taken through averaged subtraction images when subjects see sets of word-like stimuli with a control condition of simple fixation (anterior is to the top of the images, and left hemisphere is to the left of the images). The upper-left image represents activation when a series of false font stimuli are presented, the upper right when random letter strings are presented, the lower left when pseudowords are presented, and the lower right when real words are presented for all four images, there is activation in the posterior cortex in lateral extrastriate cortex. Only for the pseudoword and real word conditions is there activation in medial extrastriate cortex (*box*).

another scan, the stimuli consisted of strings of letter-like forms that were matched for many visual features to actual letters. These stimuli were called "false-font." A third set of stimuli were strings of random consonants, which were called letter-string stimuli. These stimuli were chosen to determine whether the activation was caused by processing at the level of analysis of single letters. The final set of stimuli was a series of letter strings that followed the spelling rules of English, but were not real words, e.g. "tweal" (pseudowords).

All four stimulus types produce extrastriate activation that is laterally placed in the hemispheres (figure 7.4). This activation seems most easily interpreted as related to general visual processing at the level of common visual features or characteristics shared by all of the stimulus sets.

Although the lateral activations are produced by presentation of all four types of visual word-like stimuli, left medial extrastriate cortex activation is found for real and pseudowords, but not for letter strings and false font stimuli (box in figure 7.4). These differences are probably not due to the simple visual characteristics of these stimuli, because these were carefully matched. This area (or these areas) is likely part of a system that makes distinctions based on information about the combinations of letters that English words can regularly assume, a level of analysis called ortho-graphic regularity. This level of distinction is not innate; it must have been learned as the subjects were becoming literate, and, judging by the location of the medial activation, the effects show up early in the visual processing stream. This particular level of distinction, where words and pseudowords are distinguished from other word-like stimuli, is consistent with cognitive psychology experiments about the visual processing of words. There is a well-known effect that the letters inside of words are more efficiently processed (can be seen at lower thresholds or are responded to more quickly) than the letters inside of random letter strings. This word superiority effect extends to pseudowords, such as those presented in our studies (see Carr and Pollatsek, 1985).

Temporal areas   Activation near Heschel's gyrus and the middle portion of the superior temporal gyrus has been found for a wider range of auditory stimuli than solely auditorily presented words, e.g. clicks, tones, words, orthographically regular nonwords, and real words played backwards (Lauter et al., 1985; Mazziota et al., 1982; Wise, Chollet, et al., 1991; Wise, Hadar, et al., 1991), which suggests a localization to primary and surrounding extra-primary auditory cortex.

Posterior to these areas, it appears as if activation is dependent upon more than aural presentation of a stimulus. Wise, Chollett, et al. (1991) report bilateral activa-tion of a posterior superior temporal gyral near the location of the left-lateralized temporoparietal area found by Petersen et al. (1988, 1989) during passive presenta-tion of auditory words. Wise and collaborators used three different task conditions relative to a resting baseline control. For one task, subjects listened to aurally pre-sented nonwords; in the other tasks, subjects made decisions about whether aurally presented pairs of nouns, or nouns and verbs, were appropriately matched. The presentation rate was varied across the tasks and subjects (26–60 words presented per minute). In a fourth condition, subjects were aurally presented with nouns at a rate of 15 per minute and asked to think silently of as many appropriate verbs as they could for each presented noun. Left-lateralized posterior temporal gyral activation was seen for this condition as well, despite a lack of significant activation in Heschel's gyrus and the middle superior temporal gyrus. These posterior temporal/temporoparietal regions do not appear to be activated by simple auditory stimuli, including tones, clicks, or rapidly presented synthetic syllables (Lauter et al., 1985; Mazziota et al., 1982). Temporoparietal activation has been found when subjects performed rhyme-detection tasks on visually presented words (Petersen et al., 1989).

Different interpretations have been placed upon the posterior temporal and tem-poroparietal regions of activation. Wise, Chollet, et al. (1991) have suggested that the posterior temporal area is related to word comprehension and semantic process-ing, based largely upon the observation that a left posterior region is active during a task when subjects must silently generate verbs appropriate for aurally presented nouns. Because more anterior areas were not significantly active during the silent verb generation task, and activity in the left posterior temporal region appears uncorrelated with the rate of stimulus presentations, the authors suggest that the activation probably does not represent some form of phonological processing.

Other findings present difficulties for the conclusion that posterior temporal areas are related to word comprehension in a general way. Most troubling is the lack of significant activation observed in a similar verb generation task in which subjects were asked to say aloud appropriate verbs for visually (in contrast to auditorily) presented nouns (Petersen et al., 1988), and a condition in which subjects were asked to say aloud either as many jobs or words beginning with the letter "a" as they could think of in three minutes (in which case no stimuli were presented during the scan, either auditorily or visually) (Frith et al., 1991b). For both of these generation tasks, subjects verbally, rather than silently, produced responses, and the control condition consisted of verbal output (read aloud visually presented words and count aloud, respectively).

Wise and collaborators have suggested that the failure to replicate posterior temporal activation may be accounted for on the basis of verbal output. They raise two issues. One possibility is that word repetition may activate the semantic (i.e. posterior superior gyral region) system. Thus, in the subtraction of generate verbs versus read aloud visually presented nouns, semantic activation during the verb-generation task might have been canceled out (Wise, Chollet, et al., 1991; Wise, Hadar, et al., 1991). Arguing against this interpretation is the lack of any temporal activation during any level of lexical analysis upon visually presented words, e.g. passive visual presentation-fixation point, or repeat visual word-passive visual words (Petersen et al., 1988).

Based upon the lack of superior temporal gyral activation during reading aloud of visual words versus passive viewing of visually presented words, Wise and collaborators also suggest that perhaps the process of self-vocalization causes a reduction of activity in temporal regions and, thus, might cancel out any activation produced by semantic analysis/word comprehension during any generation condition in which subjects say aloud their responses (Frith et al., 1991b; Wise, Chollet, et al., 1991; Wise, Hadar, et al., 1991). This explanation is only plausible if the additional semantic processing demands of the generate task (above the repeat task) produce no further neuronal activity.

An alternative explanation for the posterior temporal/temporoparietal results is that the activation is related to auditory or phonological processing that includes activity related to short-term storage. This would explain its ubiquitous presence and modulation in auditory tasks, its presence in a task that requires phonological processing of visual information, and its absence in a semantic association task (the generate a verb task) when the input for the task is presented visually. Such an explanation might also explain why a low presentation rate of auditory words (15 words per minute) might not produce significant activation in early auditory regions, but might produce significant activation in posterior regions when the task requires continued storage of each presented word for the interval between stimuli.

All of this is not to say that some aspects of semantic processing or word comprehension are not present in the posterior temporal lobe. The lesion literature relates damage to posterior cortex with comprehension deficits much more strongly than anterior cortex (see reviews by Damasio, 1992, and Geschwind, 1979). Regions in the temporal lobe, other than the particular temporoparietal region discussed here, might be more directly related to computations of a semantics or word meaning. Some data recently produced in PET studies (Marrett et al., 1990; Raichle et al., 1991) are consistent with this possibility.

More fundamentally, the comprehension of normal language is not likely to be subsumed in a single enclosed localization. The computation of meaning must occur above the level of the single word in its context within the grammatical structure. Because many lexical items have different meanings dependent on context (such as "play"), several types of coding must be coordinated to compute meaning. Rather than attempt to find the region of the brain related to word comprehension, an experimental analysis of the contribution of different areas to semantic processing should be the goal.

Lateral premotor areas   Simple repetition of visual and auditory words evoke activation in Sylvian-opercular and premotor regions. The left-sided regions surround the traditionally defined Broca's area. Damage to Broca's area in the left hemisphere is often thought to be associated with speech production deficits and agrammatism (Geschwind, 1965; Mohr et al., 1978). Although the lateral premotor activation could be related to some specifically linguistic functions, both simple tongue and hand movements (Fox, Pardo, et al., 1987) caused similar lateral premotor activation. Mohr et al. (1978) have shown that lesions that are confined to classically defined Broca's area produce motor and praxis deficits without specific language involvement. The full-blown syndrome of Broca's aphasia requires much larger lesions.

Similar lateral and opercular activation was also found in a study of selective and divided attention to visual features of color, speed, and shape (Corbetta et al., 1991) (figure 7.5). In the selective conditions, a subject judged changes in only one of the attributes during the scan; in the divided condition, the subjects monitored for changes in all of the attributes simultaneously. For each of the selective conditions, a lateral or opercular activation was found, but none was seen in the divided condition. The divided condition produced activation in prefrontal and cingulate cortex (see below).

In each of the conditions that produce lateral premotor-Sylvian-opercular activation, efficient response selection can be based on simple or limited processing of input. In the word repetition conditions, the relationship between a visual or auditory word, and articulation of that word, is an overlearned association that might be mediated by left-medial extrastriate cortex (for visual words) and parietotemporal cortex (for auditory words). In the selective conditions of visual attention experiments, the information on which the decision is made is limited to a single attribute, perhaps allowing much of the processing to occur in different extrastriate cortical regions.

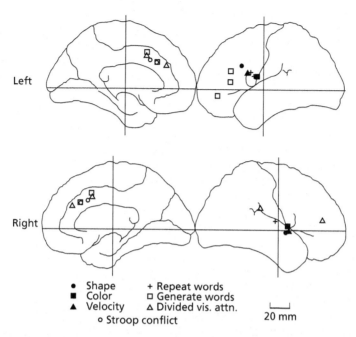

Left

Right

• Shape   + Repeat words
■ Color   □ Generate words
▲ Velocity △ Divided vis. attn.
○ Stroop conflict   20 mm

Figure 7.5 Diagrammatic representation of the location of frontal, prefrontal, and cingulate activations in different tasks. The filled symbols represent activation seen when subjects selectively responded to changes in a single visual attribute (e.g. color) of a visual array in the visual attention experiment referred to in the text. The open triangle refers to the divided condition from the same study. The open circle refers to activation seen in the Stroop conflict paradigm (Pardo et al., 1990), and the open square to regions activated in the generate verb condition with visual presentation. There is a cluster of points in the anterior cingulate across a range of conditions and a cluster of points in posterior frontal cortex across several conditions. Tentative explanations for these clusterings are presented in the text.

Anterior cingulate When the demands of a particular task do not have a strong stimulus response association, which allows for the use of posterior processing regions and lateral premotor regions, another set of networks seems to become involved. These include the use of lateral prefrontal regions (see below) as processing centers for complex associations and the use of the anterior cingulate as the response selection, or premotor stage.

The anterior cingulate region activated in the generate a verb task has now been seen in several other conditions (figure 7.5). In the conflict condition of the Stroop color naming task, subjects name the color of ink in which a word is printed, rather than say aloud the word itself, e.g. say "green" when the word "red" is presented in green ink (Pardo et al., 1990). In a divided attention condition, subjects were asked to monitor simultaneously visual arrays for changes in the color, shape, or speed of a visual stimulus (Corbetta et al., 1991). And, in another condition, subjects were asked to pick an arbitrary direction for movement when given a signal to move (Frith et al., 1991a). All of these conditions were also associated with activation of lateral prefrontal regions (see below). Although cingulate activation tends to occur in more difficult conditions, e.g. generating a word versus simple repetition, psychophysical results (Corbetta et al., 1991) provide evidence against a general task difficulty interpretation of cingulate function.

Consistent with this view of the cingulate as being related to response selection based on complex stimulus–response associations, are results from other methodologies. Large midline lesions, including SMA and anterior cingulate, often produce akinetic mutism, a syndrome in which spontaneous speech is extremely rare (Barris and Schuman, 1953; Masdau et al., 1978; Nielsen and Jacobs, 1951). Recent single-unit recording work has shown greater activation for many anterior cingulate neurons in monkeys when the animals are performing more complex sensory-motor tasks than in more simple conditions (Shima et al., 1991).

Prefrontal activations   There are two issues related to understanding the prefrontal activations in lexical processing experiments. One concerns the general nature of the task involved. The other is more specific and relates to the domain of specific information necessary to perform a task. As discussed above, both prefrontal and anterior cingulate activation appear to be critically dependent upon the nature of the relationship between presented stimuli and expected responses in a manner that cuts across many different types of tasks (figure 7.5). For lexical tasks, manipulation of the degree to which a response to a stimulus is overlearned appears to affect the degree of observed prefrontal activation strongly. Thus, when subjects are merely asked to read aloud or repeat presented words, prefrontal activation is not found. When subjects are required to generate a response to a presented noun, but the response is practiced and "automatic," prefrontal activation is not found. However, when a response is not habitual, or overlearned, prefrontal activation is found (Frith et al., 1991b; Petersen et al., 1988; Raichle et al., 1991).

The second issue is the domain of information that is used to perform the tasks in the less automatic conditions. Activation has been seen in left dorsolateral prefrontal cortex across a wide range of conditions. Petersen et al. (1988, 1989) found activation in the verb generation subtractions, for both visual and auditory input. Activation was also found in tasks where subjects silently monitored a list of words for members of a semantic category, although it was weak. Further studies with simple visual presentation of words and word-like stimuli (described in the extrastriate visual cortex section) showed that when simple visual activation is subtracted away (by subtracting the false font condition from the pseudoword and real word conditions), there is greater activation in left inferior prefrontal cortex for real words than for pseudowords (Petersen et al., 1990). The combination of results led us to suggest that some computation related to semantic processing or association between words gives rise to this activation. The difficulty has been in understanding what this computation might actually be, because as stated earlier, comprehension is much more affected by posterior than frontal lesions.

There is evidence that some aspects of semantic processing might be affected by frontal lesions, particularly in semantic priming tasks (Milberg and Blumstein, 1981; Swinney et al., 1989). The interpretation of the priming results has been that frontal lesions produce difficulties with access to semantic codes for lexical items. However, neither the PET nor the lesion results can distinguish whether the prefrontal activation represents processing the semantic code itself, or the process of accessing it for a particular task. It seems plausible that the prefrontal activation described is caused by task-specific use of semantic information, particularly when the response to a stimulus is not an automatic or overlearned association (such as simple repetition of the seen or heard item).

Goldman-Rakic (1988) has outlined a similar interpretation of prefrontal cortical function. Her work in macaques has implicated areas in anterior prefrontal cortex in monkeys as being involved in higher-order transformations or representations of information, particularly when an animal is involved in working memory tasks. An animal with lesions in this region has difficulty withholding a preponent, direct response to the stimulus when asked to hold and transform the information and act on that transformed representation. This description holds for many of the tasks that have produced prefrontal activation in PET.

Goldman-Rakic (1988) also hypothesizes that different subregions of dorsolateral prefrontal cortex may perform similar transformations on different input information. Different prefrontal areas have been associated with each of the tasks that produced cingulate activation. For example, the divided attention condition of the attention to visual features experiment produced activation in right prefrontal cortex.

Cerebellum   The activation in right lateral inferior cerebellum is anatomically distinct from activation found with the repeat words and other motor tasks, which argues for a "cognitive," rather than sensory or motor computation being related to this activation. The lateralization to the right cerebellar hemisphere is consistent with observations that each cerebellar hemisphere is anatomically and functionally related to the contralateral (in this case left, language-dominant) cerebral hemisphere. A role for the cerebellum in cognition has been advanced (e.g. see reviews by Leiner et al., 1991, and Schmahmann, 1991).

This activation motivated a study of a 49-year-old man (RC1) with right cerebellar damage on a variety of tasks involving complex nonmotor processing, including the generate verb task, which produced right cerebellar PET activation (Fiez et al., 1992). RC1's performance on standard tests of intelligence and language was excellent. He did, however, have profound deficits in two areas: practice-related learning and detection of errors that manifested itself across several tasks. In the generate verb task, he often produced nonverb associates and could not improve his performance on this and several related habit formation tasks, even though both his short- and long-term memory appeared to be above normal in function. These results suggest that some functions performed by the cerebellum can be generalized beyond a purely motor domain consistent with the earlier PET activation (Thompson, 1986).

5 Conclusion

These studies have shown that PET can contribute to our understanding of the brain mechanisms that underlie the processing of single words and, by extension, to other areas of study. However, PET should be viewed as evolutionary, rather than revolutionary. The results are most interpretable when they occur in an environment of information from other types of studies (e.g. the psychological studies of visual word processing coupled to the visual word studies of extrastriate cortex, and the use of lesion-behavior evidence to help interpret PET studies, as in the case of the cerebellar activation in the verb generation experiments). Because of the complexity underlying an understanding of even the reduced arena of the processing of single words, it is best to accept constraints from many different modalities, including the relatively recent contributions of PET activation studies.

Acknowledgments

The authors wish to thank Paula Jost for her help in preparing the manuscript and Chris Reid for reading an earlier version of the manuscript. This work was supported by the McDonnell Center for Higher Brain Function, and National Institutes of Health grants EY08775 and NS06833. J. A. Fiez received fellowship support from the National Science Foundation and the Mr and Mrs Spencer T. Olin Program for Women.

References

Baron, J. C. 1989: Depression of energy metabolism in distant brain structures: Studies with positron emission tomography in stroke patients. *Seminars in Neurology*, 9, 281–5.

Baron, J. C., Frackowiak, R. S. J., Herholz, K., Jones, T., Lammertsma, A. A., et al. 1989: Use of PET methods for measurement of cerebral energy metabolism and hemodynamics in cerebrovascular disease. *Journal of Cerebral Blood Flow and Metabolism*, 9, 723–42.

Barris, R. W., and Schuman, H. R. 1953: Bilateral anterior cingulate gyrus lesions. *Neurology*, 3, 44–52.

Caramazza, A. 1988: Some aspects of language processing revealed through the acquired aphasia: The lexical system. *Annual Review of Neuroscience*, 11, 395–421.

Carr, T. H., and Pollatsek, A. 1985: Recognizing printed words: A look at current models. In D. Besner, T. G. Weller, and G. E. MacKinnon (eds), *Reading Research*, New York: Academic, 2–73.

Corbetta, M., Miezin, F. M., Dobmeyer, S., Shulman, G. L., and Petersen, S. E. 1991: Selective and divided attention during visual discriminations of shape, color, and speed: Functional anatomy by positron emission tomography. *Journal of Neuroscience*, 11, 2383–402.

Crosson, B. 1985: Subcortical functions in language: A working model. *Brain and Language*, 25, 257–92.

Damasio, A. R. 1992: Aphasia. *New England Journal of Medicine*, 326, 531–9.

Damasio, A. R., and Damasio, H. 1983: The anatomic basis of pure alexia. *Neurology*, 33, 1573–83.

Davidson, B. J. 1986: Activation of semantic and phonological codes during reading. *Journal of Experimental Psychology: Learning, Memory, and Cognition*, 12, 201–7.

Evans, A. C., Beil, C., Marrett, S., Thompson, C. J., and Hakim, A. 1988: Anatomical-functional correlation using an adjustable MRI-based region of interest atlas with positron emission tomography. *Journal of Cerebral Blood Flow and Metabolism*, 8, 513–30.

Felleman, D. J., and Van Essen, D. C. 1991: Distributed hierarchical processing in the primate cerebral cortex. *Cerebral Cortex*, 1, 1–47.

Fiez, J. A., Petersen, S. E., Cheney, M. K., and Raichle, M. E. 1992: Impaired nonmotor learning and error detection associated with cerebellar damage: A single-case study. *Brain*, 115, 155–78.

Fox, P. T., Miezin, F. M., Allman, J. M., Van Essen, D. C., and Raichle, M. E. 1987: Retinotopic organization of human visual cortex mapped with positron emission tomography. *Journal of Neuroscience*, 7, 913–22.

Fox, P. T., Mintun, M. A., Reiman, E. M., and Raichle, M. E. 1988: Enhanced detection of focal brain responses using intersubject averaging and change-distribution analysis of subtracted PET images. *Journal of Cerebral Blood Flow and Metabolism*, 8, 642–53.

Fox, P. T., and Pardo, J. V. 1991: Does intersubject variability in cortical functional organization increase with neural "distance" from the periphery? In P. T. Fox and J. V. Pardo (eds), *Exploring Brain Functional Anatomy with Positron Emission Tomography*, Chichester: Wiley (Ciba Foundation Symposium 163), 125–44.

Fox, P. T., Pardo, J. V., Petersen, S. E., and Raichle, M. E. 1987: Supplementary motor and premotor responses to actual and imagined hand movement with positron emission tomography. *Society of Neuroscience Abstracts*, 13, 1433.

Fox, P. T., Perlmutter, J. S., and Raichle, M. E. 1985: A stereotactic method of anatomical localization for positron emission tomography. *Journal of Computer Assisted Tomography*, 9, 141–53.

Fox, P. T., and Raichle, M. E. 1986: Focal physiological uncoupling of cerebral blood flow and oxidative metabolism during somatosensory stimulation in human subjects. *Proceedings of the National Academy of Sciences (USA)*, 83, 1140–4.

Friston, K. J., Frith, C. D., Liddle, P. F., and Frackowiak, R. S. J. 1991a: Comparing functional (PET) images: The assessment of significant change. *Journal of Cerebral Blood Flow and Metabolism*, 11, 690–9.

Friston, K. J., Frith, C. D., Liddle, P. F., and Frackowiak, R. S. J. 1991b: Plastic transformation of PET images. *Journal of Computer Assisted Tomography*, 15, 634–9.

Frith, C. D., Friston, K., Liddle, P. F., and Frackowiak, R. S. J. 1991a: Willed action and the prefrontal cortex in man: A study with PET. *Proceedings of the Royal Society (London) Series B*, 244, 241–6.

Frith, C. D., Friston, K., Liddle, P. F., and Frackowiak, R. S. J. 1991b: A PET study of word finding. *Neuropsychologia*, 29, 1137–48.

Frostig, R. D., Lieke, E. E., Ts'o, D. Y., and Grinvald, A. 1990: Cortical functional architecture and local coupling between neuronal activity and the microcirculation revealed by in vivo high-resolution optical imaging of intrinsic signals. *Proceedings of the National Academy of Sciences (USA)*, 87, 6082–6.

Geschwind, N. 1965: Disconnection syndromes in animals and man, Part 1. *Brain*, 88, 237–94.

Geschwind, N. 1979: Specializations of the human brain. *Scientific American*, 283 (3), 158–68.

Glushko, R. J. 1979: The organization and activation of orthographic knowledge in reading aloud. *Journal of Experimental Psychology: Perception and Performance*, 5, 674–91.

Goldman-Rakic, P. S. 1988: Topography of cognition: Parallel distributed networks in primate association cortex. *Annual Review of Neuroscience*, 11, 137–56.

Grafton, S. T., Woods, R. P., Mazziotta, J. C., and Phelps, M. E. 1991: Somatotopic mapping of the primary motor cortex in humans: Activation studies with cerebral blood flow and positron emission tomography. *Journal of Neurophysiology*, 66, 735–43.

Henderson, V. W. 1986: Anatomy of posterior pathways in reading: A reassessment. *Brain and Language*, 29, 119–33.

Herholz, K., Pawlik, G., Wienhard, K., and Heiss, W. D. 1985: Computer assisted mapping in quantitative analysis of cerebral positron emission tomograms. *Journal of Computer Assisted Tomography*, 9, 154–61.

Herscovitch, P., Markham, J., and Raichle, M. E. 1983: Brain blood flow measured with intravenous H_2O. 1. Theory and error analysis. *Journal of Nuclear Medicine*, 24, 782–9.

Jackson, A., and Morton, J. 1984: Facilitation of auditory word recognition. *Memory and Cognition*, 12, 568–74.

Karbe, H., Herholz, K., Szelies, B., Pawlik, G., Wienhard, K., et al. 1989: Regional metabolic correlates of Token test results in cortical and subcortical left hemispheric infarction. *Neurology*, 39, 1083–8.

Karbe, H., Szelies, B., Herholz, K., and Heiss, W. D. 1990: Impairment of language is related to left parieto-temporal glucose metabolism in aphasic stroke patients. *Journal of Neurology*, 237, 19–23.

Kushner, M., Reivich, M., Alavi, A., Greenberg, J., Stern, M., et al. 1987: Regional cerebral glucose metabolism in aphemia: A case report. *Brain and Language*, 31, 201–14.

Kutas, M., and Van Petten, C. 1990: Electrophysiological perspectives on comprehending written language. In P. M. Rossini and F. Mauguière (eds), *New Trends and Advanced Techniques in Clinical Neurophysiology*, New York: Elsevier, 155–67.

Lassen, N. A., Ingvar, D. H., and Skinhoj, E. 1978: Brain function and blood flow. *Scientific American*, 239, 62–71.

Lauter, J., Herscovitch, P., Formby, C., and Raichle, M. E. 1985: Tonotopic organization in human auditory cortex revealed by positron emission tomography. *Hearing Research*, 20, 199–205.

Leiner, H. C., Leiner, A. L., and Dow, R. S. 1991: The human cerebrocerebellar system: Its computing, cognitive, and language skills. *Behavioral Brain Research*, 44, 113–28.

Marrett, S., Bub, D., Chertkow, H., Meyer, E., Gum, T., et al. 1990: Functional neuroanatomy of visual single word processing studied with PET/MRI. *Society for Neuroscience Abstracts*, 16, 27.

Masdau, J. C., Schoene, W. C., and Funkenstein, H. 1978: Aphasia following infarction of the left supplementary motor area. *Neurology*, 28, 1220–3.

Mazziotta, J. C., Pelizzari, C. C., Chen, G. T., Brookstein, F. L., and Valentino, D. 1991: Region of interest issues: The relationship between structure and function in the brain. *Journal of Cerebral Blood Flow and Metabolism*, 11, A51–6.

Mazziotta, J. C., Phelps, M. E., Carson, R. E., and Kuhl, D. E. 1982: Tomographic mapping of human cerebral metabolism: Auditory stimulation. *Neurology*, 32, 921–37.

Mazziotta, J. C., Phelps, M., Plummer, D., Schwab, R., and Halgren, E. 1983: Optimization and standardization of anatomical data in neurobehavioral investigations using positron computed tomography. *Journal of Cerebral Blood Flow and Metabolism*, 3, S266–7.

Metter, E. J., Hanson, W. R., Jackson, C. A., Kempler, D., van Lancker, D., et al. 1990: Temporoparietal cortex in aphasia. *Archives of Neurology*, 47, 1235–8.

Metter, E. J., Jackson, C., Kempler, D., Riege, W. H., Hanson, W. R., et al. 1986: Left hemisphere intracerebral hemorrhages studied by (F-18)-fluorodeoxyglucose PET. *Neurology*, 36, 1155–62.

Metter, E. J., Kempler, D., Jackson, C., Hanson, W. R., Mazziotta, J. C., et al. 1989: Cerebral glucose metabolism in Wernicke's, Broca's, and conduction aphasia. *Archives of Neurology*, 46, 27–34.

Metter, E. J., Kempler, D., Jackson, C. A., Hanson, W. R., Riege, W. H., et al. 1987: Cerebellar glucose metabolism in chronic aphasia. *Neurology*, 37, 1599–606.

Metter, E. J., Riege, W. H., Hanson, W. R., Jackson, C. A., Kempler, D., et al. 1988: Subcortical structures in aphasia: An analysis based on (F-18)-fluorodeoxyglucose, positron emission tomography, and computed tomography. *Archives of Neurology*, 45, 1229–34.

Milberg, W., and Blumstein, S. E. 1981: Lexical decision and aphasia: evidence for semantic processing. *Brain and Language*, 14, 371–85.

Mintun, M. A., Fox, P. T., and Raichle, M. E. 1989: A highly accurate method of localizing regions of neuronal activation in the human brain with positron emission tomography. *Journal of Cerebral Blood Flow and Metabolism*, 9, 96–103.

Mohr, J. P., Pessin, M. S., Finkelstein, S., Funkenstein, H. H., Duncan, G. W., et al. 1978: Broca aphasia: Pathologic and clinical. *Neurology*, 28, 311–24.

Naeser, M., Alexander, M. P., Heim-Estabrooks, N., Levine, H. L., Laughlin, S. A., et al. 1982: Aphasia with predominantly subcortical lesion sites. *Archives of Neurology*, 39, 2–14.

Neely, J. H. 1977: Semantic priming and retrieval from lexical memory: Roles of inhibitionless spreading activation and limited-capacity attention. *Journal of Experimental Psychology: General*, 106, 226–54.

Nielsen, J. M., and Jacobs, L. L. 1951: Bilateral lesions of the anterior cingulate gyri. *Bulletin of the Los Angeles Neurological Society*, 16, 231–4.

Pardo, J. V., Pardo, P. J., Janer, K. W., and Raichle, M. E. 1990: The anterior cingular cortex mediates processing selection in the Stroop attentional conflict paradigm. *Proceedings of the National Academy of Sciences (USA)*, 87, 256–9.

Petersen, S. E., Fox, P. T., Posner, M. I., Mintun, M., and Raichle, M. E. 1988: Positron emission tomographic studies of the cortical anatomy of single-word processing. *Nature*, 331, 585–9.

Petersen, S. E., Fox P. T., Posner, M. I., Mintun, M., and Raichle, M. E. 1989: Positron emission tomographic studies of the processing of single words. *Journal of Cognitive Neuroscience*, 1, 153–70.

Petersen, S. E., Fox, P. T., Snyder, A., and Raichle, M. E. 1990: Activation of extrastriate and frontal cortical areas by visual words and word-like stimuli. *Science*, 249, 1041–4.

Posner, M. I. 1978: *Chronometric Explorations of Mind*. Englewood Heights, NJ: Erlbaum.

Posner, M. I., Petersen, S. E., Fox, P. T., and Raichle, M. E. 1988: Localization of cognitive functions in the human brain. *Science*, 240, 1627–31.

Powers, W. J., and Raichle, M. E. 1985: Positron emission tomography and its application to the study of cerebrovascular disease in man. *Stroke*, 16, 361–76.

Raichle, M. E. 1989: Developing a functional anatomy of the human brain with positron emission tomography. *Current Neurology*, 9, 161–78.

Raichle, M. E., Fiez, J., Videen, T. O., Fox, P. T., Pardo, J. V., et al. 1991: Practice-related changes in human brain functional anatomy. *Society for Neuroscience Abstracts*, 17, 21.

Raichle, M. E., Martin, W. R. W., Herscovitch, P., Mintun, M. A., and Markham, J. 1983: Brain blood flow measured with intravenous $H_2^{15}O$. II. Implementation and validation. *Journal of Nuclear Medicine*, 24, 790–8.

Reivich, M., Kuhl, D., Wolf, A., Greenberg, J., Phelps, M., et al. 1979: The [^{18}F] fluorodeoxyglucose method for the measurement of local cerebral glucose utilization in man. *Circulation Research*, 44, 127–37.

Schmahmann, J. D. 1991: An emerging concept. The cerebellar contribution to higher function. *Archives of Neurology*, 48, 1178–87.

Shallice, T., McLeod, P., and Lewis, K. 1985: Isolating cognitive modules with the dual task paradigm: Are speech perception and production separate processes? *Quarterly Journal of Experimental Psychology A: Human Experimental Psychology*, 37A, 507–31.

Shima, K., Aya, K., Mushiake, H., Inase, M., Aizawa, H., et al. 1991: Two movement-related foci in the primate cingulate cortex observed in signal-triggered and self-paced forelimb movements. *Journal of Neurophysiology*, 65, 188–202.

Sokoloff, L., Reivich, M., Kennedy, C., Des Rosiers M. H., Patlak, C. S., et al. 1977: The [^{14}C] deoxyglucose method for the measurement of local cerebral glucose utilization: Theory, procedure, and normal values in the conscious and anesthetized albino rat. *Journal of Neurochemistry*, 28, 897–916.

Steinmetz, H., and Seitz, R. J. 1991: Functional anatomy of language processing: Neuroimaging and the problem of individual variability. *Neuropsychologia*, 29, 1149–61.

Stytz, M. R., and Frieder, O. 1990: Three-dimensional medical imaging modalities: An overview. *Critical Reviews in Biomedical Engineering*, 18, 1–35.

Swinney, D., Zurif, E., and Nicol, J. 1989: The effects of focal brain damage on sentence processing: An examination of the neurological organization of a mental module. *Journal of Cognitive Neuroscience*, 1, 25–37.

Talairach, J., Szikla, G., and Tournoux, P. 1967: *Atlas d'anatomie stéréotaxique du téléencephale*. Paris: Masson.

Talairach, J., and Tournoux, P. 1988: *Co-Planar Stereotaxic Atlas of the Human Brain*. New York: Thieme.

Tanenhaus, M. K., Flanigan, H. P., and Seidenberg, M. S. 1980: Orthographic and phonological activation in auditory and visual word recognition. *Memory and Cognition*, 8, 513–20.

Thompson, R. F. 1986: The neurobiology of learning and memory. *Science*, 223, 941–7.

Tyrrell, P. J., Warrington, E. K., Frackowiak, R. S. J., and Rossor, M. N. 1990: Heterogeneity in progressive aphasia due to focal cortical atrophy. *Brain*, 113, 1321–36.

Vallar, G., Perani, D., Cappa, S. F., Messa, C., Lenzi, G. L., et al. 1988: Recovery from aphasia and neglect after subcortical stroke: neuropsychological and cerebral perfusion study. *Journal of Neurology, Neurosurgery, and Psychiatry*, 51, 1269–76.

Wise, R., Chollet, F., Hadar, U., Friston, K., Hoffner, E., et al. 1991: Distribution of cortical neural networks involved in word comprehension and word retrieval. *Brain*, 114, 1803–17.

Wise, E., Hadar, U., Howard, D., and Patterson, K. 1991: Language activation studies with positron emission tomography. In *Exploring Brain Functional Anatomy with Positron Emission Tomography*, Chichester: Wiley (Ciba Foundation Symposium 163), 218–34.

Worsley, K. J., Evans, A. C., Marrett, S., and Neelin, P. 1992: A three-dimensional statistical analysis for CBF activation studies in human brain. *Journal of Cerebral Blood Flow and Metabolism*, 12, 900–18.

Yaqub, B. A., Gascon, G. G., Alnosha, M., and Whitaker, H. 1988: Pure word deafness (acquired verbal auditory agnosia) in an Arabic speaking patient. *Brain*, 111, 457–66.

Zeki, S., Watson, J. D. G., Lueck, C. J., Friston, K. J., Kennard, C., and Frackowiak, R. S. J. 1991: A direct demonstration of functional specialization in human visual cortex. *Journal of Neuroscience*, 11, 641–9.

8

Modularity, Domain Specificity and the Development of Language

Elizabeth Bates

Debates about the nature and evolution of language often shed more heat than light, because they confuse three logically separable issues: innateness, localization, and domain specificity. Proponents of innateness argue that our ability to acquire a language is determined by genetic factors, and mediated by a form of neural organization that is unique to our species. Proponents of localization argue that our ability to process language is localized to specific regions of the brain. Proponents of domain specificity build on both these points, but add the further specification that our localized language abilities are discontinuous from the rest of mind, separate and "special," constituting what Chomsky (1988) has termed a "mental organ."

The first claim has to be true at some level of analysis, because we are indeed the only species that can acquire a language in its full-fledged form (cf. Greenfield and Savage-Rumbaugh, 1991; Savage-Rumbaugh et al., 1993). The second claim is also well attested. Indeed, one of the oldest findings in cognitive neuroscience is the finding that lesions to specific regions of the left cerebral hemisphere in adults usually lead to irreversible forms of language breakdown, or aphasia – although, as we shall see, there is still considerable controversy about the nature of those symptoms (Bates and Wulfeck, 1989a, 1989b). The real debate revolves around the mental-organ claim. Are the mental structures that support language "modular," discontinuous and dissociable from all other perceptual and cognitive systems? Does the brain of the newborn child contain neural structures that are destined to mediate language, and language alone? The domain specificity view can be contrasted with an approach in which language is viewed as an innate system, but one that involves a reconfiguration of mental and neural systems that exist in other species (Deacon, 1990a, 1990b; Sereno, 1990), and which continue to serve at least some nonlinguistic functions in our own (Bates et al., 1991, 1992).

In this paper, I will provide arguments for innateness and localization but against domain specificity, in research on adult aphasia (the adult endpoint that is the source of most hypotheses about early specialization for language), and in research on normal and abnormal language development. I will begin with a brief explication of the modular approach to language, and then describe some general arguments and specific findings that support a different view, i.e. that "Language is a new machine built out of old parts" (Bates et al., 1988).

1 Modularity and Domain Specificity: What Are They?

The word "module" is used in markedly different ways by neuroscientists and behavioral scientists, a fact that has led to considerable confusion and misunderstanding in interdisciplinary discussions of brain and language. When a neuroscientist uses the word "module," s/he is usually trying to underscore the conclusion that brains are structured, with cells, columns, layers and/or regions that divide up the labor of information processing in a variety of ways. In all fairness, there are few neuroscientists or behavioral scientists who would quibble with this claim. Indeed, Karl Lashley himself probably had something similar in mind, despite his notorious claims about equipotentiality and mass action (Lashley, 1950). In cognitive science and linguistics, the term "module" refers to a stronger and more controversial claim, one that deserves some clarification before we proceed.

The strongest and clearest definition of modularity in cognitive science comes from Jerry Fodor's influential book *Modularity of Mind* (Fodor, 1983; see also Fodor, 1985). Fodor begins his book with an acknowledgment to psycholinguist Merrill Garrett, thanking him for the inspiring line "Parsing is a reflex." This is, in fact, the central theme in Fodor's book, and the version of modularity that most behavioral scientists have in mind when they use this contentious word. A module is a specialized, encapsulated mental organ that has evolved to handle specific information types of enormous relevance to the species. Following the MIT linguist Noam Chomsky (Chomsky, 1957, 1965, 1988), Fodor argues that human language fits this definition of a module. Elaborating on this argument, Fodor defines modules as cognitive systems (especially perceptual systems) that meet nine specific criteria. Five of these criteria describe the way that modules process information. These include encapsulation (it is impossible to interfere with the inner workings of a module), unconsciousness (it is difficult or impossible to think about or reflect upon the operations of a module), speed (modules are very fast), shallow outputs (modules provide limited output, without information about the intervening steps that led to that output), and obligatory firing (modules operate reflexively, providing predetermined outputs for predetermined inputs regardless of the context). As Fodor himself acknowledges (Fodor, 1985), these five characteristics can also be found in acquired skills that have been learned and practiced to the point of automaticity (Schneider and Shiffrin, 1977; Norman and Shallice, 1980). Another three criteria pertain to the biological status of modules, to distinguish these behavioral systems from learned habits. These include ontogenetic universals (i.e. modules develop in a characteristic sequence), localization (i.e. modules are mediated by dedicated neural systems), and pathological universals (i.e. modules break down in a characteristic fashion following some insult to the system). It is assumed (although this assumption may not be correct – see below) that learned systems do not display these particular regularities. The ninth and most important criterion is domain specificity, i.e. the requirement that modules deal exclusively with a single information type, albeit one of enormous relevance to the species. Aside from language, other examples might include face recognition in humans and other primates, echo location in bats, or fly detection in the frog. Of course learned systems can also be domain-specific (e.g. typing, driving, or baseball), but they lack the instinctual base that characterizes a "true" module. In the same vein, innate systems may exist that operate across domains (see below for examples). However, in Fodor's judgment such domain-general or "horizontal" modules are of much less interest and may prove intractable to study, compared with domain-specific or "vertical" modules like language and face recognition.

Fodor's version of modularity unifies the three claims that language is innate, localized, and domain-specific. This is a thoroughly reasonable proposal, but other forms of mental and neural organization are possible. In fact, all logical combinations of innateness, domain specificity, and localization may be found in the minds and brains of higher organisms. Here are a few possible examples.

1   Well-defined regions of the brain may become specialized for a particular function as a result of experience. In other words, learning itself may serve to set up neural systems that are localized and domain-specific, but not innate. A good example comes from positron emission tomography studies of brain activity showing a region of visual cortex that is specialized for words that follow the spelling rules of English (Petersen et al., 1992). Surely we would all agree that English spelling is not part of our biological heritage (and if it is, it should be clear to every teacher that such a module is not well fixed in the genome of American students). The ultimate location of a "spelling module" must be based on general facts about the organization of visual cortex, and its connections to the auditory system (in particular, the areas with primary responsibility for language – see below).

2   There may be a strong innate predisposition to set up domain-specific functions in a form that is broadly distributed across many different cortical regions, in patterns that vary widely from one individual brain to another. In other words, these systems may be innate and domain-specific, but not strongly localized. An example comes from cortical stimulation showing that many different regions of the left hemisphere can interrupt naming, although some sites are more vulnerable than others (Ojemann, 1991; Burnstine et al., 1990; Lüders, Lesser, Dinner, et al., 1991; Lüders et al., 1986; Lüders, Lesser, Hahn, et al., 1991).

3   There may be systems that are innate and highly localized, but not domain-specific. Instead, they are used to process many different kinds of information. Posner's three different attentional systems might be good candidates for this category (Posner and Driver, 1992).

In short, although evidence for localization is extremely interesting, it is simply not germane to the problems of domain specificity or innateness. Many studies of local-

Table 8.1 Proposed levels of analysis for the domain specificity, localization and innateness of language

	Domain specificity (unique to language)	*Localization (restricted to specific sites)*
Tasks/problems to be solved	Yes	No
Behaviors/skills	Yes	No
Representations/knowledge	Yes	No
Processing mechanisms	No	Yes
Genetic substrate	No	Yes

ization in adult animals (e.g. Goldman-Rakic, 1987) provide compelling evidence for regional specialization of a very intricate sort under "default" developmental conditions. On the other hand, there has been a veritable explosion of evidence for cortical plasticity in vertebrates, showing how many alternative forms of organization are possible when the default conditions do not hold (e.g. the "rewiring" results of Frost, Sur, Killackey, O'Leary, Merzenich and others – see Johnson, 1993, for a review). Indeed, some neuroscientists have argued that experience literally sculpts the brain into its final form (Merzenich et al., 1984; Rakic, 1975; Huttenlocher, 1990). Hence localization and domain specificity may be the endpoints of learning and development, but they are not necessarily the starting points (Karmiloff-Smith, 1993).

My arguments here will focus on domain specificity, but first I should clarify that domain specificity itself can apply at several different levels. A system may have unique properties at one level, while it follows general laws at another. Table 8.1 lists five levels at which a claim of domain specificity can be made: (1) the task or problem to be solved, (2) the behaviors or skills that evolve (or emerge) to solve the problem, (3) the knowledge or representations that must be present somewhere in the mind/brain of an individual who can solve the problem and produce the requisite behaviors, (4) the neural mechanisms or processors that are required to sustain those representations, and (5) the genetic substrate that makes 1–4 possible (in interaction with some environment). What level do we have in mind when we argue that language is "special"? Surely we can agree that language represents a special response to a special problem, i.e. the problem of mapping thoughts and concepts that are inherently non-linear (or atemporal) on to a channel with heavy linear (temporal) constraints. That is, symbols must be produced one at a time (one word or one sign), fast enough to fall within memory constraints but clearly and efficiently enough for successful production and comprehension. Human languages represent a very broad set of possible solutions to this special problem, but taken together (for all their similarities and differences), languages do not really look very much like anything else that we do (i.e. Turkish and tennis both take place in real time, but they do not look alike). Finally, we can all agree that the detailed and unique set of behaviors that comprise language must be supported by a detailed and unique set of mental/neural representations, i.e. knowledge of Turkish cannot look very much like knowledge of tennis.

In other words, there is no controversy surrounding the claim that language is "special" at the first three levels in table 8.1. The problem of language is unique, it is solved in a special way, and the knowledge required to solve that problem does not look like anything else we know. The real controversy revolves around the next two levels in the chart. To solve a special problem, do we really have to have a special information processor? Have we evolved new neural tissue, a new region, or a special form of computation that deals with language, and language alone? And is that new mechanism guaranteed by its own special stretch of DNA? These are the levels at which I part company with the Fodor/Chomsky view. In the words of Eric Kandel:

> The functions localized to discrete regions in the brain are not complex faculties of mind, but elementary operations. More elaborate faculties are constructed from the serial and parallel (distributed) interconnections of several brain regions.
>
> (Kandel et al., 1991, p. 15)

Our challenge is to figure out how these older, simpler neural systems have been reconfigured to solve the language problem. I will argue that language is domain-specific at levels 1–3 (the problem, its behavioral solution, the representations that support behavior), but these levels are not innate or localized. On the other hand, linguistic knowledge is acquired and supported by processors that are innate and are localized, but not domain-specific (that is, they can also process information from other domains).

2 General Arguments Against the Domain Specificity of Language

My own long-standing skepticism about the mental-organ claim is based on four kinds of evidence: (1) phylogenetic recency, (2) behavioral plasticity, (3) neural plasticity, (4) the arbitrariness of mappings from form to meaning. None of these arguments constitutes a disproof of domain specificity, but together they weaken its plausibility.

Phylogenetic recency

Bates et al. (1991) note that the species on this earth have had a great deal of time to evolve ways of dealing with light, gravity, motion, spatial organization, cause and effect, and the boundaries of common objects and events. By contrast, language is a newcomer – about 30,000 years old by current best estimates. It is hard to imagine how we could have developed elaborate, innate and domain-specific mechanisms for language in a relatively short period of time (although "poverty of the imagination" is an admittedly weak argument for any case, including my own).

Behavioral plasticity

Although one sometimes reads in textbooks that languages are based on a host of universal principles, the same everywhere, cross-linguistic research testifies to a surprising variability in structure and function across natural languages (MacWhinney and Bates, 1989, a volume based on studies of sentence processing in 15 different languages, as drastically different as Hungarian, Warlpiri and Chinese; see also Wurm, 1993). To be sure, there are some similarities (e.g. all languages have a semantics and a grammar). But the variability that has been recorded so far greatly exceeds reports for other putatively innate and domain-specific systems (including the oft-invoked example of birdsong). Oral languages present a daunting range of possibilities, from Chinese (a language with absolutely no inflections of any kind on nouns or verbs) to Greenlandic Eskimo (a language in which a sentence can consist of a single word with 8–12 prefixes, suffixes, and infixes). But an even more important lesson comes from the fact that deaf communities have developed full-blown linguistic systems in the visual-manual modality (e.g. Klima and Bellugi, 1988). If bats were suddenly deprived of echo location, would they develop an equally complex and efficient system in some other modality, within two generations? Probably not. To me, the very existence of languages like ASL argues strongly against domain specificity – although it does argue that our species has a robust and passionate urge of some kind to communicate our most complex thoughts, and a powerful set of information-processing mechanisms that permit us to solve this problem.

Neural plasticity

In contrast with the best-known examples of innate and domain-specific brain systems, the systems that support languages also show an extraordinary and perhaps unprecedented degree of neural plasticity. Research on the long-term effects of early focal brain injury suggests that children with large lesions to the classic language zones go on, more often than not, to attain levels of language ability that are indistinguishable from normal (Bates et al., 1992; Thal, Marchman, et al., 1991; Marchman et al., 1991; Stiles and Thal, 1993; Vargha-Khadem et al., 1991; Aram, 1988). As Milner and her colleagues have shown (Rasmussen and Milner, 1977; Milner, 1993), this steady state can be achieved in a variety of ways. In roughly 40 percent of the adult survivors of early local brain injury who received a sodium amytal test to determine the hemispheric specialization for speech, language production was interrupted by paralysis of the right hemisphere. Another 40 percent of this sample displayed left-hemisphere dominance for speech, suggesting that some kind of reorganization has taken place within the left hemisphere. The remaining 20 percent displayed some form of bilateral organization for speech, with some language functions controlled by the left and others by the right.

This does not mean that the two hemispheres are initially equipotential for language. For the last ten years, we have carried out prospective studies of language development in children with focal injuries to the left or right hemisphere. That is, we locate children with early focal brain injury in the prelinguistic period (before six months of age), and follow them through their first encounters with cognitive domains that are lateralized in normal adults (e.g. language, spatial cognition, facial affect). Our findings for language are largely compatible with retrospective studies of the same populations, i.e. most children go on to achieve linguistic abilities within the normal or low normal range. However, it is also clear that this reorganization takes place after an initial phase where regional biases for language are evident (whether or not those biases map on to the adult picture). Regardless of side, size, or site of lesion, most children with focal brain injury are delayed in the first stages of language production. Receptive delays are not uniquely associated with left-hemisphere injury at any point in the stages that we have studied so far, suggesting that the acquisition of receptive control over language may be a bilateral phenomenon (indeed, receptive deficits tend to be slightly greater with right-hemisphere injury). On the other hand, recovery from initial delays in expressive language does take longer (on average) in children with left-hemisphere injury. We may conclude with some confidence that the recovery of language observed in children with focal brain injury represents a true reorganization, an alternative to the default model that is discovered after an initial delay.

The same degree of plasticity is not observed in other, phylogenetically older cognitive domains (Stiles and Thal, 1993). Working with the same population of children, Stiles and colleagues (Stiles-Davis, 1988; Stiles-Davis et al., 1988; Stiles and Nass, 1991) have observed patterns of behavioral deficit along the lines that we would expect from work on spatial cognitive deficits in adults (although the childhood variants are more subtle). Reilly and colleagues have reported similar parallels to the adult model in their research on facial affect in these children (Reilly et al., 1994). Although it is difficult to compare apples and oranges, it looks as though there may be more plasticity for language than we observe in other perceptual and cognitive systems.

Arbitrariness of form–meaning mapping

This final point is a bit more difficult to summarize, but I think it is at least as important as the first three. A defining characteristic of language (indeed, one of its few universals) is the arbitrariness of the relationship between sound and meaning (and, to a surprising degree, between signs and their meanings in ASL). The words *dog, chien, perro, cane, Hund*, etc. do not in any way resemble the fuzzy four-legged creatures that they signify. The same is true for the relationship between grammatical forms and the communicative work that those forms carry out. For example, depending on the language that one speaks, basic information about "who did what to whom" can be signaled through word order (as it is in English), case inflections on nouns (e.g. Latin, Russian, Hungarian), agreement marking between subject and verb (a major source of information in Italian, but only a minor source in English), and a range of other cues. What does this have to do with modularity? If one examines all the known examples of innate and domain-specific knowledge, there is always some kind of a physical constant, a partial isomorphism between the source of information in the world to which the animal must respond, and the internal state that the animal must take (at some level in the nervous system) in order to respond correctly. Consider, for example, the "bug detector" in the retina of the frog (Lettvin et al., 1959), or the line angle detectors located in the visual cortex of kittens (Hubel and Wiesel, 1963). To evolve an innate perceptual and/or motor system, it seems that nature needs something to work with, something that holds still, something physically solid, constant, reliable. Language lacks this property, and for that reason, I find it hard to understand in concrete, material terms what an "innate language-specific acquisition device" might look like.

As I have said, none of these are knock-down arguments by themselves. They simply serve to put us on our guard, to raise an appropriate level of skepticism in the face of claims about a grammar gene or a language neuron. Let us turn now to some more specific claims about innateness, localization, and domain specificity, starting with the adult aphasia (the first test case for localization and domain specificity in the history of cognitive neuroscience).

3 Arguments Based on Adult Aphasia

Let us assume, for the moment, that there is good evidence for localization of language in our species, in high-probability default patterns that must (I agree) mean that some kind of genetic bias is at work. Exactly what is localized?

In the early stages of research on aphasia, it was generally argued that Broca's aphasia (non-fluent with spared comprehension) results from a breakdown in the motor aspects of language, while Wernicke's aphasia (comprehension deficits in the presence of fluent speech) results from injury to sensory areas. This characterization made reasonably good neuroanatomical sense, in view of the fact that Broca's aphasia correlates with frontal injury while Wernicke's aphasia is associated with posterior lesions, but its fit to the behavioral data was always fairly loose. As Freud (1953 [1891]) pointed out a hundred years ago, a sensory deficit cannot explain the severe word-finding deficits and substitution errors that characterize the fluent output observed in Wernicke's aphasia. In the 1970s, analogous problems arose for the motor account of Broca's aphasia (Zurif and Caramazza, 1976; Heilman and Scholes, 1976). In particular, careful experimental studies showed that these patients also suffer from comprehension problems when they are forced to rely on grammatical markers to interpret complex sentences (e.g. patients could interpret "The apple was eaten by the boy," but not "The boy was chased by the girl"). At this point, several investigators offered an alternative view based on modular theories of linguistic organization (e.g. Caramazza and Berndt, 1985). In particular, it was argued that Broca's aphasics have lost the ability to comprehend or produce grammar (resulting in telegraphic output, and subtle comprehension deficits that are most evident when semantic information is too ambiguous to support sentence interpretation). Conversely, the comprehension deficits and word-finding problems observed in Wernicke's aphasia could be jointly and parsimoniously explained if these patients have lost the ability to process content words. This apparent double dissociation provided support for the idea that the brain is organized into innate, domain-specific, and localized modules for grammar and semantics, respectively (see Gazzaniga, 1993, for arguments along the same lines).

But this unifying view has also fallen on hard times. More recent studies of language breakdown in aphasia have forced investigators to abandon the idea of a "grammar box," i.e. neural tissue that is devoted exclusively to grammar, and contains the representations that are necessary for grammatical processing. To offer just a few examples, there are (1) numerous studies showing that so-called agrammatic aphasics can make remarkably fine-grained judgments of grammaticality (Linebarger et al., 1983; Wulfeck, 1987; Shankweiler et al., 1989; Wulfeck and Bates, 1991), and (2) a host of cross-linguistic studies showing differences in the symptoms displayed by agrammatic patients in different language communities – differences that can only be explained if we acknowledge that the patient still retains detailed knowledge of his/her grammar (Bates et al., 1991; Menn and Obler, 1990). It begins to look as though linguistic knowledge is broadly represented in the adult brain – a conclusion that is also supported by studies of brain activity during normal language use (Petersen et al., 1992; Kutas and Kluender, 1991). Some areas do play a more important role than others in getting a particular process under way in real time, but the knowledge itself is not strictly localized.

So what is localized? The classic sensorimotor view of Broca's and Wernicke's aphasia has fallen by the wayside, and now the grammar/semantics view has fallen as well. But their successor is still unnamed. Some investigators have argued that left frontal regions are specialized for the rapid processes required for fluent use of grammar, while posterior regions play a more important role in controlled, strategic choice of words and sentence frames (e.g. Frazier and Friederici, 1991; Zurif et al., 1990; Milberg and Albert, 1991). These ideas are still distressingly vague, but they point us in a new direction.

From a developmental perspective, the default pattern of brain organization for language observed in adults can be viewed as the end product of regional differences in neural computation and processing that "attract" or "recruit" language processes under default conditions. The perisylvian areas of the left hemisphere are not "innate language tissue," any more than a tall child constitutes an "innate basketball player." However, all other things being equal, the left perisylvian areas will take over the language problem, and the tall child has a very good chance of ending up on the basketball team. This brings me to the problem of how (and where) language is acquired.

4   Arguments Based on Normal and Abnormal
Language Development

In line with Fodor's criterion for ontogenetic universals, it is well known that children go through a series of universal stages in language learning: from babbling in vowel sounds (around 3 months) to babbling in consonants (between 6 and 9 months); from first signs of word comprehension (from 8–10 months) to the onset of word production (averaging 12 months, with a substantial range of individual variability); from the single-word stage (from 12–20 months, on average) to the onset of word combinations; from simple two-word strings (so-called telegraphic speech) to complex grammar (evident in most normal children by three years of age). But can we conclude that these milestones reflect the unfolding of a domain-specific module? Probably not, at least not on the basis of the evidence that is currently available (see Bates et al., 1992, for details). First of all, there is enormous variability from one child to another in the onset and duration of these stages. Second, there are important variations in this basic pattern from one language to another (e.g. children who are exposed to a richly inflected language like Turkish often display signs of productive grammar in the one-word stage). Third, each of these milestones in early language is correlated with specific changes outside the boundaries of language (e.g. the use of familiar gestures like drinking, combing, or putting a telephone receiver to the ear as a way of "labeling" common objects – gestures that appear in the hearing child right around the time that naming takes off in the vocal modality). In other words, one cannot conclude that the universal maturational timetable for language is really universal, or that it is specific to language.

These problems of interpretation are compounded in research on abnormal language development. Two recent examples illustrate the confusion between innateness and domain specificity that has plagued this field, much like the confusion between domain specificity and localization that has characterized research on adult aphasia.

Petitto and Marentette (1991) published an influential paper demonstrating that deaf infants exposed to sign language "babble" with their hands, producing meaningless but systematic actions that are not observed in hearing children. Furthermore, this form of manual babbling occurs around 8–10 months of age, the point at which vocal babbling appears in the hearing child. The authors conclude that language learning involves innate abilities that are independent of modality (i.e. vocal or manual); they also claim that these abilities are specific to language, providing support for Chomsky's mental-organ claim. Their first conclusion is clearly supported by the evidence, but the second is not. We have known for more than 100 years that children begin to imitate novel actions (i.e. actions that are not already in their repertoire) around 8–10 months. The more systematic the adult input, the more systematic the child's imitation is likely to be. Petitto and Marentette's demonstration of babbling in the visual modality constitutes a particularly beautiful example of this interesting but well-established fact. The kind of imitation that underlies babbling is undoubtedly based upon abilities that are innate, and particularly well developed in our species (human children imitate far better and more often than any other primate – Greenfield and Savage-Rumbaugh, 1991; Chevalier-Skoinikoff, 1991), but proof of its existence does not, in itself, constitute evidence in favor of the notion that language is "special."

A somewhat different example appeared in a letter to *Nature* by Gopnik (1990; for further details see Gopnik and Crago, 1991), describing preliminary results from a study of grammatical abilities in a family of individuals suffering from some kind of genetically based disorder (see also Tallal et al., 1991). Members of this family have difficulty with particular aspects of grammar, including regular verb inflections (e.g. the *-ed* in verbs like *walked* and *kissed*). By contrast, they are reported to have less trouble with irregular forms like *came* or *gave*. This pattern is offered as an example of an innate and domain-specific disorder, termed "feature-blind dysphasia," and has been cited as evidence in favor of Pinker's claim that regular and irregular forms are handled by separate mental and perhaps neural mechanisms (Pinker, 1991). Shortly after Gopnik's letter appeared, *Nature* published a rebuttal by Vargha-Khadem and Passingham (1990; see also Fletcher, 1990), who have studied the same family for a number of years. These authors point out that the members of this family suffer from a much broader range of linguistic and non-linguistic deficits than one might conclude from Gopnik's description. Their peculiar grammatical symptoms are only the tip of an iceberg, one by-product of a disorder with repercussions in many different areas of language and cognition, providing further evidence for innateness but none for domain specificity (Marchman, 1993).

The above examples are part of a long tradition in neurolinguistics, where unusual profiles of language ability and disability are cited as events for the eccentricity and modularity of language. Some other "parade cases" include specific language impairment or SLI, and children with Williams syndrome.

By definition, specific language impairment (SLI) refers to delays in receptive and/or expressive language development in children with no other known form of neurological or cognitive impairment. However, recent studies of SLI suggest that this definition may not be accurate (Cohen et al., 1991; Tallal et al., 1985). Although these children do not suffer from global forms of mental retardation, they do show subtle impairments in aspects of cognition and/or perception that are not specific to language. For example, many children with SLI experience difficulty in processing rapid transitions in acoustic information (including non-linguistic stimuli). This may help to explain new studies comparing SLI in English, Italian, and Hebrew (Rom and Leonard, 1990; Leonard et al., in press) showing that the specific areas of grammar that are most delayed vary from one language to another, and the most vulnerable elements within each language appear to be those that are low in phonological substance (i.e. salience). The subtle deficits associated with SLI may also transcend the acoustic modality, affecting certain kinds of manual gesture (Thal, Tobias and Morrison, 1991). Taken together, these studies suggest that SLI may not be a purely linguistic (or acoustic) phenomenon.

The strongest evidence to date in favor of domain specificity comes from rare cases in which language appears to be remarkably spared despite severe limitations in other cognitive domains. Etiologies associated with this unusual profile include spina bifida and hydrocephalus, and a rare form of mental retardation called Williams syndrome, or WMS (Bellugi et al., 1991; Jernigan and Bellugi, 1990). The dissociations observed in WMS prove that language can "decouple" from mental age at some point in development. Nevertheless, recent studies of WMS place constraints on the conclusion that language is a separate mental system from the beginning. First, it is clear that language development is seriously delayed in infants and preschool children with WMS, suggesting that certain "cognitive infrastructures" must be in place before language can be acquired (Thal et al., 1989). Second, studies of older children with WMS demonstrate peculiar islands of sparing in some non-linguistic domains (e.g. face recognition, and recognition of common objects from an unfamiliar perspective), and unusual patterns of deficit in other non-linguistic domains that are not at all comparable to the patterns displayed by Down syndrome children matched for mental age. Third, the language of older children and adults with WMS includes some deviant characteristics that are not observed in normal children. For example, in a word fluency test in which WMS children and Down syndrome controls were asked to generate names for animals, Down syndrome and normal controls tend to generate high-frequency words like *dog* and *cat*; WMS individuals tend instead to generate unusual, low-frequency items like *ibex* and *brontosaurus*. In view of such findings, it seems that WMS may not represent sparing of normal language, but a completely different solution to the language problem, achieved with a deviant form of information processing.

In short, the dissociations between language and cognition observed in SLI (where language < cognition) and in Williams syndrome (where language > cognition) cannot be used to support a mental-organ view. Things are just not that simple. Instead, these unusual profiles offer further evidence for the behavioral and neural plasticity of language. There are many ways to solve the problem of language learning. Some are more efficient than others, to be sure, but the problem can be solved with several different configurations of learning, memory, perception, and cognition. This brings us to my final point: How is it that language is learnable at all?

There is a branch of language acquisition research called "learnability theory" (e.g. Lightfoot, 1991) which uses formal analysis to determine the range of conditions under which different kinds of grammars can (in principle) be learned. Until recently, most of this research has been based upon the assumption that language learning in humans is similar to language learning in serial digital computers, where a priori hypotheses about grammatical rules are tested against strings of input symbols, based on some combination of positive evidence ("here is a sentence in the target language") and negative evidence ("here is a sentence that is not permitted by the target language"). A famous proof by Gold (1967) showed that a broad class of grammars (including generative grammars of the sort described by Chomsky) could not be learned by a system of this kind unless negative evidence was available in abundance, or strong innate constraints were placed upon the kinds of hypotheses that the system would consider. Since we know that human children are rarely given explicit negative evidence, the learnability theory seems to require the conclusion that children have an extensive store of innate and domain-specific grammatical knowledge.

In the last two years, this conclusion has been challenged by major breakthroughs in the application of a different kind of computer architecture (called neural networks, connectionism, and/or parallel distributed processing) to classic problems in language learnability. Because connectionism makes a very different set of assumptions about the way that knowledge is represented and acquired, Gold's pessimistic conclusions about language learnability do not necessarily apply. This new era began in 1986 with a simulation by Rumelhart and McClelland (1986) on the acquisition of the English past tense, showing that connectionist networks go through stages that are very similar to the ones displayed by children who are acquiring English (producing and then recovering from rule-like overgeneralizations like *comed* and *wented*, in the absence of negative evidence). This simulation has been severely criticized (see especially Pinker and Prince, 1988; Kim et al., 1991). However, a number of new works have appeared that get around these criticisms, replicating and extending the Rumelhart-McClelland findings in several new directions (Elman, 1990, 1991; MacWhinney, 1991; Plunkett and Marchman, 1991; Marchman, 1993). The most recent example comes from Marchman (1993), who has "lesioned" neural networks at various points during learning of the past tense (randomly eliminating between 2 and 44 percent of the connections in the network). These simulations capture some classic "critical period" effects in language learning (e.g. smaller, earlier lesions lead to better outcomes; later, larger lesions lead to persistent problems in grammar), showing that such effects can occur in the absence of "special" maturational constraints (compare with Newport, 1990, and Elman, 1991). In addition, Marchman's damaged systems found it more difficult to acquire regular verbs (e.g. *walked*) than irregulars (e.g. *came*), proving that the specific pattern of deficits described by Gopnik and by Pinker can result from non-specific forms of brain damage in a general-purpose learning device. Such research on language learning in neural networks is still in its infancy, and we do not know how far it can go. But it promises to be an important tool, helping us to determine just how much innate knowledge has to be in place for certain kinds of learning to occur.

In short, a great deal has been learned in the last few years about the biological foundations for language development. Evidence for innateness is good, but evidence for a domain-specific "mental organ" is difficult to find. Instead, language learning appears to be based on a relatively plastic mix of neural systems that also serve other functions. I believe that this conclusion renders the mysteries of language evolution at issue in this volume somewhat more tractable. That is, the continuities that we have observed between language and other cognitive systems make it easier to see how this capacity came about in the first place.

Acknowledgments

This research was supported by the Center for the Study of the Neurological Basis of Language (NIH-NINDS NS 22343), the Crosslinguistic Studies in Aphasia Project (NIH-NIDCD DC00216) and the Center for the Study of the Neurological Basis of Disorders of Language, Learning and Behavior (NINCDS 1 P50 NS22343), with partial support from the Grammatical Abilities in Alzheimer's Disease Project (NIH No. 5 RO1 DC00348). Selected passages in this essay have appeared in E. Bates, Language Development, in E. Kandel and L. Squire (eds), Special Issue on Cognitive Neuroscience, *Current Opinion in Neurobiology*, 2, 180–5.

References

Aram, D. M. 1988: Language sequelae of unilateral brain lesions in children. In F. Plum (ed.), *Language, Communication and the Brain*. New York: Raven Press.

Bates, E., Bretherton, I., and Snyder, L. 1988: *From First Words to Grammar – Individual Differences and Dissociable Mechanisms*. New York: Cambridge University Press.

Bates, E., Thal, D., and Janowsky, J. 1992: Early language development and its neural correlates. In I. Rapin and S. Segalowitz (eds), *Handbook of Neuropsychology*, vol. 7: *Child Neuropsychology*. Amsterdam: Elsevier.

Bates, E., Thal, D., and Marchman, V. 1991: Symbols and syntax: A Darwinian approach to language development. In N. Krasnegor, D. Rumbaugh, E. Schiefelbusch, and M. Studdert-Kennedy (eds), *Biological and Behavioral Determinants of Language Development*, Hillsdale, NJ: Erlbaum, 29–65.

Bates, E., and Wulfeck, B. 1989a: Crosslinguistic studies of aphasia. In B. MacWhinney and E. Bates (eds), *The Crosslinguistic Study of Sentence Processing*. New York: Cambridge University Press.

Bates, E., and Wulfeck, B. 1989b: Comparative aphasiology: A crosslinguistic approach to language breakdown. *Aphasiology*, 3, 111–42 and 161–8.

Bates, E., Wulfeck, B., and MacWhinney, B. 1991: Crosslinguistic research in aphasia: An overview. *Brain and Language*, 41, 123–48.

Bellugi, U., Bihrle, A., Neville, H., Jernigan, T., and Doherty, S. 1991: Language, cognition and brain organization in a neurodevelopmental disorder. In W. Gunnar and C. Nelson (eds), *Developmental Behavioral Neuroscience*. Hillsdale, NJ: Erlbaum.

Burnstine, T. H., Lesser, R. P., Hart, J. Jr., Uematsu, S., Zinreich, S. J., Krauss, G. L., Fisher, R. S., Vining, E. P., and Gordon, B. 1990: Characterization of the basal temporal language area in patients with left temporal lobe epilepsy. *Neurology*, 40 (6), 966–70.

Caramazza, A., and Berndt, R. 1985: A multicomponent view of agrammatic Broca's aphasia. In M. L. Kean (ed.), *Agrammatism*. New York: Academic Press.

Chevalier-Skoinikoff, S. 1991: Spontaneous tool use and sensorimotor intelligence in Cebus compared with other monkeys and apes. *Behavioral and Brain Sciences*, 14 (2).

Chomsky, N. 1957: *Syntactic Structures*. The Hague: Mouton.

Chomsky, N. 1965: *Aspects of the Theory of Syntax*. Cambridge, MA: MIT Press.

Chomsky, N. 1988: *Language and Problems of Knowledge*. Cambridge, MA: MIT Press.

Cohen, H., Gelinas, C., Lassonde, M., and Geoffrey, G. 1991: Auditory lateralization for speech in LI children. *Brain and Language*, 41, 395–401.

Deacon, T. 1990a: Brain-language coevolution. In J. A. Hawkins and M. Gell-Mann (eds), *The Evolution of Human Languages: Proceedings of the Santa Fe Institute Studies in the Sciences of Complexity*. Reading, MA: Addison-Wesley.

Deacon, T. 1990b: Rethinking mammalian brain evolution. *American Zoologist*, 30, 629–705.

Elman, J. 1990: Finding structure in time. *Cognitive Science*, 14, 179–211.

Elman, J. 1991: Incremental learning, or the importance of starting small. *Proceedings of the Thirteenth Annual Conference of the Cognitive Science Society*. Hillsdale, NJ: Erlbaum, 443–8.

Fletcher, P. 1990: Speech and language defects. *Nature*, 346, 226.

Fodor, J. 1983: *The Modularity of Mind*. Cambridge, MA: MIT Press.

Fodor, J. A. 1985: Multiple book review of "The modularity of mind". *Behavioral and Brain Sciences*, 8, 1–42.

Frazier, L., and Friederici, A. 1991: On deriving the properties of agrammatic comprehension. *Brain and Language*, 40, 51–66.

Freud, A. 1953: *On Aphasia: A Critical Study*. New York: International Universities Press. (Original work published in 1891.)

Gazzaniga, M. 1993: *Language and the Cerebral Hemispheres*. Paper presented at the FESN Study Group on Evolution and Neurology of Language, Geneva, April.

Gold, E. 1967: Language identification in the limit. *Information and Control*, 16, 447–74.

Goldman-Rakic, P. S. 1987: Development of cortical circuitry and cognitive function. *Child Development*, 58, 601–22.

Gopnik, M. 1990: Feature-blind grammar and dysphasia. *Nature*, 344, 715.

Gopnik, M., and Crago, M. 1991: Familial aggregation of a developmental language disorder. *Cognition*, 39 (1), 1–50.

Greenfield, P., and Savage-Rumbaugh, E. 1991: Imitation, grammatical development and the invention of protogrammar by an ape. In N. Krasnegor, D. Rumbaugh, R. Schiefelbusch, and M. Studdert-Kennedy (eds), *Biological and Behavioral Determinants of Language Development*, Hillsdale, NJ: Erlbaum, 235–62.

Heilman, K. M., and Scholes, R. J. 1976: The nature of comprehension errors in Broca's, conduction and Wernicke's aphasics. *Cortex*, 12, 258–65.

Hubel, D. H., and Wiesel, T. N. 1963: Receptive fields of cells in striate cortex of very young, visually inexperienced kittens. *Journal of Neurophysiology*, 26, 944–1002.

Huttenlocher, P. R. 1990: Morphometric study of human cerebral cortex development. *Neuropsychologia*, 28 (6), 517–27.

Jernigan, T., and Bellugi, U. 1990: Anomalous brain morphology on magnetic resonance images in Williams Syndrome and Down Syndrome. *Archives of Neurology*, 47, 429–533.

Johnson, M. (ed.) 1993: *Brain Development and Cognition: A Reader*. Oxford: Blackwell.

Kandel, E. R., Schwartz, J. H., and Jessell, T. H. 1991: *Principles of Neural Science*, 3rd edn. New York: Elsevier.

Karmiloff-Smith, A. 1993: *Beyond Modularity: A Developmental Perspective on Cognitive Science*. Cambridge, MA: MIT Press.

Kim, J., Pinker, S., Prince, A., and Sandup, P. 1991: Why no mere mortal has ever flown out to center field. *Cognitive Science*, 15 (2), 173–218.

Klima, E., and Bellugi, U. 1988: *The Signs of Language*. Cambridge, MA: Harvard University Press.

Kutas, M., and Kluender, R. 1991: What is who violating? A reconsideration of linguistic violations in light of event-related potentials. *Center for Research in Language Newsletter*, 6, 1. La Jolla: University of California, San Diego, Center for Research in Language.

Lashley, K. S. 1950: In search of the engram. In *Symposia of the Society for Experimental Biology, No. 4. Physiological mechanisms and animal behavior.* New York: Academic Press.

Leonard, L., Bortolini, U., Caselli, M., McGregor, K., and Sabbadini, L. 1992: Morphological deficits in children with Specific Language Impairment: the status of features in the underlying grammar. *Language Acquisition*, 2: 151–79.

Lettvin, J. Y., Maturana, H. R., McCulloch, W. S., and Pitts, W. H. 1959: What the frog's eye tells the frog's brain. *Proceedings of the Institute of Radio Engineering of New York*, 47, 1940–51.

Lightfoot, D. 1991: The child's trigger experience – Degree-0 learnability. *Behavioral Brain Sciences*, 14 (2), 364.

Linebarger, M., Schwartz, M., and Saffran, E. 1983: Sensitivity to grammatical structure in so-called agrammatic aphasics. *Cognition*, 13, 361–92.

Lüders, H., Lesser, R., Dinner, D., Morris, H., Wyllie, E., and Godoy, J. 1991: Localization of cortical function: New information from extraoperative monitoring of patients with epilepsy. *Epilepsia*, 29 (Suppl 2), S56–S65.

Lüders, H., Lesser, R., Hahn, J., Dinner, D., Morris, H., Resor, S., and Harrison, M. 1986: Basal temporal language area demonstrated by electrical stimulation. *Neurology*, 36, 505–9.

Lüders, H., Lesser, R., Hahn, J., Dinner, D., Morris, H., Wyllie, E., and Godoy, J. 1991: Basal temporal language area. *Brain*, 114, 743–54.

MacWhinney, B. 1991: Implementations are not conceptualizations: Revising the verb-learning model. *Cognition*, 40, 121–57.

MacWhinney, B., and Bates, E. (eds) 1989: *The Cross Linguistic Study of Sentence Processing.* New York: Cambridge University Press.

Marchman, V. 1993: Constraints on plasticity in a connectionist model of the English past tense. *Journal of Cognitive Neuroscience*, 5 (21), 215–34.

Marchman, V., Miller, R., and Bates, E. 1991: Babble and first words in children with focal brain injury. *Applied Psycholinguistics*, 12, 1–22.

Menn, L., and Obler, L. K. (eds) 1990: *Agrammatic Aphasia: Cross-language Narrative Sourcebook.* Amsterdam/Philadelphia: John Benjamins.

Merzenich, M., Nelson, R., Stryker, M., Cynader, M., Schoppmann, A., and Zook, J. 1984: Somatosensory cortical map changes following digit amputation in adult monkeys. *Journal of Comparative Neurology*, 224, 591–605.

Milberg, W., and Albert, M. 1991: The speed of constituent mental operations and its relationship to neuronal representation: An hypothesis. In R. G. Lister and H. J. Weingartner (eds), *Perspectives on Cognitive Neuroscience.* New York: Oxford University Press.

Milner, B. 1993: *Carotidamytal Studies of Speech Lateralization and Gesture Control.* Paper presented at the FESN Study Group on Evolution and Neurology of Language, Geneva, April.

Newport, E. 1990: Maturational constraints on language learning. *Cognitive Science*, 14, 11–28.

Norman, D. A., and Shallice, T. 1980: *Attention to Action: Willed and Automatic Control of Behavior.* Center for Human Information Processing (Technical Report No. 99). (Reprinted in revised form in R. J. Davidson, G. E. Schwartz, and D. Shapiro [eds] [1986], *Consciousness and Self-regulation* [Vol. 4]. New York: Plenum Press.)

Ojemann, G. A. 1991: Cortical organization of language. *Journal of Neuroscience*, 11 (8), 2281–7.

Petersen, S. E., Fiez, J. A., and Corbetta, M. 1992: Neuroimaging. *Current Opinion in Neurobiology – Special Issue on Cognitive Neuroscience*, 2, 217–22.

Petitto, L., and Marentette, P. F. 1991: Babbling in the manual mode: Evidence for the ontogeny of language. *Science*, 251, 1493–9.

Pinker, S. 1991: Rules of language. *Science*, 253, 530–5.

Pinker, S., and Prince, A. 1988: On language and connectionism: An analysis of a parallel distributed processing model of language acquisition. *Cognition*, 28, 73–193.

Plunkett, K., and Marchman, V. 1991: U-shaped learning and frequency effects in a multi-layered perceptron: Implications for child language acquisition. *Cognition*, 38 (1), 43–102.

Plunkett, K., and Marchman, V. 1993: From rote learning to system building: Acquiring verb morphology in children and connectionist nets. *Cognition*, 48, 21–69.

Posner, M. I., and Driver, J. 1992: The neurobiology of selective attention. *Current Opinion in Neurobiology Special Issue on Cognitive Neuroscience*, 2, 165–9.

Rakic, P. 1975: Timing of major ontogenetic events in the visual cortex of the rhesus monkey. In N. Buchwald and M. Brazier (eds), *Brain Mechanisms in Mental Retardation*, New York: Academic Press.

Rasmussen, T., and Milner, B. 1977: The role of early left-brain injury in determining lateralization of cerebral speech functions. *Annals of the New York Academy of Sciences*, 299, 355–69.

Reilly, J., Stiles, J., Larsen, J., and Trauner, D. 1994: *Affective Facial Expression in Infants with Focal Brain Damage. Neuropsychologia*, 33, 83–99.

Rom, A., and Leonard, L. 1990: Interpreting deficits in grammatical morphology in specifically language impaired children: Preliminary evidence from Hebrew. *Clinical Linguistics and Phonetics*, 4 (2), 93–105.

Rumelhart, D., McClelland, J., and the PDP Research Group 1986: *Parallel Distributed Processing: Explorations in the Microstructure of Cognition*, Vol. 1. Cambridge, MA: MIT/Bradford Books.

Savage-Rumbaugh, S., Murphy, J., Sevcik, R., Brakke, K., Williams, S., and Rumbaugh, D. 1993: Language comprehension in ape and child. *Monographs of the Society for Research in Child Development*, Serial no. 233, Volume 58, 3–4, 222–42.

Schneider, W., and Shiffrin, R. 1977: Controlled and automatic human information processing: 1. Detection, search and attention. *Psychological Review*, 84, 321–30.

Sereno, M. 1990: Language and the primate brain. *Center for Research in Language Newsletter*, 4 (4). La Jolla: University of California, San Diego, Center for Research in Language.

Shankweiler, D., Crain, S., Gorrell, P., and Tuiler, B. 1989: Reception of language in Broca's aphasia. *Language and Cognitive Processes*, 4 (1), 1–33.

Stiles, J., and Nass, R. 1991: Spatial grouping activity in young children with congenital right- or left-hemisphere brain injury. *Brain and Cognition*, 15, 201–22.

Stiles, J., and Thal, D. 1993: Linguistic and spatial cognitive development following early focal brain injury: Patterns of deficit and recovery. In M. Johnson (ed.), *Brain Development and Cognition: A Reader*. Oxford: Blackwel.

Stiles-Davis, J. 1988: Spatial dysfunctions in young children with right cerebral hemisphere injury. In J. Stiles-Davis, M. Kritchevsky, and U. Bellugi (eds), *Spatial Cognition: Brain Bases and Development*. Hillsdale, NJ: Erlbaum.

Stiles-Davis, J., Janowsky, J., Engel, M., and Nass, R. 1988: Drawing ability in four young children with congenital unilateral brain lesions. *Neuropsychologia*, 26, 359–71.

Tallal, P., Townsend, J., Curtiss, S., and Wulfeck, B. 1991: Phenotypic profiles of language-impaired children based on genetic/family history. *Brain and Language*, 41, 81–95.

Tallal, P., Stark, R., and Mellits, D. 1985: Identification of language impaired children on the basis of rapid perception and production skills. *Brain and Language*, 25, 314–22.

Thal, D., Bates, E., and Bellugi, U. 1989: Language and cognition in two children with Williams Syndrome. *Journal of Speech and Hearing Research*, 3, 489–500.

Thal, D., Marchman, V., Stiles, J., Aram, D., Trauner, D., Nass, R., and Bates, E. 1991: Early lexical development in children with focal brain injury. *Brain and Language*, 40, 491–527.

Thal, D., Tobias, S., and Morrison, D. 1991: Language and gesture in late talkers: A one-year follow-up. *Journal of Speech and Hearing Research*, 34, 604–12.

Vargha-Khadem, F., Isaacs, E., Papaleloudi, H., Polkey, C., and Wilson, J. 1991: Development of language in six hemispherectomized patients. *Brain*, 114, 473–95.

Vargha-Khadem, F., and Passingham, R. 1990: Speech and language defects. *Nature*, 346, 226.

Wulfeck, B. 1987: *Sensitivity to Grammaticality in Agrammatic Aphasia: Processing of Word Order and Agreement Violations*. Doctoral dissertation, UCSD.

Wulfeck, B., and Bates, E. 1991: Differential sensitivity to errors of agreement and word order in Broca's aphasia. *Journal of Cognitive Neuroscience*, 3, 258–72.

Wurm, S. A. 1993: *Language Contact and Unusual Semantic Features: Some Ideas on Language and Thought*. Paper presented at the FESN Study Group on Evolution and Neurology of Language, Geneva, April.

Zurif, E., and Caramazza, A. 1976: Psycholinguistic structures in aphasia: Studies in syntax and semantics. In H. and H. A. Whitaker (eds), *Studies in Neurolinguistics*, vol. 1. New York: Academic Press.

Zurif, E. B., Swinney, D., and Garrett, M. 1990: Lexical processing and sentence comprehension in aphasia. In A. Caramazza (ed.), *Cognitive Neuropsychology and Neurolinguistics: Advances in Models of Cognitive Function and Impairment*, pp. 123–36. Hillsdale, NJ: Lawrence Erlbaum.

9

Linking Cognition and Brain: The Cognitive Neuroscience of Language

William Bechtel

The pioneering investigations of Broca (chapter 5, this volume) and Wernicke (chapter 6, this volume) provided exemplars of how to relate brain structures and psychological function in domains outside of language. Language itself, however, has been one of the most difficult domains of cognition to *understand* in terms of brain activities. As a result, the most developed approaches to the study of language are found in other disciplines – linguistics, psycholinguistics, and philosophy of language – that draw little if at all on information about the neural structures underlying language. There are several reasons for this. One is the fact that fully developed linguistic abilities are only found in humans, rendering it difficult to employ animal models in understanding language. Yet much of our understanding of the brain mechanisms involved in other psychological activities, such as seeing, resulted from studies in other species where it is possible to induce lesions or record from individual cells (see chapter 13, this volume). A second is that, until the recent emergence of neuroimaging techniques, language deficits resulting from naturally occurring lesions provided virtually the only avenue to studying how the human brain performed cognitive functions, including those involved in language. Naturally occurring lesions, however, generally do not damage single functional components of the brain. A final factor is that, for most humans, thinking is so dependent on language that it has been hard to conceive of more basic cognitive activities that might explain language itself. This motivation for seeing language as relying on special cognitive processes has been buttressed by the arguments of some linguists who argue for a special language module or language instinct (Chomsky, 1988; Pinker, 1994).

My goal in this chapter is to provide a philosophical perspective on neuroscience research on language. I begin with the question of the relation between language and thought, and lay out two proposals, one of which makes language foundational for thought, while the other makes thought foundational for language. I show that these different proposals provide different frameworks of understanding the project of relating language to the brain. I then turn to the multidisciplinary character of research on language, examining the differences in the way linguistics, psycholinguistics, and neuropsychology have approached the study of language. Next I examine how the study of neural processes has been related to various disciplinary inquiries into language. Finally, I take up three currently controversial issues concerning brain and language – whether linguistic ability (a) is a unique human adaptation, (b) results from a special module, and (c) is innate.

1 Language and Thought

As humans, we are often aware of covertly formulating our thoughts in language and phenomenologically it often seems as if we are *hearing* ourselves speak when we think to ourselves. In such situations, we use natural language sentences, privately rehearsed, as vehicles for representing information. For example, in solving problems, we privately construct sentences that identify the features of the problem, advance hypotheses for solving the problem, and identify evidence that supports or undercuts the hypotheses. (This is not to imply that our whole private life is linguistic. Most people are also aware of manipulating images, and employing such images in solving problems. There appear to be substantial individual differences with respect to how much different people rely in problem solving on talking to themselves and manipulating images.) Thus, language and thought seem closely related. But theorists differ significantly in the way in which they have envisaged language and thought to be related.

The philosopher Jerry Fodor has defended one extreme position. For Fodor (1975), thought is itself a linguistic activity, involving the formulation of hypotheses and evaluation of evidence. This would seem to have the striking consequence of denying thought to all creatures lacking a language, including human infants who are just learning language. (Other philosophers, such as Donald Davidson (1982), do not blanch at such a suggestion, arguing that unless a creature is able to make the sort of distinctions that can be represented in language, it lacks thought.) But for Fodor the language required for thought is not a *natural* language such as English or German, but an internal language which he terms *the language of thought*. Creatures who cannot learn natural languages, such as cats and dogs, can still possess a language of thought in which they can contemplate hypotheses and weigh evidence. In creatures who can learn a natural language, the language of thought provides the vehicle in which they can formulate and test hypotheses about the meanings of natural language expressions and the syntactic rules governing them. One way to think about Fodor's proposal is that for him the language of thought constitutes the innate machine language of the human information-processing system and any natural language must be compiled or interpreted in terms of it. Fodor's proposal resonates with theorists who have construed the von Neumann computer (a computer which carries out symbol manipulation using a stored program) as a model for understanding cognition.

Theorists at the opposite pole dissociate thought from language, construing thinking as a quite different sort of activity than the manipulation of linguistic items.

One alternative construal is that thinking without language involves perceptual processes, such as the ability to recognize perceptual patterns and relate them to each other. The emergence of connectionist or neural network models of information processing, where the primitive operations are not operations on language-like (symbolic) representations but excitations and inhibitions between neural type units, has provided a powerful alternative model to the von Neumann computer. Connectionist research suggests how a system could carry out intelligent processes by recognizing and transforming patterns without using the medium of language (see Rumelhart et al., 1986; Bechtel and Abrahamsen, in press). This renders the relation between thought and language more complex. One possibility is that learning to communicate in a natural language is not analogous to translating into a different language, but perhaps consists in creating a string of words as a result of complex neural activity involving interactions between a large number of units (see Churchland, 1995). Another is that acquisition of language provides a vehicle for radically transforming the thought process, allowing the thinker to take advantage of some of the special characteristics of language, such as productivity and systematicity (Bechtel, 1996; Clark, 1987). From this perspective, some human thought is indeed linguistic in nature, but the language in question is not a language of thought, but a natural language such as English. A related view is that language is first acquired as an interpersonal communication system and only later transformed into an internal representational system that can be invoked in activities such as problem solving (Vygotsky, 1962).[1]

These different proposals as to the relation of language and thought generate very different perspectives on the task of relating language to the brain. On the language of thought model, a major objective is to determine how the brain generally implements linguistic representations and operations on them. The task of explaining natural languages focuses on how the brain translates between natural language representations and representations in the language of thought. If, on the other hand, language is viewed as an acquired capacity, then the challenge is to determine how other cognitive capacities implemented in the brain are recruited to provide for the acquisition and use of language.

2   Disciplinary Perspectives on Language:
Linguistics, Psycholinguistics, and Neuropsychology[2]

One of the intriguing things about language is that it has been studied from many different perspectives in a variety of disciplines, each of which brings different tools to bear in analyzing it. Here I focus on three of the disciplinary approaches that are most likely to be influenced by results in neuroscience, but these do not exhaust the possibilities. Yet another disciplinary approach is found in philosophy, where philosophy of language was a central area of investigation during the twentieth century. A central interest of philosophers has been the meaning of linguistic structures, and it was the philosopher Gottlob Frege who introduced the important distinction between the *sense* or connotation of an expression and its *reference* or denotation. To date, however, research in philosophy of language has been theoretical in nature and has been conducted largely in isolation from more empirical investigations of language. Accordingly, it is unlikely to be influenced in the short term by results in neuroscience. In this section I review the recent contributions of three disciplines that have played a central role in the empirical study of language – linguistics, psycholinguistics, and neuropsychology – in which the impact of discoveries about brain mechanisms might be expected to have the greatest impact.

Linguistics

Linguistics itself is a multifaceted discipline and practicing linguists approach the phenomena of language from a variety of different perspectives. Some linguists focus on the diversity of languages and the specific, often distinctive, features of particular languages. Others focus on the historical relations between languages (e.g. reconstructing the long extinct Proto-Indo-European language and tracing its divergence into contemporary languages of Europe, Iran, and northern India). Yet other linguists focus more abstractly on the distinctive features of language that are found in all languages. During the first half of the twentieth century a tradition known as *structural* linguistics attempted to characterize linguistic phenomena, introducing critical concepts such as *morpheme* (the smallest unit carrying meaning) and *phoneme* (a unit of sound) to characterize linguistic structures. For structuralists, the analysis of *syntax* (the arrangement of morphological units into sentences) proved challenging. In an effort to make syntax tractable, Zellig Harris advanced the idea of normalizing complex sentences by using *transformations* to relate them to simpler kernel sentences. For example, the passive sentence *the home run was hit by McGwire* is a transformation of the kernel sentence *McGwire hit the home run*. The potent idea of transformations was further developed by Harris's student Noam Chomsky (1957, 1965), who advanced the idea of a *grammar* as a *generative system* comprising a set of rules that would generate all and only members of the infinite set of grammatically well-formed sentences of a language.

Chomsky's early grammars employed phrase structure rules. Two examples of phrase structure rules are $S \rightarrow NP\ VP$ and $NP \rightarrow Adj\ N$; the first specifies that a sentence can be composed of a noun phrase followed by verb phrase while the second states that a noun phrase can be composed of an adjective followed by a noun. Application of phrase structure rules generates what Chomsky referred to as *deep structures*. Transformation rules could then be applied to deep structures to generate *surface structures*, which constitute the grammatical structures of actual sentences. Because of the role of transformation rules, Chomsky's early grammars were known as *transformational grammars*. Over the subsequent half century of periodic revisions in his grammatical theories, Chomsky has come to minimize the role of transformational rules, replacing them with specifications in the lexicon which constrain permissible movements of lexical items within grammatical structures. Despite these changes in the actual grammars he has proposed, Chomsky's goal throughout his career has been to identify principles that could account for the well-formed sentences in any natural language. These principles would constitute a *Universal Grammar*. For Chomsky, specific natural languages, such as English and Turkish, all employ the same Universal Grammar but implement various features of it in different ways. In Chomsky's more recent grammars, this involves setting values for specific parameters identified in the Universal Grammar.

Chomsky clearly set the agenda for many linguists during the second half of the twentieth century and the idea of a generative system exercised considerable influence on the emergence of cognitive science. His approach offered an answer to Descartes, who had claimed that mere physical devices, such as the brain, could not produce human thought because they could not exhibit the flexible use of the potentially infinite set of sentences found in any natural language (this was a major argument Descartes advanced for dualism – see chapter 1, this volume). Chomsky contended that by implementing phrase structure and transformational rules a machine could generate any sentence of a language.[3] Chomsky also bestowed other very influential ideas on the emerging cognitive sciences which had a profound impact on the manner in which linguistic ability was analyzed. I introduce three of these ideas here: (1) that linguistic ability is found only in humans, (2) that it is dependent upon a specialized module, and (3) that basic grammatical knowledge is innate. I return to these in the last section of this chapter.

As I will describe in the next section, Chomskian analyses of language have had a major impact on the analysis of brain mechanisms underlying language ability. Gradually, however, other approaches to linguistic analysis that developed largely in the shadow of Chomsky are also beginning to influence that analysis of neural findings. One of the major alternatives is referred to as *cognitive linguistics*; it rejects the autonomy given to syntax in Chomskian approaches, and attempts to derive linguistic forms from semantically grounded cognitive processes that are not unique to language (Langacker, 1987; Tomasello, 1998).

Psycholinguistics

Efforts to understand the psychological processing involved in comprehending and producing linguistic structures has had a complex relation to linguistic theory. Many of the nineteenth-century pioneers in developing an experimental psychology (such as Wilhelm Wundt and Hermann Paul) proposed accounts of psychological processes involved in language (Wundt emphasized the sentence as a basic unit whereas Paul emphasized a process of construction from individual words – see Blumenthal, 1987). During the first decades of the twentieth century, this psychological interest in language was largely eclipsed as behaviorists focused on general models of learning that could equally explain animal and human behavior. Modern psycholinguistics, though, was inaugurated in the period just prior to Chomsky's appearance on the intellectual stage. In an eight-week summer seminar sponsored by the Social Science Research Council in 1953 an ambitious agenda for collaboration between psychologists and linguists was formulated (Osgood and Sebeok, 1954). A representative endeavor was the attempt to establish the *psychological reality* of linguistic constructs, such as the *phoneme*, through analysis of speech errors. This enterprise of evaluating psychological reality was naturally extended in the wake of Chomsky's proposals of transformational grammars by attempts to demonstrate, either through reaction time studies or memory studies, that sentences requiring more transformations were harder to process than sentences with fewer (Miller, 1962). This specific attempt was relatively short-lived, partly due to results indicating that not all transformations in the grammar resulted in longer processing time and partly due to the fact that Chomsky periodically changed his grammar, rendering previous psychological studies uninterpretable (Reber, 1987). Nonetheless, psycholinguistics has remained an active pursuit in which researchers, primarily psychologists, interested in processing models, have periodically drawn upon and reformulated ideas in linguistics to account for psychological processes (Abrahamsen, 1987).

In addition to adapting frameworks from linguists, psycholinguists have also relied on ideas of mental representation and processing in their research. One of the more powerful tools for analyzing language processing has been semantic networks, first proposed by Ross Quillian (1968). These employ networks of nodes to represent relations between word meanings. Each sense of a word would be represented by a type node, which would be related to token nodes for other concepts that figured in its definition, and these to nodes that figured in their definitions. By employing a process of *spreading activation* from one node to another, Quillian showed how one could compare two related concepts (figure 9.1). Other theorists have put semantic networks to other uses, such as explaining what are known as *priming effects*. Priming is exhibited when reaction times for words are shortened by prior presentation of other related words. Suppose, for example, one has to decide whether the second item in a pair is a word (this is known as a *lexical decision task*). Subjects will respond affirmatively faster to *truck* in *car-truck* than in *snake-truck*; a plausible explanation is that the subject represents meanings of words in a semantic network, and activation spreads quickly from *car* to *truck*, priming it and making access faster when *truck* was presented after *car* than after *snake*.

Neuropsychology

While the root *neuro-* suggests that *neuropsychology* refers to a general integration of neuroscience and psychology, neuropsychology has traditionally focused on what can be learned about psychological processing from patterns of deficits found in instances of brain damage (and, on the applied side, on tests for evaluating brain-damaged patients). Out of necessity in the era before neuroimaging, neuropsychology often proceeded on a purely behavioral level, advancing detailed analyses of the deficits exhibited by patients without specific information about the locus and extent of underlying brain damage. An important strategy in neuropsychology has been to try to establish the independence of cognitive processes from each other by showing that each can be damaged independently. Neuropsychologists refer to this as *disso-*

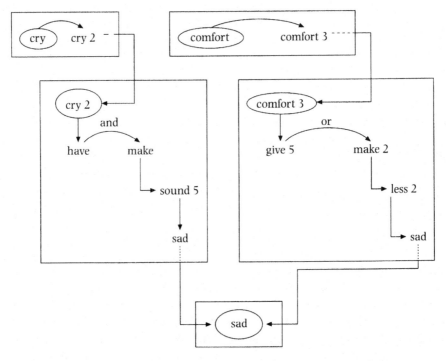

Figure 9.1 A portion of a semantic network as explored by Quillian (1968). Two concepts, *cry* and *comfort*, are shown to be related in that paths lead from both of them to *sad*.

ciating: one cognitive capacity can be abolished while another is retained. A *double dissociation* obtains when each can be separately abolished while retaining the other. A double dissociation of two capacities is often taken to be strong evidence that the two capacities rely on different resources and are carried out by different components in the brain (Shallice, 1988).

The differing patterns of deficits exhibited by Broca's and Wernicke's aphasics appeared to provide an example of a double dissociation between comprehension and production (although, as I shall discuss shortly, there have been several alternative proposals as to just what the alternative capacities are). An even more powerful example of a double dissociation is provided by different types of reading disorders or dyslexias. Individuals with *surface dyslexia*, for example, tend to regularize the pronunciation of words which have exceptional pronunciations in English. For example, they may pronounce *pint* to rhyme with the regularly pronounced words *hint* and *lint* rather than producing the irregular but correct pronunciation. Their tendency to overregularize the pronunciation of these words suggests that they are processing written words only as letter strings (not as lexical items that have predetermined pronunciations). In contrast, individuals with *phonological dyslexia* are able to pronounce words correctly (whether regular or irregular), but experience great difficulty when they attempt to pronounce non-words (such as *zat*). The ability to read regular and irregular words but not even simple non-words suggests that their reading is mediated by recognition of words as items in their lexicon. As a result of these different patterns of deficits, theorists such as Max Coltheart (1987) have proposed that there are two routes to reading, one through a lexicon and one employing grapheme-to-phoneme transition rules. Normal readers have both routes available to them, but the two forms of dyslexia just discussed result when one of these routes is disrupted, leaving only the other. Such an interpretation of double dissociations has recently been challenged by those developing connectionist models – which are designed to use only one processing pathway but which, when lesioned in different ways, exhibit double dissociations (Hinton and Shallice, 1991; van Orden et al., 2001).

3   Relating Brain Structures and Language Functions

The three approaches to understanding language identified in the previous section have historically been pursued with little input from direct investigations of the brain. Increasingly, however, researchers are integrating pursuits in these other disciplines with research on the brain. In this section I will examine several attempts to link linguistic functions, especially those having to do with syntax and semantics, with brain processes. But before doing so, though, it is important to note that these are not the only aspects of language that might be related to brain processes. Comprehending language, for example, requires analysis of auditory or visual stimuli. Like primary visual cortex, primary auditory cortex has a distinct topography, but in this case different tones are mapped on to different parts of primary auditory cortex (generating a tonotopic organization). Neuroimaging studies in humans have revealed several areas that are activated by both words and other sounds, and two areas in the anterior superior temporal cortex and temporal parietal cortex were activated with words but not other sounds (Petersen et al., 1988). This suggests a complex auditory processing system in which common processing areas interact with areas dedicated to speech processing. Thus, although I am emphasizing syntactic and semantic processing, one should not overlook the potential for brain-level research on other linguistic abilities.

The pioneering research of Broca (chapter 5, this volume) and Wernicke (chapter 6, this volume) provided compelling evidence that areas in the (left) inferior frontal lobe and in the (left) superior temporal gyrus were critically involved in language functions. It is worth drawing attention to the differences in the way they construed their contributions. In accordance with the idea that separate faculties were responsible for different psychological functions, Broca took himself to have identified the locus of articulate speech (but not of language generally, since he recognized that Tan was capable of comprehending speech even while he could not produce it). In this he offered a direct or simple localization of the faculty, and his concern was to specify more precisely the locus of this faculty (in part, by trying to identify where Tan's lesion originated). Wernicke operated out of the quite different framework of associationism, according to which cognitive performance depended upon connections between different cognitive capacities. He took the superior temporal gyrus, located adjacent to the primary auditory cortex, to be the center for acoustic speech imagery, responsible for the acoustic recognition of words. He proposed that knowledge of the actual meaning of words depended on connections to visual, auditory, and other sensory images stored elsewhere in the brain. Thus, only two of the aphasias he identified (see figure 6.4) involve damage to localized centers; the others all involve disruptions of various connection pathways.

Despite the initial promise, the efforts to find localized centers of language function were soon abandoned. An attitude decidedly opposed to seeking brain loci for any functions but the most basic sensory ones dominated research in the first decades of the twentieth century (Franz, 1917; Head, 1918; Lashley, 1929, 1950). But beginning in the 1960s, Norman Geschwind at the Boston Veterans Administration Hospital began to resurrect the Wernicke model and it soon came to provide the dominant neurological framework for understanding aphasias. To Wernicke's model, Geschwind added the idea that sensory domains other than speech provided input to the language system through projections to the angular gyrus. Geschwind then proposed a multistage processing system that figured in either speaking a written word or repeating a heard word (see figure 9.2). Geschwind describes the process of speaking a word as involving the building up of a motor program as structure is passed from Wernicke's area to Broca's and then to motor cortex:

> In this model the underlying structure of an utterance arises in Wernicke's area. It is then transferred through the arcuate fasciculus to Broca's area, where it evokes a detailed and coordinated program for vocalization. The program is passed on to the adjacent face area of the motor cortex, which activates the appropriate muscles of the mouth, the lips, the tongue, the larynx, and so on.
>
> (Geschwind, 1979)

The traditional differentiation of Wernicke's and Broca's areas characterized the primary contribution of each area in terms of what we do with language – we comprehend it and we produce it. As natural as the decomposition into comprehension and production is from one point of view, it is orthogonal to the linguist's decomposition, which is grounded in the types of knowledge one must have of one's language – knowledge of phonology, morphology, syntax, semantics, pragmatics, and so on. Both comprehension and production require competency in all of these; for example, to comprehend and to produce a sentence one must know what the units of meaning are (morphology) and how they sound (phonology). If the brain organized language processing in terms of comprehension and production, then it would seem that the brain would have separate stores for phonological, morphological, syntactic, semantic, and pragmatic knowledge for comprehension and production. On this scheme, an individual could develop completely different phonological, syntactical, and semantic processes for comprehension and production.

Finding the comprehension/production perspective problematic, a number of Geschwind's younger colleagues, including David Caplan, Mary-Louise Kean, and

Speaking a written word   Motor cortex

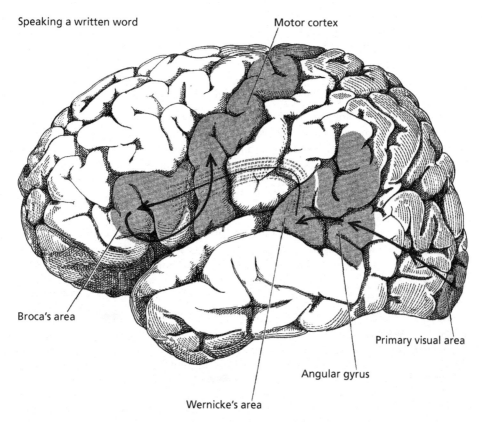

Broca's area

Primary visual area

Angular gyrus

Wernicke's area

Figure 9.2   Geschwind's adaptation of Wernicke's model of speaking a written word. Geschwind proposed that the sensation is first registered by the primary visual cortex and then relayed to the angular gyrus. The visual form of the word is then associated with the corresponding auditory pattern in Wernicke's area. A representation is then transmitted from Wernicke's area to Broca's area, through a bundle of nerve fibers called the arcuate fasciculus. In Broca's area the word evokes a detailed program for articulation, which is supplied to the face area of the motor cortex. Figure adapted from N. Geschwind, Specializations of the human brain, *Scientific American*, 238, 1979, p. 190.

Edgar Zurif, began to develop an alternative decomposition. They thought in terms of systems for processing phonology, syntax, and semantics, with these systems subserving both comprehension and production. Deficits in Wernicke's area had long been associated with deficits in both comprehension and production, since language production in Wernicke's aphasics often consisted in paraphasic speech in which syntax is preserved but the utterances do not make sense (sometimes referred to as cocktail chatter). This suggested that Wernicke's area played a role in processing word meanings in both comprehension and production. The key to prompting an overall re-examination of the Wernicke-Geschwind model was a re-analysis of the deficits found in Broca's aphasics which suggested it too played a role in both comprehension and production. This re-analysis resulted from a fine-grained examination of the speech of Broca's aphasics which revealed that they tend to omit a particular class of words, known as *closed-class* vocabulary, which often serve as grammatical markers.[4] A similar careful examination of comprehension revealed that when comprehension depended upon such closed-class markers, Broca's aphasics exhibited comprehension deficits. For example, Caramazza and Zurif employed sentences with a complex syntactic structure known as *center embedding* in which one or more relative clauses intervene between a noun and its associated verb (e.g. "The girl that the boy is chasing is tall"). To understand this sentence correctly one must attend to the relative clause marker *that*. Caramazza and Zurif (1976) found that Broca's aphasics made errors on such sentences, indicating that they were not processing the syntactic indicators. This comprehension deficit is often masked in other sentences such as "The apple that the boy is eating is red," where Broca's aphasics are able to employ semantic cues to understand the sentence correctly (that is, they rely on the semantic information that *red* is more likely to describe a fruit than a person).

These and other studies suggested that Broca's area was primarily involved in syntactic analysis and Wernicke's area in semantic processing. This decomposition is further supported by evidence of patients with more posterior damage (i.e. in Wernicke's area) who have difficulty processing word stems (*walk* in *walked*) but not the inflectional suffix (*-ed*) and patients with more anterior damage (i.e. in Broca's area) with the reverse pattern of deficits.

Despite the plausibility of this decomposition into syntax and semantics, subsequent data have not been fully consistent with the attempt to map these phenomena on to Broca's and Wernicke's areas. Particularly influential in indicating a much more complex pattern of localization of deficits was the introduction, beginning in the 1970s, of CT-scans to identify the locus of brain damage in various aphasic patients.[5] Mohr (1976), for example, found that patients with lesions restricted to Broca's area did not exhibit Broca's aphasia; rather, much more extensive damage was required to yield the symptoms of Broca's aphasia. Subsequent studies (de Bleser, 1988; Willmes and Poeck, 1993) have provided additional evidence that damage to Broca's and Wernicke's areas does not necessarily result in Broca's and Wernicke's aphasias, and that the damage in patients with those aphasias may be located elsewhere, including in the right hemisphere (Caplan et al., 1996). This may be due simply to individual variability in brain development, but it does make establishing correlations between brain activity and linguistic function more difficult.

Perhaps the central issue, though, is whether decomposition into processes such as syntactic analysis and semantic analysis is the correct decomposition. Grodzinsky (2000) argues that it is not the ability to process syntax *per se* that is damaged by lesions in Broca's area. Rather, the deficit affects very specific syntactic forms whose linguistic analysis involves transformational movements in which a constituent of a sentence is moved from one position in the sentence to another but where a trace is left at the original position. The trace continues to play a syntactic role in the sentence such as in the assignment of thematic roles, but is not phonetically voiced. For example, the structure to the right of the verb *liked* is assigned the thematic role of Theme (recipient of the action) in

(1) Mary liked t_i.

When this is transformed into a relative clause, as in (2),

(2) [*which man*]$_i$ did Mary like t_i

the trace t_i is left in its original position.

Grodzinsky proposes that Broca's aphasics delete the trace and attempt to rely on other information to interpret the sentence. The following pair of sentences illustrates his analysis:

(3) The boy who pushed the girl was tall.

(4) The boy who the girl pushed was tall.

The linguistic derivation of (4) begins with

(5) The girl pushed the boy.

The phrase *the boy* is then moved but, in normal processing, a trace is left to indicate that *the boy* is the theme of *pushed*. But in Broca's aphasics, no trace is left and there appear to be two candidates for agent of *pushed*: *the boy* and *the girl*. The result, according to this analysis, is that Broca's aphasics must guess which is the agent, and are at chance in arriving at the correct interpretation of (4). Since (3) does not involve such a transformation, they process it normally.

Chapters 7 and 8, this volume, further develop the case for rethinking the identification of semantic processing with Wernicke's area, and syntactic processing with Broca's area. Bates points to two additional psycholinguistic sources of evidence that challenge the association of grammatical capacity with Broca's area. The first are studies showing that Broca's aphasics can make fine-grained grammaticality judgments, thereby indicating that they retain much linguistic knowledge. The second are comparative psycholinguistic studies indicating that Wernicke's aphasics who speak more highly inflected languages substitute grammatical markers. From this she proposes that the more general pattern is that Broca's aphasics make errors of omission while Wernicke's aphasics make errors of commission (Elman et al., 1996). Petersen and Fiez review evidence, much of it originating with the early PET studies of Petersen and his colleagues (Petersen et al., 1989, 1988) indicating that semantic tasks such as generating verbs appropriate for particular nouns activate areas in dorsolateral prefrontal cortex, suggesting that semantic processing occurs in frontal cortex, not Wernicke's area. As we discuss in chapter 4, this volume, this alternative localization of semantic processing proved controversial, although it coheres with reports of difficulty in processing verbs as well as closed-class vocabulary in patients with left frontal lobe damage (Caramazza and Shelton, 1991).

A major direction of recent research on brain and language has been to expand the number of brain regions thought to have a role in language processing. Petersen's work not only introduces dorsolateral prefrontal cortex as an area involved in language processing, but also identifies an area in extrastriate cortex that is more active when subjects read words or word-like letter strings (i.e. strings that follow the spelling rules of English) than when they read random letter strings or false-font stimuli, indicating that that area is involved in some manner in language processing. More fine-grained analysis of patients with lexical difficulties has revealed patients with temporal lobe lesions who exhibit deficits in processing particular categories of words (e.g. names for living things, or sometimes more specifically names for plants – see Caramazza and Shelton, 1998). Neuroimaging studies on intact subjects support this decomposition by semantic category, indicating increased activation in the inferior temporal lobe and occipital lobe in response to animal pictures and increased activation in more dorsal parts of the temporal lobe and parts of the frontal lobe in response to artifacts. There is also evidence of modality-specific deficits involving particular grammatical categories (e.g. impairment in producing verbs only in speech, not writing) that suggest even more differentiation of language-related functions.

As researchers press on, they are continuing to identify more brain areas involved in language processing, including areas in the right hemisphere where damage manifests itself in deficits in comprehension above the sentence level – with jokes, etc. – and with prosodic features of speech. The picture that emerges is that language processing involves a complex, integrated system in which a host of brain areas contribute. As the analysis develops, it may prove increasingly difficult to characterize what each of these areas contributes in terms of large-scale language tasks (e.g. comprehension or syntactic processing); rather, each will carry out more specific tasks that are employed in language processing. Moreover, some investigators propose that these processes need not be exclusively linguistic; rather, language processing may recruit brain areas originally dedicated to other psychological activities:

> though language processing recruits specialized localized structures in the left cerebral hemisphere, there may be no one-to-one correlation of individual brain regions with functions defined in linguistic terms. Instead of neural modules being specialized for specific classes of linguistic functions, it appears that many language processes are distributed as component neural computations, performed in concert in many different brain structures. Specific classes of linguistic operations might better be identified with specific signature patterns of distributed activity.
>
> (Deacon, 1998, p. 216)

4 Major Issues in the Study of Language: Uniqueness, Modularity, and Nativism

I noted in section 2, above, that in addition to his specific proposals for linguistic analysis, Chomsky took stances on three issues: (1) whether language processing is a uniquely human ability, (2) whether it involves a special module, and (3) whether the capacity for language is innate. All of these themes are related and were developed in a critique Chomsky (1959) published of B. F. Skinner's book *Verbal Behavior*, a large-scale effort to explain linguistic performances using the behaviorist principles of operant conditioning. For most behaviorists, including Skinner, linguistic abilities were the product of the same underlying capacities for learning that account for all behavior. In contrast, Chomsky argued that linguistic abilities are distinctive, relying on specialized rules for constructing and modifying linguistic structures. Chomsky denied that one could train up a linguistic capacity through conditioning. Accordingly, he and his followers have been critics of claims that training programs employed with other species (e.g. chimpanzees) could yield even primitive linguistic abilities. In contrast, he attributes language to a special module that embodies the special grammatical knowledge on which real linguistic ability relies (see also Pinker, 1994). The same arguments against the potential for training language abilities in other species apply equally to humans, and thus Chomsky commits himself to the knowledge being innate, not learned.

As just indicated, one way to assess whether language is a uniquely human ability has been to focus on other species and ask whether they exhibit anything like a language or whether they can be trained to use human languages (or humanly constructed artificial languages). Ethological research has revealed that other species have complex signaling and call systems. Bees, for example, are able to communicate the location of sources of nectar to each other, but this system has limited expressive capacity and is not productive – it does not provide the capacity for continually generating novel sentences, as do human languages (von Frisch, 1967). Vervet monkeys have a set of calls which specify different types of danger (eagle, snake, etc.), but again, their capacity is not productive (Cheney and Seyfarth, 1990). Moreover, as Deacon argues, call systems rely on very different neural components, particularly ones in the limbic system, from the ones used in human language processing (Deacon, 1997).

Since the 1930s, researchers have attempted to train a variety of non-human species, including dolphins and sea lions as well as several higher primates such as common and bonobo chimpanzees, in the use of artificial linguistic systems. In one of the most successful of these efforts, Sue Savage-Rumbaugh (1986) taught two common chimpanzees to use a system of lexigrams (non-iconographic visual symbols, often mounted on a computer keyboard) to designate objects and use them in a variety of activities, including ones in which the two animals coordinate their behaviors by communicating with lexigrams. In subsequent research, she demonstrated, first with bonobos (especially Kanzi) and then with common chimpanzees, that these primates could acquire the use of lexigrams and comprehension of some spoken English without rigorous conditioning when lexigrams were made available to facilitate the animal's own objectives (e.g. negotiating subsequent activities – Savage-Rumbaugh and Lewin, 1994).

Critics of language research with animals (Pinker, 1994) note that even by the best assessment, apes only acquire the linguistic skills of very young children (Kanzi's comprehension was comparable to that of a $2\frac{1}{2}$-year-old child). Their vocabularies do not grow rapidly, as do those of human children, and they do not acquire the complex syntax of human languages. But these observations only indirectly address the core question of uniqueness, which concerns whether language acquisition relies on cognitive capacities that are shared with other species (but more developed in humans), or whether it depends on new, specialized language abilities. Thus, this question links back to the modularity question.

A distinctive approach that holds the potential for finally advancing this often stalled debate is found in Deacon's *The Symbolic Species*. On the one hand, Deacon points to a capacity that seems to be only exhibited naturally in humans: the capacity for symbolic reference. His emphasis is on *symbolic*. Reference is exhibited in bee and monkey communication – members of both species produce behaviors that are appropriately linked to a referent. What is distinctive of human languages is that words comprise a structured but open-ended semantic system. Words stand in relation to one another such that when a new word is introduced into a person's vocabulary, it immediately takes up relations to existing words:

> words refer to things indirectly, by virtue of an implicit system of relationships between them. This requires that they work in combination (even if only implicitly), referring to one another and modifying one another's reference, to produce a kind of *virtual reference* in which each is associated not so much with some specific concrete object or event, but with kinds, abstract classes, or predicates that can be applied to things.
>
> (Deacon, 1998, p. 220)

While emphasizing the distinctiveness of symbolic reference as exhibited in human languages, Deacon also maintains that it derives through evolutionary changes from capacities present in primate brains. Of particular note in the evolution of the human brain from that of primates is the disproportionate expansion of prefrontal areas. One result of increased prefrontal cells in human brains is that these cells send out more processes to cells elsewhere than do cells in these areas of primate brains, with the result that these cells win the competition for developing processing pathways to other brain areas. Research on monkeys by Goldman-Rakic (1987) shows that lesions to prefrontal areas in monkeys destroy their ability to carry out delayed non-match to sample tasks. In these tasks, monkeys must retain information about stimuli seen previously and then inhibit the natural tendency to choose those stimuli in subsequent choice situations. Humans and members of other species with damage to similar areas exhibit deficits of perseveration – they are unable to repress previous response strategies when shifts are called for. Deacon proposes that amplification of such capacities (which would result from disproportionate expansion of relevant brain areas) may be critical to establishing a network of semantically related symbols required for symbolic reference. For him, a semantic network sets word meanings off from one another by negation so that different words cover different parts of semantic space. If so, then at least this proto-linguistic capacity is grounded in cognitive capacities shared with other species (Deacon, 1997).

By grounding linguistic capacities in cognitive abilities shared with other species, Deacon's approach also rejects the modularity of language. While advocates of the Chomskian tradition still argue for the modularity of language (Fodor, 1983), the discoveries of more and more brain areas involved in language processing suggest that it is a distributed process, and less likely to be due to a segregated module (see chapter 8, this volume). But until research reveals just what brain areas are involved in language, how they are connected to other brain areas, and what kind of information processing each performs, it will not be possible to fully settle just how modular language mechanisms are.

Our final controversial issue concerns whether the capacity for language is innate. Those committed to strong nativism claims, such as Chomsky, Fodor, and Pinker, often appeal to the fact that linguistic capacities seem to be acquired in the absence of environmental inputs sufficient for learning. This is often referred to as the *poverty of the stimulus* argument. Related evidence is found in research on deaf children of hearing parents who create their own signed language (Goldin–Meadow et al., 1994) and of deaf children who develop a more systematically grammatical form of American Sign Language than their parents (Singleton and Newport, 1994). Nativists account for these abilities by proposing that the representations employed in language processing (i.e. the principles of Universal Grammar) are pre-wired into the brain and only need to be triggered by appropriate experience. For Pinker, these neural representations must in turn be specified by the genome.

Although Skinner's (1957) anti-nativism denied any specific genetic or brain structure specifically dedicated to language, contemporary opponents to nativism generally do not deny that some particular brain structures are more suited to handle language processing than others and that the human genome specifies a pattern of development highly suited to acquisition of language. What they reject is the claim that genes could specify neurocircuitry at the level of detail required for innate representations. Rather, like Elman et al. (1996), they propose that the genome provides architectural and chronological constraints that, in the course of normal development as the organism interacts with its environment, result in a brain with areas predisposed to learning specific types of information. Thus, Bates (chapter 8, this volume) proposes that the left temporal cortex is predisposed to extract perceptual detail and that this bias, under normal development, results in specialization for language. But because other brain areas can perform these functions if this area is destroyed early in development, Bates and Elman reject the nativist claim that it encodes innate representations used for language processing. Current discussion of nativism thus turns on the question of what is innately specified and how it figures in the development of normal language capacity.

Although questions about the modularity of language, its uniqueness in humans, and what aspects of it are innate remain controversial, there has been a change in how these questions are investigated. Investigators are increasingly relying on information about the brain and how it supports language function. Partly as a result of this additional potent constraint on theorizing, the positions on both sides of each issue are being refined and moderated.

Conclusions

Language remains our most fascinating cognitive capacity. As I have discussed, it has been the focus of inquiry in a number of different disciplines. The early promise of determining what specific brain areas contribute to language has not yet been fulfilled, but the near future promises to be exciting. New tools, including neuroimaging, are pointing to more brain areas involved in language processing. The challenge is to determine the different processing contributions each makes. Continuing progress in this effort will also likely reduce the differences on such polarizing issues as the uniqueness, modularity, and nativism of language.

Notes

1 A challenge from this perspective is to explain how a child first develops the intention to communicate when it lacks any prior propositional representation of what it intends to communicate. An intriguing suggestion, developed by Andrew Lock (1980), is that the child's first communications (often in the form of gestures) are not intentional from the child's perspective, but interpreted as such by the hearer, and that only after it recognizes the efficacy of its behavior does the child produce gestures and linguistic structures with communicative intent.

2 Material for this section was partly adapted from Bechtel, Abrahamsen, and Graham, (1998).

3 As enticing as this idea is, that the mind might employ the same operations as posited in a linguist's grammar, it is important to bear in mind that the tasks of the linguist in developing a grammar (accounting for all the well-formed sentences of a language) and of a psychologist in explaining linguistic behavior (accounting for linguistic production and comprehension) are different and that different ways of representing language may be required for the two pursuits (Abrahamsen, 1987).

4 Bradley, Garrett, and Zurif distinguish *open*- and *closed*-class vocabularies as follows: "The closed-class (grammatical morphemes, minor grammatical categories, nonphonological words) includes sentence elements that, by and large, are vehicles of phrasal construction rather than primary agents of reference, as is the case with open-classed words (content words, major grammatical categories)" (Bradley et al., 1980, p. 277).

5 It is important to note that while CT-scans provide reliable information as to regions of dead tissue in the brain, it is possible that processing in areas where tissue has not died is nonetheless disrupted, perhaps as a result of their connectivity to areas where the tissue has died.

References

Abrahamsen, A. A. 1987: Bridging boundaries versus breaking boundaries: Psycholinguistics in perspective. *Synthese*, 72, 355–88.

Bechtel, W. 1996: What knowledge must be in the head in order to acquire language. In B. Velichkovsky and D. M. Rumbaugh (eds), *Communicating Meaning: The Evolution and Development of Language*, Hillsdale, NJ: Erlbaum, 45–78.

Bechtel, W., and Abrahamsen, A. A. in press: *Connectionism and the Mind*, 2nd edn. Oxford: Blackwell.

Bechtel, W., Abrahamsen, A., and Graham, G. 1998: The life of cognitive science. In W. Bechtel and G. Graham (eds), *A Companion to Cognitive Science*, Oxford: Blackwell, 1–104.

Blumenthal, A. L. 1987: The emergence of psycholinguistics. *Synthese*, 72, 313–23.

Bradley, D., Garrett, M., and Zurif, E. 1980: Syntactic deficits in Broca's aphasia. In D. Caplan (ed.), *Biological Studies of Mental Processes*, Cambridge, MA: MIT Press, 269–86.

Caplan, D., Hildebrandt, N., and Makris, N. 1996: Location of lesions in stroke patients with deficits in syntactic processing in sentence comprehension. *Brain*, 119, 933–49.

Caramazza, A., and Shelton, J. 1991: Lexical organization of nouns and verbs in the brain. *Nature*, 249, 788–90.

Caramazza, A., and Shelton, J. R. 1998: Domain-specific knowledge systems in the brain: The animate-inanimate distinction. *Journal of Cognitive Neuroscience*, 10 (1), 1–34.

Caramazza, A., and Zurif, E. 1976: Dissociation of algorithmic and heuristic processes in language comprehension: Evidence from aphasia. *Brain and Language*, 3, 572–82.

Cheney, D. L., and Seyfarth, R. M. 1990: *How Monkeys See the World: Inside the Mind of Another Species*. Chicago: University of Chicago Press.

Chomsky, N. 1957: *Syntactic Structures*. The Hague: Mouton.

Chomsky, N. 1959: Review of *Verbal Behavior*. *Language*, 35, 26–58.

Chomsky, N. 1965: *Aspects of a Theory of Syntax*. Cambridge, MA: MIT Press.

Chomsky, N. 1988: *Language and the Problems of Knowledge*. Cambridge, MA: MIT Press.

Churchland, P. M. 1995: *The Engine of Reason, the Seat of the Soul*. Cambridge, MA: MIT Press.

Clark, A. 1987: *Microcognition: Philosophy, Cognitive Science, and Parallel Distributed Processing*. Cambridge, MA: MIT Press.

Coltheart, M. 1987: Cognitive neuropsychology and the study of reading. In M. I. Posner and O. S. M. Marvin (eds), *Attention and Performance*, vol. 11. Hillsdale, NJ: Lawrence Erlbaum, 3–37.

Davidson, D. 1982: Rational animals. *Dialectica*, 36, 317–27.

de Bleser, R. 1988: Localisation of aphasia: Science or fiction. In G. Denes, C. Semenza, and P. Bisiacchi (eds), *Perspectives in Cognitive Neuropsychology*, Hove, UK: Lawrence Erlbaum.

Deacon, T. W. 1997: *The Symbolic Species*. New York: Norton.

Deacon, T. 1998: Language evolution and neuromechanisms. In W. Bechtel and G. Graham (eds), *A Companion to Cognitive Science*, Oxford: Blackwell, 212–25.

Elman, J. L., Bates, E., Johnson, M., Karmiloff-Smith, A., Parisi, D., and Plunkett, K. 1996: *Rethinking Innateness: A Connectionist Perspective on Development*. Cambridge, MA: MIT Press.

Fodor, J. A. 1975: *The Language of Thought*. New York: Crowell.

Fodor, J. A. 1983: *The Modularity of Mind*. Cambridge, MA: MIT Press.

Franz, S. I. 1917: Cerebral adaptation vs. cerebral organology. *Psychological Bulletin*, 14, 137–40.

Geschwind, N. 1979: Specializations of the human brain. *Scientific American*, 238 (3), 158–68.

Goldin-Meadow, S., Butcher, C., Mylander, C., and Dodge, M. 1994: Nouns and verbs in a self-styled gesture system: What's in a name? *Cognitive Psychology*, 27 (3), 259–313.

Goldman-Rakic, P. S. 1987: Circuitry of primate prefrontal cortex and regulation of behavior by representational memory. In J. M. Brookhart, V. B. Mountcastle, and S. R. Geiger (eds), *Handbook of Physiology: The Nervous System*, vol. 5. Bethesda, MD: American Physiological Society, 373–417.

Grodzinsky, Y. 2000: The neurology of syntax: Language use without Broca's area. *Behavioral and Brain Sciences*, 23.

Head, H. 1918: Some principles of neurology. *Brain*, 41, 344–54.

Hinton, G. E., and Shallice, T. 1991: Lesioning a connectionist network: Investigations of acquired dyslexia. *Psychological Review*, 98, 74–95.

Langacker, R. 1987: *Foundations of Cognitive Grammar*, vol. 1. Stanford, CA: Stanford University Press.

Lashley, K. S. 1929: *Brain Mechanisms and Intelligence*. Chicago: University of Chicago Press.

Lashley, K. S. 1950: In search of an engram. *Symposium on Experimental Biology*, 4, 45–8.

Lock, A. 1980: *The Guided Reinvention of Language*. New York: Academic.

Miller, G. A. 1962: Some psychological studies of grammar. *American Psychologist*, 17, 748–62.

Mohr, J. P. 1976: Broca's area and Broca's aphasia. In H. Whitaker and H. A. Whitaker (eds), *Studies in Neurolinguistics*, vol. 1. New York: Academic, 201–35.

Osgood, C. E., and Sebeok, T. A. (eds) 1954: *Psycholinguistics: A Survey of Theory and Research Problems*, vol. 10. Bloomington: Indiana University Press.

Petersen, S. E., Fox, P. T., Posner, M. I., Mintun, M., and Raichle, M. E. 1988: Positron emission tomographic studies of the cortical anatomy of single-word processing. *Nature*, 331, 585–8.

Petersen, S. E., Fox, P. J., Posner, M. I., Mintun, M., and Raichle, M. E. 1989: Positron emission tomographic studies of the processing of single words. *Journal of Cognitive Neuroscience*, 1 (2), 153–70.

Pinker, S. 1994: *The Language Instinct*. New York: Morrow.

Quillian, M. R. 1968: Semantic memory. In M. Minsky (ed.), *Semantic Information Processing*. Cambridge, MA: MIT Press.

Reber, A. S. 1987: The rise and (surprisingly rapid) fall of psycholinguistics. *Synthese*, 72 (3), 325–39.

Rumelhart, D. E., McClelland, J. L., and the PDP Research Group (eds) 1986: *Parallel Distributed Processing: Explorations in the Microstructure of Cognition. Vol. 1. Foundations*, Cambridge, MA: MIT Press.

Savage-Rumbaugh, E. S. 1986: *Ape Language: From Conditioned Response to Symbol*. New York: Columbia University Press.

Savage-Rumbaugh, S., and Lewin, R. 1994: *Kanzi: The Ape at the Brink of the Human Mind*. New York: John Wiley and Sons.

Shallice, T. 1988: *From Neuropsychology to Mental Structure*. New York: Cambridge University Press.

Singleton, J., and Newport, E. 1994: When learners surpass their models: The acquisition of American Sign Language. MS., University of Rochester.

Skinner, B. F. 1957: *Verbal Behavior*. Englewood Cliffs, NJ: Prentice-Hall.

Tomasello, M. 1998: Cognitive linguistics. In W. Bechtel and G. Graham (eds), *A Companion to Cognitive Science*, Oxford: Blackwell, 477–87.

van Orden, G. C., Pennington, B. F., and Stone, G. O. 2001: What do double dissociations prove? *Cognitive Science*, 25.

von Frisch, K. 1967: *The Dance Language and Orientation of Bees*. Cambridge, MA: Harvard University Press.

Vygotsky, L. S. 1962: *Thought and Language*. Cambridge, MA: MIT Press.

Willmes, K., and Poeck, K. 1993: To what extent can aphasic syndromes be localized. *Brain*, 116, 1527–40.

Questions for Further Study and Reflection

1   What, for Broca, is the faculty of articulate speech? What makes it a faculty for him? Are there alternatives to construing it as a faculty?

2   Broca spends considerable time describing the ideas of Bouillaud concerning the localization of articulate speech, which in many ways parallel his own. He also discusses the evidence adduced by Auburtin for Bouillaud's claims. What makes Broca's work different? Why do you suppose history has credited him and not his predecessors with first localizing a mental function in the brain?

3   Why did Wernicke not adopt the view that meanings of words were stored in the region of temporal cortex which produced a deficit when lesioned? Why, in other words, did he insist on and interpret his findings in terms of an associationist framework? Could they be accommodated by a faculty framework such as Broca employed?

4   Petersen and Fiez emphasize that only "elementary operations" are likely to be localized in specific brain regions. What would count as an elementary operation for these purposes? What reasons can be offered either supporting or denying localization of function at this level?

5   Petersen and Fiez's PET studies employ the subtraction method – subtracting the activation produced during a control task from that produced in a target task. What is the motivation for this approach? When might it be misleading?

6   For many theorists, nativism and modularity are tightly linked – it is a domain-specific module that might be innately specified. What does nativism come to if the connection with modularity is broken, as Bates proposes? What kind of evidence could be brought forward for or against nativism without modularity?

7   Bates recommends the "distressingly vague" but suggestive idea that "left frontal regions are specialized for the rapid processes required for fluent use of grammar, while posterior regions play a more important role in controlled, strategic choice of words and sentence frames." Suggest a research program that could make this proposal less vague and subject it to critical experimental inquiry.

8 How have arguments about learnability, which Bates reviews, been used to establish a domain-specific language module? How does a neural network's acquisition of the past tense provide Bates with a response to these learnability arguments?

9 What is meant by establishing the psychological reality of a linguistic construct? What would it be to establish the neuroscientific reality of a linguistic construct?

10 What might be learned about human language use from studies of animal communication?

Part III

Vision

Introduction

William Bechtel

Exploration of how different brain regions contribute to the performance of mental tasks has been easiest for sensory and motor processing, and more difficult for central cognitive tasks such as reasoning, memory, and language. The reasons have to do with the means researchers have to probe the contributions of distinct brain areas. While lesions can occur anywhere in the brain, it is often very difficult to infer from the deficits in overall performance just what the damaged region contributed to performance when it was not damaged (see chapter 4, above). Thus, there has been extensive controversy, as we saw in Part II, as to what Broca's and Wernicke's areas contribute to linguistic performance. Recording the electrical activity of cells and determining what kind of stimulus will elicit activity in the cell has been a far more probative tool for discovering the contributions of various brain regions, but this procedure works best for those areas responsive to features of the stimulus – sensory processing areas.

For humans and other primates, the most developed sense is vision. One indication of this is the amount of brain tissue devoted to visual analysis (see chapter 12). Accordingly, vision is the mental process for which researchers have produced the most detailed understanding at the neural level.

As chapter 13 reviews, serious research on brain areas contributing to vision began in the late nineteenth century, but it was the landmark investigations of David Hubel and Torsten Wiesel in the 1950s that initiated the process of developing detailed accounts of what different brain regions contributed to visual processing. Chapter 10 is a paper by Hubel and Wiesel that appeared in *Scientific American* in 1979. A major portion of the paper describes the results of their own work on primary visual cortex (Brodmann's area 17) which revealed a complex architecture that seemed to extract information about edges of objects in the visual field. As Hubel and Wiesel note at the end of their chapter, they and other researchers had already in the 1970s identified other visual areas to which neurons in primary visual cortex projected which seemed to analyze different features of visual stimuli (several of these discoveries are described in chapter 13).

Thus, by the 1980s numerous visual processing areas had been identified, but there was no organizing principle by which to conceptualize the decomposition of processing in the brain. It was at this stage that Mortimer Mishkin and Leslie Unger-leider proposed that the visual system after primary visual cortex was organized into two pathways, one involved in object identification and another in specifying spatial location (accordingly, these two pathways are often referred to as the "what" and "where" pathways). They describe their proposal, and the evidence on which it is based, in chapter 11.

Although subsequent research has confirmed the broad distinction into two visual pathways, it has also revealed additional complexity. In chapter 12 van Essen and Gallant describe a complex hierarchical stream of subcortical and cortical process-ing areas involved in vision. In both subcortical and early cortical areas, they differ-entiate three processing streams, which eventually reorganize into the two streams identified by Mishkin and Ungerleider. Noting the critical role cell recording has played in determining the functions of neurons in various visual areas, van Essen and Gallant also discuss two different ways of analyzing results – treating neurons as feature detectors or as filters.

In the last chapter of this Part, Bechtel provides both an overall account of the history of research on vision and provides a framework in which to understand the reasoning which guided the discovery of contemporary accounts of vision. He describes a process in which it was the discovery of more and more visual process-ing areas that guided the further decomposition of visual processing into sub-tasks. He ends by suggesting the importance today of developing a computational analysis to couple with the details of neural processing areas and their functional contributions.

10

Brain Mechanisms of Vision

David H. Hubel and Torsten N. Wiesel

Viewed as a kind of invention by evolution, the cerebral cortex must be one of the great success stories in the history of living things. In vertebrates lower than mammals the cerebral cortex is minuscule, if it can be said to exist at all. Suddenly impressive in the lowest mammals, it begins to dominate the brain in carnivores, and it increases explosively in primates; in man it almost completely envelops the rest of the brain, tending to obscure the other parts. The degree to which an animal depends on an organ is an index of the organ's importance that is even more convincing than size, and dependence on the cortex has increased rapidly as mammals have evolved. A mouse without a cortex appears fairly normal, at least to casual inspection; a man without a cortex is almost a vegetable, speechless, sightless, senseless.

Understanding of this large and indispensable organ is still woefully deficient. This is partly because it is very complex, not only structurally but also in its functions, and partly because neurobiologists' intuitions about the functions have so often been wrong. The outlook is changing, however, as techniques improve and as investigators learn how to deal with the huge numbers of intricately connected neurons that are the basic elements of the cortex, with the impulses they carry and with the synapses that connect them. In this article we hope to sketch the present state of knowledge of one subdivision of the cortex: the primary visual cortex (also known as the striate cortex or area 17), the most elementary of the cortical regions concerned with vision. That will necessarily lead us into the related subject of visual perception, since the workings of an organ cannot easily be separated from its biological purpose.

The cerebral cortex, a highly folded plate of neural tissue about two millimeters thick, is an outermost crust wrapped over the top of, and to some extent tucked under, the cerebral hemispheres. In man its total area, if it were spread out, would be about 1.5 square feet. (In a 1963 article in *Scientific American* one of us gave the area as 20 square feet and was quickly corrected by a neuroanatomist friend in Toronto, who said he thought it was 1.5 square feet – "at least that is what

Canadians have.") The folding is presumably mainly the result of such an unlikely structure's having to be packed into a box the size of the skull.

A casual glance at cortical tissue under a microscope shows vast numbers of neurons: about 10^5 (100,000) for each square millimeter of surface, suggesting that the cortex as a whole has some 10^{10} (10 billion) neurons. The cell bodies are arranged in half a dozen layers that are alternately cell-sparse and cell-rich. In contrast to these marked changes in cell density in successive layers at different depths in the cortex there is marked uniformity from place to place in the plane of any given layer and in any direction within that plane. The cortex is morphologically rather uniform in two of its dimensions.

One of the first great insights about cortical organization came late in the nineteenth century, when it was gradually realized that this rather uniform plate of tissue is subdivided into a number of different regions that have very different functions. The evidence came from clinical, physiological, and anatomical sources. It was noted that a brain injury, depending on its location, could cause paralysis or blindness or numbness or speech loss; the blindness could be total or limited to half or less of the visual world, and the numbness could involve one limb or a few fingers. The consistency of the relation between a given defect and the location of the lesion gradually led to a charting of the most obvious of these specialized regions, the visual, auditory, somatic sensory (body sensation), speech, and motor regions.

In many cases a close look with a microscope at cortex stained for cell bodies showed that in spite of the relative uniformity there were structural variations, particularly in the layering pattern, that correlated well with the clinically defined subdivisions. Additional confirmation came from observations of the location (at the surface of the brain) of the electrical brain waves produced when an animal was stimulated by touching the body, sounding clicks or tones in the ear or flashing light in the eye. Similarly, motor areas could be mapped by stimulating the cortex electrically and noting what part of the animal's body moved.

This systematic mapping of the cortex soon led to a fundamental realization: most of the sensory and motor areas contained systematic two-dimensional maps of the world they represented. Destroying a particular small region of cortex could lead to paralysis of one arm; a similar lesion in another small region led to numbness of one hand or of the upper lip, or blindness in one small part of the visual world; if electrodes were placed on an animal's cortex, touching one limb produced a correspondingly localized series of electric potentials. Clearly the body was systematically mapped on to the somatic sensory and motor areas; the visual world was mapped on to the primary visual cortex, an area on the occipital lobe that in man and in the macaque monkey (the animal in which our investigations have mainly been conducted) covers about 15 square centimeters.

In the primary visual cortex the map is uncomplicated by breaks and discontinuities except for the remarkable split of the visual world down the exact middle, with the left half projected to the right cerebral cortex and the right half projected to the left cortex. The map of the body is more complicated and is still perhaps not completely understood. It is nonetheless systematic, and it is similarly crossed, with the right side of the body projecting to the left hemisphere and the left side projecting to the right hemisphere. (It is worth remarking that no one has the remotest idea why there should be this amazing tendency for nervous-system pathways to cross.)

An important feature of cortical maps is their distortion. The scale of the maps varies as it does in a Mercator projection, the rule for the cortex being that the regions of highest discrimination or delicacy of function occupy relatively more cortical area. For the body surface, a millimeter of surface on the fingers, the lips, or the tongue projects to more cortex than a millimeter of trunk, buttocks, or back; in vision the central part of the retina has a representation some 35 times more detailed than the far peripheral part.

Important as the advances in mapping cortical projections were, they tended for some time to divert thought from the real problem of just how the brain analyzes information. It was as though the representation could be an end in itself instead of serving a more subtle purpose – as though what the cortex did was to cater to some little green man who sat inside the head and surveyed images playing across the cortex. In the course of this article we shall show that, for vision at least, the world is represented in a far more distorted way; any little green man trying to glean information from the cortical projection would be puzzled indeed.

The first major insight into cortical organization was nonetheless the recognition of this subdivision into areas having widely different functions, with a tendency to ordered mapping. Just how many such areas there are has been a subject of wide speculation. Anatomists' estimates have on the whole been rather high – up to several hundred areas, depending on the individual worker's sensitivity to fine differences in microscopic patterns and sometimes also on his ability to fool himself. Physiologists began with lower estimates, but lately, with more powerful mapping methods, they have been revising their estimates upward. The important basic notion is that information on any given modality such as sight or sound is transmitted first to a primary cortical area and from there, either directly or via the thalamus, to successions of higher areas. A modern guess as to the number of cortical areas might be between 50 and 100.

The second major insight into cortical organization came from the work of the anatomist Santiago Ramón y Cajal and his pupil Rafael Lorente de Nó. This was the realization that the operations the cortex performs on the information it receives are local. What that means can best be understood by considering the wiring diagram that emerged from the Golgi method used by Cajal and Lorente de Nó. In essence the wiring is simple. Sets of fibers bring information to the cortex; by the time several synapses have been traversed the influence of the input has spread vertically to all cell layers; finally several other sets of fibers carry modified messages out of the area. The detailed connections between inputs and outputs differ from one area to the next, but within a given area they seem to be rather stereotyped. What is common to all regions is the local nature of the wiring. The information carried into the cortex by a single fiber can in principle make itself felt through the entire thickness in about three or four synapses, whereas the lateral spread, produced by branching trees of axons and dendrites, is limited for all practical purposes to a few millimeters, a small proportion of the vast extent of the cortex.

The implications of this are far-reaching. Whatever any given region of the cortex does, it does locally. At stages where there is any kind of detailed, systematic topographical mapping the analysis must be piecemeal. For example, in the somatic sensory cortex the messages concerning one finger can be combined and compared with an input from elsewhere on that same finger or with input from a neighboring finger, but they can hardly be combined with the influence from the trunk or from a foot. The same applies to the visual world. Given the detailed order of the input to the primary visual cortex, there is no likelihood that the region will do anything to correlate information coming in from both far above and far below the horizon, or from both the left and the right part of the visual scene. It follows that this cannot by any stretch of the imagination be the place where actual perception is enshrined. Whatever these cortical areas are doing, it must be some kind of local analysis of the sensory world. One can only assume that as the information on vision or touch or sound is relayed from one cortical area to the next the map becomes progressively more blurred and the information carried more abstract.

Even though the Golgi-method studies of the early 1900s made it clear that the cortex must perform local analyses, it was half a century before physiologists had the least inkling of just what the analysis was in any area of the cortex. The first understanding came in the primary visual area, which is now the best-understood of any cortical region and is still the only one where the analysis and consequent transformations of information are known in any detail. After describing the main transformations that take place in the primary visual cortex we shall go on to show how increasing understanding of these cortical functions has revealed an entire world of architectural order that is otherwise inaccessible to observation.

We can best begin by tracing the visual path in a primate from the retina to the cortex. The output from each eye is conveyed to the brain by about a million nerve fibers bundled together in the optic nerve. These fibers are the axons of the ganglion cells of the retina. The messages from the light-sensitive elements, the rods and cones, have already traversed from two to four synapses and have involved four other types of retinal cells before they arrive at the ganglion cells, and a certain amount of sophisticated analysis of the information has already taken place.

A large fraction of the optic-nerve fibers pass uninterrupted to two nests of cells deep in the brain called the lateral geniculate nuclei, where they make synapses (figure 10.1). The lateral geniculate cells in turn send their axons directly to the primary visual cortex. From there, after several synapses, the messages are sent to a number of further destinations: neighboring cortical areas and also several targets deep in the brain. One contingent even projects back to the lateral geniculate bodies; the function of this feedback path is not known. The main point for the moment is that the primary visual cortex is in no sense the end of the visual path. It is just one

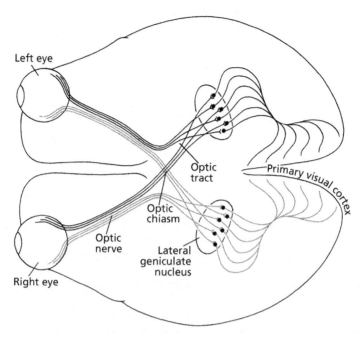

Figure 10.1   The visual pathway is traced schematically in the human brain, seen here from below. The output from the retina is conveyed, by ganglion-cell axons bundled in the optic nerves, to the lateral geniculate nuclei; about half of the axons cross over to the opposite side of the brain, so that a representation of each half of the visual scene is projected on the geniculate of the opposite hemisphere. Neurons in the geniculates send their axons to the primary visual cortex.

stage, probably an early one in terms of the degree of abstraction of the information it handles.

As a result of the partial crossing of the optic nerves in the optic chiasm, the geniculate and the cortex on the left side are connected to the two left half retinas and are therefore concerned with the right half of the visual scene, and the converse is the case for the right geniculate and the right cortex. Each geniculate and each cortex receives input from both eyes, and each is concerned with the opposite half of the visual world.

To examine the workings of this visual pathway our strategy since the late 1950s has been (in principle) simple. Beginning, say, with the fibers of the optic nerve, we record with microelectrodes from a single nerve fiber and try to find out how we can most effectively influence the firing by stimulating the retina with light. For this one can use patterns of light of every conceivable size, shape, and color, bright on a dark background or the reverse, and stationary or moving. It may take a long time, but sooner or later we satisfy ourselves that we have found the best stimulus for the cell being tested, in this case a ganglion cell of the retina. (Sometimes we are wrong!)

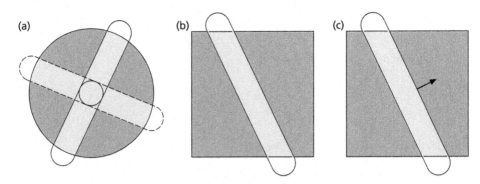

Figure 10.2   The receptive fields of various cells in the visual pathway are compared. Retinal ganglion cells and neurons in the lateral geniculate nucleus have circular fields with either an excitatory center and an inhibitory surround (a) or the opposite arrangement. A spot of light falling on the center stimulates a response from such a cell; so does a bar of light falling on the field in any orientation, provided it falls on the center. In the visual cortex there is a hierarchy of neurons with increasingly complex response properties. The cortical cells that receive signals directly from the geniculate have circularly symmetrical fields. Cortical cells farther along the pathway, however, respond only to a line stimulus in a particular orientation. A "simple" cell (b) responds to such a fine stimulus only in a particular part of its field. A "complex" cell (c) responds to a precisely oriented line regardless of where it is in its field and also to one moving in a particular direction (arrow).

We note the results and then go on to another fiber. After studying a few hundred cells we may find that new types become rare. Satisfied that we know roughly how the neurons at this stage work, we proceed to the next stage (in this case the geniculate) and repeat the process. Comparison of the two sets of results can tell us something about what the geniculate does. We then go on to the next stage, the primary cortex, and repeat the procedure.

Working in this way, one finds that both a retinal ganglion cell and a geniculate cell respond best to a roughly circular spot of light of a particular size in a particular part of the visual field (figure 10.2(a)). The size is critical because each cell's receptive field (the patch of retinal receptor cells supplying the cell) is divided, with an excitatory center and an inhibitory surround (an "on center" cell) or exactly the reverse configuration (an "off center" cell). This is the center-surround configuration first described by Stephen W. Kuffler at the Johns Hopkins University School of Medicine in 1953. A spot exactly filling the center of an on–center cell is therefore a more effective stimulus than a larger spot that invades the inhibitory area, or than diffuse light. A line stimulus (a bar of light) is effective if it covers a large part of the center region and only a small part of the surround. Because these cells have circular symmetry they respond well to such a line stimulus whatever its orientation. To sum up, the retinal ganglion cells and the cells of the lateral geniculate – the cells supplying the input to the visual cortex – are cells with concentric, center-surround receptive fields. They are primarily concerned not with assessing levels of illumination but rather with making a comparison between the light level in one small area of the visual scene and the average illumination of the immediate surround.

The first of the two major transformations accomplished by the visual cortex is the rearrangement of incoming information so that most of its cells respond not to spots of light but to specifically oriented line segments. There is a wide variety of cell types in the cortex, some simpler and some more complex in their response properties, and one soon gains an impression of a kind of hierarchy, with simpler cells feeding more complex ones. In the monkey there is first of all a large group of cells that behave (as far as is known) just like geniculate cells: they have circularly symmetrical fields. These cells are all in the lower part of one layer, called layer IV, which is precisely the layer that receives the lion's share of the geniculate input. It makes sense that these least sophisticated cortical cells should be the ones most immediately connected to the input.

Cells outside layer IV all respond best to specifically oriented line segments (figure 10.2(b)). A typical cell responds only when light falls in a particular part of the visual world, but illuminating that area diffusely has little effect or none, and small spots of light are not much better. The best response is obtained when a line that has just the right tilt is flashed in the region or, in some cells, is swept across the region. The most effective orientation varies from cell to cell and is usually defined sharply enough so that a change of 10 or 20 degrees clockwise or counterclockwise reduces the response markedly or abolishes it. (It is hard to convey the precision of this discrimination. If 10 to 20 degrees sounds like a wide range, one should remember that the angle between 12 o'clock and one o'clock is 30 degrees.) A line at 90 degrees to the best orientation almost never evokes any response.

Depending on the particular cell, the stimulus may be a bright line on a dark background or the reverse, or it may be a boundary between light and dark regions. If it is a line, the thickness is likely to be important; increasing it beyond some optimal width reduces the response, just as increasing the diameter of a spot does in the case of ganglion and geniculate cells. Indeed, for a particular part of the visual field the geniculate receptive-field centers and the optimal cortical line widths are comparable.

Neurons with orientation specificity vary in their complexity. The simplest, which we call "simple" cells, behave as though they received their input directly from several cells with center-surround, circularly symmetrical fields – the type of cells found in layer IV. The response properties of these simple cells, which respond to an optimally oriented line in a narrowly defined location, can most easily be accounted for by requiring that the centers of the incoming center-surround fields all be excitatory or all be inhibitory, and that they lie along a straight line. At present we have no direct evidence for this scheme, but it is attractive because of its simplicity and because certain kinds of indirect evidence support it. According to the work of Jennifer S. Lund of the University of Washington School of Medicine, who in the past few years [the late 1970s] has done more than anyone else to advance the Golgi-stain anatomy of this cortical area, the cells in layer IV project to the layers just above, which is roughly where the simple cells are found.

The second major group of orientation-specific neurons are the far more numerous "complex" cells. They come in a number of subcategories, but their main feature is that they are less particular about the exact position of a line. Complex cells behave as though they received their input from a number of simple cells, all with the same receptive-field orientation but differing slightly in the exact location of their fields. This scheme readily explains the strong steady firing evoked in a complex cell as a line is kept in the optimal orientation and is swept across the receptive field. With the line optimally oriented many cells prefer one direction of movement to the opposite direction (figure 10.2(c)). Several possible circuits have been proposed to explain this behavior, but the exact mechanism is still not known.

Although there is no direct evidence that orientation-sensitive cells have anything to do with visual perception, it is certainly tempting to think they represent some early stage in the brain's analysis of visual forms. It is worth asking which cells at this early stage would be expected to be turned on by some very simple visual form, say a dark blob on a light background. Any cell whose receptive field is entirely inside or outside the boundaries of such an image will be completely unaffected by the figure's presence because cortical cells effectively ignore diffuse changes in the illumination of their entire receptive fields.

The only cells to be affected will be those whose field is cut by the borders. For the circularly symmetrical cells the ones most strongly influenced will be those whose center is grazed by a boundary (because for them the excitatory and inhibitory subdivisions are most unequally illuminated). For the orientation-specific cells the only ones to be activated will be those whose optimal orientation happens to coincide with the prevailing direction of the border. And among these the simple cells will be much more exacting than the complex ones, responding optimally only when the border falls along a line separating an excitatory and an inhibitory region. It is important to realize that this part of the cortex is operating only locally, on bits of the form; how the entire form is analyzed or handled by the brain – how this information is worked on and synthesized at later stages, if indeed it is – is still not known.

The second major function of the monkey visual cortex is to combine the inputs from the two eyes. In the lateral geniculate nuclei a neuron may respond to stimulation of the left eye or of the right one, but no cell responds to stimulation of both eyes. This may seem surprising, since each geniculate receives inputs from both eyes, but the fact is that the geniculates are constructed in a way that keeps inputs from the two eyes segregated. Each geniculate body is divided into six layers, three left-eye layers interdigitated with three right-eye ones (figure 10.3). The opposite-side half of the visual world is mapped on to each layer (with the six maps in precise register, so that in a radial pathway traversing the six layers the receptive fields of all the cells encountered have virtually identical positions in the visual field). Since any

Right eye   Left eye   Right eye

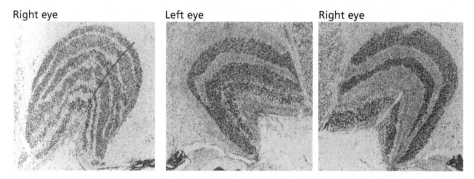

Figure 10.3 The lateral geniculate nucleus of a normal monkey (left) is a layered structure in which cells in layers 1, 4, and 6 (numbered from bottom to top) receive their input from the eye on the opposite side and those in layers 2, 3, and 5 receive information from the eye on the same side. The maps are in register, so that the neurons along any radius (black line) receive signals from the same part of the visual scene. The layered nature of the input is demonstrated in the two geniculates of an animal that had vision in the left eye only (two micrographs at right): in each geniculate cells in the three layers with input from right eye have atrophied. Geniculates are enlarged 10 diameters.

one layer has input from only one eye, the individual cells of that layer must be monocular.

Even in the visual cortex the neurons to which the geniculate cells project directly, the circularly symmetrical cells in layer IV, are all (as far as we can tell) strictly monocular; so are all the simple cells. Only at the level of the complex cells do the paths from the two eyes converge, and even there the blending of information is incomplete and takes a special form. About half of the complex cells are monocular, in the sense that any one cell can be activated only by stimulating one eye. The rest of the cells can be influenced independently by both eyes.

If one maps the right-eye and left-eye receptive fields of a binocular cell (by stimulating first through one eye and then through the other) and compares the two fields, the fields turn out to have identical positions, levels of complexity, orientation, and directional preference; everything one learns about the cell by stimulating one eye is confirmed through the other eye. There is only one exception: if first one eye and then the other are tested with identical stimuli, the two responses are usually not quantitatively identical; in many cases one eye is dominant, consistently producing a higher frequency of firing than the other eye.

From cell to cell all degrees of ocular dominance can be found, from complete monopoly by one eye through equality to exclusive control by the other eye. In the monkey the cells with a marked eye preference are somewhat commoner than the cells in which the two eyes make about equal contributions. Apparently a binocular cell in the primary visual cortex has connections to the two eyes that are qualitatively virtually identical, but the density of the two sets of connections is not necessarily the same.

It is remarkable enough that the elaborate sets of wiring that produce specificity of orientation and of direction of movement and other special properties should be present in two duplicate copies. It is perhaps even more surprising that all of this can be observed in a newborn animal. The wiring is mostly innate, and it presumably is genetically determined. (In one particular respect, however, some maturation of binocular wiring does take place mostly after birth.)

We now turn to a consideration of the way these cells are grouped in the cortex. Are cells with similar characteristics – complexity, receptive-field position, orientation, and ocular dominance – grouped together or scattered at random? From the description so far it will be obvious that cells of like complexity tend to be grouped in layers, with the circularly symmetrical cells low in layer IV, the simple cells just above them and the complex cells in layers II, III, V, and VI. Complex cells can be further subcategorized, and the ones found in each layer are in a number of ways very different.

These differences from layer to layer take on added interest in view of the important discovery, confirmed by several physiologists and anatomists during the past few decades, that fibers projecting from particular layers of the cortex have particular destinations. For example, in the visual cortex the deepest layer, layer VI, projects mainly (perhaps only) back to the lateral geniculate body; layer V projects to the superior colliculus, a visual station in the midbrain; layers II and III send their projections to other parts of the cortex. This relation between layer and projection site probably deserves to be ranked as a third major insight into cortical organization.

The next stimulus variable to be considered is the position of the receptive field in the visual field. In describing the lateral geniculate nucleus we pointed out that in each layer the opposite-half visual field forms an ordered topographical map. In the projection from lateral geniculate to primary visual cortex this order is preserved, producing a cortical map of the visual field. Given this ordered map it is no surprise that neighboring cells in this part of the cortex always have receptive fields that are close together; usually, in fact, they overlap. If one plunges a microelectrode into the cortex at a right angle to the surface and records from cell after cell (as many as 100 or 200 of them) in successively deeper layers, again the receptive fields mostly overlap, with each new field heaped on all the others (figure 10.4). The extent of the entire pile of fields is usually several times the size of any one typical field.

There is some variation in the size of these receptive fields. Some of the variation is tied to the layering: the largest fields in any penetration tend to be in layers III, V, and VI. The most important variation, however, is linked to eccentricity, or the distance of a cell's receptive field from the center of gaze. The size of the fields and the extent of the associated scatter in the part of the cortex that maps the center of gaze are tiny compared to the size and amount of scatter in the part that maps

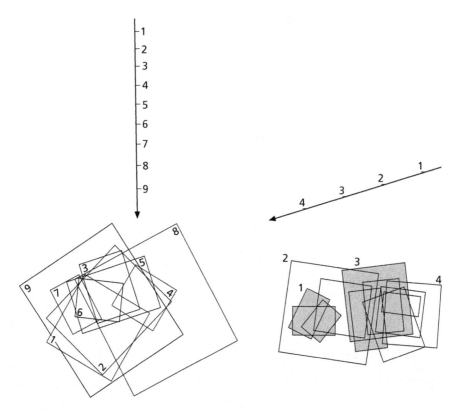

Figure 10.4 The positions of the receptive fields (numbered from 1 to 9) of cortical neurons mapped by an electrode penetrating at roughly a right angle to the surface are essentially the same (left), although the fields are different sizes and there is some scatter. In an oblique penetration (right) from two to four cells were recorded at 0.1-millimeter intervals, at each of four sites (numbered from 1 to 4) one millimeter apart. Each group includes various sizes and some scatter, but now there is also a systematic drift: fields of each successive group of cells are somewhat displaced.

the far periphery. We call the pile of superimposed fields that are mapped in a penetration beginning at any point on the cortex the "aggregate field" of that point. The size of the aggregate field is obviously a function of eccentricity.

If the electrode penetrates in an oblique direction, almost parallel to the surface, the scatter in field position from cell to cell is again evident, but now there is super-imposed on the scatter a consistent drift in field position, its direction dictated by the topographical map of the visual fields. And an interesting regularity is revealed: it turns out that moving the electrode about one or two millimeters always produces a displacement in visual field that is roughly enough to take one into an entirely new region. The movement in the visual field, in short, is about the same as the size of the aggregate receptive field. For the primary visual cortex this holds wherever the recording is made. At the center of gaze the fields and their associated scatter are tiny, but so is the displacement corresponding to a one-millimeter movement along the cortex. With increasing eccentricity (farther out in the visual field) both the field and scatter and the displacement become larger, in parallel fashion. It seems that everywhere a block of cortex about one or two millimeters in size is what is needed to take care of a region of the visual world equivalent to the size of an aggregate field.

These observations suggest the way the visual cortex solves a basic problem: how to analyze the visual scene in detail in the central part and much more crudely in the periphery. In the retina, which has the same problem, for obvious optical reasons the number of millimeters corresponding to a degree of visual field is constant. The retina handles the central areas in great detail by having huge numbers of ganglion cells, each subserving a tiny area of central visual field; the layer of ganglion cells in the central part of the retina is thick, whereas in the outlying parts of the retina it is very thin. The cortex, in contrast, seems to want to be uniform in thickness everywhere. Here there are none of the optical constraints imposed on the retina, and so area is simply allotted in amounts corresponding to the problem at hand.

The machinery in any square millimeter of cortex is presumably about the same as in any other. A few thousand geniculate fibers enter such a region, the cortex does its thing and perhaps 50,000 fibers leave – whether a small part of the visual world is represented in great detail or a larger part in correspondingly less detail. The uniformity of the cortex is suggested, as we indicated at the outset, by the appearance of stained sections. It is compellingly confirmed when we examine the architecture further, looking specifically at orientation and at ocular dominance.

For orientation we inquire about groupings of cells just as we did with field position, looking first at two cells sitting side by side. Two such cells almost invariably have the same optimal stimulus orientation. If the electrode is inserted in a direction perpendicular to the surface, all the cells along the path of penetration have identical or almost identical orientations (except for the cells deep in layer IV, which have no optimal orientation at all). In two perpendicular penetrations a millimeter or so apart, however, the two orientations observed are usually different. The cortex must therefore be subdivided by some kind of vertical partitioning into regions of constant receptive-field orientation. When we came on this system almost 20 years ago, it intrigued us because it fitted so well with the hierarchical schemes we had proposed to explain how complex cells are supplied by inputs from simple cells: the circuit diagrams involve connections between cells whose fields cover the same part of the visual world and that respond to the same line orientation. It seemed eminently reasonable that strongly interconnected cells should be grouped together.

If the cortex is diced up into small regions of constant receptive-field orientation, can one say anything more about the three-dimensional shape of the regions than that their walls are perpendicular to the surface? Are neighboring regions related in any systematic way or are regions subserving all the possible orientations scattered

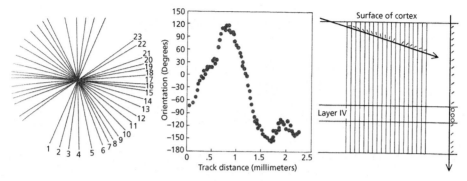

Figure 10.5   The orientation preferences of 23 neurons encountered as a microelectrode penetrated the cortex obliquely are charted (left); the most effective tilt of the stimulus changed steadily
in a counterclockwise direction. The results of a similar experiment are plotted (center); in this
case, however, there were several reversals in direction of rotation. The results of a large number
of such experiments, together with the observation that a microelectrode penetrating the cortex
perpendicularly encounters only cells that prefer the same orientation (apart from the circularly
symmetrical cells in layer IV, which have no preferred orientation), suggested that the cortex is
subdivided into roughly parallel slabs of tissue, with each slab, called an orientation column,
containing neurons with like orientation specificity (right).

over the cortex at random? We began to study these questions simply by penetrating the cortex obliquely or parallel to the surface. When we first did this experiment
in about 1961, the result was so surprising that we could hardly believe it. Instead
of a random assortment of successive orientations there was an amazing orderliness.
Each time the electrode moved forward as little as 25 or 50 micrometers (thousandths
of a millimeter) the optimal orientation changed by a small step, about 10 degrees
on the average; the steps continued in the same direction, clockwise or counterclockwise, through a total angle of anywhere from 90 to 270 degrees. Occasionally
such a sequence would reverse direction suddenly, from a clockwise progression to
a counterclockwise one or vice versa. These reversals were unpredictable, usually
coming after steady progressions of from 90 to 270 degrees (figure 10.5).

Since making this first observation we have seen similar order in almost every
monkey. Either there is a steady progression in orientation or, less frequently, there
are stretches in which orientation stays constant. The successive changes in orientation are small enough so that it is hard to be sure that the regions of constant
orientation are finite in size; it could be that the optimal orientation changes in
some sense continuously as the electrode moves along the cortex.

We became increasingly interested in the three-dimensional shape of these
regional subdivisions. From considerations of geometry alone the existence of small
or zero changes in every direction during a horizontal or tangential penetration
points to parallel slabs of tissue containing cells with like orientation specificity, with
each slab perpendicular to the surface. The slabs would not necessarily be planar,
like slices of bread; seen from above they might well have the form of swirls, which

could easily explain the reversals in the direction of orientation changes. Recording large numbers of cells in several parallel electrode penetrations seemed to confirm this prediction, but it was hard to examine more than a tiny region of brain with the microelectrode.

Fortunately an ideal anatomical method was invented at just the right time for us. This was the 2-deoxyglucose technique for assessing brain activity, devised by Louis Sokoloff and his group at the National Institute of Mental Health. . . . The method capitalizes on the fact that brain cells depend mainly on glucose as a source of metabolic energy and that the closely similar compound 2-deoxyglucose can to some extent masquerade as glucose. If deoxyglucose is injected into an animal, it is taken up actively by neurons as though it were glucose; the more active the neuron, the greater the uptake. The compound begins to be metabolized, but for reasons best known to biochemists the sequence stops with a metabolite that cannot cross the cell wall and therefore accumulates within the cell.

The Sokoloff procedure is to inject an animal with deoxyglucose that has been labeled with the radioactive isotope carbon 14, stimulate the animal in a way calculated to activate certain neurons and then immediately examine the brain for radioactivity, which reveals active areas where cells will have taken up more deoxyglucose than those in quiescent areas. The usual way of examining the brain for this purpose is to cut very thin slices of it (as one would for microscopic examination) and press them against a photographic plate sensitive to the radioactive particles. When the film is developed, any areas that were in contact with radioactive material are seen as dark masses of developed silver grains. Together with Michael P. Stryker we adapted the Sokoloff method to our problem, injecting an anesthetized animal with deoxyglucose and then moving a pattern of black and white vertical stripes back and forth 1.5 meters in front of the animal for 45 minutes. We then cut the brain into slices, either perpendicular to the surface of the cortex or parallel to it.

The autoradiographs quickly confirmed the physiological results. Sections cut perpendicular to the surface showed narrow bands of radioactivity about every 570 micrometers (roughly half a millimeter), extending through the full thickness of the cortex. Evidently these were the regions containing cells responsive to vertical lines. The deep part of layer IV was uniformly radioactive, as was expected from the fact that the cells in the layer have circularly symmetrical receptive fields and show no orientation selectivity.

Sections cut parallel to the surface showed an unexpectedly complex set of periodically spaced bands, often swirling, frequently branching and rejoining, only here and there forming regular parallel slabs. What was particularly striking was the uniformity of the distance from one band to the next over the entire cortex. This fitted perfectly with the idea of a uniform cortex. Moreover, the distance between stripes fitted well with the idea that the cortical machinery must repeat itself at least every millimeter. If the distance were, for example, 10 millimeters from vertical through 180 degrees and back to vertical, sizable parts of the visual field would lack cells sensitive to any given orientation, making for a sketchy and extremely bizarre representation of the visual scene.

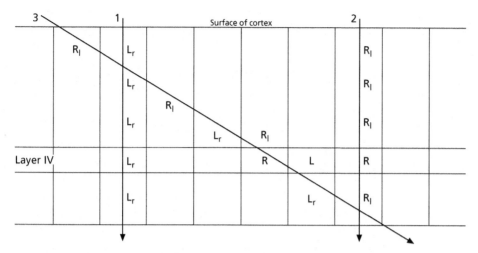

Figure 10.6 Grouping of cells according to ocular dominance was revealed by physiological studies. In one typical vertical penetration of the cortex (1) a microelectrode encounters only cells that respond preferentially to the left eye (L_r) and, in layer IV, cells that respond only to the left eye (L); in another vertical penetration (2) the cells all have right-eye dominance (R_l) or, in layer IV, are driven exclusively by the right eye (R). In an oblique penetration (3) there is a regular alternation of dominance by one eye or the other eye. Repeated penetrations suggest that the cortex is subdivided into regions with a cross-sectional width of about 0.4 millimeter and with walls perpendicular to the cortical surface and layers: the ocular-dominance columns.

The final variable whose associated architecture needs to be considered is eye preference. In microelectrode studies neighboring cells proved almost invariably to prefer the same eye. If in vertical penetrations the first cell we encountered preferred the right eye, then so did all the cells, right down to the bottom of layer VI; if the first cell preferred the left eye, so did all the rest. Any penetration favored one eye or the other with equal probability. (Since the cells of layer IV are monocular, there it was a matter not of eye preference but of eye monopoly.) If the penetration was oblique or horizontal, there was an alternation of left and right preferences, with a rather abrupt switchover about every half millimeter. The cortex thus proved to be diced up into a second set of regions separated by vertical walls that extend through the full cortical thickness. The ocular-dominance system was apparently quite independent of the orientation system, because in oblique or tangential penetrations the two sequences had no apparent relation to each other (figure 10.6).

The basis of these ocular-dominance columns, as they have come to be called, seems to be quite simple. The terminals of geniculate fibers, some subserving the left eye and others the right, group themselves as they enter the cortex so that in layer IV there is no mixing. This produces left–eye and right–eye patches at roughly half-millimeter intervals. A neuron above or below layer IV receives connections from that layer from up to about a millimeter away in every direction. Probably the strongest connections are from the region of layer IV closest to the neuron, so that it is presumably dominated by whichever eye feeds that region.

Again we were most curious to learn what these left-eye and right-eye regions might look like in three dimensions; any of several geometries could lead to the cross-sectional appearance the physiology had suggested. The answer first came from studies with the silver-degeneration method for mapping connections, devised by Walle J. H. Nauta of the Massachusetts Institute of Technology. Since then we have found three other independent anatomical methods for demonstrating these columns.

A particularly effective method (because it enables one to observe in a single animal the arrangement of columns over the entire primary visual cortex) is based on the phenomenon of axonal transport. The procedure is to inject a radioactively labeled amino acid into an area of nervous tissue. A cell body takes up the amino acid, presumably incorporates it into a protein and then transports it along the axon to its terminals. When we injected the material into one eye of a monkey, the retinal ganglion cells took it up and transported it along their axons, the optic-nerve fibers. We could then examine the destinations of these fibers in the lateral geniculate nuclei by coating tissue slices with a silver emulsion and developing the emulsion; the radioactive label showed up clearly in the three complementary layers of the geniculate on each side.

This method does not ordinarily trace a path from one axon terminal across a synapse to the next neuron and its terminals, however, and we wanted to follow the path all the way to the cortex. In 1971 Bernice Grafstein of the Cornell University Medical College discovered that after a large enough injection in the eye of a mouse some of the radioactive material escaped from the optic-nerve terminals and was taken up by the cells in the geniculate and transported along their axons to the cortex. We had the thought that a similarly large injection in a monkey, combined with autoradiography, might demonstrate the geniculate terminals from one eye in layer IV of the visual cortex.

Our first attempt yielded dismayingly negative results, with only faint hints of a few silver grains visible in layer IV. It was only after several weeks that we realized that by resorting to dark-field microscopy we could take advantage of the light-scattering properties of silver grains and so increase the sensitivity of the method. We borrowed a dark-field condenser, and when we looked at our first slide under the microscope, there shining in all their glory were the periodic patches of label in layer IV.

The next step was to try to see the pattern face on by sectioning the cortex parallel to its surface. The monkey cortex is dome-shaped, and so a section parallel to the surface and tangent to layer IV shows that layer as a circle or an oval, while a section below layer IV shows it as a ring. By assembling a series of such ovals and rings from a set of sections one can reconstruct the pattern over a wide expanse of cortex.

From the reconstructions it was immediately obvious that the main overall pattern is one of parallel stripes representing terminals belonging to the injected eye, separated by gaps representing the other eye. The striping pattern is not regular like wallpaper. (We remind ourselves occasionally that this is, after all, biology!) Here and there a stripe representing one eye branches into two stripes, or else it ends blindly at a point where a stripe from the other eye branches. The irregularities are commonest near the center of gaze and along the line that maps the horizon. The stripes always seem to be perpendicular to the border between the primary visual cortex and its neighbor, area 18, and here the regularity is greatest. Such general rules seem to apply to all macaque brains, although the details of the pattern vary from one individual to the next and even from one hemisphere to the other in the same monkey.

The width of a set of two stripes is constant, about 0.8 millimeter, over the entire primary visual cortex, once more emphasizing the uniformity of the cortex. Again the widths fit perfectly with the idea that all of the apparatus needed to look after an area the size of an aggregate field must be contained within any square millimeter of cortex. The two techniques, deoxyglucose labeling and amino acid transport, have the great advantage of being mutually compatible, so that we have been able to apply both together, one to mark orientation lines and the other to see the ocular-dominance columns. The number of brains examined so far is too small to justify any final conclusions, but the two systems appear to be quite independent, neither parallel nor at right angles but intersecting at random.

The function served by ocular-dominance columns is still a mystery. We know there are neurons with all grades of eye preference throughout the entire binocular part of the visual fields, and it may be that a regular, patterned system of converging inputs guarantees that the distribution will be uniform, with neither eye favored by accident in any one place. Why there should be all these grades of eye preference everywhere is itself not clear, but our guess is that it has something to do with stereoscopic depth perception.

Given what has been learned about the primary visual cortex, it is clear that one can consider an elementary piece of cortex to be a block about a millimeter square and two millimeters deep (figure 10.7). To know the organization of this chunk of tissue is to know the organization for all of area 17; the whole must be mainly an iterated version of this elementary unit. Of course the elementary unit should not be thought of as a discrete, separable block. Whether the set of orientation slabs begins with a slab representing a vertical orientation, an oblique one or a horizontal one is completely arbitrary; so too is whether an ocular-dominance sequence begins with a left-plus-right pair of dominance slabs or a right-plus-left pair. The same thing is true for a unit crystal of sodium chloride or for any complex repetitive pattern such as is found in wallpaper.

What, then, does the visual scene really look like as it is projected on to the visual cortex? Suppose an animal fixes its gaze on some point and the only object in the visual field is a straight line above and a bit to the left of the point where the gaze is riveted. If each active cell were to light up, and if one could stand above the cortex and look down at it, what would the pattern be? To make the problem more interesting, suppose the pattern is seen by one eye only. In view of the architecture just

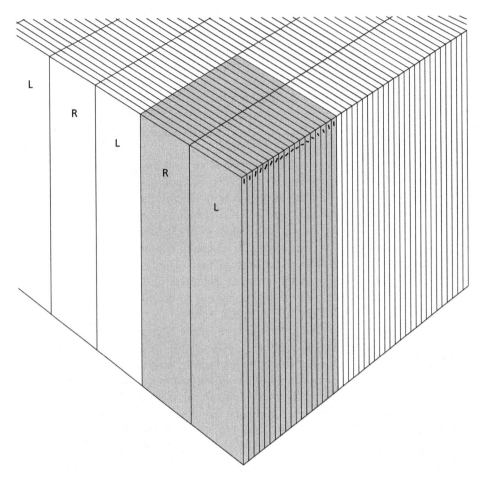

Figure 10.7   A block of cortex about a millimeter square and two millimeters deep (shaded) can be considered an elementary unit of the primary visual cortex. It contains one set of orientation slabs subserving all orientations and one set of ocular-dominance slabs subserving both eyes. The pattern is reiterated throughout the primary visual area. The placing of the boundaries (at the right or the left eye, at a vertical, horizontal or oblique orientation) is arbitrary; representation of the slabs as flat planes intersecting at right angles is an oversimplification.

described the pattern turns out to be not a line but merely a set of regularly spaced patches. The reasoning can be checked directly by exposing a monkey with one eye closed to a set of vertical stripes and making a deoxyglucose autoradiograph. The resulting pattern should not be a great surprise: it is a set of regularly spaced patches, which simply represents the intersection of the two sets of column systems. Imagine the surprise and bewilderment of a little green man looking at such a version of the outside world!

Why evolution has gone to the trouble of designing such an elaborate architecture is a question that continues to fascinate us. Perhaps the most plausible notion is that the column systems are a solution to the problem of portraying more than two dimensions on a two-dimensional surface. The cortex is dealing with at least four sets of values: two for the x and y position variables in the visual field, one for orientation and one for the different degrees of eye preference. The two surface coordinates are used up in designating field position; the other two variables are accommodated by dicing up the cortex with subdivisions so fine that one can run through a complete set of orientations or eye preferences and meanwhile have a shift in visual-field position that is small with respect to the resolution in that part of the visual world.

The strategy of subdividing the cortex with small vertical partitions is certainly not limited to the primary visual area. Such subdivisions were first seen in the somatic sensory area by Vernon B. Mountcastle of the Johns Hopkins University School of Medicine about 10 years before our work in the visual area. In the somatic sensory area, as we pointed out above, the basic topography is a map of the opposite half of the body, but superimposed on that there is a twofold system of subdivisions, with some areas where neurons respond to the movement of the joints or pressure on the skin and other areas where they respond to touch or the bending of hairs. As in the case of the visual columns, a complete set here (one area for each kind of neuron) occupies a distance of about a millimeter. These subdivisions are analogous to ocular-dominance columns in that they are determined in the first instance by inputs to the cortex (from either the left or the right eye and from either deep receptors or receptors in the upper skin layers) rather than by connections within the cortex, such as those that determine orientation selectivity and the associated system of orientation regions.

The columnar subdivisions associated with the visual and somatic sensory systems are the best-understood ones, but there are indications of similar vertical subdivisions in some other areas: several higher visual areas, sensory parietal regions recently studied by Mountcastle and the auditory region, where Thomas J. Imig, H. O. Adrián, and John F. Brugge of the University of Wisconsin Medical School and their colleagues have found subdivisions in which the two ears seem alternately to add their information or to compete.

For most of these physiologically defined systems (except the visual ones) there are so far no anatomical correlates. On the other hand, in the past few years several anatomists, notably Edward G. Jones of the Washington University School of Medicine and Nauta and Patricia Goldman at MIT, have shown that connections from one region of the cortex to another (for example from the somatic sensory area on one side to the corresponding area on the other side) terminate in patches that have a regular periodicity of about a millimeter. Here the columns are evident morphologically, but one has no idea of the physiological interpretation. It is clear, however, that fine periodic subdivisions are a very general feature of the cerebral cortex. Indeed, Mountcastle's original observation of that feature may be said to supply a fourth profound insight into cortical organization.

It would surely be wrong to assume that this account of the visual cortex in any way exhausts the subject. Color, movement, and stereoscopic depth are probably all dealt with in the cortex, but to what extent or how is still not clear. There are indications from work we and others have done on depth and from work on color by Semir Zeki of University College London that higher cortical visual areas to which the primary area projects directly or indirectly may be specialized to handle these variables, but we are a long way from knowing what the handling involves.

What happens beyond the primary visual area, and how is the information on orientation exploited at later stages? Is one to imagine ultimately finding a cell that responds specifically to some very particular item? (Usually one's grandmother is selected as the particular item, for reasons that escape us.) Our answer is that we doubt there is such a cell, but we have no good alternative to offer. To speculate broadly on how the brain may work is fortunately not the only course open to investigators. To explore the brain is more fun and seems to be more profitable.

There was a time, not so long ago, when one looked at the millions of neurons in the various layers of the cortex and wondered if anyone would ever have any idea of their function. Did they all work in parallel, like the cells of the liver or the kidney, achieving their objectives by pure bulk, or were they each doing something special? For the visual cortex the answer seems now to be known in broad outline: Particular stimuli turn neurons on or off; groups of neurons do indeed perform particular transformations. It seems reasonable to think that if the secrets of a few regions such as this one can be unlocked, other regions will also in time give up their secrets.

11

Object Vision and Spatial Vision: Two Cortical Pathways

Mortimer Mishkin, Leslie G. Ungerleider, and Kathleen A. Macko

Thirty-five years ago Lashley concluded that visual mechanisms do not extend beyond the striate cortex. He was led to this view after finding that "None of the lesions in the prestriate region of the monkey has produced symptoms resembling object agnosia as described in man. . . . Uncomplicated destruction of major portions of the prestriate region . . . has not been found to produce any disturbances in sensory or perceptual organization" (Lashley, 1948).

We now know, of course, that Lashley's conclusion was wrong. Tissue essential for vision extends far beyond striate cortex to include not only the prestriate region of the occipital lobe but also large portions of the temporal and parietal lobes. Neurobehavioral studies since Lashley's (Cowey and Gross, 1970; Gross, 1973; Mishkin, 1972, 1982; Mishkin et al., 1982; Pohl, 1973), together with converging evidence from physiological (Allman et al., 1981; Gross, 1973; Hyvärinen, 1981; Mountcastle et al., 1975; Robinson et al., 1978; Zeki, 1978) and anatomical studies (Desimone et al., 1980; Rockland and Pandya, 1981; Ungerleider and Mishkin, 1979; Van Essen et al., 1981; Weller and Kaas, 1981; Zeki, 1969), indicate that these extrastriate regions contain numerous visual areas that can be distinguished both structurally and functionally. Moreover, recent work from our own laboratory (Ungerleider and Mishkin, 1982) suggests that these multiple visual areas are organized hierarchically into two separate cortical visual pathways, one specialized for "object" vision, the other for "spatial" vision.

1   Two Pathways

The two cortical visual pathways are schematized in figure 11.1. One of them consists of a multisynaptic occipitotemporal projection system that follows the course of the inferior longitudinal fasciculus. This pathway, which interconnects the striate, prestriate, and inferior temporal areas, is crucial for the visual identification of

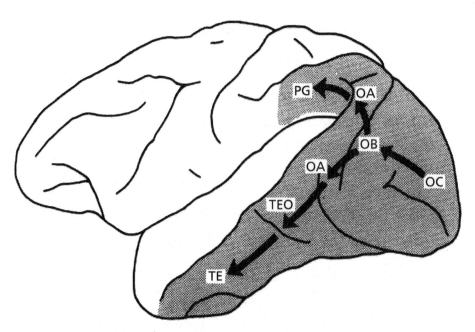

Figure 11.1   Lateral view of the left hemisphere of a rhesus monkey. The shaded area defines the cortical visual tissue in the occipital, temporal, and parietal lobes. Arrows schematize two cortical visual pathways, each beginning in primary visual cortex (area OC), diverging within prestriate cortex (areas OB and OA), and then coursing either ventrally into the inferior temporal cortex (areas TEO and TE) or dorsally into the inferior parietal cortex (area PG). Both cortical visual pathways are crucial for higher visual function, the ventral pathway for object vision, and the dorsal pathway for spatial vision.

objects (Mishkin, 1982). Subsequent links of the occipitotemporal pathway with limbic structures in the temporal lobe (Turner et al., 1980) and with ventral portions of the frontal lobe (Kuypers et al., 1965) may make possible the cognitive association of visual objects with other events, such as emotions and motor acts.

The other pathway consists of a multisynaptic occipitoparietal projection system that follows the course of the superior longitudinal fasciculus. This pathway, which interconnects the striate, prestriate, and inferior parietal areas, is critical for the visual location of objects (Ungerleider and Mishkin, 1982). Subsequent links of the occipitoparietal pathway with dorsal limbic (Pandya and Kuypers, 1968) and dorsal frontal cortex (Kuypers et al., 1965; Pandya and Kuypers, 1968) may enable the cognitive construction of spatial maps, as well as the visual guidance of motor acts (Haaxma and Kuypers, 1975) that were initially triggered by activity in the ventral pathway. In contrast to the ventral pathway, which remains modality-specific throughout its course, the later stations in the dorsal pathway appear to receive convergent input from other modalities and so may constitute polysensory areas (Hyvärinen, 1981; Seltzer and Pandya, 1980).

The notion that separate neural systems mediate object and spatial vision is not new (Ingle et al., 1967; Newcombe and Russell, 1969). In previous formulations, however, these two types of visual perception were attributed to the geniculostriate and tectofugal systems, respectively, rather than to separate cortical pathways diverging from a common striate origin. The shift to the present view is in keeping with the cumulative evidence that, in primates at least, all forms of visual perception, as distinguished from visuomotor functions, are more heavily dependent on the geniculostriate than on the tectofugal system.

2 Object Vision

The anterior part of inferior temporal cortex, or area TE in Bonin and Bailey's (1947) terminology, is the last exclusively visual area in the pathway that begins in the striate cortex, or area OC, and continues through the prestriate and posterior temporal areas, OB, OA, and TEO (figure 11.1). This ventrally directed chain of cortical visual areas appears to extract stimulus-quality information from the retinal input to the striate cortex (Mishkin, 1972), processing it for the purpose of identifying the visual stimulus and ultimately assigning it some meaning through the mediation of area TE's connections with the limbic and frontal-lobe systems (Jones and Mishkin, 1972). According to this view, the analysis of the physical properties of a visual object (such as its size, color, texture, and shape) is performed in the multiple subdivisions of the prestriate–posterior temporal complex (Zeki, 1978), and may even be completed within this tissue. Such a proposal gains support from the striking loss in pattern-discrimination ability that follows damage to the posterior temporal area (Mishkin, 1972). But the synthesis of all the physical properties of the particular object into a unique configuration appears to entail the funneling of the outputs from the prestriate–posterior temporal region into area TE (Mishkin, 1982). This postulated integration of the coded visual properties of an object within area TE would make TE especially well suited to serve not only as the highest-order areas for the visual perception of objects but also as the storehouse for their central representations and, hence, for their later recognition.

That area TE is important for the retention of some form of visual experience has been suspected for decades (Mishkin, 1954). Numerous behavioral studies (Gross, 1973) have demonstrated that bilateral removal of inferior temporal cortex in monkeys yields marked improvement both in the retention of visual discrimination habits acquired prior to surgery and in the postoperative acquisition of new ones. This impairment, which is exclusively visual, appears in the absence of any sensory loss and thus has long been considered a higher-order, or "visuopsychic," dysfunction.

But that the impairment is in fact a visual retention disorder was demonstrated only later when it was found that area TE lesions impair performance on visual tests that tax memory even more than they do on visual tests that tax perceptual ability (Cowey and Gross, 1970). Now, having examined the ability of monkeys with TE

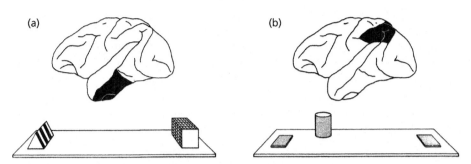

Figure 11.2   Behavioral tasks sensitive to cortical visual lesions in monkeys. (a) Object discrimination. Bilateral removal of area TE in inferior temporal cortex produces severe impairment on
object discrimination. A simple version of such a discrimination is a one-trial object-recognition
task based on the principle of non-matching to sample, in which monkeys are first familiarized
with one object of a pair in a central location (familiarization trial not shown) and are then rewarded
in the choice test for selecting the unfamiliar object. (b) Landmark discrimination. Bilateral
removal of posterior parietal cortex produces severe impairment on landmark discrimination. On
this task, monkeys are rewarded for choosing the covered foodwell closer to a tall cylinder, the
"landmark," which is positioned randomly from trial to trial closer to the left cover or closer to
the right cover, the two covers being otherwise identical.

lesions simply to remember the visual appearance of newly presented objects, we
have uncovered what is perhaps the most dramatic impairment of all (Mishkin,
1982). After just a few days of training, normal monkeys shown an object only
once will demonstrate that they recognize that object when it is presented several
minutes later (figure 11.2(a)). Thus, somewhere in the visual system the single presentation of a complex stimulus leaves a trace against which a subsequently presented
stimulus can be matched. If it does match, i.e. if the original neural trace is reacti-
vated, there is immediate recognition, as demonstrated by the monkey's highly accurate performance. The area in which the neural trace appears to be preferentially
established is area TE, since lesions here – but not lesions elsewhere in the cortical
visual system – nearly abolish the monkey's ability to perform the recognition task.
Apparently, area TE contains the traces laid down by previous viewing of stimuli,
and these serve as stored central representations against which incoming stimuli are
constantly being compared. In the process, old central representations may either
decay, be renewed, or even be refined, while new representations are added to the
store.

It is significant that by virtue of the extremely large visual receptive fields of inferior temporal neurons (Gross, 1973) this area seems to provide the neural basis for
the phenomenon of stimulus equivalence across retinal translation (Gross and
Mishkin, 1979), i.e. the ability to recognize a stimulus as the same, regardless of its
position in the visual field. But a necessary consequence of this mechanism for
stimulus equivalence is that within the occipitotemporal pathway itself there is a loss
of information about the visual location of the objects being identified.

3   Spatial Vision

The neural mechanism that enables the visual location of objects also entails the transmission of information from striate through prestriate cortex; however, the pre-striate route in this case, as well as the rest of the pathway for spatial vision, appears to be quite separate from the pathway for object vision (figure 11.1). Evidence in support of this dichotomy of cortical visual pathways has come from our studies of posterior parietal cortex.

In the initial study of the series, Pohl (1973) demonstrated a dissociation of visual deficits after inferior temporal and parietal lesions. That is, whereas the temporal but not the parietal lesion produced severe impairment on an object-discrimination learning task, just the reverse was found on tests in which the monkey had to learn to choose a response location on the basis of its proximity to a visual "landmark" (figure 11.2(b)). These results provided compelling evidence that

> the inferior temporal cortex participates mainly in the acts of noticing and remember-ing an object's qualities, not its position in space. Conversely, the posterior parietal cortex seems to be concerned with the perception of the spatial relations among objects, and not their intrinsic qualities.
>
> (Mishkin, 1972)

The effective lesions in Pohl's study were large, since they included not only infe-rior parietal cortex, or area PG, but also dorsal prestriate tissue within area OA. To test for the possibility of a further localization of function within this region, addi-tional experiments were performed with more restricted lesions (Mishkin et al., 1982). The results, however, failed to reveal any evidence of a cortical focus serving spatial vision; rather, the severity of impairment on the landmark task was found to depend on the amount of tissue included in the lesion, completely independent of the lesion site. Since damage to the same region, no matter how extensive, failed to produce any impairment in the acquisition of a visual pattern discrimination, it appears that the entire posterior parietal region, including dorsal OA cortex, partic-ipates selectively in the processing of visuospatial as distinguished from visual object-quality information.

Our findings support the accumulating neurobiological evidence that parietal area PG, rather than being a purely tactual association area as was once thought, is a poly-sensory area to which both the visual and tactual modalities contribute (Hyvärinen, 1981; Mountcastle et al., 1975; Robinson et al., 1978). The findings are thus consis-tent with the proposal (Semmes, 1967) that area PG serves a supramodal spatial ability that subsumes both the macrospace of vision and the microspace encompassed by the hand. According to this proposal, visuospatial and tactual discrimination deficits, as well as the inaccuracies in reaching that also follow inferior parietal damage, are dif-ferent reflections of a single, supramodal disorder in spatial perception.

Polysensory area PG is presumed to depend for its visual input on the modality-specific prestriate area OA, which appears to serve visual spatial functions selectively.

Such a hierarchical model for spatial perception suggests, in turn, that the source of the critical visual input for the entire dorsal prestriate-parietal region is, again, the striate cortex. The alternative possibility, namely, that the source of the critical input is the superior colliculus, found no support in a study of the effects of tectal lesions on performance of the landmark task; even complete bilateral destruction of the superior colliculus failed to produce a reliable loss in retention. We therefore examined the contribution of striate inputs to the visuospatial functions of posterior parietal cortex (Mishkin and Ungerleider, 1982), using a disconnection technique analogous to the one used originally to examine the contribution of striate inputs to the object-vision functions of inferior temporal cortex (Mishkin, 1966). Our results suggested that the posterior parietal cortex, like the inferior temporal, is totally dependent on striate input for its participation in vision; but unlike the inferior temporal, the posterior parietal cortex does not seem to receive a heavy visual input via the corpus callosum. It therefore appears that each posterior parietal area may be organized largely as a substrate for contralateral spatial function, which could account in part for the symptom of contralateral spatial neglect that has so often been reported after unilateral parietal injury in man (Denny-Brown and Chambers, 1958; Heilman and Watson, 1977; Mesulam, 1981).

A second difference in the organization of visual inputs to posterior parietal and inferior temporal cortex was uncovered in an experiment that compared the effects of selective removals of striate cortex (Mishkin and Ungerleider, 1982). In this experiment monkeys received bilateral lesions of the striate areas representing either central vision (lateral striate) or peripheral vision (medial striate). The results indicated that while inputs from central vision are the more important ones for the object-recognition functions of inferior temporal cortex, inputs from central and peripheral vision are equally important for the visuospatial functions of posterior parietal cortex.

In summary, interactions with striate cortex are critical for the parietal just as they are for the temporal area, but the striate inputs to these two cortical targets are organized differently: relative to inferior temporal cortex, posterior parietal cortex receives a greater contribution from inputs representing both the contralateral and the peripheral visual fields. These differences, which are seen also in the visual receptive field topography of inferior temporal vs. posterior parietal neurons (Gross, 1973; Robinson et al., 1978), presumably reflect differences in the sensory processing required for object vs. spatial vision.

4 Metabolic and Anatomical Mapping

The evidence from our behavioral work demonstrates that the neural mechanisms underlying object and spatial vision depend on the relay of information from striate cortex through prestriate cortex to targets in inferior temporal and inferior parietal areas, respectively. We have now mapped the full extent of both cortical visual pathways combined, using the 2-[^{14}C]deoxyglucose method (Macko et al., 1982). By comparing a blinded and a seeing hemisphere in the same monkey we have found that the entire visual system can be outlined on the basis of differential hemispheric glucose utilization during visual stimulation. Reduced glucose utilization in the blind as compared with the seeing hemisphere was seen cortically throughout the entire expanse of striate and prestriate cortex (areas OC, OB, and OA), inferior temporal cortex as far forward as the temporal pole (areas TEO and TE), and the posterior part of the inferior parietal lobule (area PG). These results, which are in remarkably close agreement with our neurobehaviorally derived model of the two cortical visual pathways, have allowed us to delineate the exact limits of the entire system (Macko et al., 1981) (figure 11.1).

To trace the flow of visual information within each system we undertook a series of studies using autoradiographic and degeneration tracing techniques. Our goal in these anatomical investigations was to identify the multiple visual areas within the prestriate cortex, explore their organization, and map their projections forward into both the temporal and parietal lobes.

The findings indicated that the striate cortex is indeed the source of two major cortical projection systems. The first system begins with the known striate projection to the second visual area, V2 (Rockland and Pandya, 1981; Tigges et al., 1981; Weller and Kaas, 1981; Zeki, 1969). We found that V2 in turn projects to areas V3 and V4 (Ungerleider et al., 1983). These three prestriate areas are arranged in adjacent "belts" that nearly surround the striate cortex, and, like striate cortex, each belt contains a topographic representation of the visual field. Area V2 corresponds to prestriate area OB, while V3 and V4 are both contained within prestriate area OA, exclusive of its dorsal part. Area V4 in turn projects to both areas TEO and TE in the inferior temporal cortex (Desimone et al., 1980).

The second major system begins with both striate and V2 projections to visual area MT (Rockland and Pandya, 1981; Tigges et al., 1981; Ungerleider and Mishkin, 1979; van Essen et al., 1981; Weller and Kaas, 1981; Zeki, 1969), which is located in the caudal portion of the superior temporal sulcus, mainly within dorsolateral OA. Area MT in turn projects to four additional areas in the upper superior temporal and the intraparietal sulci (Ungerleider et al., 1982). Although the total extents of these four areas are not yet completely established, the more anterior one in the intraparietal sulcus clearly falls within area PG. Thus, one major system of projections out of striate cortex is directed ventrally into the temporal lobe, while a second is directed dorsally into the parietal lobe. Furthermore, the divergence between these two systems appears to begin almost immediately after striate cortex, i.e. in its initial projections.

The two multisynaptic projection systems that we have traced provide not only the anatomical substrate for our two functionally defined visual pathways but also a partial solution to the puzzle that was presented at the outset, namely, why extensive removals of prestriate cortex in monkeys have repeatedly failed to yield the expected losses in either object or spatial vision (Lashley, 1948; Pribram et al., 1969; Ungerleider and Mishkin, 1982). If prestriate cortex constitutes an essential relay in both a striate–temporal and a striate–parietal pathway, then damage to this relay should yield effects at least as severe as damage to both its target areas. Yet such dramatic effects have not been found. The reason appears to be that no prestriate lesion to date has produced a total visual disconnection of the temporal and parietal lobes, since all removals have spared varying extents of prestriate tissue that could continue to relay visual information. Comparison with our anatomical maps indicates that the portions of prestriate cortex that have consistently escaped damage are those parts of both the belt areas and the MT-related areas that represent the peripheral visual fields. Thus, just as we had found from sparing in striate cortex, sparing of peripheral-field representations in prestriate cortex will protect both object and spatial vision from serious losses.

5   Objects in Spatial Locations

A major question posed by the present analysis is how object information and spatial information, initially carried together in the geniculostriate projections but then analyzed separately in the two cortical visual pathways, are eventually reintegrated. As already noted, both pathways have further connections to the limbic system and the frontal lobe, and each of these target areas therefore constitutes a potential site of convergence and synthesis for object and spatial information. This theoretical possibility has not yet been sufficiently tested. Preliminary work does indicate, however, that one such site of reintegration may be the hippocampal formation and that one of its functions may be to enable the rapid memorization of the particular locations occupied by particular objects (Parkinson and Mishkin, 1982; Smith and Milner, 1981). Further application of this concept of reintegration to research on the limbic system and the frontal lobe could throw new light on some old questions of local cerebral function.

References

Allman, J. M., Baker, J. F., Newsome, W. T., and Petersen, S. E. 1981: Visual topography and function: Cortical visual areas in the owl monkey. In C. N. Woolsey (ed.), *Cortical Sensory Organization, Vol 2: Multiple Visual Areas*, Clifton, NJ: Humana Press, 171–85.

Bonin, G. von, and Bailey, P. 1947: *The Neocortex of Macaca Mulatta.* Urbana, IL: University of Illinois Press.

Cowey, A., and Gross, C. G. 1970: Effects of prestriate and inferotemporal lesions on visual discrimination by Rhesus monkey. *Experimental Brain Research*, 11, 128–44.

Denny-Brown, D., and Chambers, R. A. 1958: *Research Publications of the Association for Research in Nervous and Mental Health*, 36, 35–117.

Desimone, R., Fleming, J., and Gross, C. G. 1980: Prestriate afferents to inferior temporal cortex: An HRP study. *Brain Research*, 184, 41–55.

Gross, C. G. 1973: Visual functions of inferotemporal cortex. In R. Jung (ed.), *Handbook of Sensory Physiology VII/3*, Berlin: Springer-Verlag, 451–82.

Gross, C. G., and Mishkin, M. 1979: In S. Hatnad, R.W. Doty, L. Goldstein, J. Jaynes, and G. Krauthamer (eds), *Lateralization in the Nervous System*, New York: Academic Press, 109–22.

Haaxma, R., and Kuypers, H. G. J. M. 1975: Intrahemispheric cortical connections and visual guidance of hand and finger movements in the rhesus monkey. *Brain*, 98, 239–60.

Heilman, K. M., and Watson, R. T. 1977: Mechanisms underlying the unilateral neglect syndrome. In E. A. Weinstein and R. P. Friedland (eds), *Advances in Neurology*, vol. 18, New York: Raven Press, 93–106.

Hyvärinen, J. 1981: Regional distribution of functions in parietal association area 7 of the monkey. *Brain Research*, 206, 287–303.

Ingle, D., Schneider, G. E., Trevarthan, G. B., and Held, R. 1967: Locating and identifying: Two modes of visual processing. *Psychologische Forschung*, 31, 42–3.

Jones, B., and Mishkin, M. 1972: Limbic lesions and the problem of stimulus-reinforcement associations. *Experimental Neurology*, 36, 352–77.

Kuypers, H. G. J. M., Szwarcbart, M. K., Mishkin, M., and Rosvold, H. E. 1965: Occipitotemporal corticocortical connections in the rhesus monkey. *Experimental Neurology*, 11, 245–62.

Lashley, K. S. 1948: The mechanism of vision: XVIII. Effects of destroying the visual "associative areas" of the monkey. *Genetic Psychological Monographs*, 37, 107–66.

Macko, K. A., Jarvis, C. D., Kennedy, C., Miyaoka, M., Shinohara, M., Sokoloff, L., and Mishkin, M. 1982: Mapping the primate visual system with [2-^{14}C] deoxyglucose. *Science* 218, 394–7.

Macko, K. A., Kennedy, C., Sokoloff, L., and Mishkin, M. 1981: Limits of visually related cortex in the monkey's parietal and temporal lobes as delineated by the 2-DG technique. *Society of Neuroscience Abstracts*, 7, 832.

Mesulam, M.-M. 1981: A cortical network for directed attention and unilateral neglect. *Annals of Neurology*, 10, 309–25.

Mishkin, M. 1954: Visual discrimination performance following partial ablations of the temporal lobe: I. Ventral vs. lateral. *Journal of Comparative and Physiological Psychology*, 47, 187–93.

Mishkin, M. 1966: In R. Russell (ed.), *Frontiers of Physiological Psychology*, New York: Academic Press, 93–119.

Mishkin, M. 1972: In A. G. Karczmar and J. C. Eccles (eds), *Brain and Human Behavior*, Berlin: Springer-Verlag, 187–208.

Mishkin, M. 1982: *Philosophical Transactions of the Royal Society of London, Series B*, 298, 85–95.

Mishkin, M., Lewis, M. E., and Ungerleider, L. G. 1982: Equivalence of parieto-preoccipital subareas for visuospatial ability in monkeys. *Behavioral Brain Research*, 6, 41–55.

Mishkin, M., and Ungerleider, L. G. 1982: Contribution of striate inputs to the visuospatial functions of parieto-preoccipital cortex in monkeys. *Behavioral Brain Research*, 6, 57–77.

Mountcastle, V. B., Lynch, J. C., Georgopoulos, A., Sakata, H., and Acuna, C. 1975: Posterior parietal association cortex of the monkey: command functions for operations within extrapersonal space. *Journal of Neurophysiology*, 38, 871–908.

Newcombe, F., and Russell, W. R. 1969: Dissociated visual perceptual and spatial deficits in focal lesions of the right hemisphere. *Journal of Neurology, Neurosurgery, and Psychiatry*, 32, 73–81.

Pandya, D. N., and Kuypers, H. G. J. M. 1968: Cortico-cortical connections in the rhesus monkey. *Brain Research*, 13, 13–36.

Parkinson, J. K., and Mishkin, M. 1982: A selective mnemonic role for the hippocampus in monkeys: memory for the location of objects. *Society of Neuroscience Abstracts*, 8, 23.

Pohl, W. 1973: Dissociation of spatial discrimination deficits following frontal and parietal lesions in monkeys. *Journal of Comparative and Physiological Psychology*, 82, 227–39.

Pribram, K. H., Spinelli, D. N., and Reitz, S. L. 1969: The effects of radical disconnexion of occipital and temporal cortex on visual behavior of monkeys. *Brain*, 92, 301–12.

Robinson, D. L., Goldberg, M. E., and Stanton, G. B. 1978: Parietal association cortex in the primate: Sensory mechanisms and behavioral modulations. *Journal of Neurophysiology*, 41, 910–32.

Rockland, K. S., and Pandya, D. N. 1981: Cortical connections of the occipital lobe in the rhesus monkey: Interconnections between areas 17, 18, 19 and the superior temporal sulcus. *Brain Research*, 212, 249–70.

Seltzer, B., and Pandya, D. N. 1980: Converging visual and somatic sensory cortical input to the intraparietal sulcus of the rhesus monkey. *Brain Research*, 192, 339–51.

Semmes, J. 1967: In J. G. Bosma (ed.), *Symposium on Oral Sensation and Perception*, Springfield, IL: Thomas, 137–48.

Smith, M. L., and Milner, B. 1981: The role of the right hippocampus in the recall of spatial location. *Neuropsychologia*, 19, 781–93.

Tigges, J., Tigges, M., Anschel, S., Cross, N. A., Letbetter, W. D., and McBride, R. L. 1981: Areal and laminar distribution of neurons interconnecting the central visual cortical areas 17, 18, 19, and MT in squirrel monkey (*saimiri*). *Journal of Comparative Neurology*, 202, 539–60.

Turner, B. H., Mishkin, M., and Knapp, M. 1980: Organization of the amygdalopetal projections from modality-specific cortical association areas in the monkey. *Journal of Comparative Neurology*, 191, 515–43.

Ungerleider, L. G., Desimone, R., and Mishkin, M. 1982: Cortical projections of area MT in the macaque. *Society of Neuroscience Abstracts*, 8, 680.

Ungerleider, L. G., Gattass, R., Sousa, A. P. B., and Mishkin, M. 1983: Projections of area V2 in the macaque. *Society of Neuroscience Abstracts*, 9, 152.

Ungerleider, L. G., and Mishkin, M. 1979: The striate projection zone in the superior temporal sulcus of *macaca mulatta*: Location and topographic organization. *Journal of Comparative Neurology*, 188, 347–66.

Ungerleider, L. G., and Mishkin, M. 1982: Two cortical visual systems. In D. J. Ingle, M. A. Goodale, and R. J. W. Mansfield (eds), *Analysis of Visual Behavior*, Cambridge, MA: MIT Press, 549–86.

Van Essen, D. C., Maunsell, J. H. R., and Bixby, J. L. 1981: The middle temporal visual area in the macaque monkey: Myeloarchitecture, connections, functional properties, and topographic organization. *Journal of Comparative Neurology*, 199, 293–326.

Weller, R. E., and Kaas, J. H. 1981: Cortical and subcortical connections of visual cortex in primates. In C. N. Woolsey (ed.), *Cortical Sensory Organization, Vol. 2: Multiple Visual Areas*, Clifton, NJ: Humana Press, 121–55.

Zeki, S. M. 1969: Representation of central visual fields in prestriate cortex monkey. *Brain Research*, 14, 271–91.

Zeki, S. M. 1978: Functional specialisation in the visual cortex of the rhesus monkey. *Nature* (London), 274, 423–8.

12

Neural Mechanisms of Form and Motion Processing in the Primate Visual System

David C. Van Essen and Jack L. Gallant

Our visual system uses the information contained in the two-dimensional images on the retina to generate a rich set of perceptions about the characteristics of our external three-dimensional world. These percepts arise in a split second and require little conscious effort, yet they are the result of sophisticated information processing strategies that entail many stages of analysis. A major aim of visual neuroscience is to explain our subjective perceptions in terms of the properties of single neurons at these different processing stages. Much of our current understanding of higher visual processing comes from studies of the macaque monkey, whose visual system is in many respects similar to that of humans. The macaque visual system includes dozens of distinct areas within the cerebral cortex, plus numerous subcortical centers. The pattern of anatomical connections among these components suggests two basic organizational principles. First, the visual system is hierarchically organized but with extensive two-way information flow between levels. Second, each hierarchical stage contains multiple subdivisions, which are embodied as separate areas, separate anatomical compartments within an area, or (at subcortical levels) different morphological cell classes. Subdivisions at different levels are connected in specific patterns, giving rise to distinct processing streams which can be traced through many hierarchical stages.

This anatomical organization raises two broad questions about functional specialization within the visual system. First, what are the functions of the different processing streams? Second, how do the response characteristics of individual neurons differ at successive stages of the anatomical hierarchy? These questions have been tackled most successfully in relation to two key aspects of vision: the perception of form (the shapes of objects) and of motion (where things are moving). These explorations mainly involve physiological and behavioral techniques that are quite different from the cellular and molecular techniques most familiar to this journal's

[*Neuron*] readership. Nonetheless, we hope that a review of recent progress in understanding visual cortex will interest a broad spectrum of neuroscientists who share the ultimate objective of attaining a continuum of explanations of brain function, from the most molecular to the most cognitive levels.

We will begin with a brief overview of the anatomical organization of the visual system. The remainder of the review will discuss the functional distinctions among the different processing streams, starting in the retina but emphasizing the early and intermediate stages of cortical processing.

1   The Visual Cortex is a Distributed Hierarchical System

The general layout of the macaque visual system can best be visualized on a two-dimensional unfolded representation of its different subcortical and cortical centers (figure 12.1). The major subcortical centers (lower left) include the two retinae, the lateral geniculate nucleus (LGN), the superior colliculus (SC), and the pulvinar complex. In the cortex, 32 distinct areas associated with vision have been identified, which collectively occupy more than half of the total surface area of the neocortex in the macaque (Felleman and Van Essen, 1991; see also Desimone and Ungerleider, 1989). Areas specifically mentioned in this review are indicated by different shading in figure 12.1; the remaining visual areas are unlabeled and shown in gray.

At the earliest stage of cortical processing are the two largest areas, V1 and V2, each of which occupies about 10 percent of the neocortex. At intermediate stages, there are more than a dozen visual areas, but only four of them will be discussed here: area V4, the middle temporal area (MT), and the dorsal and lateral subdivisions of the medial superior temporal complex (MSTd and MSTℓ, respectively). The highest stages of processing include two distinct clusters of areas – one located in the inferotemporal (IT) cortex and the other in the posterior parietal (PP) cortex.

Many visual areas contain internal compartments, or modules, that have been revealed by a combination of anatomical and physiological techniques. Figure 12.1 illustrates the spatial pattern and the relative sizes of these modules in five different areas. In V1, modularity is most clearly manifested by a fine-grained pattern of "blobs" and "interblobs" that can be identified histochemically and by their pattern of connections (Livingstone and Hubel, 1984). The modules that have been identified in higher areas (V2, V4, MT, and posterior IT; see below) are larger in size and differ in their topology. For example, in V2 they form stripes rather than the spotted pattern found in V1.

The 32 visual areas of the macaque are interconnected by more than 300 distinct cortico-cortical pathways. The great majority of these pathways are reciprocal; for example, V1 projects to V2, and V2 projects back to V1. In most cases, there are pronounced asymmetries in the specific cortical layers in which reciprocal pathways originate and terminate. On this basis, the entire collection of visual areas can be arranged in a hierarchy that contains ten distinct levels of cortical processing (Felleman and Van Essen, 1991). Figure 12.2 illustrates a simplified hierarchical

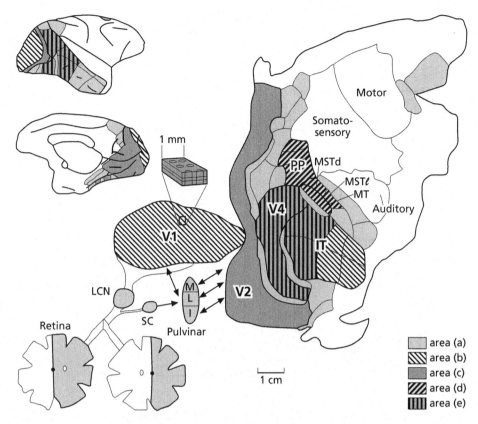

Figure 12.1 Two-dimensional map of the cerebral cortex and major subcortical visual centers in the macaque monkey. The flattened cortical map encompasses the entire right hemisphere. Shaded areas are discussed in the text; remaining visual areas are shown in gray. The cortical map also shows schematic representations of the various compartmental subdivisions that have been identified within individual visual areas, including the blobs, interblobs, and layer 4B of V1; the thick stripes, thin stripes, and interstripes of V2; and the recently described subdivisions of V4, MT, and posterior IT. The spatial configuration of compartments is illustrated schematically but approximately to scale. Partitioning within and between areas is based on information and references provided in Felleman and Van Essen, 1991; Van Essen et al., 1992; Van Essen and DeYoe, 1994; and DeYoe et al., submitted. Modified, with permission, from Felleman and Van Essen, 1991; and Van Essen et al., 1992.

scheme that includes only the subset of visual areas shown by shading in figure 12.1, but that also includes the various compartments within these areas which contribute to the different processing streams.

Three major processing streams have been identified at the subcortical level. In the lateral geniculate nucleus, these streams are represented by distinct populations of cells residing in different layers. The most prominent of these are the parvocellular (P) and magnocellular (M) layers. In addition, there is a small population of koniocellular (K) neurons whose function remains poorly understood (Casagrande

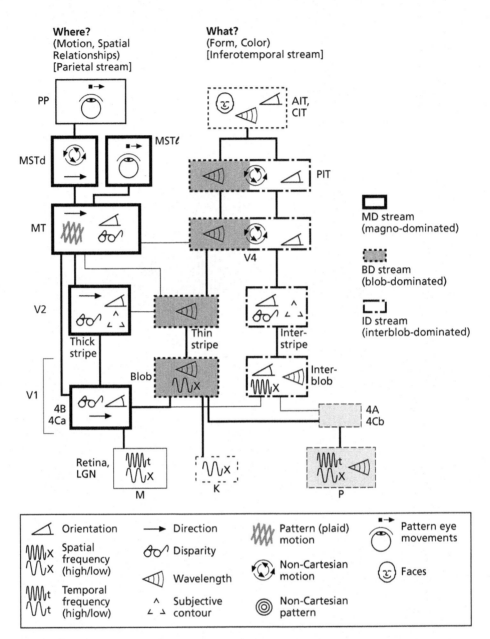

Figure 12.2 Hierarchical organization of concurrent processing streams in the macaque monkey. Boxes represent visual areas, compartments within an area, and subcortical centers; solid lines represent major connections between structures (usually reciprocal pathways); and icons represent characteristic neurophysiological properties. Subcortical streams in the retina and lateral geniculate nucleus (LGN) include the M, K, and P streams. Cortical streams at early and intermediate stages include the MD, BD, and ID streams. The IT complex includes posterior inferotemporal areas (PIT), which are components of the BD and ID streams, and central and anterior areas (CIT and AIT). Pathways are based on information cited in Felleman and Van Essen, 1991; and Van Essen and DeYoe, 1994. Physiological specializations are based on information cited in Felleman and Van Essen, 1987; DeYoe and Van Essen, 1988; and in the present text.

and Norton, 1991; Hendry and Yoshioka, 1994). As they converge in V1, the three subcortical streams are reorganized into three new streams, which can be traced through several subsequent stages of the extrastriate cortex (see Van Essen and DeYoe, 1994). The blob-dominated (BD) and interblob-dominated (ID) streams are named after the compartments in V1 from which they originate. The magnodominated (MD) stream is named after its dominant source of subcortical inputs. At higher stages of the visual hierarchy, there is another reorganization into separate dorsal and ventral streams (Ungerleider and Mishkin, 1982; Desimone and Ungerleider, 1989). The dorsal stream is largely an extension of the MD stream, and it projects to the PP cortex, which is involved in the analysis of spatial relations (where things are). The ventral stream includes the BD and ID streams, and it projects to the IT complex, which is involved in pattern recognition (what things are). The outputs of the visual cortex are distributed to various other cortical areas in the temporal and frontal lobes and to numerous subcortical structures, which are all outside the scope of this review.

Overall, these anatomical findings constitute a progress report that represents a major revision of our understanding of the organization of the primate visual cortex. Until the 1970s, it was widely presumed that the visual cortex contained only three subdivisions (areas 17, 18, and 19), which were linked by a strictly serial (unidirectional) flow of information, from area 17 to 18 to 19. Starting with the work of a few early pioneers (see Allman and Kaas, 1974; Zeki, 1975), this classic picture has evolved into our current scheme, which includes dozens of visual areas, many of which have an internal modular structure and a principle of distributed hierarchical organization that includes bidirectional information flow and multiple processing streams.

2 Neural Function Can Be Analyzed by Lesions and Single-neuron Recordings

The hierarchical scheme illustrated in figure 12.2 provides a basic structural framework for addressing questions about the specific functions associated with different streams and different stages of processing. The various icons within the figure summarize much of what is known about the distribution of different physiological and functional characteristics in the macaque visual system. This information derives mainly from two complementary approaches: lesion studies and single-unit recording.

Lesion studies assess the behavioral deficits resulting from ablation or inactivation of a specific cortical or subcortical structure. This approach can suggest how the information processed within the lesioned structure actually contributes to different perceptual and visuomotor tasks. However, it has inherent limitations, one of which is that simply identifying a deficit says little about the detailed neural processing carried out within the lesioned region.

Single-unit neurophysiological recordings are used to determine the specific stimulus characteristics that are effective in eliciting responses from neurons in any given region. Typically, the first step is to determine the region of the visual field from which a cell can be directly activated (the classical receptive field). Various stimuli are then presented within this receptive field, and the firing rate of the cell is determined in response to each stimulus. There are an infinite number of possible visual stimuli, but usually no more than a few hundred can be presented to any given cell during the limited period it is available for study. Choosing stimuli appropriate for investigation of any particular issue is a major challenge, and the choice often depends on the investigator's assumptions about the way information is encoded by individual neurons. Some tend to regard visual neurons as feature detectors that convey information about the likelihood that a particular trigger feature is present in the image (Barlow, 1972; Tanaka, 1993). From this perspective, a key objective is to determine what particular stimulus is most effective in driving any given cell. Others tend to regard visual neurons as filters, which are typically tuned along more than one stimulus dimension (orientation, wavelength, etc.). This perspective puts greater emphasis on determining which stimulus dimensions are most relevant to each cell and on characterizing the tuning curves along each of these dimensions. The underlying assumption is that differences in firing rate convey information useful for discriminating among stimuli that lie on the slopes of each cell's multidimensional tuning surface. Our emphasis will be on the filtering point of view, which has been very useful in computational vision (see Marr, 1982) and is becoming more widely accepted in neurophysiology.

3 Low-level Analysis Occurs in the Subcortical Streams

The different subcortical processing streams carry out a basic division of labor that has been revealed by neurophysiological recordings and by behavioral tests after selective lesions (see Merigan and Maunsell, 1993; Van Essen et al., 1992). This division is best understood for the P and M streams, which differ in the information they carry along several low-level stimulus dimensions.

P cells are the most numerous cell type, and they provide a higher-resolution (finer-grained) spatial representation of the visual world than the less numerous M cells. In the time domain, P cells convey information mainly about relatively slow changes (low temporal frequencies, corresponding to patterns that are static or moving slowly), whereas M cells emphasize relatively rapid changes (high temporal frequencies, corresponding to rapid motion or sudden transients). In the color domain, P cells convey most of the information about the colors of different objects, and M cells play only a minor role.

Given that the P and M streams are selective for distinct yet partially overlapping aspects of the information contained in images, it should not be surprising that each stream contributes to many different aspects of perception (Schiller et al., 1990; Merigan and Maunsell, 1993). For example, if P cells are selectively lesioned, a monkey cannot discriminate colors, but it can discriminate different shapes, textures, depths, and motion patterns, as long as the stimuli contain the appropriate spatial and temporal frequencies for activating cells in the surviving M stream. Likewise, if M cells are selectively lesioned, a monkey can discriminate different shapes, textures, depths, and even motion patterns, as long as the stimuli contain the appropriate spatial and temporal frequencies for activating cells in the surviving P stream.

4   Motion Analysis is Emphasized in the MD Stream

The MD stream (figure 12.2) includes portions of V1 and V2 plus several higher extrastriate areas (MT, MSTd, and MSTℓ). At all stages of the MD stream, there is a relatively high incidence of cells selective for the direction of stimulus motion (arrow icon). The highest incidence of direction selectivity occurs in MT, whose involvement in motion analysis has been established more directly by lesion and microstimulation experiments. In particular, lesions of MT lead to specific deficits in motion perception and also in the smooth eye movements used to track a moving target (Newsome et al., 1985; Newsome and Paré, 1988). Electrical stimulation of a small region of MT (by passing current through a microelectrode) can markedly perturb a monkey's perception of the direction of stimulus motion, causing the monkey to report, for example, leftward movement for a stimulus that is actually moving to the right (Salzman et al., 1992; Salzman and Newsome, 1994).

When tested with simple stimuli (such as moving spots, bars, or gratings), most cells in MT are tuned relatively broadly for the direction of stimulus motion and also for stimulus speed. On the presumption that MT is likely to be involved in higher-order motion analysis as well, a number of investigators have studied the responses of MT neurons to more complex motion patterns. One such example involves plaid patterns, which consist of two superimposed gratings that differ in orientation and direction of movement. Under some conditions, human observers perceive the two gratings as a pair of transparent gratings moving in different directions. Under other conditions, the percept is of a single coherent plaid pattern moving in a direction different from either of the component gratings. Some cells in MT have responses that correlate with the subjective reports of human observers, suggesting that they play a role in the representation of complex motion (see figure 12.2; Movshon et al., 1985; Stoner and Albright, 1993).

Our visual system uses motion information for a variety of purposes. One useful distinction is between the local motion signals used to analyze the trajectory of an individual object and the global motion flow fields that arise when we move our eyes or navigate through the environment. Neurophysiological studies suggest that these two types of motion analysis are segregated into distinct subregions within MT (see figure 12.1), whose outputs are then distributed to separate visual areas, MSTd and MSTℓ, at the next stage of the hierarchy (Born and Tootell, 1992; Komatsu and Wurtz, 1988a, 1988b).

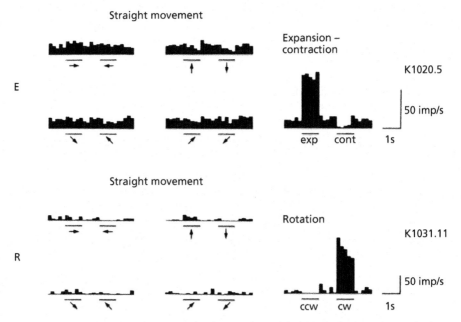

Figure 12.3   Responses of two cells in MSTd to different Cartesian and non-Cartesian patterns of motion. As indicated by the histograms of neural firing rate versus time, the cell illustrated on the top (E) responded vigorously to a pattern of random dots when it was expanding outward (exp) but not when it was contracting inward (cont) or when it moved uniformly in different directions (arrows). The cell illustrated on the bottom (R) responded vigorously to clockwise (cw) rotation of a random dot pattern but not to counterclockwise (ccw) rotation of the same pattern or to uniform motion in different directions (arrows). Reprinted, with permission, from Tanaka and Saito, 1989.

The motion flow fields that arise during navigation typically include components of image expansion (when moving forward), contraction (when moving backward), rotation (when tilting the head), and shear (when moving past objects at different distances). The characteristics of cells in MSTd suggest a specific involvement in the analysis of these types of flow fields (Saito et al., 1986; Duffy and Wurtz, 1991a; Orban et al., 1992). This is illustrated in figure 12.3 for two different neurons (Tanaka and Saito, 1989). The neuron on the top responded vigorously to an expanding motion field but not to contraction or ordinary linear motion. The neuron on the bottom responded best to clockwise rotation but not to counterclockwise rotation or linear motion. We refer to cells of this general type as non-Cartesian motion cells (rotation icon in figure 12.2) because they are tuned along dimensions that are described most simply in polar or hyperbolic coordinates rather than conventional Cartesian (x,y) coordinates. However, motion analysis in MSTd involves more than just extracting the components of motion along the cardinal non-Cartesian axes (pure rotation, pure expansion, etc.). In particular, some MSTd neurons are selective for spiral motion patterns, which have components of both rotation and expansion (Graziano et al., 1994; see also Duffy and Wurtz, 1991a, 1991b). Thus, it appears that MSTd contains a population of motion filters whose peaks are scattered throughout a multidimensional stimulus space. This strategy for representing higher-order motion information is presumably useful for mediating the variety of behavioral tasks (avoiding obstacles, guiding eye and limb movements, etc.) that are carried out during navigation through the environment.

The major outputs of the MD stream are directed dorsally to areas in the parietal lobe. Lesion studies and physiological recordings indicate that the PP cortex is involved in a variety of high-level functions. These include analyzing spatial relations (where things are located), controlling eye movements, and determining where visual attention is to be directed (see Mountcastle et al., 1975; Ungerleider and Mishkin, 1982; Andersen, 1989). Although many details remain to be worked out, the various areas within the PP complex appear to be specialized for handling distinct (but partially overlapping) subsets of this constellation of visuomotor and attentional tasks (see Andersen, 1989; Colby et al., 1993).

5   Form and Color Analysis Are Emphasized in the Ventral Stream

The ventral stream encompasses the BD and ID streams (see figure 12.2), which parallel one another over four stages of the cortical hierarchy, from V1 to the posterior IT complex (DeYoe, Felleman, Van Essen, and McClendon, 1994). Originally, both streams were thought to be dominated by inputs from the subcortical P stream. More recent evidence indicates a much greater degree of convergence, with the ID stream receiving inputs from two subcortical streams and the BD stream from all three subcortical streams (figure 12.2; see Nealey and Maunsell, 1994; Van Essen and DeYoe, 1994).

Physiological recordings and lesion studies implicate the ventral stream in the analysis of both form and color, but it has proven difficult to determine the specific role of the individual streams and individual areas in these different aspects of perception. There is some physiological evidence for a functional dichotomy in which the BD stream mediates color perception, and the ID stream mediates form perception (Livingstone and Hubel, 1984, 1988). However, there are difficulties with this hypothesis because there is significant intermixing of receptive field characteristics in different streams (see DeYoe and Van Essen, 1988; Peterhans and von der Heydt, 1993) and because blobs in V1 (and other structures in the BD stream) are present in the owl monkey, a primate species that has no color vision (Tootell et al., 1985). An alternative hypothesis is that the BD stream analyzes the surface characteristics of objects (including their texture and brightness as well as their color), whereas the ID stream analyzes the shapes of objects, based on their boundaries rather than their interiors (see Grossberg, 1987; Allman and Zucker, 1990; Van Essen and DeYoe, 1994).

At present, little is known about functional differences between the BD and ID streams at higher stages of the hierarchy (V4 and beyond) that would help to distinguish between these or other hypotheses. It has been suggested that V4 as a whole is primarily involved in color vision, both in the macaque (Zeki, 1973, 1983) and in a putative homologous area in humans (Zeki et al., 1991). However, lesions of V4 in the macaque have much greater effects on pattern discrimination than on color discrimination tasks (Schiller et al., 1990; Schiller and Lee, 1991; Heywood et al., 1992; Merigan, 1993). This suggests that V4 is more important for form vision than for color vision.

Physiological recordings from V4 have provided additional evidence in support of its involvement in form vision (Desimone and Schein, 1987). To obtain a more systematic analysis of how form information is represented in V4, we have used a novel class of visual stimuli that are more complex than conventional bars and gratings (Gallant et al., 1993; Gallant et al., 1996). These include concentric, radial, spiral, and hyperbolic gratings, which we collectively refer to as non–Cartesian gratings. Although the stimuli we used were stationary rather than moving, they are mathematically analogous to the non–Cartesian motion stimuli successfully used in the studies of MSTd mentioned above (concentric gratings are analogous to rotation, radial gratings to expansion, etc.). As illustrated in figure 12.4, we found many cells in V4 that responded better to certain non–Cartesian gratings than to any conventional sine wave (Cartesian) grating. The stimuli in this figure are grouped into three subsets.

Cartesian gratings (left) vary in orientation and spatial frequency; hyperbolic gratings (center) also vary in orientation and spatial frequency; and polar gratings (concentric, radial, and spiral; right) vary in concentric and radial frequency. Each stimulus is color-coded according to the mean response that it elicited from the cell. Blue corresponds to stimuli that were relatively ineffective, and red to stimuli that were maximally effective (spiral patterns for the cell in figure 12.4A and hyperbolic patterns for the cell in figure 12.4B). Cells preferring polar gratings were more common than cells preferring hyperbolic gratings, and within this subpopulation, concentric preferences were more common than radial preferences. Many cells responded to a wide range of stimuli within both the Cartesian and non–Cartesian stimulus spaces. We suspect that these cells would also respond to a variety of other patterns containing appropriate combinations of curvature cues (e.g. curved edges, intersections, and other patterns that are not pure gratings). We consider the most significant aspect of non–Cartesian form cells to be their orderly tuning across a set of higher-order stimulus dimensions rather than their best stimulus *per se*. An interesting possibility is that V4 might include cells tuned along additional higher-order dimensions distinct from the particular non–Cartesian dimensions illustrated here. Testing this hypothesis will require the development of new classes of visual stimuli that capture the complex aspects of shape information contained in natural images.

The IT complex (figure 12.2) includes posterior, central, and anterior subdivisions that have been collectively implicated in pattern recognition. Lesions of the

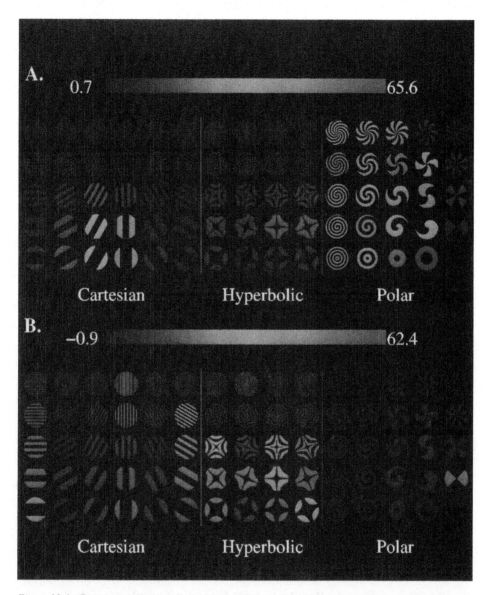

Figure 12.4 Responses of two cells in V4 to different Cartesian and non-Cartesian spatial patterns. Each icon represents a particular visual stimulus, and its color represents the mean response to that stimulus, relative to the spontaneous background rate, using the color scale shown above. (A) A neuron that responded maximally to polar stimuli, particular spirals, and much less to hyperbolic and Cartesian gratings. Modified, with permission from Gallant et al., 1993. (B) A neuron that responded maximally to hyperbolic stimuli of low to moderate spatial frequencies. J. G. and D. V. E., unpublished data.

IT cortex impair the ability to recognize previously learned patterns (Ungerleider and Mishkin, 1982). Many cells in the IT cortex respond best to complex shapes, including various geometrical stimuli (Fujita et al., 1992; Tanaka, 1993) and natural shapes such as faces or hands (Perrett et al., 1982; Desimone et al., 1984). These cells generally respond to a range of stimuli (e.g. several different faces or hands when tested with a large number of stimuli), suggesting that they are tuned along dimensions associated with certain classes of natural stimuli but not for unique examples within that class. It is unlikely that the complex receptive field characteristics encountered in higher visual areas are genetically hardwired. Instead, these properties are probably related to visual experience so that the specific pattern of higher-order receptive fields present in any particular brain depends upon the individual's particular sensory history. Indeed, recent experiments have shown that the selectivity of cells in the anterior IT cortex can be markedly altered by repeated exposure to a limited set of behaviorally relevant stimuli (Miyashita, 1993; Sakai and Miyashita, 1993).

6   Concluding Remarks

Functional specialization reflects the diversity of visual cues and visual tasks

To illustrate the nature of functional specialization in the visual cortex, we have emphasized the roles of the MD stream in motion analysis and the ID and BD streams in form and color analysis. However, each stream contains a diversity of response characteristics suggestive of involvement in a range of visual tasks (Peterhans and von der Heydt, 1993; Sary et al., 1993). These characteristics make sense, given that high-level visual functions generally require access to a variety of low-level cues, and low-level cues can contribute to a variety of perceptual and visuomotor tasks. Viewed from this perspective, the extensive convergence, divergence, and cross talk between processing streams provide an anatomical substrate that enables the visual system to perform a wide variety of tasks in a complex visual environment (DeYoe and Van Essen, 1988; Van Essen et al., 1992).

There are parallels between form and motion analysis

We have noted a parallel between the non-Cartesian spatial dimensions linked to form analysis in V4 and the non-Cartesian spatiotemporal dimensions linked to motion analysis in MSTd. The specific perceptual functions of non-Cartesian selectivity in V4 and MSTd are surely quite different, and there are many differences in the specific receptive field properties of each area. Yet, the neural computations underlying higher-level form and motion analysis may have important features in common, as has been suggested for the neural circuitry underlying conventional orientation and direction selectivity in V1 (Douglas and Martin, 1991). If such commonalities indeed exist, it may help to explain how the cerebral neocortex, which is characterized by a remarkable degree of structural uniformity, is able to carry out such a wide diversity of functionally distinct tasks.

Visual neurons are multidimensional filters

At every stage of the visual hierarchy, neurons act as filters that are tuned along several distinct dimensions. At early cortical levels, the dimensions, such as orientation and spatial frequency, tend to be relatively simple. At higher levels, inputs are combined in ways that lead to selectivity along more complex dimensions, such as the non-Cartesian dimensions relevant for V4 and MSTd. In addition, as one moves up the visual hierarchy, responses become progressively less dependent on exactly where stimuli are placed in the visual field and progressively more dependent on attention and other aspects of behavioral state (see Maunsell and Newsome, 1987; Miyashita, 1993; Connor et al., 1997; Motter, 1994). Treating neurons as filters rather than feature detectors has inherent analytic advantages even when dealing with these highly nonlinear aspects of neural selectivity.

Biological systems are complex but decipherable

The organization and function of the primate visual system is far more complex than most neuroscientists appreciated as recently as a decade ago. Similar statements can be made for many important topics in cellular, molecular, systems, and developmental neuroscience. Processes that were once discussed in relatively simple mechanistic terms have been shown to involve complex interacting components, whether they be transmitters, channels, and trophic factors (in studies focusing on the cellular/molecular level), or cell classes, receptive field types, and cortical areas (in studies focusing on the systems level). Whatever the level of analysis, it remains a great challenge to understand how these diverse components interact dynamically with one another to produce coordinated, reliable performance in a living organism.

Acknowledgments

We thank Drs J. Sanes, J. Lichtman, D. Gottlieb, and C. E. Connor for insightful comments on the manuscript. Work from the authors' laboratory was supported by grants from NIH (EY02091) and ONR (NO0014-89-J-1192).

References

Allman, J. M., and Kaas, J. H. 1974: A crescent-shaped cortical visual area surrounding the middle temporal area (MT) in the owl monkey (*Aotus trivirgatus*). *Brain Research*, 81, 199–213.

Allman, J., and Zucker, S. 1990: Cytochrome oxidase and functional coding in primate striate cortex: A hypothesis. *Cold Spring Harbor Symposia on Quantitative Biology*, 55, 979–82.

Andersen, R. A. 1989: Visual and eye movement functions of the posterior parietal cortex. *Annual Review of Neuroscience*, 12, 377–403.

Barlow, H. B. 1972: Single units and sensation: A neuron doctrine for perceptual psychology? *Perception*, 1, 371–94.

Born, R. T., and Tootell, R. B. H. 1992: Segregation of global and local motion processing in primate middle temporal visual area. *Nature*, 357, 497–9.

Casagrande, V. A., and Norton, T. T. 1991: Lateral geniculate nucleus: A review of its physiology and function. In A. G. Leventhal (ed.), *Vision and Visual Dysfunction*, New York: Macmillan Press, 41–84.

Colby, C. L., Duhamel, J. R., and Goldberg, M. E. 1993: Ventral intraparietal area of the macaque: Anatomic location and visual response properties. *Journal of Neurophysiology*, 3, 902–14.

Connor, C. E., Preddie, D. G., Gallant, J. L., and Van Essen, D. C. 1997: Spatial attention effects in macaque area V4. *Journal of Neurophysiology*, 17, 3201–14.

Desimone, R., Albright, T. D., Gross, C. G., and Bruce, C. 1984: Stimulus-selective properties of inferior temporal neurons in the macaque. *Journal of Neuroscience*, 4, 2051–62.

Desimone, R., and Schein, S. J. 1987: Visual properties of neurons in area V4 of the macaque: Sensitivity to stimulus form. *Journal of Neurophysiology*, 57, 835–68.

Desimone, R., and Ungerleider, L. 1989: Neural mechanisms of visual processing in monkeys. In F. Boller and J. Graman (eds), *Handbook of Neuropsychology*, Amsterdam: Elsevier, 267–99.

DeYoe, E. A., and Van Essen, D. C. 1988: Concurrent processing streams in monkey visual cortex. *Trends in Neuroscience*, 11, 219–26.

DeYoe, E. A., Felleman, D. J., Van Essen, D. C., and McClendon, E. 1994: Multiple processing streams in occipito-temporal visual cortex. *Nature*, 371, 151–4.

Douglas, R. J., and Martin, K. A. C. 1991: A functional microcircuit for cat visual cortex. *Journal of Physiology*, 440, 735–69.

Duffy, C. J., and Wurtz, R. H. 1991a: Sensitivity of MST neurons to optic flow stimuli. 1. A continuum of response selectivity to large-field stimuli. *Journal of Neurophysiology*, 65, 1329–45.

Duffy, C. J., and Wurtz, R. H. 1991b: Sensitivity of MST neurons to optic flow stimuli. II. Mechanisms of response selectivity revealed by small-field stimuli. *Journal of Neurophysiology*, 65, 1346–59.

Felleman, D. J., and Van Essen, D. C. 1987: Receptive field properties of neurons in area V3 of macaque monkey extrastriate cortex. *Journal of Neurophysiology*, 57, 889–920.

Felleman, D. J., and Van Essen, D. C. 1991: Distributed hierarchical processing in primate visual cortex. *Cerebral Cortex*, 1, 1–47.

Fujita, I., Tanaka, K., Ito, M., and Cheng, K. 1992: Columns for visual features of objects in monkey inferotemporal cortex. *Nature*, 360, 343–6.

Gallant, J. L., Braun, J., and Van Essen, D. C. 1993: Selectivity for polar, hyperbolic, and Cartesian gratings in macaque visual cortex. *Science*, 259, 100–3.

Gallant, J. L., Connor, C. E., Rakshit, S., Lewis, J., and Van Essen, D. C. 1996: Neural responses to polar, hyperbolic, and Cartesian gratings in area V4 of the macaque monkey. *Journal of Neurophysiology*, 76, 2718–37.

Graziano, M. S. A., Andersen, R. A., and Snowden, R. J. 1994: Tuning of MST neurons to spiral motions. *Journal of Neuroscience*, 14, 54–67.

Grossberg, S. 1987: Cortical dynamics of three-dimensional form, color, and brightness perception: 1. Monocular theory. *Perceptual Psychophysiology*, 41, 87–116.

Hendry, S. H. C., and Yoshioka, T. 1994: A neurochemically distinct third channel in the macaque dorsal lateral geniculate nucleus. *Science*, 264, 575–7.

Heywood, C. A., Gadotti, A., and Cowey, A. 1992: Cortical area V4 and its role in the perception of color. *Journal of Neuroscience*, 12, 4056–65.

Komatsu, H., and Wurtz, R. H. 1988a: Relation of cortical areas MT and MST to pursuit eye movements: 1. Localization and visual properties of neurons. *Journal of Neurophysiology*, 60, 580–603.

Komatsu, H., and Wurtz, R. H. 1988b: Relation of cortical areas MT and MST to pursuit eye movements: Ill. Interaction with full field visual stimulation. *Journal of Neurophysiology*, 60, 621–44.

Livingstone, M. S., and Hubel, D. H. 1984: Anatomy of physiology of a color system in the primate visual cortex. *Journal of Neuroscience*, 41, 309–56.

Livingstone, M., and Hubel, D. 1988: Segregation of form, color, movement, and depth: anatomy, physiology, and perception. *Science*, 240, 740–9.

Marr, D. 1982: *Vision*. San Francisco: Plenum.

Maunsell, J. H. R., and Newsome, W. T. 1987: Visual processing in monkey extrastriate cortex. *Annual Review of Neuroscience*, 10, 363–402.

Merigan, W. H. 1993: Human V4? Lesion studies of the macaque and human visual cortices again fail to demonstrate a simple parallel between the two. *Current Biology*, 3, 226–9.

Merigan, W. H., and Maunsell, J. H. R. 1993: How parallel are the primate visual pathways? *Annual Review of Neuroscience*, 16, 369–402.

Miyashita, M. 1993: Inferior temporal cortex: Where visual perception meets memory. *Annual Review of Neuroscience*, 16, 245–64.

Motter, B. C. 1994: Neural correlates of attentive selection for color or luminance in extrastriate area V4. *Journal of Neuroscience*, 14, 2178–89.

Mountcastle, V. B., Lynch, J. C., Georgopoulos, A., Sakata, H., and Acuna, C. 1975: Posterior parietal association cortex of the monkey: Command functions for operations within extrapersonal space. *Journal of Neurophysiology*, 38, 871–908.

Movshon, J. A., Adelson, E. H., Gizzi, M. S., and Newsome, W. T. 1985: The analysis of moving visual patterns. In C. Chagas, R. Gattass, and C. Gross (eds), *Pattern Recognition Mechanisms*, Rome: Vatican Press, 117–51.

Nealey, T. A., and Maunsell, J. H. R. 1994: Magnocellular and parvocellular contributions to the responses of neurons in macaque striate cortex. *Journal of Neuroscience*, 14, 2069–79.

Newsome, W. T., and Paré, E. B. 1988: A selective impairment of motion perception following lesions of the middle temporal visual area (MT). *Journal of Neuroscience*, 8, 2201–11.

Newsome, W. T., Wurtz, R. H., Dursteler, M. R., and Mikami, A. 1985: Deficits in visual motion processing following ibotenic acid lesions of the middle temporal visual area of the macaque monkey. *Journal of Neuroscience*, 5, 825–40.

Orban, G. A., Lagae, L., Verri, A., Raiguel, S., Xiao, D., Maes, H., and Torre, V. 1992: First-order analysis of optical flow in monkey brain. *Proceedings of the National Academy of Sciences (USA)*, 89, 2595–9.

Perrett, D. I., Rolls, E. T., and Caan, W. 1982: Visual neurons responsive to faces in the monkey temporal cortex. *Experimental Brain Research*, 47, 329–42.

Peterhans, E., and von der Heydt, R. 1993: Functional organization of area V2 in the alert macaque. *European Journal of Neuroscience*, 5, 509–24.

Saito, H.-A., Yuki, M., Tanaka, K., Hikosaka, K., Fukada, Y., and Iwai, E. 1986: Integration of direction signals of image motion in the superior temporal sulcus of the macaque monkey. *Journal of Neuroscience*, 6, 145–57.

Sakai, K., and Miyashita, Y. 1993: Memory and imagery in the temporal lobe. *Current Biology*, 3, 166–70.

Salzman, C. D., and Newsome, W. T. 1994: Neural mechanisms for forming a perceptual decision. *Science*, 264, 231–7.

Salzman, C. D., Murasagi, C. M., Britten, K. H., and Newsome, W. T. 1992: Microstimulation in visual area MT: Effects of direction discrimination performance. *Journal of Neuroscience*, 12, 2331–85.

Sary, G., Vogels, R., and Orban, G. A. 1993: Cue-invariant shape selectivity of macaque inferior temporal neurons. *Science*, 260, 995–7.

Schiller, P. H., and Lee, K. 1991: The role of primate extrastriate area V4 in vision. *Science*, 251, 1251–3.

Schiller, P. H., Logothetis, N. K., and Charles, E. R. 1990: Role of the color-opponent and broad-band channels in vision. *Visual Neuroscience*, 5, 321–46.

Stoner, C. R., and Albright, T. D. 1993: Image segmentation cues in motion processing: Implications for modularity in vision. *Journal of Cognitive Neuroscience*, 5, 129–49.

Tanaka, K. 1993: Neuronal mechanisms of object recognition. *Science*, 262, 685–8.

Tanaka, K., and Saito, H. A. 1989: Analysis of motion of the visual field by direction, expansion/contraction, and rotation cells clustered in the dorsal part of the medial superior temporal area of the macaque monkey. *Journal of Neurophysiology*, 62, 626–41.

Tootell, R., Hamilton, S., and Silverman, M. 1985: Topography of cytochromeoxidase activity in owl monkey cortex. *Journal of Neuroscience*, 51, 2786–800.

Ungerleider, L. G., and Mishkin, M. 1982: Two cortical visual systems. In D. G. Ingle, M. A. Goodale, and R. J. Q. Mansfield (eds), *Analysis of Visual Behavior*, Cambridge, MA: MIT Press, 549–86.

Van Essen, D. C., and DeYoe, E. A. 1994: Concurrent processing in the primate visual cortex. In M. S. Gazzaniga (ed.), *The Cognitive Neurosciences*. Cambridge, MA: MIT Press.

Van Essen, D. C., Anderson, C. H., and Felleman, D. J. 1992: Information processing in the primate visual system: An integrated systems perspective. *Science*, 255, 419–23.

Zeki, S. M. 1973: Colour coding in rhesus monkey prestriate cortex. *Brain Research*, 53, 422–7.

Zeki, S. M. 1975: The functional organization of projections from striate to prestriate visual cortex in the rhesus monkey. *Cold Spring Harbor Symposium on Quantitative Biology*, 40, 591–600.

Zeki, S. 1983: Colour coding in the cerebral cortex: The responses of wavelength-selective and colour-coded cells in monkey visual cortex to changes in wavelength composition. *Neuroscience*, 9, 767–81.

Zeki, S., Watson, J. D. G., Lueck, C. J., Friston, K. J., Kennard, C., and Frackowiak, R. S. J. 1991: A direct demonstration of functional specialization in human visual cortex. *Journal of Neuroscience*, 11, 641–9.

Decomposing and Localizing Vision: An Exemplar for Cognitive Neuroscience

William Bechtel

To date the greatest successes in developing a brain-based model of a cognitive process have been in the domain of vision. As chapter 12, above, by Van Essen and Gallant, makes clear, by now at least 33 different cortical areas involved in visual processing have been identified in primates, with a highly complex pattern of connectivity between them. Their figure (12.2, this volume) strongly conveys the idea that different visual areas each carry out different types of processing, and that the ability to see the world is the product of a complex system in which information processing tasks are divided up, but in which the specialized components collaborate. Although not all the component operations and brain areas involved in vision have been discovered, and the account offered so far is subject to revision in the face of new research, it is nonetheless relatively complete and well supported. Moreover, from studying figure 12.2, one develops an understanding of how processes in the brain make it possible for us to see the world. As such, it offers what Thomas Kuhn refers to as an exemplar – a example of successful research which provides a model to be emulated by other domains of cognitive science.

The account of visual processing that has emerged fits the framework of *mechanistic explanation* we introduced in chapter 1, this volume – it decomposes seeing into a number of different operations and localizes each of them in different parts of the brain. Analyzing the history of research that led to this account provides us an exemplar in a second sense – an exemplar of the path cognitive neuroscience research is likely to take. The previous three chapters in this section represent major stages in the development of current mechanistic explanations of vision. But they represent only some of the steps in the overall process by which researchers arrived at this understanding of how the visual system works. In this chapter I will relate additional components of that history to illustrate important steps in developing mechanistic explanations.

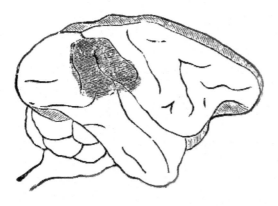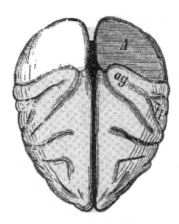

Figure 13.1 Competing proposed localizations of visual processing based on lesions made in monkeys. David Ferrier claimed that lesions to the angular gyrus (left) resulted in blindness whereas Hermann Munk argued that lesions to the occipital lobe (right) generated blindness.

Our experience of vision is simply one of opening our eyes and seeing what is before us. We are not aware of intermediate operations that the brain is performing. But this should not be surprising. Natural systems typically work very smoothly and fail to reveal their components. Thus, it takes active intervention to reveal the component operations. One especially important way of revealing component operations is to break the system, or find instances of broken system, where we can identify the operation that fails and the operations that can still be performed without the broken elements. This is the strategy of lesion research we described in chapter 4, above. While many of the cleavages in the visual system were first identified through lesion studies, researchers did not widely accept the decomposition until lesion research was complemented by single-cell recording research which could reveal more directly what the individual components were doing. Thus, a common pattern that we will observe is lesion research suggesting how the visual system is decomposed, and single-cell recording providing additional evidence and revealing in greater detail the component operations.

1   Getting Started: Identifying the Locus of Control

A common starting point in developing explanations of how a complex system carries out a particular function such as vision is the attempt to localize *that* function as a whole in a particular part of the system. Competing hypotheses as to the locus are often advanced, and the first stage in the inquiry is an attempt to adjudicate between them. In the case of vision, two loci were advanced in the nineteenth century: the angular gyrus located in the posterior parietal lobe and the occipital lobe (figure 13.1). Of these, the more popular locus was the occipital lobe. It was proposed as the cortical center for vision by the Italian investigator Bartolomeo Panizza (1855) on the basis of a study of patients who experienced blindness after strokes damaged the occipital lobe, and lesion studies on several other species in which lesions to the occipital lobe produced blindness. His publications were largely ignored, perhaps because they only appeared in local Italian journals, but neuroanatomical studies in the same period by Pierre Gratiolet (1854) and Theodor Meynert (1870) indicated that the optic tract, which first projected to an area of the thalamus known as the *lateral geniculate nucleus* (LGN), projected on to the occipital lobe (Meynert traced the projections more specifically to the area surrounding the calcarine fissure), thereby supporting the occipital lobe locus.

The occipital lobe locus was opposed by one of the leading neurologists of the time, David Ferrier (1876). He was a pioneer in the use of mild electrical stimulation to identify loci in the brain associated with different functions, and found that stimulation of the angular gyrus (a region in the posterior parietal cortex) caused monkeys to move their eyes toward the opposite side. He further supported the angular gyrus locus through lesion studies in which he reported that bilateral lesions to the angular gyrus resulted in blindness, but that large lesions in the occipital lobe produced little impairment. Ferrier (1881) later moderated his claims, holding that both the angular gyrus and the occipital lobe figured in vision and that only lesions to both could produce complete and enduring blindness, but he continued to emphasize the angular gyrus.

In retrospect, it appears that the reason Ferrier's lesions of the angular gyrus produced deficits in vision was that his incisions cut deeply and severed the nerve pathways from the thalamus to the occipital cortex (Finger, 1994). Moreover, his failure to eliminate vision with occipital lobe lesions was due to incomplete removal of the visual processing areas in the occipital lobe. But these shortcomings in his technique were only established much later and did not figure in settling the conflict. Moreover, one should not just infer that Ferrier misapplied the lesion techniques because he cut too deeply since the functional difference between the underlying white matter and the gray matter first had to be appreciated and standards for conducting lesions research developed. Standardized methods are often the outcome of such scientific controversies – they cannot be appealed to in settling the controversies.

What seems to have established the occipital lobe as the *locus of control* for vision was the accumulation of reports from numerous investigators much like those originally put forward by Panizza. Many experimentalists (e.g. Munk, 1881; Schäfer, 1888b) described visual deficits after occipital lobe lesions in animals while clinicians (Henschen, 1893; Wilbrand, 1890) reported on patients who suffered visual deficits after damage to the occipital lobe.

During the same period as researchers were identifying a locus for vision at a macro level, neuroanatomists were making great progress in discovering the microstructure of the brain. Chapter 3, above, related the process by which the neuron doctrine was established and how researchers began to map the cortex in terms of neuroanatomical features such as the thickness of particular layers in various parts of the cortex. Even earlier, in 1776 Francesco Gennari, in the course of examining frozen sections of human brains, had identified a white stripe that was especially prominent in the posterior part of the brain (Glickstein and Rizzolatti,

1984). Subsequently Paul Flechsig (1896) identified this stripped area as the target of the projections from the LGN, and Grafton Elliot Smith (1907) named it the *area striata*; the area is now often referred to as the *striate cortex*. The area was also distinguished on cytoarchitectural grounds in a wide variety of species by Korbinian Brodmann (1909/1994), who assigned this area the number 17 since it was the seventeenth cortical area he had examined. Much later the terms *primary visual cortex* and *V1* also came to be applied to this area which, by the beginning of the twentieth century, was generally accepted as the locus of visual processing.

2   From Localization to Mechanistic Explanation

As controversial as it often is to establish, a proposal of a locus of control for a function is generally only a preliminary step toward explaining it. A simple identification of a function with a structure does not really explain the function since it provides no account of how the function is accomplished. Explanation requires decomposition into component functions, which usually results either from (1) discovering other structures that are involved in carrying out the same function (thereby revealing that the first site was not the sole locus of control and provoking the question of what more specific contribution that location makes), or (2) discovering components within the structure in question (and then asking what activities each of these performs). In twentieth-century research on visual processing, both of these played a critical role – the discovery of structure within the striate cortex provided one clue to the division of labor in processing visual stimuli and the discovery of a variety of other visual centers in the brain provided another clue.

Beyond direct localization: complexity within striate cortex

A first step beyond simply identifying striate cortex as the locus of control was already made in a detailed study Salomen Henschen (1893) made of lesion sites which produced vision deficits in humans. He showed that deficits in different parts of the occipital lobe produced blindness in different parts of the visual field and proposed that the occipital lobe must be topographically organized so that different parts of the retina projected on to different areas of the visual cortex (leading him to refer to it as the *cortical retina*). The occipital lobe map that Henschen proposed lays the projections out in the reverse manner of what is now accepted. While it might seem surprising that someone could discover a topological structure, and yet get all the locations reversed, such developments are surprisingly common in the history of science. It is indicative of how difficult it is to extend beyond individual, highly suggestive findings to generate a systematic account.

Discovering the correct topographical layout of striate cortex resulted from amassing many more data. Tatsuji Inouye was able to study 29 individuals who sustained highly focal damage to the occipital lobe during the Russo-Japanese war (as a result of new bullets introduced by the Russians; see Glickstein, 1988), and with the

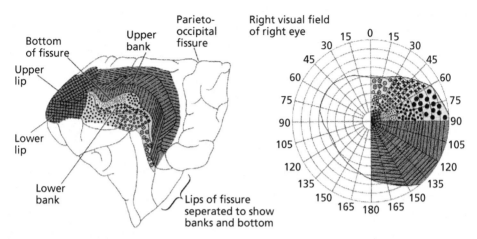

Figure 13.2   Gordon Holmes's (1918) map of how locations in the visual field (right) project onto parts of the primary visual cortex, based on studies of soldiers injured during World War I.

additional data points was able to determine that the central part of the visual field projects to the rear of the occipital lobe and the peripheral parts to the front. A similar study by Gordon Holmes and William Tindall Lister during World War I generated an even more detailed and accessible diagram of the topographical projection of parts of the visual field on to the visual cortex (see figure 13.2).

Micro-lesion studies could reveal topographical organization, but not the actual function performed by cells in striate cortex since the result of lesions was complete blindness. To make additional progress, researchers needed to determine what cells in striate cortex actually contributed to visual processing. For this, the key technique was single-cell recording. Here the strategy is to determine what kinds of stimuli cause a cell to fire most actively and then to assume that the cell is representing that stimulus (see chapter 18, this volume).[1] The pioneer for this approach was Steven Kuffler, who employed dark and light circles as stimuli while he recorded from ganglion cells in the retina; from this he discovered that the receptive fields of these cells were organized so that a cell might respond when a stimulus was in the center of its receptive field but not in the surrounding area (an *on-center* cell) or the reverse (an *off-center* cell).

Kuffler extended his research to the LGN, but it was two researchers in his laboratory at Johns Hopkins, David Hubel and Torsten Wiesel, who succeeded in extending this technique to striate cortex cells. While working at Walter Reed Army Institute, Hubel had developed a tungsten microelectrode and began using it to record from both sleeping and waking animals. He had succeeded in finding a few cells that responded when he moved his hands back and forth in front of the cat, but found he could not influence most cortical cells. After Hubel and Wiesel began their collaboration in the spring of 1958 they soon made the discovery that cells in striate cortex responded most vigorously not to spots of light but to oriented lines

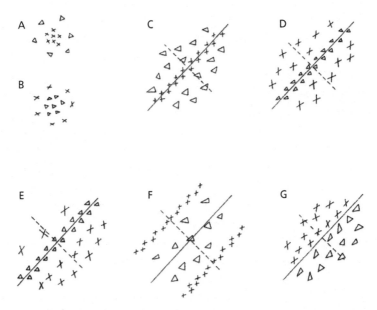

Figure 13.3 Examples of Hubel and Wiesel's (1962) mapping of receptive fields of cells in the lateral geniculate (A–B) and simple cortical cells (C–G) in the cat. X indicates an area producing excitatory responses and Δs an area producing inhibitory responses. Reprinted with permission of the Physiological Society.

or bars. Like many important scientific discoveries, theirs exhibited a bit of serendipity. They began by trying a variety of circular stimuli comparable to those Kuffler had used to elicit responses in retinal ganglion cells, but failed to produce any strong results. But as they were inserting a glass slide into their projecting ophthalmoscope, Hubel reports that "over the audiomonitor the cell went off like a machine gun" (Hubel, 1982, p. 438). They soon figured out that it was not the dot on the slide that was having an effect, but the fact that "as the glass slide was inserted its edge was casting on to the retina a faint but sharp shadow, a straight dark line on a light background" (p. 439).[2]

Over the first ten years of their collaboration, Hubel and Wiesel probed the striate cortex of both cats (Hubel and Wiesel, 1962) and monkeys (Hubel and Wiesel, 1968) and discovered a rich organization of cells with different response patterns. What they termed *simple cells* had receptive fields with spatially distinct *on* and *off* areas along a line at a particular orientation (most typically, they had a long, narrow *on* area sandwiched between two more extensive *off* areas) (see figure 13.3). Hubel and Wiesel suggested how several cells with center-surround receptive fields (such as those found in the LGN) might all send excitatory input to a single simple cell. In this regard, it is salient that simple cells predominate in layer 4, which is the input layer to cortex. Whereas simple cells were sensitive to stimuli only at a given retinal location, what Hubel and Wiesel termed *complex cells* were responsive to bars of light at a particular orientation anywhere within their receptive fields (figure 13.4). Many

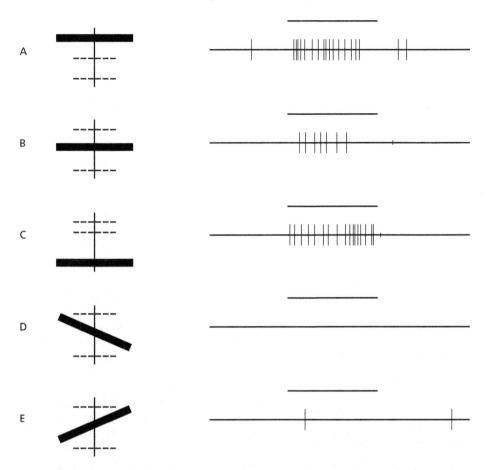

Figure 13.4 Response of complex cells in primary visual cortex to stationary, horizontally oriented black rectangles placed at different locations in the cell's receptive field (a) or moving through its receptive field (b). From Hubel and Wiesel, 1962, p. 111. Reprinted with permission of the Physiological Society.

complex cells were also sensitive to the direction of movement of bars within their receptive field. Hubel and Wiesel identified these as *complex cells* since their response pattern could be explained if they received input from several simple cells, any of which would be sufficient to cause the complex cell to fire. Complex cells are found primarily in layers 2 and 3, and 5 and 6.[3] In their papers from this period Hubel and Wiesel also distinguished *hypercomplex cells* which responded maximally only to bars extending just the width of their receptive field.[4]

Having identified three types of cells with different response properties, Hubel and Wiesel, proposed a decomposition of processing within striate cortex, with one type of cell supplying information to other cells and each carrying out its own information processing (although they also acknowledged that all three types of cells sent processes to other parts of the cortex). They also proposed the discovery of the

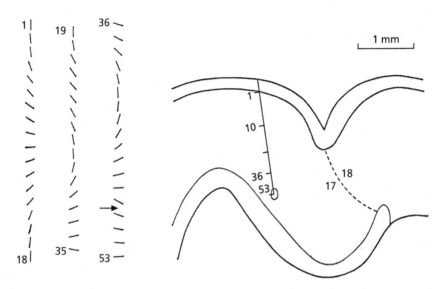

Figure 13.5   Hubel and Wiesel's reconstruction of a penetration through striate cortex about 1 mm from the 17–18 border, near the occipital pole of a spider monkey. On the left they show the preferred stimulus orientation at successive locations as electrode was gradually inserted at an angle of 45°. From Hubel and Wiesel, 1968, p. 231. Reprinted with permission of the Physiological Society.

primary function of striate cortex, but with a prophetic caveat: "The elaboration of simple cortical fields from geniculate concentric fields, complex from simple, and hypercomplex from complex is probably the prime function of the striate cortex – unless there are still other as yet unidentified cells there" (Hubel and Wiesel, 1968, p. 239).

By inserting electrodes gradually and recording from cells at different depths in the cortex, Hubel and Wiesel also discovered two additional features of the organization of striate cortex. First, when they inserted an electrode at an angle of 30° and recorded at successive locations, the preferred stimulus orientation for cells gradually changed (figure 13.5). Over the first 18 locations (approximately 1 mm) the preferred orientation varied through a full 180°. As penetration continued, a point was reached (arrow) where the variation in preferred orientation suddenly reversed. Second, they discovered that complex cells in striate cortex generally received binocular input, although they tended to be more responsive to input from one eye than the other. If the electrode were inserted perpendicularly rather than at an angle, all the cells encountered would respond to the same orientation with the same eye dominance, leading Hubel and Wiesel to adopt Vernon Mountcastle's proposal of a columnar organization of cortex. They proposed that in one direction successive columns (each 0.5 mm wide) were dominated by alternate eyes (ocular dominance columns) while in the other direction successive columns were responsive to different orientations of the stimulus.

Chapter 10, by Hubel and Wiesel, provides a summary of the major discoveries they and others made in the 1960s and 1970s about the organization of primary visual cortex. In this research we see clearly the emergence of a mechanistic analysis in which different operations are identified in the processing of visual inputs, each localized in a different cell type. Although one consequence of this research was to reveal complexity within striate cortex, another, perhaps more important consequence, was to demonstrate that it was not the sole locus of visual processing, since detecting oriented bars of light is not yet perception. This focused a question for further research: where else is visual information processed, and what does each of these areas contribute. Accordingly, Hubel and Wiesel conclude their 1968 paper with the prophetic comment:

> Specialized as the cells of 17 are, compared with rods and cones, they must, nevertheless, still represent a very elementary stage in the handling of complex forms, occupied as they are with a relatively simple region-by-region analysis of retinal contours. How this information is used at later stages in the visual path is far from clear, and represents one of the most tantalizing problems for the future.
>
> (Hubel and Wiesel, 1968, p. 242)

Beyond direct localization: identifying prestriate visual areas

The second means of moving beyond direct localization is to discover additional components that contribute to the function. However, although in the nineteenth century there were suggestions of additional visual processing areas,[5] there was an influential factor working against the identification of other brain areas involved in visual processing. In the first half of the twentieth century brain research was dominated by an anti-localizationist sentiment that construed most of the cortex as jointly subserving cognitive capacities, without any particular part playing a specialized role. This view was supported by experiments, some performed by Pierre Flourens in the early nineteenth century in response to phrenology and others performed in the early twentieth century in response to neo-phrenological localizationists, which suggested that individual parts of the cortex could be removed without loss of any particular cognitive ability. What mattered was how much was removed; cognitive performance seemed to decline roughly proportionately to the amount removed. Karl Lashley termed this the principle of mass action and applied it in particular to the area immediately surrounding the striate cortex, an area for which he coined the term *prestriate region*. He denied that prestriate cortex played a specifically visual function, insisting: "visual habits are dependent upon the striate cortex and upon no other part of the cerebral cortex" (Lashley, 1950). Prestriate cortex could be involved, along with other association areas, in higher processes resulting from visual perception, but were not involved, according to Lashley, in visual processing *per se*.

For many researchers, one sign of the lack of differentiated function beyond striate cortex was the lack of evidence that these areas were topographically organized in the manner of striate cortex. The very lack of a topographical organization was construed as support for the idea that these areas played a holistic, integratory role. Thus, one of the first indications of visual processing beyond striate cortex was Alan Cowey's (1964) discovery, using surface electrodes to record evoked responses, of a second topographically organized area in Brodmann's area 18 (which immediately adjoins area 17); this area came to be known as V2, with striate cortex being designated V1. Using single-cell recording, Hubel and Wiesel (1965) confirmed the topographical organization of this area and identified yet a third area, V3, in Brodmann's area 19. By tracing degeneration of fibers from discrete lesions in striate cortex to areas in surrounding cortex, Semir Zeki (1969) offered collaborative evidence for the existence of these additional areas. Zeki (1971) then extended this approach by creating lesions in V2 and V3 and tracing degeneration forward into areas on the anterior bank of the lunate sulcus in which "the organized topographic projection, as anatomically determined, gradually breaks down" (p. 33).[6] Zeki labeled these areas V4 and V4a.[7]

For the discovery of these additional areas to advance the functional decomposition of vision, it was necessary to link them with functions distinct from those associated with V1. As with V1, single-cell recording played the major role. Zeki (1973) recorded from cells in V4 and found "in every case the units have been colour coded, responding vigorously to one wavelength and grudgingly, or not at all, to other wavelengths or to white light at different intensities" (p. 422). Using a similar procedure as Hubel and Wiesel did in studying V1, Zeki recorded from successively encountered cells in a perpendicular penetration and found they responded to the same wavelength while successively encountered cells in an oblique penetration responded to different wavelengths. Zeki interpreted this as evidence of a columnar organization. The next year Zeki (1974) reported on a study recording from cells on the posterior bank of the superior temporal sulcus, an area he would later label V5 and others would designate MT. He found that cells in this area responded primarily to movement, with some firing in response to movements in any direction, but with most being sensitive to the direction, and sometimes the shape of the moving stimulus. As with V5, he found evidence of a columnar organization of movement-sensitive cells, with adjacent cells exhibiting slight changes in their preferred orientation (figure 13.6). Soon after the topography of these areas was ascertained through single-cell recording, neuroanatomical staining studies revealed that the connections to area V4 were primarily from V2 (Zeki, 1978) and those to area V5 were from V1 (van Essen et al., 1981).

Zeki's discovery of a color area made explicable the earlier reported clinical cases of achromatopsia (the inability to see objects as colored; see note 5). These patients had presumably suffered lesions in V4. In 1983 Zihl et al. (1983) reported on a patient who, as a result of vascular damage, could not perceive motion. To the patient activities such as coffee being poured into a cup appeared as contiguous shapes, like a glacier. Zeki's discovery of motion detection by V5 or MT could likewise explain this patient's deficit as due to damage to that area. The advent of neuroimaging techniques (chapter 4, this volume) has made it possible to identify the areas of increased

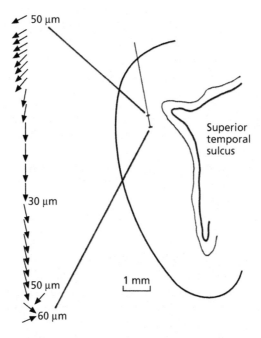

Figure 13.6   Zeki's (1974, p. 563) reconstruction of a penetration into the posterior bank of the superior temporal sulcus showing the preferred direction of movement of successive cells from which recordings were made.

blood flow when humans are presented with colored and moving stimuli; in confirmation of the decomposition suggested by single-cell recording in monkeys, Zeki et al. (1991), using PET, found distinct loci of increased blood flow in color- and motion-processing tasks. Since it is not yet feasible to differentiate areas such as V4 and V5 in humans on purely neuroanatomical grounds, these neuroimaging studies provide one of the best indications as to where these areas are located in the human brain.

Beyond direct localization: expanding visual analysis into temporal and parietal cortexes

The discovery of visual processing areas surrounding V1 which analyzed distinct visual properties such as color and motion both significantly advanced the functional decomposition of vision and posed a major question: where is the information about edges, colors, and motion put to use to permit the recognition of objects and events in the world? To address this question researchers had to expand the quest for specialized visual processing areas into more anterior parts of the temporal and parietal lobes. The first suggestions that areas in the temporal lobe played a specific role in visual processing was a study by Edward Schäfer (1888a) ostensibly devoted to showing that, contrary to Ferrier's claims, the temporal cortex was not the locus of an auditory center. In monkeys in which either the superior temporal gyrus or nearly all the temporal lobes were removed, Schäfer reports no detectable loss of hearing but describes a deficit in recognizing visually presented stimuli:

> the condition was marked by loss of intelligence and memory, so that the animals, although they received and responded to impressions from all the senses, appeared to understand very imperfectly the meaning of such impressions. This was not confined to any one sense, and was most evident with visual impressions. For even objects most familiar to the animals were carefully examined, felt, smelt and tasted exactly as a monkey will examine an entirely strange object, but much more slowly and deliberately. And on again, after only a few minutes, coming across the same object, exactly the same process of examination would be renewed, as if no recollection of it remained.
>
> (p. 375)

Little attention was paid to Schäfer's observations until after a study by Heinrich Klüver and Paul Bucy in the late 1930s, in which removal of the temporal lobe in monkeys resulted in a condition they described as *psychic blindness* or *visual agnosia* in which "the ability to recognize and detect the meaning of objects on visual criteria alone seems to be lost although the animal exhibits no or at least no gross defects in the ability to discriminate visually" (Klüver, 1948). The effects of the lesions induced by Klüver and Bucy were referred to as a syndrome since the monkeys exhibited a variety of other behavioral changes, including loss of emotional responsiveness and increased sexual behavior. Pribram and Brashaw (1953) addressed the question of whether these different deficits were due to a common process in the same brain area. By demonstrating that different lesions in temporal cortex could generate one or another deficit they showed that the various deficits were due to interrupting different processes. In particular, they traced visual agnosia to lesions of the amygdala and adjacent cortex. Subsequently, Pribram collaborated with Mortimer Mishkin in localizing visual agnosia specifically to lesions in inferotemporal cortex (Mishkin and Pribram, 1954). Subsequently, through a complex set of lesions involving the striate cortex in one hemisphere and the inferotemporal cortex in the other and the sectioning of the forebrain commissures, Mishkin (1966), succeeded in separating striate and inferotemporal cortex, and demonstrated that the deficits in visual learning and memory result when inferotemporal cortex is cut off from earlier visual processing. He also demonstrated that TE and TEO, areas within inferotemporal cortex that von Bonin and Bailey (1951) had distinguished on cytoarchitectonic grounds, produced differential deficits, with TEO lesions producing greater deficits in single-pattern discrimination tasks and TE lesions generating greater deficits on learning to perform multiple discriminations in parallel.

Again, the lesion studies indicating separate processing areas were complemented by single-cell recording studies that sought to determine what stimuli generated specific responses in inferotemporal cortex. Charles Gross, together with Carlos Eduardo Rocha-Miranda and David Bender (1972), found cells in the inferotemporal cortex of the macaque which responded most vigorously to shapes such as hands. (Like Hubel and Wiesel's, their discovery resulted from serendipity: after failing to find a light source that would drive a particular cell, they waved a hand in front of the stimulus screen and produced a vigorous response.) Although nearly a decade passed before further research was published confirming different areas where individual cells were responsive to specific shapes,[8] in the 1990s there was an explosion of reports of specific areas in inferotemporal cortex responsive to different specific shapes (see Tanaka, 1996, for a review).

A similar pattern of first lesion studies, then single-cell recording studies, emerged in research on the parietal cortex.[9] Ettlinger and Kahlsbeck (1962) analyzed deficits in monkeys with lesions in posterior parietal cortex and revealed deficits in visual orientation and reaching, indicating that these areas are involved in analysis of the location of objects in the visual field. In the early 1970s Hyvärinen and Poranen began recording from neurons in posterior parietal cortex, where they found cells which they interpreted as involved in visuospatial guidance of movement. These cells fired

> . . . when a sensory stimulus which interested the animal was placed in a specific location in space where it became the target of the monkey's gaze or manual reaching, tracking or manipulation. . . . Some cells were clearly related to eye movements whereas others appeared to discharge in response to visual sensory stimuli.
> (Hyvärinen and Poranen, 1974; quoted in Gross, 1998, p. 203)

The link Hyvärinen and Poranen found between activity of parietal cells and eye movement suggested a motor function for parietal cortex cells, a suggestion that was further developed by Vernon Mountcastle and his colleagues, who identified parietal cells linked not just to eye movement and visual tracking of objects, but to arm and hand manipulation (Mountcastle et al., 1975). Mountcastle interpreted these cells as involving motor commands linked to selective attention. Other research, however, suggested that the posterior parietal cortex was primarily involved in visual analysis since some cells are responsive in the absence of any motor activity (Goldberg and Robinson, 1980). But importantly, Richard Andersen and his colleagues demonstrated that cells in posterior parietal cortex mapped stimuli in terms of spatial location, a feature to which temporal lobe cells are relatively unresponsive (Andersen et al., 1985).[10]

3 Proposing a Complex, Organized System

The research described in the last two sections clearly advanced the efforts to functionally decompose and localize visual processing. Whereas initially only V1 seemed to be involved, it now appeared that much of the back half of the brain was devoted to analyzing visual inputs, with different areas analyzing different aspects of visual scenes. As discoveries piled up, a possible outcome was a theoretical morass – the discovery that the brain consisted of many special-purpose processing areas with no systematic organization. But Mortimer Mishkin and Leslie Ungerleider introduced what proved to be a powerful organizing principle (see chapter 11, this volume). They proposed that visual processing beyond V1 was organized into two pathways, one progressing dorsally into posterior parietal cortex that was involved in analyzing *where* objects are in the visual field, the other, progressing ventrally down into inferotemporal cortex, that is involved in analyzing *what* objects are present in the visual scene. The distinction between *what* and *where* processing had been advanced previously by Schneider (1967) and Trevarthen (1968) for subcortical areas, but by proposing it for cortical areas Mishkin and Ungerleider offered a macro-level organizing principle for visual areas that integrated the findings related in previous sections.

For Mishkin and Ungerleider, the separation of two the pathways began in pre-striate cortex. Other researchers soon proposed extending the scheme back into V1, LGN, and the retina, generating a model of two processing streams from the very earliest visual input. An important piece of evidence for projecting the two streams further back was a distinction between two different cell types in the retina and the LGN. Enroth-Cugell and Robson (1966) had differentiated two types of cells in the cat retina, which they named X and Y *cells*. X cells had small receptive fields (hence, they were sensitive to high spatial frequencies), medium conductance velocities, and responded as long as the stimulus was present. In contrast, Y cells had large receptive fields, rapid conductance velocities, and responded transiently. A similar distinction of retinal cell-types was advanced for primates. $P\alpha$ (or P ganglion) cells correspond to the X cells in the cat while the $P\beta$ (or M ganglion) cells correspond to the Y cells in the cat. Research on old-world monkeys revealed that this scheme is maintained in the LGN where the cells in the two inner layers have large cell bodies (the layers are thus known as *magnocellular* or *M layers*) while those in the outer four have small cell bodies (thus called *parvocellular* or *P layers*). The M layers of the LGN receive projections from the M ganglion cells, while the P layers receive input from the P ganglion cells (Dreher et al., 1976).

One challenge was how to link the two precortical pathways with the two cortical pathways. The early studies of Hubel and Wiesel and others had suggested that V1 had a homogeneous cytoarchitecture; if this were the case, the two precortical pathways would converge in V1 and then two other pathways would diverge beyond V1. But, in accord with the caveat in the passage quoted above from Hubel and Wiesel, a new technique, involving the application of cytochrome oxidase stains (developed by Margaret Wong-Riley, 1979), revealed additional complexity in V1. Cytochrome oxidase is an enzyme critical to the oxidative metabolism of the cell; staining for it reveals areas of high metabolic activity. In layers 2 and 3, and 5 and 6 of V1 these showed up as "blobs"[11] which indicated regions of particularly high metabolic activity. Recording separately from cells in the blob regions and in the interblob regions, Livingstone and Hubel (1984) found orientation-selective cells only in the interblob regions, and wavelength-sensitive cells in the blobs, indicating a separation of processing within V1. On the basis of this differentiation, Livingstone and Hubel proposed extending Mishkin and Ungerleider's two pathways to account for all visual processing from the retina on.

As Van Essen and Gallant's chapter (12, this volume) makes clear, the integrating scheme of two processing streams receives support from the neuroanatomy. The M layers of the LGN project to layer 4B in V1, where there are no blobs, whereas the P layers of the LGN project, via layers 4A and 4Cb, to layers 2 and 3 of V1, where there are both blob and interblob regions. Cytochrome oxidase stain also revealed a differentiation in V2 of alternating thick and thin stripes with interstripe areas between them. The differentiation in V1 is maintained, with the thick-stripe regions receiving their input from layer 4B, the thin-stripe regions from the blobs of layers 2 and 3, and the interstripe regions from the interblob regions in V1. From the differentiated areas in V1 and V2, processing largely separates into the *what* and *where* pathways originally distinguished by Mishkin and Ungerleider (see figure 12.2).

The proposal of two processing streams has provided an organizing framework for research on visual processing in the brain and, as we shall see, has influenced computational and psychological investigations of perception. But it is important to note that, like most integrating schemes, this one is subject to a variety of qualifications. Van Essen and Gallant draw attention to the fact that the two streams are not entirely independent – there are neural connections between areas such as MT and V4, which appear in different streams, and processing in the later parts of one stream continues even when its primary input is removed. Moreover, there is considerable interaction between the two precortical streams so that processing in each stream can continue even if the supposed precortical input is removed. Furthermore, the characterization of the two streams as processing *what* and *where* information has been questioned. Milner and Goodale (1995) argue that the dorsal stream receives information about the identity of objects (revealed in the ability of individuals with temporal lobe lesions to grasp objects appropriately for their use) and propose that what is distinctive about it is that it is primarily concerned with coordinating information about visual stimuli for action. In their view, the ventral stream is principally involved in extracting information about visual stimuli required for higher cognitive processing. Even with such qualifications, though, the idea of two visual streams plays an important integratory role in theorizing about visual processes, providing for a relatively coherent and graspable account of how the brain processes visual information.

4   What is Still Needed: A Computational Analysis

I have characterized the model of vision presented in figure 12.2 as providing an exemplar of successful explanation in cognitive neuroscience, and have used its history to provide an exemplar of how mechanistic models of cognitive processes can be developed. However, it is important to bear in mind that the account is still incomplete. One way it can be filled in is to discover more processing components and figure out what operations they seem to be performing. But another way is to develop computational models of visual processes that specify the operations to be performed by each component in the proposed pathways of visual processing – models that specify how the firing patterns of neurons in one area encode specific information about distal objects and how later neurons are able to extract yet other information by operating on the firing patterns of the first set of neurons.

Many computational modelers, including many who now refer to themselves as computational neuroscientists, are currently engaged in developing just such models. But the effort is not new. One of the most influential proponents of applying computational modeling to vision was the British physiological psychologist turned MIT AI researcher David Marr. Marr began his career as an enthusiastic supporter and active contributor (Marr, 1969, 1970) to the approach of attempting to understand the visual system by discovering the responsiveness of individual cells, but he gradually became disenchanted with it. He was dissatisfied not only because the exciting initial discoveries, such as Hubel and Wiesel's discovery of edge detectors, had not yet been followed by similar discoveries of higher-level correlates of perception (his negative assessment here might have been premature), but also because of a recognition that such discoveries would not suffice to *explain* perception. As he worked out his final position in his posthumously published book, *Vision* (Marr, 1982), an explanation of perception, as opposed to a description of how the visual system behaved, required figuring out what the visual system was doing and how the various processes contributed to doing it:

> The message was clear. There must exist an additional level of understanding at which the character of the information processing tasks carried out during perception are analyzed and understood in a way that is independent of the particular mechanisms and structures that implement them in our heads. This was what was missing – the analysis of the problem as an information processing task. Such analysis does not usurp an understanding at other levels – of neurons or of computer programs – but it is a necessary complement to them, since without it there can be no real understanding of the function of all of those neurons.
>
> (p. 19)

In fact, Marr went on to argue that there are "three levels at which any machine carrying out an information processing task must be understood." The highest of these levels is what he called the *computational theory* where one specified both (a) "what is computed and why" and (b) "that the resulting operation is defined uniquely by the constraints it has to satisfy" (p. 23). Marr illustrated what he was seeking by providing the computational theory for a cash register: *what* a cash register does is arithmetic, and the reason it does it is because the operations of arithmetic are those we "intuitively feel to be appropriate" for combining prices (p. 22). The constraints it must satisfy are principles such as associativity and commutivity. In the case of vision, he contends, "the underlying task is to reliably derive properties of the world from images of it; the business of isolating constraints that are both powerful enough to allow a process to be defined and generally true of the world is a central theme of our inquiry" (p. 23).

The second level focuses on how the machine carries out its information-processing task by specifying the *representations* it uses at the input and output level, and the *algorithms* it uses to transform input representations into output representations. Marr notes that the representations and algorithms must be appropriately coupled since particular kinds of operations are suited for specific kinds of representations, but that there often are many kinds of representations that can be selected for a given computation (one can represent the number one with binary or digital numerals) and, even given a choice of representations, different algorithms can perform the computation.

Finally, the third level addresses the question of how the representation and algorithm are physically realized. It is here that appeal to neural mechanisms is relevant, but Marr wants to stress two points: (1) any explanation of what the neurons are doing requires specifying the computation being performed and the representational system and algorithms employed; and (2) there is only a loose coupling between these levels so that details about the neural architecture constrain but do not determine the representations and algorithm, and the computation being performed.

As Marr noted in a passage quoted above, much of his concern was with identifying the constraints operating on vision, as well as with specifying a representational system and sets of algorithms that could carry out the computation. Computer programming provided him one means of examining various representational systems and exploring what algorithms might operate on them, and he identified psychophysics as a major source for evaluating the empirical adequacy of such accounts. The task Marr set for himself was to determine how, from low-level feature detection, a system could come to recognize objects. His positive proposal was that the system constructs three successive representations of each visual scene which he termed the primal sketch, a 2½-D sketch, and a 3-D model representation. The primal sketch is a two-dimensional representation in which lines correspond to intensity changes in light reaching the retina and larger configurations are constructed by grouping the resulting lines. The 2½-D sketch is a viewer-centered representation that specifies the depth and orientation of visible surfaces in the world. Finally, the 3-D model representation is an object-centered representation of volumetric primitives which facilitates recognition of the objects present. Marr specified computational procedures (e.g. algorithms in his three-level account) for producing each of these. For example, especially important for creating the primal sketch is a procedure for identifying zero-crossings (areas in the scene in which the second derivative of the intensity function changes sign), and Marr proposed a procedure whereby this could be done through local computation by neurons such as Hubel and Wiesel's simple cells in V1. To create a 2½-D sketch from a flat image, Marr proposed independent procedures that relied on information in the primal sketch about, for example, motion, stereopsis, optic flow, and surface texture, to determine properties of surfaces and their relation to the viewer.

As he proceeds to the 3-D model representation and the process of identifying objects, Marr's account becomes more sketchy and he offers fewer proposals as to the neural substrates involved in the hypothesized processes. More recently, psychologist Irving Biederman (1995) has put forward a proposal as to how object recognition could proceed quickly from an object-oriented representation such as Marr's 3-D model representation. He proposes that primitive volumes correspond to one of 24 basic shapes known as *geons* that are created by varying four attributes of a cone shape. He argues that combinations of two or three geons (produced by varying their attributes, e.g. orientation, and relations to each other) can uniquely specify objects. Although the details of the geon theory extend beyond what can be evaluated at the neural level, Biederman does appeal to evidence by Tanaka (1993) that cells in area TE in the inferotemporal cortex respond to complex object features.

One thing that is striking about Marr's work is, once he became dissatisfied with traditional neuroscience approaches, how much he dissociated the development of computational models from the attempt to develop neural grounding. His approach was rather to propose algorithms that could compute a 3-D model from the low-level information available to the nervous system, evaluating his proposals largely by whether they could compute a 3-D model. The wealth of information now available or soon to be forthcoming about how the brain decomposes the computational process, however, can provide an additional important constraint on computational modeling. A further important consideration that Marr did not address was whether real organisms ever compute a 3-D model of their environment. Recently a number of theorists have in fact questioned this, arguing that in fact visual processing extracts only partial information from a scene as it is needed (Ballard, 1991; Churchland et al., 1994).

Despite these shortcomings in the way he executed his project, Marr's insistence on an information-processing analysis as a critical component of an explanation is well taken. Currently, many of the computational models being developed employ connectionist or artificial neural networks. These are computational models in which the primitive elements (units) are simple processors modeled loosely on neurons. Instead of exhibiting a pattern of spiking, units take on activation values. They are connected to each other by weighted connections so that activation in one unit can excite or inhibit activation in other units. Programs running on standard computers can determine how activation values of units in a network will change in response to inputs and interactions within the network. Such programs can also implement a variety of rules for changing the strengths of the connections between units; such changes alter how the network will respond to inputs and provide a way of implementing learning. In many of the early neural network models processing is only in one direction from input to output units (such networks are spoken of as *feedforward*); increasingly, though, researchers are exploring models in which there are also backwards or recurrent connections and collateral connections (Bechtel and Abrahamsen, in press).

Building neural networks can often provide insight into the reasons for and importance of decomposition found in the brain. An example of the utility of this approach is found in two network models that were developed to understand the reasons for the brain's decomposition into *what* and *where* pathways. Rueckl, Cave, and Kosslyn

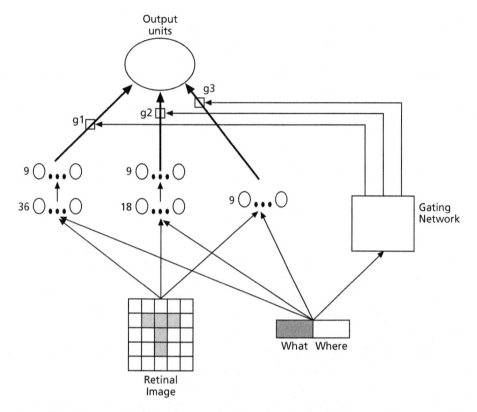

Figure 13.7 Jacobs et al.'s (1991) network that learned to employ different modular networks to determine the identity or location of the object specified on the artificial retina. The modular networks differed in whether they employed hidden units or not and the number of hidden units. The gating network learned to recognize which network produced the best answer for *where* and *what* questions and gated the output from the modular networks so that only the best-performing network would be allowed to control the output units.

(1989) designed feedforward networks whose input was presented on an artificial retina comprising a 5 × 5 grid of cells, each of which supplied an input of 1 if the input pattern covered the cell and 0 otherwise (see figure 13.7). The input from the 25 cells was processed through a hidden layer[12] and projected on to an output layer which consisted of two sets of units, on one of which the network was to identify, by the pattern of activation produced, the shape of the input and on the other of which it was to specify the location of the shape on the artificial retina. The network was trained through an error correction procedure known as *backpropagation* in which information about how far off an output unit is from its target activation is propagated backwards through the network, and weights on connections are changed in a direction that will reduce that error in the future. Rueckl et al. compared two network designs. In the more distributed design, activation was processed through a single set of hidden units; potentially, every hidden unit could contribute to both the *what* and *where* responses. In the more modular design that was inspired by the research of Mishkin and Ungerleider, the hidden units were partitioned into two sets: one set sent activation only to the *what* units, and the other set sent activation only to the *where* units. Rueckl et al. demonstrated that the dual task was learned more readily when there were dedicated sets of hidden units supplying the two sets of output units, a finding which suggests that the brain might have separated the processing of the two sorts of information for reasons of computational efficiency. Using a somewhat more complex network model, Jacobs, Jordan, and Barto (1991) showed that networks could discover how to decompose the task on their own (figure 13.7).

These neural network models represent early attempts to employ computational models in the effort to understand the operations performed by the different components in the visual system. As we have seen, the processing of *what* and *where* information in the brain is much more complex, involving many different brain areas, and different constituents within brain areas. Current computational research is often sensitive to these details, and attempts to develop biologically realistic neural networks. As successful computational models are developed, they will provide a critical complement to the details about the decomposition and localization of visual processing by allowing us to understand how a set of components found in the brain are able to jointly perform the tasks needed for us to see.

Conclusions

In this chapter I have treated neuroscience research on vision as an exemplar in two senses. First, vision research has reached a such level of maturity that it provides other domains of neuroscience with an exemplar which their research might emulate. Second, it provides a exemplar for philosophers of how mechanistic models that decompose and localize cognitive functions develop through time. The trajectory from direct localization, to discovering multiple components performing different functions, to developing integrated models supported by computational models, is one also exhibited in other disciplines of biology (Bechtel and Richardson, 1993). In each case there will be differences, in part as a result of the particular investigatory tools and information available to researchers at a time. But the challenges of figuring out what the parts of a system are, what they do, and how they interact, are common features of the development of mechanistic models.

Notes

1   One of the first positive results from recording from cells in striate cortex was actually simply to confirm its topographical organization. Talbot and Marshall (1941), working with lightly anesthetized cats and rhesus monkeys, recorded the responses evoked by bright light either from moist threads on the exposed pia or from insulated needles inserted into the cortex. They described their procedure as follows:

If the visual pattern is narrowed down to a band and then to a small square, a position can be found in the field where a movement of a degree or less will reduce the response at the cortical point. This response is observed periodically on a cathode-ray tube, and the stimulus moved by manipulating crossed slits, until the primary response shows maximum amplitude and minimum latency for the least stimulus area.   (p. 1255)

2   Hubel describes some of the sense of surprise at the finding that individual cells responded to a bar of light at a particular orientation: "This was unheard of. It is hard, now, to think back and realize just how free we were from any idea of what cortical cells might be doing in an animal's daily life" (1982, p. 439).

3   An important difference between the different layers is that they generally project to different brain areas: layers 2 and 3 to other cortical areas, layer 5 to the superior colliculus, pons, and pulvinar, and layer 6 back to the LGN.

4   Hubel and Wiesel (1965) identified such cells only in areas 18 and 19 of the cat and assumed that these cells received their inputs from complex cells. Later, though, they found them in area 17 in both cat and monkey. After Dreher (1972) found cells in cats that were location-specific like simple cells but whose response dropped off as the length of the stimulus exceeded an optimum length, they dropped the assumption that they received their inputs from complex cells.

5   By analyzing patients with cortical achromatopsia (the inability to see colors) whose lesions could be traced to the fusiform gyrus adjacent to the striate cortex, both Verrey (1888) and MacKay and Dunlop (1899) had provided evidence of a second visual area, one devoted to color perception, but most nineteenth century researchers dismissed these claims in favor of the supposition of one cortical center for vision in the striate cortex, which might produce achromatopsia with mild lesions and full blindness with more serious lesions. One finding supporting this interpretation was that most cases of achro-matopsia also manifested scotomas or areas of total blindness, suggesting that one lesion produced both effects.

6   Zeki ends the paper with the following comment about projections to other brain areas: "How the prestriate cortex is organized in regions beyond (central to) V4 and V4a remains to be seen. It is perhaps sufficient to point out at present that the organization of the prestriate areas would seem to be far more complicated than previously envisaged and that the simplistic wiring diagram from area 17 to area 18, from area 18 to area 19 and from area 19 to the so-called "interior temporal" area will have to be abandoned. At any rate, we were not able in this study to find any projections to the "inferior temporal" areas from areas 18 and 19 (V2 and V3). (p. 34)

7   During the same period John Allman and Jon Kaas, through single-cell recording in squirrel monkeys, traced topographically organized visual areas not only into extra-striate regions but also into temporal and parietal cortexes.

8   Gross (1998, pp. 199–200) reports on the slowness of response:

for more than a decade there were no published attempts to confirm or deny these and our other early basic results, such as that IT cells have large bilateral fields that include the fovea and are not visuotopically organized. And unlike Panizza, the discoverer of visual cortex in the nineteenth century, we did not publish in obscure journals or from an unknown institution. Perhaps because of the general skepticism, we did not ourselves publish a full account of a face-selective neuron until 1981.

9　Both Ferrier and Yeo (1884) and their opponents Brown and Schäfer (1888) reported deficits from lesions to the angular gyrus in the posterior parietal cortex which fit the pattern of deficit in spatial localization identified in the 1960s. Ferrier and Yeo report that the lesioned monkey was "evidently able to see its food, but constantly missed laying hold of it" and Brown and Schäfer report that their monkey "would evidently see and run up to [a raisin], but then often fail to find it . . ." (both quotations from Gross, 1998, pp. 200 and 201). Based on studies of brain injuries in World War I veterans, Gordon Holmes (1918) identified deficits in spatial localization of objects that the veterans could easily identify visually.

10　Subsequent research has confirmed a close relation between parietal cells and motor action and investigations into whether these cells are directly involved in planning action or in maintaining attention on visual stimuli (Snyder et al., 1997; Batista et al., 1999).

11　Livingstone and Hubel introduced the term *blobs* to characterize their appearance, citing the *Oxford English Dictionary* for the term. These blobs are "oval, measure roughly 150 × 200 μm, and in the macaque monkey lie centered along ocular dominance columns, to which their long axes are parallel" (1984, p. 310).

12　These units are referred to as *hidden* since they neither receive external input nor constitute output units.

References

Andersen, R. A., Essick, G. K., and Siegel, R. M. 1985: Encoding of spatial location by posterior parietal neurons. *Science*, 230, 456–8.

Ballard, D. H. 1991: Animate vision. *Artificial Intelligence*, 48, 57–86.

Batista, A. P., Buneo, C. A., Snyder, L. H., and Andersen, R. A. 1999: Reach plans in eye-centered coordinates. *Science*, 285, 257–60.

Bechtel, W., and Abrahamsen, A. (in press): *Connectionism and the Mind: Parallel Processing, Dynamics, and Evolution in Neural Networks*. Oxford: Blackwell.

Bechtel, W., and Richardson, R. C. 1993: *Discovering Complexity: Decomposition and Localization as Scientific Research Strategies*. Princeton, NJ: Princeton University Press.

Biederman, I. 1995: Visual object recognition. In S. M. Kosslyn and D. N. Osherson (eds), *Visual Cognition*, 2nd edn, vol. 2. Cambridge, MA: MIT Press, 121–65.

Brodmann, K. 1909/1994: *Vergleichende Lokalisationslehre der Grosshirnrinde* (L. J. Garvey, Trans.). Leipzig: J. A. Barth.

Brown, S., and Schäfer, E. A. 1888: An investigation into the functions of the occipital and temporal lobes of the monkey's brain. *Philosophical Transactions of the Royal Society of London*, 179, 303–27.

Churchland, P. S., Ramachandran, V. S., and Sejnowski, T. J. 1994: A critique of pure vision. In C. Koch and J. L. Davis (eds), *Large-scale Neuronal Theories of the Brain*. Cambridge, MA: MIT Press.

Cowey, A. 1964: Projection of the retina on to striate and prestriate cortex in the squirrel monkey *Saimiri Sciureus*. *Journal of Neurophysiology*, 27, 366–93.

Dreher, B. 1972: Hypercomplex cells in the cat's striate cortex. *Investigative Ophthalmology*, 11, 355–6.

Dreher, B., Fukada, Y., and Rodieck, R. W. 1976: Identification, classification and anatomical segregation of cells with X-like and Y-like properties in the lateral geniculate nucleus of old-world primates. *Journal of Physiology*, 258, 433–52.

Elliot Smith, G. 1907: New studies on the folding of the visual cortex and the significance of the occipital sulci in the human brain. *Journal of Anatomy*, 41, 198–207.

Enroth-Cugell, C., and Robson, J. G. 1966: The contrast sensitivity of retinal ganglion cells of the cat. *Journal of Physiology*, 187, 517–52.

Ettlinger, G., and Kahlsbeck, J. E. 1962: Changes in tactual discrimination and in visual reaching after successive and simultaneous bilateral posterior parietal ablations in the monkey. *Journal of Neurological and Neurosurgical Psychiatry*, 25, 256–68.

Ferrier, D. 1876: *The Functions of the Brain*. London: Smith, Elder.

Ferrier, D. 1881: Cerebral amblyopia and hemiopia. *Brain*, 11, 7–30.

Ferrier, D., and Yeo, G. F. 1884: A record of the experiments on the effects of lesions of different regions of the cerebral hemispheres. *Philosophical Transactions of the Royal Society of London*, 175, 479–564.

Finger, S. 1994: *Origins of Neuroscience*. Oxford: Oxford University Press.

Flechsig, P. E. 1896: *Gehirn und Steele*. Leipzig: Veit.

Glickstein, M. 1988: The discovery of the visual cortex. *Scientific American*, 259 (3), 118–27.

Glickstein, M., and Rizzolatti, G. 1984: Francesco Gennari and the structure of the cerebral cortex. *Trends in Neuroscience*, 7, 464–7.

Goldberg, M. E., and Robinson, D. L. 1980: The significance of enhanced visual responses in posterior parietal cortex. *Behavioral and Brain Sciences*, 3, 503–5.

Gratiolet, P. 1854: Note sur les expansions des racines cérébrales du nerf optique et sur leur terminaison dans une région déterminée de l'écorce des hémisphères. *Comptes Rendus Hebdomadaires des Séances de l'Académie des Sciences de Paris*, 29, 274–8.

Gross, C. G. 1998: *Brain, Vision, and Memory*. Cambridge, MA: MIT Press.

Gross, C. G., Rocha-Miranda, C. E., and Bender, D. B. 1972: Visual properties of neurons in inferotemporal cortex of the macaque. *Journal of Neurophysiology*, 35, 96–111.

Henschen, S. E. 1893: On the visual path and centre. *Brain*, 16, 170–80.

Holmes, G. 1918: Disturbances of visual orientation. *British Journal of Ophthalmology*, 2, 449–68.

Hubel, D. H. 1982: Evolution of ideas on the primary visual cortex, 1955–1978: A biased historical account. *Bioscience Reports*, 2, 435–69.

Hubel, D. H., and Wiesel, T. N. 1962: Receptive fields, binocular interaction and functional architecture in the cat's visual cortex. *Journal of Physiology (London)*, 160, 106–54.

Hubel, D. H., and Wiesel, T. N. 1965: Binocular interaction in striate cortex of kittens reared with artificial squint. *Journal of Neurology*, 28, 1041–59.

Hubel, D. H., and Wiesel, T. N. 1968: Receptive fields and functional architecture of monkey striate cortex. *Journal of Physiology (London)*, 195, 215–43.

Hyvärinen, J., and Poranen, A. 1974: Function of the parietal associative area 7 as revealed from cellular discharges in alert monkeys. *Brain*, 97, 673–92.

Jacobs, R. A., Jordan, M. I., and Barto, A. G. 1991: Task decomposition through competition in a modular connectionist architecture: The what and where vision tasks. *Cognitive Science*, 15, 219–50.

Klüver, H. 1948: Functional differences between the occipital and temporal lobes with special reference to the interrelations of behavior and extracerebral mechanisms. In L. Jeffress (ed.), *Cerebral Mechanisms in Behavior*, New York: Wiley, 147–99.

Lashley, K. S. 1950: In search of an engram. *Symposium on Experimental Biology*, 4, 45–8.

Livingstone, M. S., and Hubel, D. H. 1984: Anatomy and physiology of a color system in the primate visual cortex. *Journal of Neuroscience*, 4, 309–56.

MacKay, G. and Dunlop, J. C. 1899: The cerebral lesions in a case of complete acquired colourblindness. *Scot. Med. Surg. J.*, 5, 503–12.

Marr, D. 1969: A theory of the cerebellar cortex. *Journal of Physiology*, 202, 437–70.

Marr, D. 1970: A theory of the cerebral cortex. *Proceedings of the Royal Society of London*, B 176, 161–234.

Marr, D. C. 1982: *Vision: A Computation Investigation into the Human Representational System and Processing of Visual Information*. San Francisco: Freeman.

Meynert, T. 1870: Beiträge zur Kenntniss der centralen Projection der Sinnesoberflächen. *Sitzungberichte der Kaiserlichten Akademie der Wissenschaften, Wien. Mathematisch-Naturwissenschaftliche Classe*, 60, 547–62.

Milner, A. D., and Goodale, M. G. 1995: *The Visual Brain in Action*. Oxford: Oxford University Press.

Mishkin, M. 1966: Visual mechanisms beyond the striate cortex. In R. W. Russel (ed.), *Frontiers in Physiological Psychology*. New York: Academic.

Mishkin, M., and Pribram, K. H. 1954: Visual discrimination performance following partial ablations of the temporal lobe: I. Ventral vs. lateral. *Journal of Comparative and Physiological Psychology*, 47, 14–20.

Mountcastle, V. B., Lynch, J. C., Georgopoulos, A., Sakata, H., and Acuna, C. 1975: Posterior parietal association cortex of the monkey: Command functions for operations within extrapersonal space. *Journal of Neurophysiology*, 38, 871–908.

Munk, H. 1881: *Über die Funktionen der Grosshirnrinde*. Berlin: A. Hirschwald.

Panizza, B. 1855: Osservazioni sul nervo ottico. *Memoria, Istituto Lombardo di Scienze, Lettere e Arte*, 5, 375–90.

Pribram, K. H., and Bagshaw, M. 1953: Further analysis of the temporal lobe syndrome utilizing fronto-temporal ablations. *Journal of Comparative Neurology*, 99, 347–75.

Rueckl, J. G., Cave, K. R., and Kosslyn, S. M. 1989: Why are "what" and "where" processed by separate cortical visual systems? A computational investigation. *Journal of Cognitive Neuroscience*, 1, 171–86.

Schäfer, E. A. 1888a: Experiments on special sense localisations in the cortex cerebri of the monkey. *Brain*, 10, 362–80.

Schäfer, E. A. 1888b: On the functions of the temporal and occipital lobes: A reply to Dr Ferrier. *Brain*, 11, 145–66.

Schneider, G. E. 1967: Contrasting visuomotor functions of tectum and cortex in the golden hamster. *Psychologische Forschung*, 31, 52–62.

Snyder, L. H., Batista, A. B., and Andersen, R. A. 1997: Coding of intention in the posterior parietal cortex. *Nature*, 386: 167–70.

Talbot, S. A., and Marshall, W. H. 1941: Physiological studies on neural mechanisms of visual localization and discrimination. *American Journal of Ophthalmology*, 24, 1255–63.

Tanaka, K. 1993: Neuronal mechanism for object recognition. *Science*, 262, 685–8.

Tanaka, K. 1996: Inferotemporal cortex and object vision. *Annual Review of Neuroscience*, 19, 109–40.

Trevarthen, C. 1968: Two mechanisms of vision in primates. *Psychologische Forschung*, 31, 299–337.

Van Essen, D. C., Maunsell, J. H. R., and Bixby, J. L. 1981: The middle temporal visual area in the macaque monkey: Myeloarchitecture, connections, functional properties, and topographic organization. *Journal of Comparative Neurology*, 248, 164–89.

Verrey, L. 1888: Hemiachromatopsie droite absolute. *Archs. Ophthalmol.*, 8, 289–301.

von Bonin, G., and Bailey, P. 1951: *The Isocortex of Man*. Urbana: University of Illinois Press.

Wilbrand, H. 1890: *Die hemianopischen Gesichtsfeld-Formen und das optische Wahrnehmungszentrum*. Wiesbaden: J. F. Bergmann.

Wong-Riley, M. 1979: Changes in the visual system of monocularly sutured or enucleated cats demonstratable with cytochrome oxidase histochemistry. *Brain Research*, 171, 11–28.

Zeki, S. M. 1969: Representation of central visual fields in prestriate cortex monkey. *Brain Research*, 14, 271–91.

Zeki, S. M. 1971: Cortical projections from two prestriate areas in the monkey. *Brain Research*, 34, 19–35.

Zeki, S. M. 1973: Colour coding of the rhesus monkey prestriate cortex. *Brain Research*, 53, 422–427.

Zeki, S. M. 1974: Functional organization of a visual area in the posterior bank of the superior temporal sulcus of the rhesus monkey. *Journal of Physiology*, 236, 549–73.

Zeki, S. M. 1978: Functional specialisation in the visual cortex of the rhesus monkey. *Nature*, 274, 423–8.

Zeki, S. M., Watson, J. D. G., Lueck, C. J., Friston, K. J., Kennard, C., and Frackowiak, R. S. J. 1991: A direct demonstration of functional specialization in human visual cortex. *Journal of Neuroscience*, 11, 641–9.

Zihl, J., von Cramon, D., and Mai, N. 1983: Selective disturbance of movement vision after bilateral brain damage. *Brain*, 106, 313–40.

Questions for Further Study and Reflection

1 Describe the procedure by which Hubel and Wiesel proposed to discover the contributions of cells at different stages in the pathway of processing visual information. Why, for areas beyond the first level of processing, does their method require examination of cells in at least two successive areas of processing?

2 Hubel and Wiesel reject the proposal that there is a homunculus (little green man) observing the pattern of activity of cells in the visual cortex. What is wrong with this conception of how the brain processes visual information, and what alternative accounts are there?

3 Most of Mishkin, Ungerleider, and Macko's evidence for two visual pathways appeals to deficits animals exhibit after damage to one or the other pathway. Near the end of their paper, though, they introduce additional metabolic and anatomical evidence. How important is this additional evidence for supporting their claim?

4 Mishkin, Ungerleider, and Macko describe the processes in the two pathways as focusing on the object's identity versus its spatial location. Are the other ways one might characterize the activities of the two pathways that are compatible with the evidence they review?

5 Explain the distinction Van Essen and Gallant draw between neurons as feature detectors and as filters. What difference does it make if neurons employ one or the other means of coding information?

6 What might be the reason for the degree of interaction between the visual processing streams that Van Essen and Gallant identify?

7 Would it have been possible to discover cortical areas involved in vision if they had not involved a topological map of the visual field?

8 Why is a computational analysis of visual processing important? What does it add beyond finding the different brain areas involved and figuring out what sorts of information each seems to process?

Part IV

Consciousness

Introduction

Pete Mandik

In some circles, consciousness poses the ultimate and insurmountable challenge to a scientific understanding of minds and brains. In other circles, consciousness poses only ill-formed pseudo-problems not worthy of serious consideration. The contributions in this section occupy a middle ground between mystery-mongering and pessimistic dismissal. These chapters are united in the view that scientific approaches to consciousness have shone and will continue to shine much light on the topic.

Much of what is problematic about consciousness stems from the fact that the terms *conscious* and *consciousness* are applied to so many diverse phenomena. To name just a few, one might ask whether a recently pummeled boxer is conscious, or whether a driver talking on the car phone is conscious of the other cars on the road, or whether Americans have raised their consciousness of international and domestic human rights violations. Here the discussion of consciousness is restricted to sensory consciousness (consciousness of something that may be perceived by the senses), with a large, but not total, emphasis on visual consciousness. In chapter 14, neuroscientists Crick and Koch review their own theory of consciousness as it arises in vision. Continuing the interest in vision, in chapter 15 Prinz articulates a neuroscientifically informed account of consciousness with special attention paid to philosophical issues. Attention turns from vision to pain viewed as a sensory system in Hardcastle's chapter 16. Hardcastle examines philosophical issues that, when the topic is pain, take on enormous moral and medical significance. For instance, the question arises as to whether pain is subjective or objective. Differing answers to this question have led to differing answers to the possibility of (and the obligation to find) treatments of pain. In the final chapter, 17, Mandik focuses on the alleged subjectivity of conscious experience: again with an emphasis on vision, but also drawing on discussion of the sensory experience of temperature (discussed at great length in Akins's chapter 20). Mandik argues for a conception of the objective/subjective distinction that draws on a neuroscientific understanding of egocentric and allocentric representations.

14

Consciousness and Neuroscience

Francis Crick and Christof Koch

When all's said and done, more is said than done.

Anon

The main purposes of this review are to set out for neuroscientists one possible approach to the problem of consciousness and to describe the relevant ongoing experimental work. We have not attempted an exhaustive review of other approaches.

1 Clearing the Ground

We assume that when people talk about "consciousness," there is something to be explained. While most neuroscientists acknowledge that consciousness exists, and that at present it is something of a mystery, most of them do not attempt to study it, mainly for one of two reasons:

1 they consider it to be a philosophical problem, and so best left to philosophers;
2 they concede that it is a scientific problem, but think it is premature to study it now.

We have taken exactly the opposite point of view. We think that most of the philosophical aspects of the problem should, for the moment, be left on one side, and that the time to start the scientific attack is now.

We can state bluntly the major question that neuroscience must first answer: It is probable that at any moment some active neuronal processes in your head correlate with consciousness, while others do not; *what is the difference between them?* In particular, are the neurons involved of any particular neuronal type? What is special (if anything) about their connections? And what is special (if anything) about their way of firing? The neuronal correlate of consciousness is often referred to as the

NCC. Whenever some information is represented in the NCC it is represented in consciousness.

In approaching the problem, we made the tentative assumption (Crick and Koch, 1990) that all the different aspects of consciousness (for example, pain, visual awareness, self-consciousness, and so on) employ a basic common mechanism or perhaps a few such mechanisms. If one could understand the mechanism for one aspect, then, we hope, we will have gone most of the way towards understanding them all.

We made the personal decision (Crick and Koch, 1990) that several topics should be set aside or merely stated without further discussion, for experience had shown us that otherwise valuable time can be wasted arguing about them without coming to any conclusion.

1   Everyone has a rough idea of what is meant by being conscious. For now, it is better to avoid a precise definition of consciousness because of the dangers of premature definition. Until the problem is understood much better, any attempt at a formal definition is likely to be either misleading or overly restrictive, or both. If this seems evasive, try defining the word "gene." So much is now known about genes that any simple definition is likely to be inadequate. How much more difficult, then, to define a biological term when rather little is known about it.

2   It is plausible that some species of animals – in particular the higher mammals – possess some of the essential features of consciousness, but not necessarily all. For this reason, appropriate experiments on such animals may be relevant to finding the mechanisms underlying consciousness. It follows that a language system (of the type found in humans) is not essential for consciousness – that is, one can have the key features of consciousness without language. This is not to say that language does not enrich consciousness considerably.

3   It is not profitable at this stage to argue about whether simpler animals (such as octopus, fruit flies, nematodes) or even plants are conscious (Nagel, 1997). It is probable, however, that consciousness correlates to some extent with the degree of complexity of any nervous system. When one clearly understands, both in detail and in principle, what consciousness involves in humans, then will be the time to consider the problem of consciousness in much simpler animals. For the same reason, we will not ask whether some parts of our nervous system have a special, isolated, consciousness of their own. If you say, "Of course my spinal cord is conscious but it's not telling me," we are not, at this stage, going to spend time arguing with you about it. Nor will we spend time discussing whether a digital computer could be conscious.

4   There are many forms of consciousness, such as those associated with seeing, thinking, emotion, pain, and so on. Self-consciousness – that is, the self-referential aspect of consciousness – is probably a special case of consciousness. In our view, it is better left to one side for the moment, especially as it would be difficult to study self-consciousness in a monkey. Various rather unusual states,

such as the hypnotic state, lucid dreaming, and sleepwalking, will not be considered here, since they do not seem to us to have special features that would make them experimentally advantageous.

Visual consciousness

How can one approach consciousness in a scientific manner? Consciousness takes many forms, but for an initial scientific attack it usually pays to concentrate on the form that appears easiest to study. We chose visual consciousness rather than other forms, because humans are very visual animals and our visual percepts are especially vivid and rich in information. In addition, the visual input is often highly structured yet easy to control.

The visual system has another advantage. There are many experiments that, for ethical reasons, cannot be done on humans but can be done on animals. Fortunately, the visual system of primates appears fairly similar to our own (Tootell et al., 1996), and many experiments on vision have already been done on animals such as the macaque monkey.

This choice of the visual system is a personal one. Other neuroscientists might prefer one of the other sensory systems. It is, of course, important to work on alert animals. Very light anesthesia may not make much difference to the response of neurons in macaque V1, but it certainly does to neurons in cortical areas like V4 or IT (inferotemporal).

Why are we conscious?

We have suggested (Crick and Koch, 1995a) that the biological usefulness of visual consciousness in humans is to produce the best current interpretation of the visual scene in the light of past experience, either of ourselves or of our ancestors (embodied in our genes), and to make this interpretation directly available, for a sufficient time, to the parts of the brain that contemplate and plan voluntary motor output, of one sort or another, including speech.

Philosophers, in their carefree way, have invented a creature they call a "zombie," who is supposed to act just as normal people do but to be completely *un*conscious (Chalmers, 1995). This seems to us to be an untenable scientific idea, but there is now suggestive evidence that part of the brain does behave like a zombie. That is, in some cases, a person uses the current visual input to produce a relevant motor output, without being able to say what was seen. Milner and Goodale (1995) point out that a frog has at least two independent systems for action, as shown by Ingle (1973). These may well be unconscious. One is used by the frog to snap at small, prey-like objects, and the other for jumping away from large, looming discs. Why does not our brain consist simply of a series of such specialized zombie systems?

We suggest that such an arrangement is inefficient when very many such systems are required. Better to produce a single but complex representation and make it available for a sufficient time to the parts of the brain that make a choice among many different but possible plans for action. This, in our view, is what seeing is about. As pointed out to us by Ramachandran and Hirstein (1997), it is sensible to have a *single* conscious interpretation of the visual scene, in order to eliminate hesitation.

Milner and Goodale (1995) suggest that in primates there are two systems, which we shall call the on-line system and the seeing system. The latter is conscious, while the former, acting more rapidly, is not. The general characteristics of these two systems and some of the experimental evidence for them are outlined below in the section on the on-line system. There is anecdotal evidence from sports. It is often stated that a trained tennis player reacting to a fast serve has no time to see the ball; the seeing comes afterwards. In a similar way, a sprinter is believed to start to run before he consciously hears the starting pistol.

The nature of the visual representation

We have argued elsewhere (Crick and Koch, 1995a) that to be aware of an object or event, the brain has to construct a multilevel, explicit, symbolic interpretation of part of the visual scene. By multilevel, we mean, in psychological terms, different levels such as those that correspond, for example, to lines or eyes or faces. In neurological terms, we mean, loosely, the different levels in the visual hierarchy (Felleman and Van Essen, 1991).

The important idea is that the representation should be explicit. We have had some difficulty getting this idea across (Crick and Koch, 1995a). By an explicit representation, we mean a smallish group of neurons which employ coarse coding, as it is called (Ballard et al., 1983), to represent some *aspect* of the visual scene. In the case of a particular face, all of these neurons can fire to somewhat face-like objects (Young and Yamane, 1992). We postulate that one set of such neurons will be all of one type (say, one type of pyramidal cell in one particular layer or sublayer of cortex), will probably be fairly close together, and will all project to roughly the same place. If all such groups of neurons (there may be several of them, stacked one above the other) were destroyed, then the person would not see a face, though he or she might be able to see the parts of a face, such as the eyes, the nose, the mouth, etc. There may be other places in the brain that explicitly represent other aspects of a face, such as the emotion the face is expressing (Adolphs et al., 1994).

Notice that while the *information* needed to represent a face is contained in the firing of the ganglion cells in the retina, there is, in our terms, no explicit representation of the face there.

How many neurons are there likely to be in such a group? This is not yet known, but we would guess that the number to represent one aspect is likely to be closer to 100–1,000 than to 10,000–1,000,000.

A representation of an object or an event will usually consist of representations of many of the relevant aspects of it, and these are likely to be distributed, to some degree, over different parts of the visual system. How these different representations are bound together is known as the binding problem (von der Malsburg, 1995).

Much neural activity is usually needed for the brain to construct a representation. Most of this is probably unconscious. It may prove useful to consider this unconscious activity as the computations needed to find the best interpretation, while the interpretation itself may be considered to be the *results* of these computations, only some of which we are then conscious of. To judge from our perception, the results probably have something of a winner-take-all character.

As a working hypothesis we have assumed that only some types of specific neurons will express the NCC. It is already known (see the discussion under "Bistable percepts") that the firing of many cortical cells does not correspond to what the animal is currently seeing. An alternative possibility is that the NCC is necessarily global (Greenfield, 1995). In one extreme form this would mean that, at one time or another, any neuron in cortex and associated structures could express the NCC. At this point, we feel it more fruitful to explore the simpler hypothesis – that only particular types of neurons express the NCC – before pursuing the more global hypothesis. It would be a pity to miss the simpler one if it were true. As a rough analogy, consider a typical mammalian cell. The way its complex behavior is controlled and influenced by its genes could be considered to be largely global, but its genetic instructions are localized, and coded in a relatively straightforward manner.

Where is the visual representation?

The conscious visual representation is likely to be distributed over more than one area of the cerebral cortex and possibly over certain subcortical structures as well. We have argued (Crick and Koch, 1995a) that in primates, contrary to most received opinion, it is not located in cortical area V1 (also called the striate cortex or area 17). Some of the experimental evidence in support of this hypothesis is outlined below. This is not to say that what goes on in V1 is not important, and indeed may be crucial, for most forms of vivid visual awareness. What we suggest is that the neural activity there is not directly correlated with what is seen.

We have also wondered (Crick, 1994) whether the visual representation is largely confined to certain neurons in the lower cortical layers (layers 5 and 6). This hypothesis is still very speculative.

What is essential for visual consciousness?

The term "visual consciousness" almost certainly covers a variety of processes. When one is actually looking at a visual scene, the experience is very vivid. This should be contrasted with the much less vivid and less detailed visual images produced by trying to remember the same scene. (A vivid recollection is usually called a hallucination.) We are concerned here mainly with the normal vivid experience. (It is possible that our dimmer visual recollections are mainly due to the back pathways in the visual hierarchy acting on the random activity in the earlier stages of the system.)

Some form of very short-term memory seems almost essential for consciousness, but this memory may be very transient, lasting for only a fraction of a second. Edelman (1989) has used the striking phrase, "the remembered present," to make this point. The existence of iconic memory, as it is called, is well established experimentally (Coltheart, 1983; Gegenfurtner and Sperling, 1993).

Psychophysical evidence for short-term memory (Potter, 1976; Subramaniam et al., 1997) suggests that if we do not pay attention to some part or aspect of the visual scene, our memory of it is very transient and can be overwritten (masked) by the following visual stimulus. This probably explains many of our fleeting memories when we drive a car over a familiar route. If we do pay attention (e.g. a child running in front of the car) our recollection of this can be longer-lasting.

Our impression that at any moment we see all of a visual scene very clearly and in great detail is illusory, partly due to ever-present eye movements and partly due to our ability to use the scene itself as a readily available form of memory, since in most circumstances the scene usually changes rather little over a short span of time (O'Regan, 1992).

Although working memory (Baddeley, 1992; Goldman-Rakic, 1995) expands the time frame of consciousness, it is not obvious that it is *essential* for consciousness. It seems to us that working memory is a mechanism for bringing an item, or a small sequence of items, into vivid consciousness, by speech, or silent speech, for example. In a similar way, the episodic memory enabled by the hippocampal system (Zola-Morgan and Squire, 1993) is not essential for consciousness, though a person without it is severely handicapped.

Consciousness, then, is enriched by visual attention, though attention is not essential for visual consciousness to occur (Rock et al., 1992; Braun and Julesz, 1998). Attention is broadly of two types: bottom-up, caused by the sensory input; and top-down, produced by the planning parts of the brain. This is a complicated subject, and we will not try to summarize here all the experimental and theoretical work that has been done on it.

Visual attention can be directed to either a location in the visual field or to one or more (moving) objects (Kanwisher and Driver, 1992). The exact neural mechanisms that achieve this are still being debated. In order to interpret the visual input, the brain must arrive at a *coalition* of neurons whose firing represents the best interpretation of the visual scene, often in competition with other possible but less likely interpretations; and there is evidence that attentional mechanisms appear to bias this competition (Luck et al., 1997).

Recent experimental results

We shall not attempt to describe all the various experimental results of direct relevance to the search for the neuronal correlates of visual consciousness in detail but rather outline a few of them and point the reader to fuller accounts.

Action without seeing

Classical blindsight   This will already be familiar to most neuroscientists. It is discussed, along with other relevant topics, in an excellent book by Weiskrantz (1997). It occurs in humans (where it is rare) when there is extensive damage to cortical area V1 and has also been reproduced in monkeys (Cowey and Stoerig, 1995). In a typical case, the patient can indicate, well above chance level, the direction of movement of a spot of light over a certain range of speed, while denying that he sees anything at all. If the movement is less salient, his performance falls to chance; if more salient (that is, brighter or faster), he may report that he had some ill-defined visual percept, considerably different from the normal one. Other patients can distinguish large, simple shapes or colors. (For Weiskrantz's comments on Gazzaniga's criticisms, see pages 152–3; and on Zeki's criticisms, see pages 247–8.)

The pathways involved have not yet been established. The most likely one is from the superior colliculus to the pulvinar and from there to parts of visual cortex; several other known weak anatomical pathways from the retina and bypassing V1 are also possible. Recent functional magnetic resonance imaging of the blindsight patient G.Y. directly implicates the superior colliculus as being active *specifically* when G.Y. correctly discriminates the direction of motion of some stimulus without being aware of it at all (Sahraie et al., 1997 – this paper should be consulted for further details of the areas involved).

The on-line system   The broad properties of the two hypothetical systems – the on-line system and the seeing system – are shown in table I of Milner and Goodale's book *The Visual Brain in Action* (1995), to which the reader is referred for a more extended account. For a recent review, see Boussaoud et al. (1996). The on-line system may have multiple subsystems (e.g. for eye movements, for arm movements, for body posture adjustment, and so on). Normally, the two systems work in parallel, and indeed there is evidence that in some circumstances the seeing system can interfere with the on-line system (Rossetti, 2000).

One striking piece of evidence for an on-line system comes from studies on patient D.F. by Milner, Perrett, and their colleagues (1991). Her brain has diffuse damage produced by carbon monoxide poisoning. She is able to see color and texture very well but is very deficient in seeing orientation and form. In spite of this, she is very good at catching a ball. She can "post" her hand or a card into a slot without difficulty, though she could not report the slot's orientation.

It is obviously important to discover the difference between the on-line system, which is unconscious, from the seeing system, which is conscious. Milner and Goodale (1995) suggest that the on-line system mainly uses the dorsal visual stream. They propose that rather than being the *where* stream, as suggested by Ungerleider and Mishkin (1982), it is really the *how* stream. This might imply that all activity in the dorsal stream is unconscious. The ventral stream, on the other hand, they

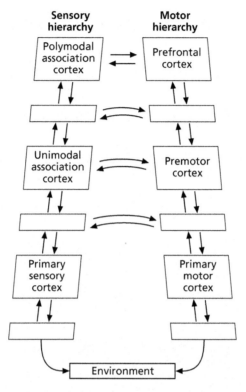

Figure 14.1 Fuster's figure (reproduced with permission by Lippincott-Raven Publishers) showing the fiber connections between cortical regions participating in the perception-action cycle. Empty rhomboids stand for intermediate areas or subareas of the labeled regions. Notice that there are connections between the two hierarchies at several levels, not just at the top level.

consider to be largely conscious. An alternative suggestion, due to Steven Wise (personal communication, and Boussaoud et al., 1996), is that direct projections from parietal cortex into premotor areas are unconscious, whereas projections to them via prefrontal cortex are related to consciousness.

Our suspicion is that while these suggestions about two systems are on the right lines, they are probably over-simple. The little that is known of the neuroanatomy would suggest that there are likely to be *multiple* cortical streams, with numerous anatomical connections between them (Distler et al., 1993). This is implied in figure 14.1, a diagram often used by Fuster (Fuster, 1997: see his figure 8.4). In short, the neuroanatomy does not suggest that the sole pathway goes up to the highest levels of the visual system, and from there to the highest levels of the prefrontal system and then down to the motor output. There are numerous pathways from most intermediate levels of the visual system to intermediate frontal regions.

We would therefore like to suggest a general hypothesis: that the brain always tries to use the quickest *appropriate* pathway for the situation at hand. Exactly how this idea works out in detail remains to be discovered. Perhaps there is competition, and the fastest stream wins. The postulated on–line system would be the quickest of these hypothetical cortical streams. This would be the zombie part of you.

Bistable percepts

Perhaps the present most important experimental approach to finding the NCC is to study the behavior of single neurons in the monkey's brain when it is looking at something that produces a bistable percept. The visual input, apart from minor eye movements, is constant; but the subject's percept can take one of two alternative forms. This happens, for example, when one looks at a drawing of the well-known Necker cube.

It is not obvious where to look in the brain for the two alternative views of the Necker cube. Allman suggested a more practical alternative: to study the responses in the visual system during binocular rivalry (Myerson et al., 1981). If the visual input into each eye is different, but perceptually overlapping, one usually sees the visual input as received by one eye alone, then by the other one, then by the first one, and so on. The input is constant, but the percept changes. Which neurons in the brain mainly follow the input, and which the percept?

This approach has been pioneered by Logothetis and his colleagues, working on the macaque visual system. They trained the monkey to report which of two rival-rous inputs it saw. The experiments are difficult, and elaborate precautions had to be taken to make sure the monkey was not cheating. The fairly similar distribution of switching times strongly suggests that monkeys and humans perceive these bistable visual inputs in the same way.

The first set of experiments (Logothetis and Schall, 1989) studied neurons in cortical area MT (medial temporal, also called V5), since they preferentially respond to movement. The stimuli were vertically drifting horizontal gratings. Only the first response was recorded. Of the relevant neurons, only about 35 percent were modulated according to the monkey's reported percept. Surprisingly, half of these responded in the opposite direction to the one expected.

The second set of experiments (Leopold and Logothetis, 1996) used stationary gratings. The orientation was chosen in each case to be optimal for the neuron studied, and orthogonal to it in the other eye. They recorded how the neuron fired during several alterations of the reported percept. The neurons were in foveal V1/V2 and in V4. The fraction following the percept in V4 was similar to that in MT, but a rather smaller fraction of V1/V2 neurons followed the percept. Also, here, but not in V4, none of the cells were anticorrelated with the stimulus.

The results of the third set of experiments (Sheinberg and Logothetis, 1997) were especially striking. In this case the visual inputs tried included images of humans, monkeys, apes, wild animals, butterflies, reptiles and various man-made objects. The

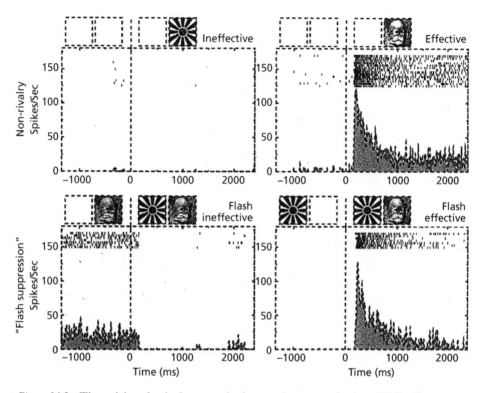

Figure 14.2 The activity of a single neuron in the superior temporal sulcus (STS) of a macaque monkey in response to different stimuli presented to the two eyes (taken from Sheinberg and Logothetis, 1997). In the upper left panel a sunburst pattern is presented to the right eye without evoking any firing response ("ineffective" stimulus). The same cell will fire vigorously in response to its "effective" stimulus, here the image of a monkey's face (upper right panel). When the monkey is shown the face in one eye for a while, and the sunburst pattern is flashed on to the monitor for the other eye, the monkey signals that it is "seeing" this new pattern and that the stimulus associated with the rivalrous eye is perceptually suppressed ("flash suppression"; lower left panel). At the neuronal level, the cell shuts down in response to the ineffective yet perceptual dominant stimulus following stimulus onset (at the dotted line). Conversely, if the monkey fixates the sunburst pattern for a while, and the image of the face is flashed on, it reports that it perceives the face, and the cell will now fire strongly (lower right panel). Neurons in V4, earlier in the cortical hierarchy, are largely unaffected by perceptual changes during flash suppression.

rivalrous image was usually a sunburst-like pattern (see figure 14.2). If a new image was flashed into one eye while the second eye was fixating another pattern, the new stimulus was the one that was always perceived ("flash suppression"). Recordings were made in the upper and lower banks of the superior temporal sulcus (STS) and inferior temporal cortex (IT). Overall, approximately 90 percent of the recorded neurons in STS and IT were found to reliably predict the perceptual state of the animal. Moreover, many of these neurons responded in an almost all-or-none fashion, firing strongly for one percept, yet only at noise level for the alternative one.

More recently, Bradley et al. (1998) have studied a different bistable percept in macaque MT, produced by showing the monkey, on a TV screen, the 2-D projection of a transparent, rotating cylinder with random dots on it, without providing any stereoscopic disparity information. Human subjects exploit structure-from-motion and see a 3-D cylinder rotating around its axis. Without further clues, the direction of rotation is ambiguous and observers first report rotation in one direction, a few seconds later, rotation in the other direction, and so on. The trained monkey responds as if it saw the same alteration. In their studies on the monkey, about half the relevant MT neurons Bradley et al. recorded from followed the percept (rather than the "constant" retinal stimulus).

These are all exciting experiments, but they are still in the early stages. Just because a particular neuron follows the percept, it does not automatically imply that its firing is part of the NCC. The NCC neurons may be mainly elsewhere, such as higher up in the visual hierarchy. It is obviously important to discover, for each cortical area, *which* neurons are following the percept (Crick, 1996). That is, what type of neurons are they, in which cortical layer or sublayer do they lie, in what way do they fire, and, most important of all, *where do they project?* It is, at the moment, technically difficult to do this, but it is essential to have this knowledge, or it will be almost impossible to understand the neural nature of consciousness.

Electrical brain stimulation

An alternate approach, with roots going back to Penfield (1958), involves directly stimulating cortex or related structures in order to evoke a percept or behavioral act. Libet and his colleagues (Libet, 1993) have used this technique to great advantage on the somatosensory system of patients. They established that a stimulus, at or near threshold, delivered through an electrode placed on to the surface of somatosensory cortex or into the ventrobasal thalamus required a minimal stimulus duration (between 0.2 and 0.5 sec) in order to be consciously perceived. Shorter stimuli were not perceived, even though they could be detected with above-chance probability, using a two-alternative forced choice procedure. In contrast, a skin or peripheral sensory-nerve stimulus of very short duration could be perceived. The difference appears to reside in the amount and type of neurons recruited during peripheral stimulation versus direct central stimulation. Using sensory events as a marker, Libet also established (1993) that events caused by direct cortical stimulation were back-dated to the beginning of the stimulation period.

In a series of classical experiments, Newsome and colleagues (Britten et al., 1992) studied the macaque monkey's performance in a demanding task involving visual motion discrimination. They established a quantitative relationship between the performance of the monkey and the neuronal discharge of neurons in its medial temporal cortex (MT). In 50 percent of all the recorded cells, the psychometric curve – based on the behavior of the entire animal – was statistically indistinguishable from the neurometric curve – based on the averaged firing rate of a single MT cell. In a second series of experiments, cells in MT were directly stimulated via an extracellular electrode (Salzman et al., 1990; MT cells are arranged in columnar structure for direction of motion). Under these conditions, the performance of the animal shifted in a predictable manner, compatible with the idea that the small brain stimulation caused the firing of enough MT neurons, encoding for motion in a specific direction, to influence the final decision of the animal. It is not clear, however, to what extent visual consciousness for this particular task is present in these highly overtrained monkeys.

The V1 hypothesis

We have argued (Crick and Koch, 1995a) that one is not directly conscious of the features represented by the neural activity in primary visual cortex. Activity in V1 may be necessary for vivid and veridical visual consciousness (as is activity in the retinae), but we suggest that the firing of none of the neurons in V1 directly correlates with what we consciously see (for a critique of our hypothesis, see Pollen, 1995, and our reply, Crick and Koch, 1995b).

Our reasons are that at each stage in the visual hierarchy the explicit aspects of the representation we have postulated is always recoded. We have also assumed that any neurons expressing an aspect of the NCC must project directly, without recoding, to at least some of the parts of the brain that plan voluntary action – that is what we have argued seeing is for. We think that these plans are made in some parts of frontal cortex (see below).

The neuroanatomy of the macaque monkey shows that V1 cells do not project directly to any part of frontal cortex (Crick and Koch, 1995a). Nor do they project to the caudate nucleus of the basal ganglia (Saint-Cyr et al., 1990), the intralaminar nuclei of the thalamus (L. G. Ungerleider, personal communication), the claustrum (Sherk, 1986) nor to the brain stem, with the exception of a small projection from peripheral V1 to the pons (Fries, 1990). It is plausible, but not yet established, that this lack of connectivity is also true for humans.

The strategy to verify or falsify this and related hypotheses is to relate the receptive field properties of individual neurons in V1 or elsewhere to perception in a quantitative manner. If the structure of perception does not map to the receptive field properties of V1 cells, it is unlikely that these neurons directly give rise to consciousness. In the presence of a correlation between perceptual experience and the receptive field properties of one or more groups of V1 cells, it is unclear whether these cells just correlate with consciousness or directly give rise to it. In that case, further experiments need to be carried out to untangle the exact relationship between neurons and perception.

A possible example may make this clearer. It is well known that the color we perceive at one particular visual location is influenced by the wavelengths of the light entering the eye from surrounding regions in the visual field (Land and McCann, 1971; Blackwell and Buchsbaum, 1988). This form of (partial) color constancy is often called the Land effect. It has been shown in the anesthetized monkey (Zeki, 1980, 1983; Schein and Desimone, 1990) that neurons in V4, but *not* in V1, exhibit the Land effect. As far as we know, the corresponding information is lacking for alert monkeys. If the same results could be obtained in a behaving monkey, it would follow that it would not be *directly* aware of the "color" neurons in V1.

Some experimental support In the last two years [1995–7], a number of psychophysical, physiological, and imaging studies have provided some support for our hypothesis, although this evidence falls short of proving it (He et al., 1996; Cumming and Parker, 1997; Kolb and Braun, 1995; summarized in Koch and Braun, 1996; but see Morgan et al., 1997). Let us briefly discuss two other cases.

When two isoluminant colors are alternated at frequencies beyond 10 Hz, humans perceive only a single fused color with a minimal sensation of brightness flicker. In spite of the perception of color fusion, color opponent cells in primary visual cortex of two alert macaque monkeys follow high-frequency flicker well above heterochromatic fusion frequencies (Gur and Snodderly, 1997). In other words, neuronal activity in V1 can clearly represent certain retinal stimulation yet is not perceived. This is supported by recent fMRI studies on humans by Engel et al. (1997).

The study by He et al. (1996) is based on a common visual aftereffect (see figure 14.3a). If a subject stares for a fraction of a minute at a horizontal grating, and is then tested with a faint grating at the same location to decide whether it is oriented vertically or horizontally, the subject's sensitivity for detecting a horizontal grating will be reduced. This adaptation is orientation-specific – the sensitivity for vertical gratings is almost unchanged – and disappears quickly. He and colleagues projected a single patch of grating on to a computer screen some 25 degrees from the fixation point. It was clearly visible and their subjects showed the predictable orientation-selective adaptation effect. Adding one or more similar patches of gratings to either side of the original grating – which remained exactly as before – removed the lines of the grating from visibility; it was now "masked." Subjectively, one still sees "something" at the location of the original grating, but one is unable to make out its orientation, even when given unlimited viewing time. Yet despite this inability to "see" the adapting stimulus, the aftereffect was as strong and as specific to the orientation of the "invisible" grating as when the grating was visible (see figure 14.3b). What this shows, foreshadowed by earlier experiments (Blake and Fox, 1974), is that visual awareness in such cases must occur at a higher stage in the visual hierarchy than orientation-specific adaptation. This aftereffect is thought to be mediated by oriented neurons in V1 and beyond, implying that at least in this case the neurons which mediate visual awareness must be located past this stage.

Our ideas regarding the absence of the NCC from V1 are not disproven by PET experiments showing that in at least some people V1 is activated during visual imagery tasks (Kosslyn et al., 1995), though severe damage to V1 is compatible with visual imagery in patients (Goldenberg et al., 1995). There is no obvious reason why such top-down effects should not reach V1. Such V1 activity would not, by itself, prove that we are *directly* aware of it, any more than the V1 activity produced there

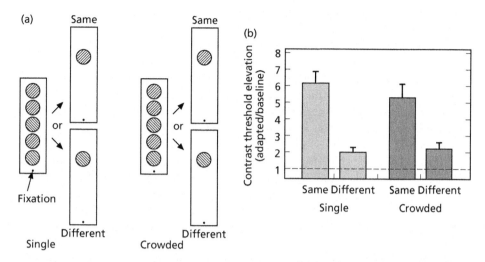

Figure 14.3 Psychophysical displays (schematic) and results pertaining to an orientation-dependent aftereffect induced by "crowded" grating patches (reproduced with permission by He, Cavanagh, and Intriligator). (a) Adaptation followed by contrast threshold measurement for a single grating (left) and a crowded grating (right). In each trial, the orientation of the adapting grating was either the same or orthogonal to the orientation of the test grating. Observers fixated at a distance of approximately 25 degrees from the adapting and test gratings. (b) Threshold contrast elevation after adaptation relative to baseline threshold contrast before adaptation. Data are averaged across four subjects. The difference between same and different adapt-test orientations reflects the orientation-selective aftereffect of the adapting grating. The data show that this aftereffect is comparable for a crowded grating (whose orientation is not consciously perceived) and for a single grating (whose orientation is readily perceived).

when our eyes are open proves this. We hope that further neuroanatomical work will make our hypothesis plausible for humans, and that further neurophysiological studies will show it to be true for most primates. If correct, it would narrow the search to areas of the brain farther removed from the sensory periphery.

The frontal lobe hypothesis

As mentioned several times, we hypothesize that the NCC must have access to explicitly encoded visual information and directly project into the planning stages of the brain, associated with the frontal lobes in general and with prefrontal cortex in particular (Fuster, 1997). We would therefore predict that patients unfortunate enough to have lost their entire prefrontal cortex on both sides (including Broca's area) would not be visually conscious, although they might still have well-preserved, but unconscious, visual-motor abilities. No such patient is known to us (not even Brickner's famous patient; for an extensive discussion of this, see Damasio and Anderson, 1993). The visual abilities of any such "frontal lobe" patient need to be carefully evaluated using a battery of appropriate psychophysical tests.

The fMRI study of the blindsight patient G.Y. (Sahraie et al., 1997) provides direct evidence for our view by revealing that prefrontal areas 46 and 47 are active when G.Y. is visually aware of a moving stimulus.

The recent findings of neurons in the inferior prefrontal cortex (IPC) of the macaque that respond selectively to faces – and that receive direct input from regions around the superior temporal sulcus and the inferior temporal gyrus that are well known to contain face-selective neurons – is also very encouraging in this regard (Scalaidhe et al., 1997). This raises the question of why face cells would be represented in *both* IT and IPC. Do they differ in some important aspect?

Large-scale lesion experiments carried out in the monkey suggest that the absence of frontal lobes leads to complete blindness (Nakamura and Mishkin, 1980, 1986). One would hope that future monkey experiments reversibly inactivate specific prefrontal areas and demonstrate the specific loss of abilities linked to visual perception while visual-motor behaviors – mediated by the on-line system – remain intact.

It will be important to study the pattern of connections between the highest levels of the visual hierarchy – such as inferotemporal cortex – and premotor and prefrontal cortex. In particular, does the anatomy reveal any feedback loops that might sustain activity between IT and prefrontal neurons (Crick and Koch, 1998)? There is suggestive evidence (Webster et al., 1994) that projections from prefrontal cortex back into IT might terminate in layer 4, but these need to be studied directly.

Gamma oscillations

Much has been made of the presence of oscillations in the gamma range (30–70 Hz) in the local-field potential and in multi-unit recordings in the visual and sensory-motor system of cats and primates (Singer and Gray, 1995). The existence of such oscillations remains in doubt in higher visual cortical areas (Young et al., 1992). We remain agnostic with respect to the relevance of these oscillations to conscious perception. It is possible that they subserve figure-ground in early visual processing.

2   Philosophical Matters

There is, at the moment, no agreed philosophical answer to the problem of consciousness, except that most living philosophers are not Cartesian dualists – they do not believe in an immaterial soul which is distinct from the body. We suspect that the majority of neuroscientists do not believe in dualism, the most notable exception being the late Sir John Eccles (1994).

We shall not describe here the various opinions of philosophers, except to say that while philosophers have, in the past, raised interesting questions and pointed to possible conceptual confusions, they have had a very poor record, historically, at arriving at valid scientific answers. For this reason, neuroscientists should listen to the questions philosophers raise but should not be intimidated by their discussions. In recent years the amount of discussion about consciousness has reached absurd proportions compared to the amount of relevant experimentation.

The problem of qualia

What is it that puzzles philosophers? Broadly speaking, it is qualia – the blueness of blue, the painfulness of pain, and so on. This is also the layman's major puzzle. How can you possibly explain the vivid visual scene you see before you in terms of the firing of neurons? The argument that you cannot explain consciousness by the action of the parts of the brain goes back at least as far as Leibniz (1686; see the translation 1965). But compare an analogous assertion: that you cannot explain the "livingness" of living things (such as bacteria, for example) by the action of "dead" molecules. This assertion sounds extremely hollow now, for a number of reasons. Scientists understand the enormous power of natural selection. They know the chemical nature of genes and that inheritance is particulate, not blending. They understand the great subtlety, sophistication, and variety of protein molecules, the elaborate nature of the control mechanisms that turn genes on and off, and the complicated way that proteins interact with, and modify, other proteins. It is entirely possible that the very elaborate nature of neurons and their interactions, far more elaborate than most people imagine, is misleading us, in a similar way, about consciousness.

Some philosophers (Searle, 1984; Dennett, 1996) are rather fond of this analogy between "livingness" and "consciousness," and so are we; but, as Chalmers (1995) has emphasized, an analogy is only an analogy. He has given philosophical reasons why he thinks it is wrong. Neuroscientists know only a few of the basics of neuroscience, such as the nature of the action potential and the chemical nature of most synapses. Most important, there is not a comprehensive, overall theory of the activities of the brain. To be shown to be correct, the analogy must be filled out by many experimental details and powerful general ideas. Much of these are still lacking.

This problem of qualia is what Chalmers (1995) calls "the hard problem": a full account of the manner in which subjective experience arises from cerebral processes. As we see it, the hard problem can be broken down into several questions, of which the first is the major problem: How do we experience anything at all? What leads to a particular conscious experience (such as the blueness of blue)? What is the function of conscious experience? Why are some aspects of subjective experience impossible to convey to other people (in other words, why are they private)?

We believe we have answers to the last two questions (Crick and Koch, 1995c). We have already explained, in the section "Why are we conscious," what we think consciousness is for. The reason that visual consciousness is largely private is, we consider, an inevitable consequence of the way the brain works. (By "private," we mean that it is inherently impossible to communicate the exact nature of what we are conscious of.) To be conscious, we have argued, there must be an explicit representation of each aspect of visual consciousness. At each successive stage in the visual cortex, what is made explicit is recoded. To produce a motor output, such as speech, the information must be recoded again, so that what is expressed by the motor neurons is related, but not identical, to the explicit representation expressed by the firing of the neurons associated with, for example, the color experience at some level in the visual hierarchy.

It is thus not possible to convey with words the exact nature of a subjective experience. It is possible, however, to convey a *difference* between subjective experiences – to distinguish between red and orange, for example. This is possible because a difference in a high-level visual cortical area can still be associated with a difference at the motor stage. The implication is that we can never explain to other people the nature of any conscious experience, only, in some cases, its relation to other ones.

Is there any sense in asking whether the blue color you see is subjectively the same as the blue color I see? If it turns out that the neural correlate of blue is exactly the same in your brain as in mine, it would be scientifically plausible to infer that you see blue as I do. The problem lies in the word "exactly." How precise one has to be will depend on a detailed knowledge of the processes involved. If the neural correlate of blue depends, in an important way, on my past experience, and if my past experience is significantly different from yours, then it may not be possible to deduce that we both see blue in exactly the same way (Crick, 1994).

Could this problem be solved by connecting two brains together in some elaborate way? It is impossible to do this at the moment, or in the easily foreseeable future. One is therefore tempted to use the philosopher's favorite tool, the thought experiment. Unfortunately, this enterprise is fraught with hazards, since it inevitably makes assumptions about how brains behave, and most of these assumptions have so little experimental support that conclusions based on them are valueless. For example, how much is a person's percept of the blue of the sky due to early visual experiences?

The problem of meaning

An important problem neglected by neuroscientists is the problem of meaning. Neuroscientists are apt to assume that if they can see that a neuron's firing is roughly correlated with some aspect of the visual scene, such as an oriented line, then that firing must be part of the neural correlate of the seen line. They assume that because they, as outside observers, are conscious of the correlation, the firing must be part of the NCC. This by no means follows, as we have argued for neurons in V1.

But this is not the major problem, which is: How do other parts of the brain know that the firing of a neuron (or of a set of similar neurons) produces the conscious percept of, say, a face? How does the brain know what the firing of those neurons represents? Put in other words, how is meaning generated by the brain?

This problem has two aspects. How is meaning expressed in neural terms? And how does this expression of meaning arise? We suspect (Crick and Koch, 1995c) that meaning derives both from the correlated firing described above and from the linkages to related representations. For example, neurons related to a certain face might be connected to ones expressing the name of the person whose face it is, and to others for her voice, memories involving her and so on, in a vast associational network, similar to a dictionary or a relational database. Exactly how this works in detail is unclear.

But how are these useful associations derived? The obvious idea is that they depend very largely on the consistency of the interactions with the environment, especially during early development. Meaning can also be acquired later in life. The usual example is a blind man with a stick. He comes to feel what the stick is touching, not merely the stick itself. For an ingenious recent demonstration along similar lines, see Ramachandran and Hirstein (1997).

3   Future Experiments

Although experiments on attention, short-term and working memory, the correlated firing of neurons, and related topics may make finding the NCC easier, at the moment the most promising experiments are those on bistable percepts. These experiments should be continued in numerous cortical and thalamic areas and need extending to cover other such percepts. It is also important to discover *which* neurons express the NCC in each case (e.g. which neuronal subtype, in what layer, and so on?), how they fire (e.g. do they fire in bursts?) and, especially, to where they project. To assist this, more detailed neuroanatomy of the connectivity will be needed. This is relatively easy to do in the macaque but difficult in humans (Crick and Jones, 1993). It is also important to discover how the various on-line systems work, so that one can contrast their (unconscious) neuronal activity with the NCC.

To discover the exact role (if any) of the frontal cortex in visual perception, it would be useful to inactivate it reversibly by cooling and/or the injection of GABA agonists, perhaps using the relatively smooth cortex of an owl monkey.

Inevitably, it will be necessary to compare the studies on monkeys with similar studies on humans, using both psychophysical experiments as well as functional imaging methods such as PET or fMRI. Conversely, functional imaging experiments on normal subjects or patients, showing for instance the involvement of prefrontal areas in visual perception (Weiskrantz, 1997; Sahraie et al., 1997), can provide a rationale for appropriate electrophysiological studies in monkeys. It would help considerably if there were more detailed architectonic studies of cortex and thalamus, since these can be done post-mortem on monkeys, apes, and humans. The extremely rapid pace of molecular biology should soon provide a wealth of new markers to help in this endeavor.

To understand a very complex nonlinear system, it is essential to be able to interfere with it both specifically and delicately. The major impact of molecular biology

is likely to be the provision of methods for the inactivation of all neurons of a particular type. Ideally, this should be done reversibly on the mature animal (see, for
example, No et al., 1996; Nirenberg and Meister, 1997). At the moment this is only
practical on mice, but in future one may hope for methods that can be used on mature
monkeys (perhaps using a viral vector), as such methods are also needed for the
medical treatment of humans.

As an example, consider the question of whether the cortical feedback pathways
– originating in a higher visual area (in the sense of Felleman and Van Essen, 1991)
and projecting into a lower area – are essential for normal visual consciousness.
There are at least two distinct types of back pathways (Salin and Bullier, 1995):
one, from the upper cortical layers, goes back only a few steps in the visual
hierarchy; the other, from the lower cortical layers, can also go back over longer
distances. We would like to be able to selectively inactivate these pathways, both
singly and collectively, in the mature macaque. Present methods are not specific
enough to do this, but new methods in molecular biology should, in time, make this
possible.

It will not be enough to show that certain neurons embody the NCC in certain –
limited – visual situations. Rather, we need to locate the NCC for all types of visual
inputs, or at least for a sufficiently large and representative sample of them. For
example, when one blinks, the eyelids briefly (30–50 msec) cover the eyes, yet the
visual percept is scarcely interrupted (blink suppression; Volkmann et al., 1980). We
would therefore expect the NCC to be also unaffected by eye blinks (e.g. the firing
activity should not drop noticeably during the blink) but not to blanking out of the
visual scene for a similar duration due to artificial means. Another example is the
large number of visual illusions. For instance, humans clearly perceive, under appropriate circumstances, a transient motion aftereffect. On the basis of fMRI imaging
it has been found that the human equivalent of cortical area MT is activated by the
motion aftereffect (in the absence of any moving stimuli; Tootell et al., 1995). The
time course of this illusion parallels the time course of activity as assayed using
fMRI. In order to really pinpoint the NCC, one would need to identify individual
cells expressing this, and similar, visual aftereffects. We have assumed that the visual
NCC in humans is very similar to the NCC in the macaque, mainly because of
the similarity of their visual systems. Ultimately, the link between neurons and
perception will need to be made in humans.

The problem of meaning and how it arises is more difficult, since there is, as yet,
not even an outline formulation of this problem in neural terms. For example, do
multiple associations depend on transient priming effects? Whatever the explanation, it would be necessary to study the developing animal to show how meaning
arises; in particular, how much is built in epigenetically and how much is due to
experience.

In the long run, finding the NCC will not be enough. A complete theory of consciousness is required, including its functional role. With luck this might illuminate
the hard problem of qualia. It is likely that scientists will then stop using the term
"consciousness" except in a very loose way. After all, biologists no longer worry

whether a seed or a virus is "alive." They just want to know how it evolved, how it develops, and what it can do.

4 Finale

We hope we have convinced the reader that the problem of the neural correlate of consciousness (the NCC) is now ripe for direct experimental attack. We have suggested a possible framework for thinking about the problem, but others may prefer a different approach; and, of course, our own ideas are likely to change with time. We have outlined the few experiments that directly address the problem and mentioned briefly other types of experiments that might be done in the future. We hope that some of the younger neuroscientists will seriously consider working on this fascinating problem. After all, it is rather peculiar to work on the visual system and not worry about exactly what happens in our brains when we "see" something. The explanation of consciousness is one of the major unsolved problems of modern science. After several thousand years of speculation, it would be very gratifying to find an answer to it.

Acknowledgments

We thank the J. W. Kieckhefer Foundation, the National Institute of Mental Health, the Office of Naval Research and the National Science Foundation. For helpful comments we thank David Chalmers, Leslie Orgel, John Searle, and Larry Weiskrantz.

Correspondence should be addressed to Dr Francis Crick, The Salk Institute, 10010 North Torrey Pines Road, La Jolla, California 92037, USA.

References

Adolphs, R., Tranel, D., Damasio, H., and Damasio, A. 1994: Impaired recognition of emotion in facial expressions following bilateral damage to the human amygdala. *Nature*, 372, 669–72.

Baddeley, A. 1992: Working memory. *Science*, 255, 556–9.

Ballard, D. H., Hinton, G. E., and Sejnowski, T. J. 1983: Parallel visual computation. *Nature*, 306, 21–6.

Blackwell, K. T., and Buchsbaum, G. 1988: Quantitative studies of color constancy. *Journal of the Optical Society of America*, A5, 1772–80.

Blake, R., and Fox, R. 1974: Adaptation to invisible gratings and the site of binocular rivalry suppression. *Nature*, 249, 488–90.

Boussaoud, D., di Pellegrino, G., and Wise, S. P. 1996: Frontal lobe mechanisms subserving vision-for-action versus vision-for-perception. *Behavioral Brain Research*, 72, 1–15.

Bradley, D. C., Chang, G. C., and Andersen, R. A. 1998: Activities of motion-sensitive neurons in primate visual area MT reflect the perception of depth. *Nature*, 392, 714–17.

Braun, J., and Julesz, B. 1998: Withdrawing attention at Little or no cost: detection and discrimination tasks. *Perception and Psychophysics*, 60, 1–23.

Britten, K. H., Shadlen, M. N., Newsome, W. T., and Movshon, J. A. 1992: The analysis of visual motion: A comparison of neuronal and psychophysical performance. *Journal of Neuroscience*, 12, 4745–65.

Chalmers, D. 1995: *The Conscious Mind: In Search of a Fundamental Theory*. Oxford: Oxford University Press.

Coltheart, M. 1983: Iconic memory. *Philosophical Transactions of the Royal Society of London, Series B*, 302, 283–94.

Cowey, A., and Stoerig, P. 1995: Blindsight in monkeys. *Nature*, 373, 247–9.

Crick, F. 1994: *The Astonishing Hypothesis*. New York: Scribners.

Crick, F. 1996: Visual perception: Rivalry and consciousness. *Nature*, 379, 485–6.

Crick, F., and Jones, E. 1993: Backwardness of human neuroanatomy. *Nature*, 361, 109–10.

Crick, F., and Koch, C. 1990: Towards a neurobiological theory of consciousness. *Seminars in the Neurosciences*, 2, 263–75.

Crick, F., and Koch, C. 1995a: Are we aware of neural activity in primary visual cortex? *Nature*, 375, 121–3.

Crick F., and Koch, C. 1995b: Cortical areas in visual awareness – Reply. *Nature*, 377, 294–5.

Crick, F., and Koch, C. 1995c: Why neuroscience may be able to explain consciousness. *Scientific American*, 273, 84–5.

Crick, F., and Koch, C. 1998: Constraints on cortical and thalamic projections. The no-strong-loops hypothesis. *Nature*, 391, 245–50.

Cumming, B. G., and Parker, A. J. 1997: Responses of primary visual cortical neurons to binocular disparity without depth perception. *Nature*, 389, 280–3.

Damasio, A. R., and Anderson, S. W. 1993: The frontal lobes. In K. M. Heilman and E. Valenstein (eds), *Clinical Neuropsychology*, 3rd edn. Oxford: Oxford University Press, 409–60.

Dennett, D. 1996: *Kinds of Minds: Toward an Understanding of Consciousness*. New York: Basic Books.

Distler, C., Boussaoud, D., Desimone, R., and Ungerleider, L. G. 1993: Cortical connections of inferior temporal area TEO in macaque monkeys. *Journal of Comparative Neurology*, 334, 125–50.

Eccles, J. C. 1994: *How the Self Controls its Brain*. Berlin: Springer-Verlag.

Edelman, G. M. 1989: *The Remembered Present: A Biological Theory of Consciousness*. New York: Basic Books.

Engel, S., Zhang, X., and Wandell, B. 1997: Colour tuning in human visual cortex measured with functional magnetic resonance imaging. *Nature*, 388, 68–71.

Felleman, D. J., and Van Essen, D. 1991: Distributed hierarchical processing in the primate cerebral cortex. *Cerebral Cortex*, 1, 1–47.

Fries, W. 1990: Pontine projection from striate and prestriate visual cortex in the macaque monkey: An anterograde study. *Vision Neuroscience*, 4, 205–16.

Fuster, J. M. 1997: *The Prefrontal Cortex: Anatomy, Physiology, and Neuropsychology of the Frontal Lobe*, 3rd edn. Philadelphia: Lippincott-Raven.

Gegenfurtner, K. R., and Sperling, G. 1993: Information transfer in iconic memory experiments. *Journal of Experimental Psychology: Human Perception and Performance*, 19, 845–66.

Goldenberg, G., Müllbacher, W., and Nowak, A. 1995: Imagery without perception – a case study of anosognosia for cortical blindsight. *Neuropsychologia*, 33, 1373–82.

Goldman-Rakic, P. S. 1995: Cellular basis of working memory. *Neuron*, 14, 477–85.

Greenfield, S. A. 1995: *Journey to the Centers of the Mind*. New York: W.H. Freeman.

Gur, M., and Snodderly, D. M. 1997: A dissociation between brain activity and perception: Chromatically opponent cortical neurons signal chromatic flicker that is not perceived. *Vision Research*, 37, 377–82.

He, S., Cavanagh, P., and Intriligator, J. 1996: Attentional resolution and the locus of visual awareness. *Nature*, 383, 334–7.

He, S., Smallman, H., and MacLeod, D. 1995: Neural and cortical limits on visual resolution. *Investigative Ophthalmology and Visual Science*, 36, 2010.

Ingle, D. 1973: Two visual systems in the frog. *Science*, 181, 1053–5.

Kanwisher, N., and Driver, J. 1992: Objects, attributes, and visual attention: Which, what, and where. *Current Directions in Psychologyical Science*, 1, 26–31.

Koch, C., and Braun, J. 1996: On the functional anatomy of visual awareness. *Cold Spring Harbor Symposium on Quantitative Biology*, 61, 49–57.

Kolb, F. C., and Braun, J. 1995: Blindsight in normal observers. *Nature*, 377, 336–9.

Kosslyn, S. M., Thompson, W. L., Kim, I. J., and Alpert, N. M. 1995: Topographical representations of mental images in primary visual cortex. *Nature*, 378, 496–8.

Land, E. H., and McCann, J. J. 1971: Lightness and retinex theory. *Journal of the Optical Society of America*, 61, 1–11.

Leibniz, G. W. 1686/1965: *Monadology and Other Philosophical Essays*. (Trans. P. Schrecker and A. M. Schrecker.) Indianapolis: Bobbs-Merrill.

Leopold, D. A., and Logothetis, N. K. 1996: Activity changes in early visual cortex reflect monkeys' percepts during binocular rivalry. *Nature*, 379, 549–53.

Libet, B. 1993: *Neurophysiology of Consciousness: Selected Papers and New Essays by Benjamin Libet*. Boston: Birkhäuser.

Logothetis, N., and Schall, J. 1989: Neuronal correlates of subjective visual perception. *Science*, 245, 761–3.

Luck, S. J., Chelazzi, L., Hillyard, S. A., and Desimone, R. 1997: Neural mechanisms of spatial selective attention in areas V1, V2, and V4 of macaque visual cortex. *Journal of Neurophysiology*, 77, 24–42.

Milner, D., and Goodale, M. 1995: *The Visual Brain in Action*. Oxford: Oxford University Press.

Milner, A. D., Perrett, D. I., Johnston, R. S., Benson, P. J., Jordan, T. R., Heeley, D. W. et al. 1991: Perception and action in "visual form agnosia." *Brain*, 114, 405–28.

Morgan, M. J., Mason, A. J. S., and Solomon, J. A. 1997: Blindsight in normal subjects? *Nature*, 385, 401–2.

Myerson, J., Miezin, F., and Allman, J. 1981: Binocular rivalry in macaque monkeys and humans: A comparative study in perception. *Behaviour Analysis Letters*, 1, 149–56.

Nagle, A. H. M. 1997: Are plants conscious? *Journal of Consciousness Studies*, 4, 215–30.

Nakamura, R. K., and Mishkin, M. 1980: Blindness in monkeys following nonvisual cortical lesions. *Brain Research*, 188, 572–7.

Nakamura, R. K., and Mishkin, M. 1986: Chronic blindness following lesions of nonvisual cortex in the monkey. *Experimental Brain Research*, 62, 173–84.

Nirenberg, S., and Meister, M. 1997: The higher response of retinal ganglion cells is truncated by a displaced amacrine circuit. *Neuron*, 18, 637–50.

No, D., Yao, T. P., and Evans, R. M. 1996: Ecdysone-inducible gene expression in mammalian cells and transgenic mice. *Proceedings of the National Academy of Sciences, USA*, 93, 3346–51.

O'Regan, J. K. 1992: Solving the "real" mysteries of visual perception: The world as an outside memory. *Canadian Journal of Psychology*, 46, 461–88.

Penfield, W. 1958: *The Excitable Cortex in Conscious Man*. Liverpool: Liverpool University Press.

Pollen, D. A. 1995: Cortical areas in visual awareness. *Nature*, 377, 293–4.

Potter, M. C. 1976: Short-term conceptual memory for pictures. *Journal of Experimental Psychology: Human Learning and Memory*, 2, 509–22.

Ramachandran, V. S., and Hirstein, W. 1997: The biological functions of consciousness and qualia: Clues from neurology. *Journal of Consciousness Studies*, in press.

Rock, I., Linnett, C. M., Grant, P., and Mack, A. 1992: Perception without attention: Results of a new method. *Cognitive Psychology*, 24, 502–34.

Rossetti, Y. 2000: Implicit perception in action: Short-lived motor representations of space. In P. G. Grossenbacher (ed.), *Finding Consciousness in the Brain*, Amsterdam: J. Benjamins, pp. 131–79.

Sahraie, A., Weiskrantz, L., Barbur, J. L., Simmons, A., Williams, S. C. R., and Brammer, M. J. 1997: Pattern of neuronal activity associated with conscious and unconscious processing of visual signals. *Proceedings of the National Academy of Sciences, USA*, 94, 9406–11.

Saint-Cyr, J. A., Ungerleider, L. G., and Desimone, R. 1990: Organization of visual cortex inputs to the striatum and subsequent outputs to the pallidonigral complex in the monkey. *Journal of Comparative Neurology*, 298, 129–56.

Salin, P. A., and Bullier, J. 1995: Corticocortical connections in the visual system: Structure and function. *Physiological Reviews*, 75, 107–54.

Salzman, C. D., Britten, K. H., and Newsome, W. T. 1990: Cortical microstimulation influences perceptual judgements of motion direction. *Nature*, 346, 174–7.

Scalaidhe, S. P. O., Wilson, F. A. W., and Goldman-Rakic, P. S. 1997: Areal segregation of face-processing neurons in prefrontal cortex. *Science*, 278, 1135–8.

Schein, S. J., and Desimone, R. 1990: Spectral properties of V4 neurons in the macaque. *Journal of Neuroscience*, 10, 3369–89.

Searle, J. 1984: *Minds, Brains, and Science*. Cambridge, MA: Harvard University Press.

Sheinberg, D. L., and Logothetis, N. K. 1997: The role of temporal cortical areas in perceptual organization. *Proceedings of the National Academy of Sciences, USA*, 94, 3408–13.

Sherk, H. 1986: The claustrum and the cerebral cortex. In E. G. Jones and A. Peters (eds), *Cerebral Cortex vol 5: Sensory-motor Areas and Aspects of Cortical Connectivity*, New York: Plenum Press, 467–99.

Singer, W., and Gray, C. M. 1995: Visual feature integration and the temporal correlation hypothesis. *Annual Review of Neuroscience*, 18, 555–86.

Subramaniam, S., Biederman, I., and Madigan, S. A. 1997: Highly accurate identification, but chance forced-choice recognition of RSVP pictures. *Journal of Experimental Psychology*, submitted.

Tootell, R. B. H., Dale, A. M., Sereno, M. I., and Malach, R. 1996: New images from human visual cortex. *Trends in Neuroscience*, 19, 481–9.

Tootell, R. B. H., Reppas, J. B., Dale, A. M., Look, R. B., Sereno, M. I., Malach, R., Brady, T. J., and Rosen, B. R. 1995: Visual motion aftereffect in human cortical area MT revealed by functional magnetic resonance imaging. *Nature*, 375, 139–41.

Ungerleider, L. G., and Mishkin, M. 1982: Two cortical visual systems. In D. J. Ingle, M. A. Goodale, and R. J. W. Mansfield (eds), *Analysis of Visual Behavior*, Cambridge, MA: MIT Press, 549–86.

Volkmann, F. C., Riggs, L. A., and Moore, R. K. 1980: Eye-blinks and visual suppression. *Science*, 207, 900–2.

von der Malsburg, C. 1995: Binding in models of perception and brain function. *Current Opinion in Neurobiology*, 5, 520–6.

Webster, M. J., Bachevalier, J., and Ungerleider, L. G. 1994: Connections of inferior temporal areas TEO and TE with parietal and frontal cortex in macaque monkeys. *Cerebral Cortex*, 5, 470–83.

Weiskrantz, L. 1997: *Consciousness Lost and Found*. Oxford: Oxford University Press.

Young, M. P., Tanaka, K., and Yamane, S. 1992: On oscillating neuronal responses in the visual cortex of the monkey. *Journal of Neurophysiology*, 67, 1464–74.

Young, M. P., and Yamane, S. 1992: Sparse population coding of faces in the inferotemporal cortex. *Science*, 256, 1327–31.

Zeki, S. 1980: The representation of colours in the cerebral cortex. *Nature*, 284, 412–18.

Zeki, S. 1983: Colour coding in the cerebral cortex: The reaction of cells in monkey visual cortex to wavelengths and colours. *Neuroscience*, 9, 741–65.

Zola-Morgan, S., and Squire, L. R. 1993: Neuroanatomy of memory. *Annual Review of Neuroscience*, 16, 547–63.

15

Functionalism, Dualism, and the Neurocorrelates of Consciousness

Jesse Prinz

There is a long-standing rift in the philosophy of mind between psychophysical identity theorists and functionalists (see chapter 1, this volume). Ironically, neither side has taken the brain terribly seriously. In illustrating their position, psychophysical identity theorists have relied on one pet example: pain is identical with C-fiber stimulation.[1] Functionalists have ignored neuroscience on the grounds that different brain state types can play the same functional role; only the functional roles, they say, are of interest. Some more sober-minded functionalists have recognized that neuroscience can help us discern the functional roles that our conscious mental states actually play, but few philosophers of this bent have gotten their hands dirty.[2] Recently, philosophers have been paying a bit more attention to neuroscience. One reason for this is that a number of neuroscientists have begun to research the nature of consciousness. Conscious states, those that have a phenomenological feel, have been especially important in evaluating the debate between identity theorists and functionalists, and in evaluating materialism, more broadly. They are often regarded as the states that make the mind–body problem a problem. Neuroscience can contribute to the philosophical investigation of the mind–body problem by offering examples of psychophysical correlations that are more up to date than the C-fiber case. But what more can neuroscience teach us about consciousness?

To answer this question, I will consider two such hypotheses. The first I'll call *Anti-functionalism*. There is a highly influential line of argument, associated with Block (1978), according to which phenomenally conscious states are more likely to find a reductive analysis in neural, rather than functional terms. For any functional analysis of phenomenal states that does not mention neural inputs and outputs, there is a potential realizer that intuitively lacks phenomenology. This intuition has supported the thesis that phenomenal states must be identified with neural states if they are identified with material states at all.

For a number of philosophers, the retreat from functionalism to the brain offers little improvement. Phenomenal states are widely thought to be the kinds of mental states for which a materialist account is least likely to succeed, even if stated in neural terms (e.g. Nagel, 1974; Jackson, 1982; Kripke, 1980; Chalmers, 1996; see also Levine, 1983). For any neural analysis of phenomenal states, there is an intuitive possibility of an automaton, with a brain like ours, but no phenomenology. This intuition has supported the thesis that phenomenal states will never be given an adequate materialist analysis. I'll call this thesis *Anti-materialism*.

In section 1, I sketch a neurally informed theory of visual consciousness. In the sections that follow, I argue against Anti-functionalism and Anti-materialism. Those arguments do not hinge on specific details of the theory that precedes them, but their central claims are illustrated by that theory. In this way, I hope to show how the neuroscience can offer perspective on theses that have been advanced by philosophers.

1   The Neural Correlates of Visual Experience

To develop a neurally informed theory of consciousness, one must first find neural correlates of conscious states. Following Crick and Koch (1990; chapter 14, this volume), I think the best way to do this is to begin with a single class of conscious states. They choose states of visual consciousness. I am largely sympathetic to the account they present in this volume, but will present several points of dissent.

Crick and Koch point out that the primate visual system involves a hierarchy of processing levels. At the lowest cortical level, we have activity in V1. This is followed by intermediate-level processing in a number of extrastriate regions (e.g. V2–V5/MT), and high-level processing in the inferior temporal cortex (IT), where recognition is normally achieved (for details, see chapters 12 and 13, this volume). Given this hierarchy, one might ask: which level of processing is most likely to be the locale of conscious experience? Crick and Koch argue persuasively that V1 should be excluded from consideration (Crick and Koch, 1995; Crick and Koch, chapter 14, this volume; Koch and Braun, 1996). For example, cells in V1 do not seem to detect illusory contours, attain color constancy, or follow the time course of motion illusions. Extrastriate areas seem to be better candidates. Illusory contours, color constancy, and motion illusions have been associated with V2, V4, and MT activity in monkeys. If visual awareness can be correlated with activity in extrastriate areas, then we will have neurophysiological support for Jackendoff's (1987) intermediate-level hypothesis, according to which visual consciousness resides in intermediate-level processing systems.

Crick and Koch seem to depart from Jackendoff's intermediate-level hypothesis in a couple of ways. First, they think that regions of prefrontal cortex associated with our ability to engage in voluntary action are essential to visual consciousness as well. They suspect that this is a primary function of visual consciousness. They also cite empirical support for the claim that prefrontal areas are involved. Prefrontal area 46 is active under conditions in which a patient with blindsight reports awareness of a stimulus in his impaired field and sectioning frontal cortex in monkeys results in a lack of visual responsiveness (see Crick and Koch, chapter 14, this volume, for references). Second, Crick and Koch are swayed by evidence that activity in high-level visual areas (IT) contributes to consciousness. In particular, they appeal to the work of Logothetis and his colleagues, who have been recording from cells in monkeys' visual pathways (see Logothetis, 1999, for review). Logothetis and colleagues found that about 40 percent of recorded cells in various extrastriate regions changed their responses when the monkeys reported changes in experience. In contrast, they found that 90 percent of the neurons recorded in IT were correlated with the experience reported by the monkeys.

I am not convinced by these considerations. Take, first, the suggestion that conscious visual experience depends on prefrontal regions involved in control of action. I dissent for several reasons (see Block, 1998, for further reasons). First, to say consciousness contributes to action control does not imply that control centers are necessary for consciousness. Conscious states may make their contribution indirectly by producing outputs to systems that input to control centers. Second, there is evidence that consciousness receives the information about deliberate actions only after those actions have been initiated (Libet, 1982). Third, as Crick and Koch acknowledge, visual systems involved in fine-tuned control of reaching behavior are thought to be unconscious, raising further questions about the role of visual consciousness in action control (Milner and Goodale, 1995). Finally, the empirical evidence Crick and Koch cite can be challenged. The fact that area 46 is active when we are visually aware is no surprise, because this area has been associated with working memory. Thus, the blindsight results only show that we typically store a temporary record of consciously perceived stimuli, not that such records are essential. Indeed, when area 46 is damaged in monkeys, the result seems to be a working memory impairment without any corresponding deficit in visual abilities (Funahashi et al., 1993). Even more strikingly, there is recent evidence that visual abilities remain after frontal cortex has been sectioned more globally (Gilman et al., 1998). This reverses the results of the earlier attempts to section frontal cortex, which Crick and Koch cite in defending their case. All this suggests that prefrontal areas are not essential for visual awareness.

Now, I want to challenge the claim that IT is necessary for visual consciousness. One reason for doubting that IT is necessary is that representations in this region seem to be more abstract than the contents of experience. Cells have large receptive fields, that make them relatively size- and position-invariant. A single cell in IT can respond to a complex stimulus, such as a face, and it may respond to faces at a range of orientations. In contrast, we experience objects as richly structured and highly specific in orientation, position, and size. Another reason for doubting that IT is necessary involves the deficits caused by damage to this region. Most notably, IT damage has been implicated in cases of apperceptive agnosia. Patients with this condition cannot recognize visually perceived objects, but they can, with great care, faithfully copy a line drawing. This suggests that they are fully visually conscious. What they lack is an ability to associate the contents of their experience with stored object representations in memory. It could turn out that their visual experiences are just like those of normal subjects, but their access to visual memory is impaired. It is more likely, however, that their experiences are abnormal. Specifically, it is likely that they fail to see the many lines and edges making up an object as part of a unified whole. Does this prove that IT is necessary for visual awareness? That IT is a neural correlate of consciousness? I think not. Instead, I suspect that IT plays a role in organizing information in earlier, extrastriate areas through efferent connections. IT may be like the conductor of an orchestra. The sound comes from the orchestra, but the conductor ensures that the players are coordinated and coherent.

The evidence cited in favor of IT can also be challenged. The Logothetis results do not entail that IT is a neural correlate of consciousness. Here, an expansion of the orchestra analogy may help. Instead of one orchestra, imagine that a group of orchestras are all playing at the same time in extrastriate regions. Each one is competing for the attention of IT. Imagine further that the loudest orchestra counts as the victor of this competition. If victory consists in being loudest, the victorious orchestra can count as victorious even if it is playing at the same time as the other orchestras. By parity of reasoning, some subset of firing cells in extrastriate cortex could correspond to the conscious percept, even though competing cells are active. The fact that only a limited percentage of cells in extrastriate areas change with changes in the percept, does not show that these areas are not correlated with experience. It may only suggest that we should identify conscious percepts with a subset of the active cells in these areas. Likewise, the fact that 90 percent of the cells in IT track the percept does not show that IT is a correlate of visual consciousness. It may only suggest that IT responds to just those cells that have won the battle of consciousness.

If this story is correct, visual consciousness can be correlated with a subset of active cells in extrastriate regions. Subsequent regions may receive outputs from these cells and may help orchestrate their activity through efferent connections. This is a reasonable starting place, but the story cannot be complete without saying what makes this subset so privileged. How do we distinguish the cells that comprise our visual experiences from the many other active, neighboring cells that do not? One possibility is that they fire at a special frequency. Crick and Koch (1990) say that firing patterns contribute to our best explanation of how separate cell populations representing different parts of a common stimulus are bound to each other. In particular, binding might be achieved by having cells fire in synchrony. They then reason that binding is intimately related to consciousness; after all, we generally experience the components of our percepts as bound. Perhaps we will find that certain cells within extrastriate cortex fire in this special way. I am not convinced that this will explain how such cells become conscious. I agree that synchronized firing could explain binding, but the connection to consciousness seems tenuous. It is hard to see why being bound is necessary or sufficient for consciousness. Must some binding not occur to process a word beneath the conscious threshold, for example? Conversely, might we not consciously perceive a pure color patch without binding it to anything else?

A more promising suggestion is that the cells underwriting conscious experience are marked by their distinctive interaction with the neural mechanisms underpinning selective attention. Winning the competition for consciousness, on this view, is a direct consequence of attention. Playing loudest amounts to attentional modulation. If this suggestion is right, then attention is necessary for consciousness. This suggestion is not incompatible with the oscillation view. Indeed, Niebur and Koch (1994) defend a model of attention that works by temporal binding. Despite this link, Crick and Koch (chapter 14, this volume) have denied that attention is necessary for consciousness. This seems untenable. I think that binding and attention can come apart, and, when they do, it is attention rather than binding that seems to usher percepts into awareness. We can imagine someone who can consciously see features without being able to bind them together (this may accurately describe what goes on in associative agnosia), but it is hard to imagine a person who can consciously see stimuli without being able to pay any attention to them. We can imagine consciousness without the ability to sustain focused attention, but it is hard to imagine consciousness without any attention at all.

The claim that attention is necessary for consciousness is hardly surprising. It is a commonplace that we often do not experience the things we do not attend to. Distraction and intensive focusing can prevent us from noticing many things in our surroundings. Certain objects seem to enter awareness by demanding our attention (a bright flash or sudden sound). Such mundane observations have gained scientific support from a recent body of psychological research on a phenomenon dubbed "inattentional blindness" by Mack and Rock (1998). In their research, an unexpected stimulus is presented, often foveally, as subjects perform a demanding line-length comparison task. Many subjects fail to see the stimulus. Mack and Rock conclude that attention is necessary for conscious perception.

Further evidence for the necessity of attention comes from cases of unilateral neglect. Patients with neglect are apparently unconscious of stimuli presented in their contralesional fields (or unconscious of one-half of visually perceived objects). This deficit seems to be attentional rather than perceptual. For one thing, high-level, categorical information, such as words or pictures, can be processed in the blind field of a patient with neglect (Bisiach and Rusconi, 1990). Perception is taking place, but the subject is unaware of it, showing only implicit signs of recognition. Moreover, the lesions that give rise to neglect are typically in regions associated with attention. The majority of cases derive from lesions in parietal cortex of the temporal/parietal junction.[3]

At this point, one might be tempted to object that the notion of attention is in as much need of analysis as consciousness; the two terms are sometimes used as synonyms. Explaining consciousness by appeal to attention is, arguably, not a good way to make progress. It may even be circular. Fortunately, this worry can be addressed. The notion of attention I am invoking is not the unanalyzed posit of folk psychology, but the well-studied posit of cognitive psychology and cognitive neuroscience. As intimated above, neuroscientists do not construe attention as a single process or mechanism. For example, Posner and Peterson (1990) have postulated three distinct attentional systems. One determines the level of vigilance, or maintained focus, on a given task. Another is involved with effortful control, as when we suppress stimulus features that ordinarily guide our responses automatically. The third, which I am implicating here, is involved in selecting some aspect of a stimulus. Selection itself may involve a number of processes. Selection can be focused on a feature, on an object, or on a region of space. It can be driven by stimulus properties or by templates in memory, as when we search for a specific object (Desimone and Duncan, 1995). Researchers debate whether selective attention works through suppression, excitation, receptive field modification, gain increases, connectivity changes, or filtering (see, e.g. Moran and Desimone, 1985; Connor et al., 1996). Perhaps these work together. For example, a filter may be created by suppressing outputs from some cells, while amplifying output of others. Filtering might also occur by modifying selective fields; the cells in one level of processing may be mapped on to a collection of cells in the next level that is larger than the collection they would ordinarily map on to.

My purpose here is not to provide a theory of attention, but to hint at some of the forms such theories have taken. It should be clear that "attention" is not being used as a covert synonym for "consciousness" on such theories. It can be characterized in non-experiential terms (excitation, suppression, filtering, etc.). Moreover, many cognitive scientists assume that selective attention can impact processing outside of awareness. There is evidence that visual selective attention has an effect on early visual processing regions, like V1, which, I have suggested, are not conscious. This would be a contradiction if the terms were synonymous. Therefore, the claim that selective attention is necessary for consciousness is a substantive claim.

The picture that I have been presenting can be summarized as follows. Visual consciousness arises when intermediate-level visual areas are connected to systems of attention. Elsewhere, I have called this the AIR theory, for *attended intermediate representations* (Prinz, 2000). I have also argued that it can be extended to encompass all forms of conscious experience (Prinz, under review). I will not rehearse that story here, but one example will serve to illustrate, and it will drive home an implication of the AIR theory that will be of use below.

It is often claimed that we can distinguish two important species of consciousness, introspective and non-introspective. Armstrong (1981) illustrates this contrast with the case of a long-distance truck driver, who has been driving so long that he is functioning on something like autopilot. The driver consciously perceives the road in some sense; he is perceptually aware of the curves, the lanes, and the other vehicles. But he is not introspectively aware. He might suddenly pop into this deeper state of awareness if, for example, another vehicle suddenly cuts him off. Armstrong suggests that this deeper form of consciousness requires a higher-order state: the driver becomes conscious of his first-order perceptual states. Before the unexpected event, he merely perceives the road, and, right after the event, he is aware that he perceives the road.

A natural strategy for explaining the contrast between these states is to postulate two distinct forms of consciousness, underwritten, perhaps, by different neural mechanisms. I prefer a more conservative strategy. There are at least two ways to accommodate introspective consciousness without departing from the AIR theory.

The shift in awareness might simply reflect a change in the *degree* of attention allocated to the situation. The unexpected event causes an attentional increase, resulting in a more vivid experience. Another possibility is that introspection involves a silent narrative. When we introspect, we typically experience verbal imagery describing the contents of our thoughts and experiences. Such silent narratives are higher-order in that they represent other mental states, but they may still qualify as attended intermediate-level perceptual representations. In particular, they may be constituted by verbal imagery housed in intermediate-level auditory systems and modulated by attention. If so, our introspective narratives are higher-order without occurring at a higher level of processing than standard perceptual events. Before the unexpected event, the truck driver may have verbal imagery concerning his plans for the next day, or no verbal imagery at all. After the event, his narrative describes the here-and-now contents of his other sensory channels. Those experiences do not become any more conscious than they were before. They do not depend on the verbal narrative or any other higher-order mental state. No qualitative change in the kind of consciousness occurs. Instead, the contents of consciousness are altered. Now, in addition to certain visual appearances, there are conscious words describing those appearances.

If either of these proposals pans out, seemingly distinct species of consciousness may be unified under a common explanation. Each involves attentionally modulated activity in intermediate processing systems in one of our sensory modalities. The apparent qualitative difference between introspective and non-introspective consciousness turns out to be a difference in quantity and content, rather than kind.

2   Anti-functionalism

Having outlined a theory of visual consciousness (which may generalize), I can now turn to the two philosophical theses described in the introduction. The first of these is Anti-functionalism. Before returning to this thesis, it is useful to say a bit about what functionalism is.

Functionalists say that mental state types can be individuated by functional roles. Outside of psychology, there are many kinds of things that are widely believed to be functional. Internal organs provide one class of examples. Arguably, something is a heart just because it plays a particular role within a circulatory system. My heart is a hunk of muscle made up of certain materials, but it qualifies as a heart in virtue of pumping blood through my body. A thing made of different stuff that played the same role, such as an artificial heart, would qualify as a heart too. Artifacts provide another class of example. For example, a battery is a device that plays the functional role of converting chemical energy into electricity. Batteries contain electrolytes, which are also functional kinds, identified by their ability to serve as ionic conductors. Electrolytes and, hence, batteries are multiply realizable, because the same function can be underwritten by different substances, including liquids, pastes, and solids.[4] Such examples support the claim that functional kinds are not ontologically peculiar or suspect. Functionalists say mental states belong to such a category.

Anti-functionalists claim that certain mental states are not functional kinds. As remarked above, Block's (1978) "absent qualia" arguments have been especially influential in defending this view. He has us imagine cases where the functional role played by a mental state is instantiated by a system that intuitively lacks such states. His most convincing examples involve phenomenal states. Most famously, he proposes that the entire Chinese population could temporarily get together to implement the functional role played by my mental states without feeling anything I feel. The large functional structure comprised by the Chinese population would not feel anything at all.

Rather than addressing Block's thought experiment head-on, I want to draw attention to a second thesis that he defends. Those who are swayed by arguments against functionalism are often also swayed by arguments against the psychophysical identity theory, which identifies mental state types with brain state types. Just as we can imagine a functional analog of my mind that lacks experience, it seems we can imagine a duplicate of my brain that lacks experience. Anti-materialism threatens. Block resists this slippery slope by arguing that there is an important difference between the epistemic status of functionalism and the identity theory. Block says we *know* that brains give rise to conscious experience. It is quite mysterious *how* they do this, but *that they do* is a thesis that has tremendous support. The truth of functionalism, in contrast, is alleged to be an open question. With this contrast in hand, he appeals to the following principle:

> If a doctrine has an absurd conclusion which there is no independent reason to believe, and if there is no way of explaining away the absurdity or showing it to be misleading or irrelevant, and if there is no good reason to believe the doctrine in the first place, then don't accept the doctrine.
>
> (Block, 1978: section 1.4)

Block thinks that this principle allows us to conclude that a mere functional analog of the mind (such as the nation of China) lacks phenomenal states, without also concluding that a functioning brain can lack phenomenal states. After all, there is an independent reason to embrace the seemingly "absurd conclusion" that brains can give rise to phenomenology, namely the fact that we are brain-headed creatures, and we have phenomenal states. We have good evidence for a materialist theory that correlates brain states with mentality, but no comparable evidence for functionalism, so functionalism is vulnerable to arguments that would not threaten the identity theory. Although it is not Block's overt aim to defend the identity theory, this argument can be regarded as aiding the identity theory by insulating it from his arguments against functionalism. I will interpret Block as defending the identity theory as against functionalism when it comes to phenomenal states.

If this reconstruction is correct, Block's argument is flawed. First, our independent reason for thinking that brains engender consciousness is also a reason for thinking that functional roles engender consciousness. We are, after all, also functional-role-headed creatures. The suggestion that brain states could engender consciousness without playing some role is absurd. Brain states are never idle. Thus we never have grounds for saying a brain state can be conscious independent of any role it plays. Moreover, Block sets up a false dichotomy. His argument implicitly assumes that the distinction between the identity theory and functionalism is principled. One problem with this view is that neural kinds are often individuated functionally (see Mundale, 1997). The account of consciousness developed in the last section illustrates nicely. It appeals to neural systems and mechanisms, but identifies these by their functional roles. Attentional mechanisms are identified by their ability to excite, inhibit, and filter, and intermediate-level systems are identified by their sequence in a hierarchy of systems that are responsive to stimuli presented through the eyes (which, like hearts, are functional kinds). Here we have a neural theory of consciousness, defended by evidence from neuroscience, stated in a language neuroscientists use, but overtly functionalist in nature.

One might reply by arguing that there are other kinds of neural properties that preserve the distinction between function and structure. Suppose you think that consciousness requires activity in a certain kind of cell. This seems to be the view behind the example of pain and C-fiber stimulation. If we identify mental states by appeal to particular cell types, one might claim, we have referred ineliminably to structure rather than function.

There is a danger of turning materialism into just another magic theory of the mind. What is it about particular kinds of brain stuff that matters for the mental? Why should a cell of a particular kind support consciousness? Would that cell produce consciousness if it were all alone, idling in a petri dish? Functional roles can be identified at every level of analysis, at the level of cells, molecules, and atoms (see Lycan, 1981). Wherever we have a structure, we can find a function that it plays. Sometimes neural structures are identified by their functions (e.g. opponent cells may be an example). In cases where they are identified in some other way, there is no reason to think that their contribution to mentality depends on something other than the roles they play. In either case, materialists who identify mental states with neural states can be often be construed as advancing functionalist hypotheses. The identity theory can be a special case of functionalism.

Ultimately, I suspect that the main distinction between the identity theory and other forms of functionalism is that the former identify functional roles in the vocabulary of some branch of neuroscience. Notice, however, that some vocabularies are not exclusive to a single discipline. The AIR theory illustrates this. People in artificial intelligence and psychology talk of attention, intermediate-level processing systems, filtering, and so on, but these terms also exist in neuroscience. "Identity theory" names a class whose membership depends on boundaries between disciplines, and these can be quite blurry. We could stipulate that a bona fide identity theory requires a vocabulary *unique* to neuroscience, but this would be imprudent because disciplinary boundaries can be unstable and artificial.

All of this adds up to a reply to Block. If absent qualia arguments work against any functional hypothesis, then they may work against many identity theorists as well. The suggestion that identity theories may be specially immune, because we are brain-headed creatures is misleading. Our brain states can be individuated functionally, and, thus, could be implemented by things made out of different stuff (including the population of China). The best materialist accounts of consciousness may turn out to be stated using terms that are at home in neuroscience, but this does not entail the rejection of functionalism.

3   Anti-materialism

In the previous section, I argued that Block is unable to establish a sharp difference between functionalism and the identity theory. This can be viewed, optimistically, as diluting the case against functionalism or, pessimistically, as strengthening the case against materialism. To diffuse the latter interpretation, I turn to arguments of contemporary dualists. These have taken a variety of forms in recent years, but I will focus on a paradigm case: Kripke's argument in *Naming and Necessity* (1980). Kripke's goal is not necessarily to prove that dualism is true, but to prove that standard forms of materialism face a major difficulty. That difficulty can be viewed as a sophisticated form of an argument used by Descartes in his defense of dualism.

Descartes famously argued that mind and body are distinct on the grounds that one can clearly and distinctly conceive of them existing apart from each other. Ever since, critics have objected that you can not infer actual distinctness from conceivable distinctness. Conceivability is not a good guide to metaphysics. For example, the fact that we can conceive of a situation in which Washington was not the first US president does not show that Washington was not the first US president. It looks as if the Cartesian argument can be easily dismissed.

Kripke thinks that this is too hasty. It is incumbent on a person who wants to resist an inference from conceivable distinctness to actual distinctness to explain why the inference fails. When we conceive of two things existing apart from each other, he claims, there must be some possible situation we are conceiving. But if there is a possible situation we are conceiving, then the two things are possibly distinct. This would cut no metaphysical ice if the identities in question were contingent. For example, in the case just mentioned, we can explain the conceivable distinctness of Washington and the first US president by saying that we are conceiving of a situation where they actually are distinct. The first US president might have been someone else. Unfortunately, such an explanation is unavailable for garden-variety theoretical identities, like "water is H_2O" or "temperature is mean molecular kinetic energy." Kripke persuasively argues that such identities are necessarily true if they are true at all. Water cannot be anything other than H_2O. Therefore, if we could really conceive of a situation where water and H_2O were distinct, that would show that they are not identical. If there is some situation where they are not identical, then they are not identical in any situation. The same, Kripke claims, is true of alleged psychophysical identities.

But, one might object, surely we can conceive of water and H_2O coming apart. Consider people who are presented with this identity for the first time. They might insist that water and H_2O are conceivably distinct. Kripke says this is an illusion. We cannot really conceive of water apart from H_2O, because no such situation is possible. Why, then, do we think we can conceive of them apart? The answer involves a split between appearance and reality. We think about water by its appearances; we conceive of a clear, tasteless liquid, found in rivers and streams. Kripke calls these attributes "contingent reference fixers." Being a clear, tasteless liquid is neither necessary nor sufficient for being water, but these attributes allow us to reliably pick out water, because most water samples (and few other things) exhibit them in this world. We are duped into thinking that water and H_2O can come apart, because we conceive of water via contingent appearances. These appearances can be manifested by things other than H_2O, and H_2O can manifest other appearances. When we seem to conceive of water and H_2O coming apart, we are really conceiving of water-like appearances and H_2O coming apart. If we could really conceive of water apart from H_2O, it would cut metaphysical ice.

The problem, Kripke argues, is that no parallel explanation is available in the case of alleged psychophysical identities. Consider the claim that an experience of red is a neural activation in extrastriate area V4 that is modulated through gain increases by selective attention mechanisms in the inferior parietal cortex.[5] I will abbreviate this claim as "red experience is V4 activation." We seem to readily conceive of these two coming apart. We can conceive of red experience without V4 activation and V4 activation without red experience. If we follow the model of garden-variety theoretical identities, this fact should be explained as follows. When we seem to conceive of red experience without V4 activation, we are really imagining something that has all the superficial properties of a red experience, an experience that seems red, but is constituted by some other kind of physical state. And when we seem to conceive of V4 activation without red experience, we are conceiving of a situation where V4 activation produces superficial properties other than those normally used to pick out red experiences. When we seem to conceive of red experiences and V4 activation coming apart, we are really conceiving of red-experience-like appearances and V4 activation coming apart. The obvious problem, Kripke tells us, is that red experience lacks contingent reference fixers. Red experience's appearance is its reality. A red-experience-like appearance just is a red experience. Therefore, when we conceive of a world where red appearances come apart from V4 activation, we are conceiving of a world where red experience and V4 activation come apart. If we cannot explain away our intuition that red experience can come apart from V4 activation, Kripke argues, we must conclude that they can come apart. If they can come apart, they are not identical.

This argument can be summarized in three premises:

1   Our ability to conceive of red experience apart from V4 activation cannot be explained the way we explain our ability to conceive of water and H_2O apart.

2   If (1), then red experience and V4 activation can come apart.
3   If red experience and V4 activation can come apart, then no red experience can
    be identical with V4 activation.

Together, these premises entail that red experience is not identical to V4 activation. The same argument scheme can be used against any alleged psychophysical
identity. This would spell great trouble for materialism. Fortunately, Kripke's
premises can be challenged. Loar (1990) develops an argument that can be taken
to challenge premise (2). Lewis (1983) argues for the contingency of psychophysical identities, thereby challenging versions of premise (3). I will challenge
premise (1).

   To refute (1), it must be shown that red experiences (and other conscious mental
states) are represented using contingent reference fixers. We must show that there is
an appearance/reality distinction of some kind in the case of conscious mental states.
At first blush, this sounds like a tall order, but it turns out to be rather simple. First,
we need to distinguish phenomenal states from the states we use to represent them.
A red experience is not the same as a representation of that experience. The latter
is a higher-order state, a state that refers to another mental state. Red experiences
refer to distal properties (if they refer to anything at all). Higher-order states can
come in the form of unconscious thoughts or, following the long-distance truck
driver example, conscious verbal images. I might represent a current red experience
by subvocally expressing the thought that I am experiencing red. When I think about
red, when I conceptualize red experiences, I must use a higher-order state. The
vehicles of my thoughts about red experiences are such higher-order states, not the
experiences themselves. The experiences themselves are only vehicles in my thoughts
about red things, things out there in the world.

   Now, we only need to establish that the relationship between red experiences and
representations of those experiences is contingent. This is, again, fairly easy. First,
one can surely have a red experience without having a higher-order representation
of that experience. The long-distance truck driver can illustrate. Before switching
out of autopilot, the truck driver is in exactly such a state. Traffic signals, street signs,
and brake lights may cause him to experience red without causing him to represent
that he is having such an experience. The experience can go unnoticed. Second, one
can have a higher-order representation of a red experience without having a red
experience. In such a situation, we might be deceived into thinking we are seeing
red, when we are not. This sounds odd, but clinical evidence suggests that it is,
tragically, quite possible. When a cognitive system is functioning normally, we form
higher-order representations of our phenomenal states only when we are in such
states. But cognitive systems do not always function normally. Patients with Anton's
syndrome are blind, but they believe they can see. The natural explanation is that
they are having representations of experiences without experiences. Ordinarily,
blindness disrupts both phenomenal states and representations of those states, but
in this syndrome only the former are disrupted. Patients are deluded into thinking
they have visual phenomenology when they do not.

All of this invites a reply to Kripke. When we judge that we are having phenomenal states, we do so by means of higher-order representations that are only contingently connected to those states. In Kripke's idiom, the representations of our phenomenal states are contingent reference fixers. If this is correct, then we can explain the psychophysical cases exactly the way that Kripke explains garden-variety theoretical identities. When we seem to conceive of red experiences apart from their correlated brain states, we are unwittingly conceiving of a situation in which our higher-order representations of red experiences occur without those experiences. When we seem to conceive of V4 activation occurring without red experience, we are unwittingly conceiving of a situation in which V4 activation (and hence a red experience) occurs, but no higher-order representation of that experience is formed. We never actually conceive of red experience and V4 stimulation coming apart. Any thought about one is a thought about the other. But there are situations where the state used to pick out our red experiences comes apart from those experiences. This allows for a conceivable division with no metaphysical cost. Conceivability misleads about metaphysics because the vehicles of conception can come apart from the things they represent.

In response to this proposal, a dualist would have to argue that phenomenal states are necessarily linked to their representations. Call this the "necessity claim." If phenomenal states are necessarily linked to representations of them, then the present attempt to explain the conceivability of mind/body separation on the model of garden-variety theoretical identities will fail. I believe there is a principled reason why a dualist cannot make the necessity claim.

Let us suppose that materialism is false. If so, it is a theoretical possibility that a creature with a brain just like mine could exist without any phenomenal states whatsoever. Philosophers call such creatures "zombies" (e.g. Chalmers, 1996). Zombies differ from us phenomenally, but they are just like us functionally, since function supervenes on physical states. Consequently, my zombie doppelgänger would form judgments very much like my judgments under the same situations. He would form a judgment that he would express by, "I am experiencing red," when he saw a ripe tomato. Now consider the representations comprising this judgment. There are two possibilities. First, we might say that the zombie has the (false) belief that he is having a red experience. This would require that he has a higher-order representation of such experiences, even though he is incapable of having them. If that is possible, then the relationship between red experiences and representations of such appearances is contingent. A creature can have the latter without the former. That would undermine the necessity claim.

The second possibility is that a zombie who claims, "I am experiencing red," cannot be expressing the thought that he is experiencing red, because he cannot have a higher-order representation of red experiences. More specifically, one might argue that the mental representation expressed by the zombie's claim cannot represent red experiences, because it is not causally related to such experiences. My zombie doppelgänger has a higher-order state that is functionally and physically equivalent to my higher-order representation of red experiences, but semantically distinct. This seems perfectly reasonable, but it offers little help to the dualist. If my zombie doppelgänger and I have higher-order mental representations that are functionally and physically alike but semantically different, then we will have reason to say that we have equivalent reference fixers. This is exactly the way reference fixers have been individuated in the literature (Putnam, 1975). If my higher-order representation of red experiences consists of a reference fixer that could have referred to something else, then the relationship between that higher-order representation and my red experiences is contingent. The dualist necessity claim is false.

In short, dualists face a fatal dilemma. If they admit that the relationship between phenomenal states and higher-order representations of those states is contingent, then the conceivable separation of mind and body can be explained in a way that parallels garden-variety theoretical identities. This would undermine any inference from conceivable separation to metaphysical separation. If they claim that the relationship between phenomenal states and higher-order representations of those states is necessary, then they must deny the possibility of zombies, because that possibility is incompatible with the necessity claim. That would be incoherent, because dualists are committed to the theoretical possibility of zombies. Therefore, dualists cannot appeal to Kripke's argument in trying to undermine materialism.

The account of visual experience outlined in section 1 was based on evidence from neuroscience. In section 2, I argued that this account is consistent with functionalism, but still counts as a version of the psychophysical identity theory. I have shown that it is invulnerable to a leading argument for dualism. Neuroscience helps us see that the line between functionalism and the identity theory can be blurred and that conscious experiences can differ from our representations of conscious experiences. These points do not depend on any details of the account defended in section 1, but they insulate that account from some possible sources of philosophical opposition.

Acknowledgments

I would like to thank Bill Bechtel and Pete Mandik for many very valuable comments on an earlier draft of this paper.

Notes

1   This identity claim turns out to be false. For one thing, C fibers belong to the peripheral nervous system, and pain sensations are likely to reside in the neocortex (probably somewhere in somatosensory cortex). C fibers are just detectors for painful stimuli. Indeed, they are not the sole pain detectors. Faster $A\delta$ fibers also participate in pain detection.
2   Some other functionalists ignore neuroscience, because they think that the nature of mental states can be determined by looking at folk psychological platitudes, which say

nothing about the brain. I will ignore this view, focusing instead on what Block (1978) has dubbed psychofunctionalism. Psychofunctionalists say that the functional roles used to individuate mental states must be discovered empirically.

3   Frontal cortex lesions, including damage to area 46, can also cause unilateral neglect (Swick and Knight, 1998). At first blush, this seems to be at odds with two points presented above. First, if lesions to area 46 can cause neglect, then why do other lesions to 46 only impair working memory? The answer may be that various different things happen in this region. It may house a working memory store and an attention mechanism. Second, if lesions to area 46 can cause neglect, then how can global frontal sections spare visual functioning? I speculate that parietal attention mechanisms and frontal attention mechanisms interact in such a way that local damage to the former can inhibit functioning in the latter, but global damage to the former cannot. Perhaps inhibitory signals are disrupted only when frontal damage is global. In any case, implicating frontal attention mechanisms does not concede anything to Crick and Koch, because the mechanisms I am adverting to are not obviously essential for the voluntary control of action.

4   Organs and artifacts are special functional kinds, in that their functions are partially determined teleologically. That means they are *designed* to play the roles they do, rather than merely playing those roles. Artifacts are designed by intentional agents to play the causal roles they do and organs are designed by nature. This may be why we are so inclined to say a heart is still a heart when it has stopped beating and a battery is still a battery when it stops generating electricity. Functionalists about mental states need not invoke functions in this teleological sense; they need not assume mental states are *designed* to serve any roles. They can identify mental states with functional roles without assuming that the roles are selected for. Functional kinds of this variety are also plentiful outside the mental realm. Toxins, tragedies, riverbeds, conductors of electricity, and drainage clogs might all sometimes (though not always) fall into this category.

5   This is meant as an illustration. V4 may turn out to only be a color area in monkeys and attention may work in some other way.

References

Armstrong, D. M. 1981: What is consciousness? In *The Nature of Mind*. Ithaca, NY: Cornell University Press.

Bisiach, E., and Rusconi, M. L. 1990: Break-down of perceptual awareness in unilateral neglect. *Cortex*, 26, 643–9.

Block, N. 1978: Troubles with functionalism. *Minnesota Studies in the Philosophy of Science*, 9, 261–325.

Block, N. 1998: How not to find the neural correlate of consciousness. In S. Hameroff, A. W. Kaszniak, and A. C. Scott (eds), *Toward a Science of Consciousness II*. Cambridge, MA: MIT Press.

Chalmers, D. 1996: *The Conscious Mind*. Oxford: Oxford University Press.

Churchland, P. S. 1986: *Neurophilosophy*. Cambridge, MA: MIT Press.

Connor, C. E., Gallant, J. L., Preddie, D. C., and Van Essen, D. C. 1996: Responses in area V4 depend on the spatial relationship between stimulus and attention. *Journal of Neurophysiology*, 75, 1306–8.

Crick, F., and Koch, C. 1990: Towards a neurobiological theory of consciousness. *Seminars in the Neurosciences*, 2, 263–75.

Crick, F., and Koch, C. 1995: Are we aware of activity in primary visual cortex? *Nature*, 375, 121–3.

Crick, F., and Koch, C. 1998: Consciousness and neuroscience. *Cerebral Cortex*, 8, 97–107.

Desimone, R., and Duncan, J. 1995: Neural mechanisms of selective visual attention. *Annual Review of Neuroscience*, 18, 193–222.

Funahashi, S., Chafee, M. V., and Goldman-Rakic, P. S. 1993: Prefrontal activity in rhesus monkey performing a delayed anti-saccade task. *Nature*, 365, 753–6.

Gilman, D. J., Saunders, R. C., Rickrode, D., and Mishkin, M. 1998: Effects of frontal cortex lesions on visual awareness in the rhesus monkey. Under review.

Jackendoff, R. 1987: *Consciousness and the Computational Mind*. Cambridge, MA: MIT Press.

Jackson, F. 1982: Epiphenomenal qualia. *Philosophical Quarterly*, 32, 127–36.

Koch, C., and Braun, J. 1996: Towards a neuronal correlate of visual awareness. *Current Opinion in Neurobiology*, 6, 158–64.

Kripke, S. 1980: *Naming and Necessity*. Cambridge, MA: Harvard University Press.

Levine, J. 1983: Materialism and qualia: the explanatory gap. *Pacific Philosophical Quarterly*, 64, 354–61.

Lewis, D. 1983: Mad pain and martian pain. In *Philosophical Papers*, vol. 1. Oxford: Oxford University Press.

Libet, B. 1982: Brain stimulation in the study of neuronal functions for conscious sensory experiences. *Human Neurobiology*, 1, 235–42.

Loar, B. 1990: Phenomenal states. In J. Tomberlin (ed.), *Philosophical Perspectives, Vol. 4: Action Theory and Philosophy of Mind*. Atascadero, CA: Ridgeview.

Logothetis, N. K. 1999: Vision: A window on consciousness. *Scientific American*, 281 (5), 68–75.

Lycan, W. G. 1981: Form, function, feel. *Journal of Philosophy*, 78, 24–50.

Mack, A., and Rock, I. 1998: *Inattentional Blindness*. Cambridge, MA: MIT Press.

Milner, A. D., and Goodale, M. A. 1995: *The Visual Brain in Action*. Oxford: Oxford University Press.

Moran, J., and Desimone, R. 1985: Selective attention gates visual processing in the extra-striate cortex. *Science*, 229, 782–4.

Mundale, J. 1997: *How Do You Know a Brain Area when You "See" One?: A Philosophical Approach to the Problem of Mapping the Brain and its Implications for the Philosophy of Mind and Cognitive Science*. Doctoral dissertation, Washington University in St Louis.

Nagel, T. 1974: What is it like to be a bat? Reprinted in *Mortal Questions*. Cambridge: Cambridge University Press.

Niebur, E., and Koch, C. 1994: A model for the neuronal implementation of selective visual attention based on temporal correlation among neurons. *Journal of Computational Neuroscience*, 1, 141–58.

Place, U. T. 1956: Is consciousness a brain process? *British Journal of Psychology*, 47, 44–50.

Posner, M. I., and Peterson, S. E. 1990: The attention system of the human brain. *Annual Review of Neuroscience*, 13, 25–42.

Prinz, J. J. under review: Toward a unified theory of consciousness.

Prinz, J. J. 2000: A neurofunctional theory of visual consciousness. *Consciousness and Cognition*, 9, 243–59.

Putnam, H. 1975: The meaning of "meaning." *Minnesota Studies in the Philosophy of Science*, 7, 131–93.

Seguin, E. G. 1886: A contribution to the pathology of hemianopsis of central origin (cortex-hemianopsia). *Journal of Nervous and Mental Diseases*, 13, 1–38.

Swick, D., and Knight, R. T. 1998: Cortical lesions and attention. In R. Parasuraman (ed.), *The Attentive Brain*, Cambridge, MA: MIT Press.

Tipper, S. P., and Behrmann, M. 1996: Object-centered not scene-based visual neglect. *Journal of Experimental Psychology: Human Perception and Performance*, 22, 1261–78.

16

The Nature of Pain

Valerie Hardcastle

Philosophers regularly take pain to be an unproblematic and simple example of a conscious experience. Nevertheless, the fact remains that we do not know exactly how pain processing works in the brain. Hence, there is much conceptual space available for wild and rampant speculation. Indeed, a little digging into the philosophical literature uncovers just about every conceivable position regarding what pain is. Some philosophers and neurophysiologists argue that pain is completely objective; it is either intrinsic to the injured body part, a functional state, a set of behavioral reactions, or a type of perception. Some philosophers and psychologists argue that pain is completely subjective; it is either essentially private and completely mysterious, or it does not correlate with any biological markers but is completely nonmysterious. A few philosophers disagree with both conceptions and hold that pain is not a state at all; it either does not exist as we commonly conceive of it or it is an attitudinal relation. Furthermore, each of these positions has become grist for someone's mill in arguing either that pain is a paradigm instance of a simple conscious state or that pain is a special case and should not be included any general theory of the mind.

Here I want to do more than merely carve out my niche among the myriad of positions: I diagnose why we have so little agreement concerning the nature of our pain states. I begin, though, by outlining a few facts regarding our other sensory systems. I do this because many of the facts of perceptual processing regarded as commonplace (even among philosophers of mind) are the same sort of facts that confuse philosophers and psychologists when theorizing about pain. In other words, I believe that we have a system that uses incoming nociceptive information to figure out which events in the world are noxious, just as we have systems that use impinging photons to compute the size and orientation of things in our environment or that use air compression waves to compute the location of objects. Understanding that we have a pain processing *system* goes a long way to alleviating some of the philosophical muddles in which we can inadvertently find ourselves ensnared.

1 Our Sensory Systems

Our visual system is quite complex, spans many areas in the brain, and comprises several subsystems whose interactions remain a mystery. It is widely known that different aspects of visual processing occur in different processing streams, at least from the feature detection perspective. For example, color is processed in the intralaminar stream, while motion is processed in the magnocellular stream (see chapters 12 and 13, this volume, for further details). The auditory system works in an analogous fashion (though the interactions of its subsystems are not as mysterious). The medial superior olive of our auditory system probably computes sound location using interaural time differences. The lateral superior olive, on the other hand, computes sound location by using differences in interaural frequency.

What is important to notice is that it is quite all right for there to exist more than one processing stream in each modality. We might be mystified how color gets joined with shape and motion so that we have a unified visual experience of particular objects. But we are not confused about whether the neuronal paths involved in computing an object's color are visual, or whether computing interaural time differences is auditory. We are perfectly happy to have each modality involved in several, maybe ultimately unrelated computations. We have decided that the parts of the brain which normally respond to impinging photons are part of the visual system and the parts of the brain which are normally sensitive to air compression trains are part of the auditory system.

Naturally, this is a gross oversimplification of how our sensory modalities are actually individuated: without unpacking what is meant by "normal functioning," the definitions are virtually unworkable. By way of partially rectifying this gloss, let me briefly touch upon the top-down and bottom-up investigative methodologies in neuroscience, for these analytic tools help disambiguate what counts as normal functioning. More importantly, they allow us to make claims about which computational algorithms and cell assemblies are and are not included in our brain systems and subsystems.

First, scientists use the method of double dissociation to isolate the processing streams that make up our subsystems. Ideally, if we can get X to occur without Y and also Y to occur without X, then scientists take this as grounds to claim that X and Y function as independent units. (Of course, nature is rarely ideal; often scientists settle for substantial loss of X with little loss of Y, and vice versa.) For example, patients with Broca's aphasia lose the meaning of certain words without losing syntax; patients with lesions in Wernicke's area lose syntax without losing meaning (see chapters in Part II, this volume). I call this a top-down strategy because we start with a crude parsing of our system writ large (e.g. linguistic processing) and then divide that system into its component pieces (syntactic processing, semantics).[1]

Second, scientists rely on a teleological analysis to unite the various and sundry parts into wholes as a bottom-up strategy. Breaking down larger pieces into smaller ones is not enough to get the explanatory job done, especially when several of our systems overlap inside the head. Our brain houses lots of individual processors; knowing all the pieces does not tell us which component piece belongs to which larger system. Why do scientists believe that color and motion processing belong to the same system but that echolocation belongs to something else? This is not a trivial question since each of the subsystems is (mainly) anatomically and physiologically distinct from the other, and since individual neurons do not know the sort of signal they are responding to. The information contained in an atmospheric compression wave or a photon wave triplet is transmitted as electrical and chemical energy once one moves inside the body. This fact makes brain physiology quite difficult. It is an even more intriguing question how the brain unites its components, if it even does.

Scientists use three converging strategies to isolate and construct systems from the component-dissociable subsystems. First, they look for correlations between neural firing patterns and events in the external world.[2] Neurophysiologists take the smallest pieces of the puzzles, usually individual neurons, or the extracellular spaces around small groups of neurons, and record what they do under a variety of circumstances. They conclude that our color and shape detectors belong together because they are active under similar circumstances, namely, when the organism's retinas are bombarded by photons. Auditory cells are active in different contexts. (This account assumes that we have already decided what counts as a different context. I am going to leave that issue aside for now.) Luckily for the scientists and their correlation project, true polymodal cells are relatively rare.

Second, scientists look at the neural connections fore and aft. Aside from knowing how the activity of a cell or an area is correlated with the environment, they also need to know what this cell or area is connected to – where the information the cell or area lights up to goes – and what is connected to the cell or area – how it gets the information it does respond to. Determining the processing algorithm of any cell group is not as easy as it might sound; it is not a matter of merely recording all the stimuli it likes and then deciding what all the stimuli have in common. (See chapter 4, this volume, for elaboration of this point.)

Finally, scientists consider historical and evolutionary facts whenever possible. We are biological organisms equipped to move through our environment. We evolved that way because (roughly speaking) those who can move most effectively through their environment succeed in reproducing the most. When thinking about our perceptual systems, especially when worrying about what functions various components serve, it is important to keep in mind how the hypothesized system or subsystem is supposed to function with regard to motor assembling. For most, if not all, information processing in the brain is related to the motor system in one way or another. For example, the visual areas all have at least some indirect contact with some motor structure or other, either the basal ganglia, or the motor cortex, or the tectum, or something (Kandel and Schwartz, 1985). Motor information needs to be "siphoned off" the visual pathways at all stages along the ascending route so that the visual input can be used for motor output (Churchland, 1986). Quite often what seems strange or curious from a psychological point of view seems quite natural from an evolutionary standpoint.

If we can group subsystems together into larger systems via their evolutionary function, then so much the better. The brain puts great emphasis on the priority of motor tasks, and we should pay attention to this emphasis. Whatever purpose we ultimately propose for a group of brain components has to fit with our biological natures. (Often, however, such considerations are not possible or are little better than just-so stories, for the details of the advantages have been lost over evolutionary time.)

I call this collection of research strategies "bottom-up" because we begin with the smaller units in the brain and then arrange them into nested hierarchies. Based on gross similarities in response patterns, connections to other systems and organs, and putative selective advantages, we group the double-dissociated subsystems into hierarchically arranged classes. The process is not cut and dried by any means, but it is the best we have at the moment.

Both approaches are required for a complete explanation of psychobiological phenomena. By breaking cognitive engines into interacting component pieces, the top-down strategy helps explain why organisms behave the way they do, and by categorizing and grouping the isolated parts, the bottom-up strategy helps explain what purpose the analyzed behavior serves. Reminding ourselves that we use both strategies in understanding our neural systems will rid us of the tendency to make our pain system into a cartoon, and reminding ourselves of biological heritage will aid in justifying a counterintuitive system that prevents our pains from occurring. Our system for perceiving pain works in exactly the same fashion as our visual and auditory systems: it is a complex system with dissociable subsystems. Furthermore, it is a system that appears quite natural when considered against an evolutionary backdrop.

2   The Classic Pain System Doctrine

The classic view of our basic pain system is of two three-neuron subsystems (see Cross, 1994; Kandel and Schwartz, 1985; Roland, 1992). (See figure 16.1.) Each subsystem has a set of neurons that resides in the dorsal root ganglion of the spinal column. These neurons extend their axons to whatever tissue they innervate and receive external input there. They also have a second axon that projects across to the dorsal horn. The axon in the dorsal horn connects with a second set of neurons housed in the dorsal horn whose axons run out of the spinal column and up to the thalamus. The third set of neurons projects from the thalamus to the post-central gyrus in cerebral cortex.

In 1911, Head and Holmes proposed a dual system of afferent projections in our pain sensory system: an epicritic system that processes information regarding intensity and precise location, and a protopathic system that delivers the actual pain sensations. Almost nine decades later, we still believe they were fundamentally

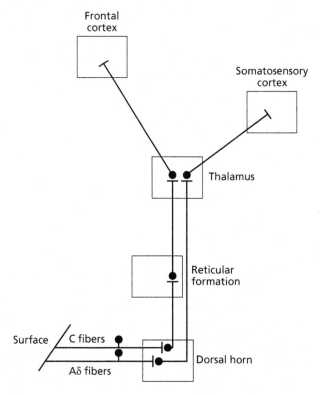

Figure 16.1 Diagram showing our pain sensory system. The first set of neurons take in information from the periphery and then synapse with a second set of neurons in the dorsal horn. These neurons ascend, some terminating in the reticular formation of the brain stem, others traveling to the thalamus. Axons that terminate medially in the thalamus synapse with a third set of neurons which project to the frontal cortex. Those that terminate laterally synapse with neurons that project to the somatosensory cortex.

correct. We now know that we have a "sensory discriminative" subsystem that computes the location, intensity, duration, and nature (stabbing, burning, prickling) of the stimuli. This subsystem is subserved by the Aδ fibers. These mechanoreceptive neurons are mylinated, so information can travel quite quickly along them (approximately 5–30 m/sec, as opposed to 0.5–2 m/sec for information traveling along unmylinated pathways). Consequently, they transmit what is known as "first pain" or "fast pain." The threshold for activation is constant from person to person, and this subsystem remains active (assuming no other defects in the organism) only as long as the raw nerve endings are stimulated.

We also have an "affective-motivational" subsystem responsible for the unpleasant part of painful sensations. This system feeds directly into our motor response systems and is considered to be phylogenetically older than other aspects of our multifaceted pain system. This polymodal subsystem begins with the well-known unmylinated C fibers. Once they are activated, they will continue to fire for some time, even after the noxious event has ceased. This subsystem gives rise to what is known as "slow pain" or "second pain," so-called because this is what we feel second whenever we are injured – a diffuse and persistent burning pain.

Similar to the color and form processors in the visual system, the Aδ-fiber and C-fiber pathways remain largely segregated. For example, generally speaking, they terminate in different layers on the dorsal horn. However, there is more interaction than the amount we find in either the visual or auditory system. The dorsal horn contains "wide-dynamic range" (WDR) neurons that respond to both Aδ and C neurons, as well as to other peripheral stimuli.

Once pain information exits the dorsal horn, it travels either to the reticular formation in the brain stem or to the thalamus. Laminae I and V project to the lateral nuclei in the thalamus (Craig et al., 1994), and laminae I, V, and VI project to the medial nuclei. Each type of nuclei underwrites a different sort of information; the lateral nuclei process discriminative information (fast pain), while the medial nuclei and reticular connections process affective-motivational information (slow pain). The two thalamic streams remain separate on their trip to cortex as well. Pain neurons in the lateral nuclei synapse in somatosensory cortex, which then can compute the location and characteristics of the pain; those in the medial nuclei synapse in the anterior cingulate gyrus in the frontal lobe, which figures in our emotional reactions to pain. The frontal lobe (and its connections) process our actual suffering.

3   Philosophy's Error

Now we can see how and why several philosophers are mistaken in their conclusions that there are no such things as pains (Churchland, 1985; Dennett, 1978), or that pains are located in our limbs (Armstrong, 1981; Newton, 1989; Pritcher, 1970), or that pains are purely subjective (Grahek, 1991; McGinn, 1983), or that pains are reactive behaviors (Gillett, 1991; Wittgenstein, 1953). Each of these positions identifies pain with one of the neuronal groups within the classic pain system, while failing to recognize that our pain system is complex and contains at least a duality of subsystems, each of which processes a different sort of information (see figure 16.2). In general, philosophers make these mistakes because they misunderstand the double-dissociation methodology. We can, either through purposeful intervention or accidents of Nature, dissociate our discriminative pain processing from our affective-motivational pain processing. Ingestion of morphine (or other opiates), lesions to the medial thalamus, and prefrontal lobotomies all result in sensations of pain without a sense of suffering and without producing characteristic pain behaviors (wincing, moaning, complaining, etc.). In these cases, patients can localize their pains but are not upset by the fact that they are in pain. We can also get reverse

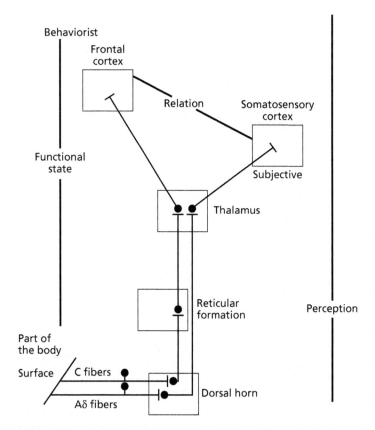

Figure 16.2   Part of the pain system identified with different philosophical views regarding the basic nature of pain. Most views identify pain with one subsystem or with one sort of neural processor. Two exceptions are functional state and perceptual views of pain, though perceptual views of pain often overlook or underestimate the motor component of pain processing.

effects, to a degree. Fentanyl causes one to react in pain, yet inhibits our discriminatory abilities for the pain (Gracely et al., 1982). Lesion studies and studies using hemispherectomies show that even with cortex completely missing, we can still have a pain sensation; we simply lack fine localization and intensity discrimination (Davis et al., 1995; Head and Holmes, 1911; Roland, 1992). Patients with Parkinson's disease and Huntington's disease often have pain sensations but are unable to indicate where they feel the pains (Chudler and Dong, 1995).

We also find instances of the pain centers in the thalamus and cortex being activated without corresponding activations of $A\delta$ or C fibers ("nociception"). Fully 80 percent of lower back pain sufferers present no external or internal injury (Wall, 1989). Phantom limbs and phantom pains in phantom limbs are quite common experiences in new amputees. Stimulating the medial periaqueductal gray region, tectum, or thalamus directly can also result in painful experiences (Davis et al., 1995; Keay et al., 1994). Finally, our emotional states heavily influence the degree of pain we feel, quite independent of actual injury. Indeed, psychogenic pains have been documented since the late nineteenth century, when D. H. Tuke reported the case of a butcher who got fouled up on a meat hook and appeared to be in agony. However, when examined by the local chemist, it was discovered that the meat hook had only penetrated his jacket sleeve and, even though the butcher was screaming in "excessive pain," he was completely unharmed.

Correlatively, there are also examples of our peripheral pain processors being activated without this information proceeding on to the thalamus or cortex. About 37 percent of emergency room visitors felt no pain at the time of their injury (Melzack et al., 1982). Athletes and soldiers can continue performing free of pain, even though they have been severely injured. Hypnosis allows some subjects to engage in what would otherwise be painful activities without being in pain. Placebos are notoriously helpful in relieving pain. And of course, some lesions to the thalamus and cortex can result in the cessation of pain experiences, even though the peripheral neurons continue to operate normally.

However, each of these double dissociations only individuates neuronal groups or subsystems within our overarching pain system. I make this claim by analogy with our other perceptual systems. Blindsight patients can discriminate shapes and figures they claim not to be able to see consciously. Sufferers of Anton's syndrome insist that they can see perfectly well, even though they are completely blind as a result of severe bilateral damage to their visual association areas. However, no one uses these facts to argue that vision does not exist, or that vision is located in our eyeballs, or that vision is purely subjective, or that vision is behavioral. We may not know exactly what to say about blindsight or Anton's syndrome, but no one claims that blindsight is not a disorder of the visual system or that patients with Anton's syndrome are not having a visual experience of some sort.

By misunderstanding what a perceptual system in the brain encompasses, many philosophers miss the boat regarding the basic nature and structure of pain. Double dissociation alone does not individuate our basic systems; the technique is rather used to isolate the subsystems that operate within the larger system. We then need to build our different systems out of the component pieces. Teleological considerations help us to do so. To wit: the neurons in our pain system all respond to roughly the same sort of information; they increase their rate of firing in the presence of noxious stimuli on skin or deep organs. Moreover, the connection among the six-neuron tract is a stable, common, and isolable pathway. Connections fore and aft show a stream of information flowing from the nociceptors on the skin up through the cortex. Finally, a pain-sensory system tied to the somatosensory processors makes good evolutionary sense. As creatures eking out lives in a hostile environment, having a system that could warn us when damage occurred and could force us to protect damaged parts until they healed would be tremendously beneficial. (Indeed, persons who cannot feel any pain at all often live a nasty, brutish, and short life.)

Neither our conscious experience of pain, the damaged tissue itself, nor our bodily or emotional reactions are fundamental to pain processing. Each is but one component of a larger processor. Hence, it is a mistake to try and claim one or the other as pain *simpliciter*. And it is equally erroneous to conclude that since we cannot identify one or the other with pain, there is no such thing. The entire pain-sensory system functions in largely the same way as any of our sensory systems. Pieces are united by our best guess of their function, based on the three types of converging evidence discussed above. Hence, we have concluded that the components of our visual system take the information contained in photons bouncing around in the world and use it to compute the location, orientation, texture, color, and movement of objects in the environment. The components of our auditory system take the information contained in atmospheric compression waves and use it to compute the placement of things. And the components of our pain system take pressure, temperature, and chemical readings of our surface (and interior) and use this information to track what is happening to our tissues. The Aδ cells and the C fibers do this, as do the spinothalamic tract and its connections to the cortex. In short, it appears we have a complex but well-defined sensory system which monitors our tissues in order to promote the welfare of our bodies.

4   The Brain in Pain: A New View of Pain Processing

Not surprisingly, though, appearances are deceiving. With the classic story of pain processing, we can determine how philosophers and others go astray in their thinking about pain: in assuming that pain is simpler than it is, they end up identifying pain with some component or subset of the entire complicated process. However, the classic story of pain *qua* sensory system only scratches the surface of the complexity of pain, and, in so doing, gives the illusion that our pain system is well defined. What counts as pain processing as opposed to an emotional reaction to pain or a belief that one is in pain, for example, is difficult to determine, for they all shade into one another. Once we move beyond the spinal column, discrete computational streams become difficult to identify and trace. What counts as the process proper and what counts as merely an influence on the process? What marks the end of one process and the beginning of another? There are no principled answers for pain.

But again, this problem appears in our other sensory systems as well (though I shall not take the time to do the comparative analysis here), so pain is not special in this regard. Perception, interpretation, judgment, and reaction are all tightly bound up with one another, so much so that separating cognitive "events" from one another becomes an artificial laboratory exercise. Pain is a far cry from the simple event that philosophers assume.

Knowledge of the higher brain processes involved in pain comes largely from imaging studies, which have only recently become cheap and easy enough to use fairly extensively in basic science research. However, imaging studies of pain are still relatively rare and replication is the exception, not the rule. (I should note, though, that true replication in any imaging study is highly unusual.) Experimental paradigms are not codified across laboratories, so definite results are hard to come by. Nevertheless, we can see general trends and patterns developing across the different studies. At the least, it is clear that the classic view of pain processing is woefully over-simplified and quite limited.

The first thing we learn from looking at the imaging studies is that blood flow or other measures of cerebral activity often decrease during painful stimulation. (We see deactivations most consistently in contralateral posterior cingulate, contralateral somatosensory cortex, and orbital gyrus; see figure 16.3.) The decrease appeared in the very first imaging study in the mid-1917s and has since remained a constant datum (Ingvar, 1975; Ingvar et al., 1976; Apkarian et al., 1992; Casey et al., 1996; Coghill et al., 1994). Though what the decreases mean is still under discussion, one good explanation is that they indicate neural inhibition. Of course, interneurons inhibiting the firing of other neurons would consume energy (and so their activity should show in any scan). However, these neurons are quite small relative to excitatory neurons and so their activity relative to excitatory cells might show up as an overall decrease in the area (cf. Apkarian, 1995).

In addition to the decreases, brain scans of induced acute pains indicate activity in the anterior cingulate cortex, anterior insula, primary and secondary somatosensory cortex, supplementary motor cortex, thalamus, and basal ganglion (Casey et al., 1994a; Coghill et al., 1994; Davis et al., 1997; Derbyshire et al., 1993; Jones et al., 1991; Talbot et al., 1991; Vogt et al., 1996). Excluding the primary somatosensory system, thalamus, and orbital gyrus, the brain regions responding to painful stimuli do not respond to non-noxious thermal stimuli or to vibrotactile stimuli. That is, the activated areas are specifically devoted to sensing and processing information about pain itself and not to analyzing any inputs which might merely resemble something painful, such as warmth or pressure. In general, parietal, insular, and cingulate cortical regions, plus the thalamus and the striatum respond significantly to painful stimuli. Figure 16.4 provides a sense of where these areas reside in the brain. Other regions may be involved as well, although they may not show any change in activity relative to baseline (because they are active during the baseline circumstances as well).

However, one danger in leaning too hard on these studies is that there is no good way to differentiate between processing the experience of pain on the one hand, and startle reflexes, anticipatory motor programming to withdraw from the pain, changes in alertness, and so forth, on the other. All the brain scans can tell us is which areas of the brain differentially respond to the task; they cannot tell us what role the activated regions are playing in the cognitive or behavioral economy of the subject. We have many responses to pain; some include the experience of pain itself, but others include our evaluation of the incident, our decisions about how to react, our bodies getting ready to respond, our planning for the next event – sorting which is which is not trivial.

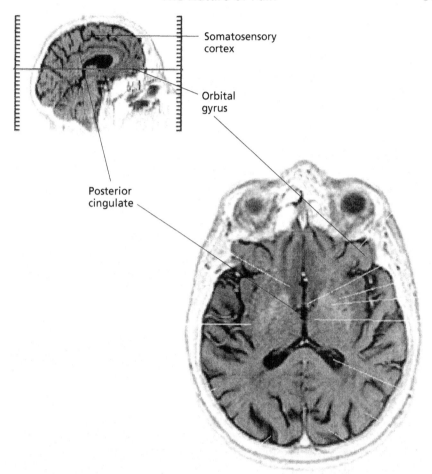

Figure 16.3 Brain image schematic showing the approximate locations of the areas that consistently indicate decreased blood flow during a painful experience. We assume that decreased blood flow correlates with neural inhibition in the brain.

Nonetheless, we can infer some things about the functional roles of the activated areas, particularly when we compare acute instances of pain with tonic ones. When the stimuli are repeated continuously over time, the ipsilateral insula, thalamus, and cerebellum also became active (Casey et al., 1994b). Here we see the first glimpse of a general arousal that might be tied to pain as the brain activity becomes bilateral and more distributed. Indeed, it looks as though tonic and chronic pains show a very different activity pattern than brief acute pains (Apkarian et al., 1992; Backonja et al., 1991).

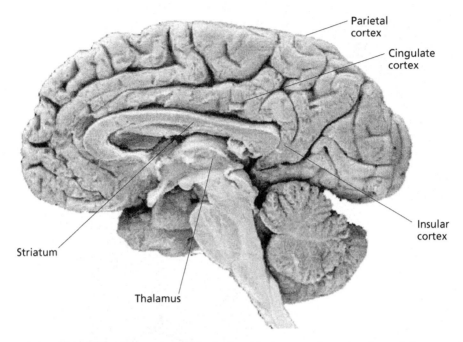

Figure 16.4 Midsagittal section of the brain with the approximate locations of the areas that respond significantly to painful stimuli marked.

In particular, some argue for a biphasic pattern of pain processing in the cortex which closely parallels the two sorts of lower-level pain processing assumed in the classic view. Apkarian (1992, 1995) suggests that the cortical activity seen in the somatosensory regions during acute pain reflects the subject's decision to remain in the painful situation for the duration of the experiment. This decision requires information concerning where the painful stimuli impinge on the body and how intense or disturbing the phenomenological experience is. This process is a relative of the Aδ fiber–based sensory discriminative subsystem, and now sometimes is called the "lateral pain system" (Guyton, 1991). The difference between the classic perspective and the new view is that Apkarian argues early pain processing to be largely motoric and to turn on one's local motivations in behavior. In fact, studies comparing what happens in the brain when a finger is heated beyond the pain threshold versus when the same finger is vibrated or moved, show about 90 percent of the activated cortical areas in common. The overlap tells us that what happens in the brain in the early pain-processing phase is very similar to what happens during motor programming and execution.

Following these early decisions, though, one can still perceive pain. This aspect of pain processing does not depend upon our somatosensory regions, which is why activity there decreases even though painful stimuli continue to impinge on the nociceptors. Despite the emphasis of the classic view on the centrality of the thalamus as a pain relay station, this later non-motoric processing probably has extrathalamic sources, for we see a decrease in thalamic activity in conjunction with increases in activity in anterior cingulate, supplementary motor, and anterior insular regions (but decreases in primary and secondary somatosensory regions, and primary motor cortex). This second phase, known as the "medial pain system," corresponds to some sort of central reaction. Pain still persists, even though the immediate behavioral response has been plotted and executed. Something else has to be done and perhaps what we are seeing is the brain's alternative strategy for dealing with the stimulus (Davis et al., 1995; Evans et al., 1992; Roland, 1992; Stea and Apkarian, 1992).

Similar decreases in brain response are seen when subjects anticipate a painful or unpleasant stimulus, and its magnitude is associated with the level of anxiety the subjects reported experiencing before the input (Drevets et al., 1995). (This correlation is striking because there is no correlation between the activity of primary somatosensory cortex and pain or intensity ratings.) The brain responds in the same way when we expect a pain to be forthcoming or when we continue to experience pain after our motor decisions have been made; the brain clearly has other coping mechanisms for pain beyond simple and quick motor responses.

"Coping mechanism" should be understood in scare quotes, however, for often what our brain decides to do to "cope" is not terribly helpful. We can see this in particular in our susceptibility to Pavlovian classical conditioning. We can easily condition sensory processing – including pain processing – to produce concomitant negative emotions, such as fear, depression, or anxiety (LeDoux, 1993; LeDoux et al., 1988, 1990). This "conditioning" is really a by-product of thalamic activity, and subcortical conditioning strongly resists extinction. One painful experience in a particular setting can induce the same emotions in the same setting later. Fear and loathing of dental procedures is perhaps the most common example of this.

Tying fear to pain intensifies the affective dimensions of pain when what is normally considered to be a minor pain is paired with intense negative emotion (Chapman, 1996). It also means that the thalamus no longer needs to be active to continue the emotional response. Once a cognitive or emotive reaction is programmed into a system, it will kick in full-strength with appropriate triggers. And it will continue to run, even if the trigger is brief and then turned off. Such is the nature of classical conditioning, and unfortunately, we have very little higher-level control over those circuits once they get set up.

Classical conditioning is just one way in which pain can become interwoven with affect. Extrathalamic connections may be even more powerful, though I shall not discuss these here. My point is simply that our pain system is intimately related to our perceptual, cognitive, and affective apparatuses. Except under unusual circumstances, we cannot react to pain without also bringing forth a lot of additional baggage. There is no such thing as a simple pain state, nor a simple pain.

5 A Return to Ancient Philosophy

Aristotle considered pain a "passion" of the soul, an affect distinct from our five senses. It is important to understand, though, that Aristotle's "passions" are wider in scope than ours. He meant "passion" (*pathe*) to refer to anything we perceive or feel that comes to us unbidden. Things we cannot control intellectually are passions. These would include our emotions as well as our perceptions. Consequently, the dividing line between the emotions and our sensory experiences is not as clear for Aristotle as it is for many contemporary philosophers. A second theme in this chapter is that an Aristotelian view is right on this score. Distinguishing among emotional, sensory, and cognitive responses is quite difficult to do, for they all run together in the brain.

Philosophers should remember this fact when theorizing about the mind/brain. We cannot have simple sensory experiences, most especially pain experiences, for these experiences are intertwined with and filtered through our emotional and cognitive lives. Contemporary resurrections of old sense data doctrines (e.g. Dretske, 1988; Fodor, 1987) are thus wrong-headed and should die a quick death.

In any event, contemporary scientists are concluding what the ancient Greek philosophers already knew: "Emotion is not simply a consequence of pain sensation that occurs after a noxious sensory message arrives at somatosensory cortex. Rather, it is a fundamental part of the pain experience. . . . [The experience of pain] is above all else a powerful and demanding feeling state" (Chapman, 1996, p. 63). The brain's reaction to pain information is complicated and multidimensional. It involves large chunks of our cognitive and emotional lives and utilizes processing resources distributed throughout our neural tissue. So far as the brain is concerned, pain is an important and all-encompassing event.

Acknowledgments

The first half of this chapter comes from "When a Pain is Not," *Journal of Philosophy* (1997). Expanded versions of these ideas can be found in *What Pain Is* (1999), MIT Press. The figures printed here come from the MIT Press book; my thanks to the publishers for allowing their reproduction.

Notes

1   This method of investigation forms the backbone of Daniel Dennett's (1987) notion of "functional decomposition."
2   Fred Dretske (1981, 1988) has the best-developed philosophical account of how correlations between environmental and internal events provide meanings for our neural firing patterns (or whatever the meaningful unit is in the brain). Over the past decade, this sort

of "informational semantics" has become quite a cottage industry in philosophy of mind and philosophy of language.

References

Apkarian, A. V. 1995: Functional imaging of pain: New insights regarding the role of the cerebral cortex in human pain perception. *Seminars in the Neurosciences*, 7, 279–93.

Apkarian, A. V., Stea, R. A., Mangos, S. H., Szeverenyi, N. M., King, R. B., and Thomas, F. D. 1992: Persistent pain inhibits contralateral somatosensory cortical activity in humans. *Neuroscience Letters*, 140, 141–7.

Armstrong, D. 1981: *The Nature of Mind and Other Essays*. Ithaca, NY: Cornell University Press.

Backonja, M., Howland, E. W., Wang, J., Smith, J., Salinsky, M., and Cleeland, C. S. 1991: Tonic changes in alpha power during immersion of the hand in cold water. *Electroencephalography and Clinical Neurophysiology*, 79, 192–203.

Casey, K. L., Minoshima, S., Berger, K. L., Koeppe, R. A., Morrow, T. J., and Frey, K. A. 1994a: Positron emission tomographic analysis of cerebral structures activated specifically by repetitive noxious heat stimuli. *Journal of Neurophysiology*, 71, 802–7.

Casey, K. L., Minoshima, S., Koeppe, R. A., Weeder, J., and Morrow, T. J. 1994b: Temporal-spatial dynamics of human forebrain activity during noxious heat stimulation. *Society of Neuroscience Abstracts*, 20, 1573.

Casey, K. L., Minoshima, S., Morrow, T. J., and Koeppe, R. A. 1996: Comparison of human cerebral activation pattern during cutaneous warmth, heat pain, and deep cold pain. *Journal of Neurophysiology*, 76, 571–81.

Chapman, C. R. 1996: Limbic processes and the affective dimension of pain. In G. Carli and M. Zimmerman (eds), *Progress in Brain Research*, vol. 10, New York: Elsevier.

Chudler, E. H., and Dong, W. K. 1995: The role of the basal ganglia in nociception and pain. *Pain*, 60, 3–85.

Churchland, P. M. 1985: Reduction, qualia, and the direct introspection of brain states. *Journal of Philosophy*, 82, 2–28.

Churchland, P. S. 1986: Epistemology in the age of neuroscience. *Journal of Philosophy*, 83, 544–53.

Coghill, R. C., Talbot, J. D., Evans, A. C., Meyer, E., Gjedde, A., Bushnell, M. C., and Duncan, G. H. 1994: Distributed processing of pain and vibration by the human brain. *Journal of Neuroscience*, 14, 4095–108.

Craig, A. D., Bushnell, M. C., Zhang, E. T., and Blomqvist, A. 1994: A thalamic nucleus specific for pain and temperature sensation. *Nature*, 372, 770–3.

Cross, S. A. 1994: Pathophysiology of pain. *Mayo Clinic Proceedings*, 69, 375–83.

Davis, K. D., Tasker, R. R., Kiss, Z. H. T., Hutchison, W. D., and Dostrovsky, J. O. 1995: Visceral pain evoked by thalamic microstimulation in humans. *Neuroreport*, 6, 369–74.

Davis, K. D., Taylor, S. J., Crawley, A. P., Wood, M. L., and Mikulis, D. J. 1997: Functional MRI of pain- and attention-related activation in the human cingulate cortex. *Neuroreport*, 77, 3370–80.

Dennett, D. C. 1978: Why you can't make a computer that feels pain. *Synthese*, 38, 449.

Dennett, D. C. 1987: *The Intentional Stance*. Cambridge, MA: MIT Press.

Derbyshire, S. W. G., Jones, A. K. P., Brown, W. D., Devani, P., Friston, K. J., Qi, L. Y., Pearce, S., Frackowiak, R. S. J., and Jones, T. 1993: Cortical and subcortical responses to pain in male and female volunteers. *7th World Congress on Pain*, 7, 500.

Dretske, F. 1981: *Knowledge and the Flow of Information*. Cambridge, MA: MIT Press.

Dretske, F. 1988: *Explaining Behavior: Reasons in a World of Causes*. Cambridge, MA: MIT Press.

Drevets, W. C., Burton, H., Videen, T. O., Snyder, A. Z., Simpson, J. R. Jr., and Raichle, M. E. 1995: Blood flow changes in human somatosensory cortex during anticipated stimulation. *Nature*, 373: 249–52.

Evans, A. C., Meyer, E., and Marret, S. 1992: Pain and activation in the thalamus. *Trends in Neuroscience*, 15, 252.

Fodor, J. A. 1987: *Psychosemantics: The Problem of Meaning in the Philosophy of Mind*. Cambridge, MA: MIT Press.

Gillett, G. R. 1991: The neurophilosophy of pain. *Philosophy*, 66, 191–206.

Gracely, R. H., Dubner, R., and McGrath, P. A. 1982: Fentanyl reduces the intensity of painful tooth pulp sensations: Controlling for detection of active drugs. *Anesthesia and Analgesia*, 61, 751–5.

Grahek, N. 1991: Objective and subjective aspects of pain. *Philosophical Psychology*, 4, 249–66.

Guyton, A. C. 1991: *Basic Neuroscience: Anatomy and Physiology*. Philadelphia: W. B. Saunders.

Head, H., and Holmes, G. 1911: Sensory disturbances from cerebral lesions. *Brain*, 34, 102–254.

Ingvar, D. H. 1975: Patterns of brain activity revealed by measurements of regional cerebral blood flow. In D. H. Ingvar and N. A. Lassen (eds), *Brain Work: The Coupling of Function, Metabolism, and Blood Flow in the Brain*, Copenhagen: Munksgaard.

Ingvar, D. H., Rosén, I., Eriksson, M., and Elmqvist, D. 1976: Activation patterns induced in the dominant hemisphere by skin stimulation. In Y. Zotterman (ed.), *Sensory Functions of the Skin*, London: Pergamon Press.

Jones, A. K. P., Brown, W. D., Friston, K. J., Qi, L. Y., and Frackowiak, R. S. J. 1991: Cortical and subcortical localization of response to pain in man using positron emission tomography. *Proceedings of the Royal Society London*, 244, 39–44.

Kandel, E. R., and Schwartz, J. H. 1985: *Principles of Neural Science*, 2nd edn. New York: Elsevier.

Keay, K. A., Clement, C. I., Owler, B., Depaulis, A., and Bandelr, R. 1994: Convergence of deep somatic and visceral nociceptive-information onto a discrete ventrolateral midbrain periqueductal gray region. *Neuroscience*, 61, 727–32.

LeDoux, J. E. 1993: Emotional memory: In search of systems and synapses. In F. M. Crinella and J. Yu (eds), *Brain Mechanisms*, vol. 233, pp. 149–57. New York: New York Academy of Sciences.

LeDoux, J. E., Farb, C., and Ruggiero, D. A. 1990: Topographic organization of neurons in the acoustic thalamus that project to the amygdala. *Journal of Neuroscience*, 10, 1043–54.

LeDoux, J. E., Iwata, J., Cicchetti, P., and Reis, D. J. 1988: Different projections of the central amygdaloid nucleus mediate autonomic and behavioral correlates of conditioned fear. *Journal of Neuroscience*, 8, 2517–29.

McGinn, C. 1983: *The Subjective View*. Oxford: Oxford University Press.

Melzack, R., Wall, P. D., and Ty, T. C. 1982: Acute pain in an emergency clinic: Latency of onset and descriptor patterns related to different injuries. *Pain*, 14, 33–43.

Newton, N. 1989: On viewing pain as a secondary quality. *Nous*, 23, 569–98.

Pritcher, G. 1970: Pain perception. *Philosophical Review*, 79, 368–93.

Roland, P. E. 1992: Cortical representations of pain. *Trends in Neuroscience*, 15, 3–5.

Stea, R. A., and Apkarian, A. V. 1992: Pain and somatosensory activation. *Trends in Neuro-science*, 15, 250–1.

Talbot, J. D., Marrett, S., Evans, A. C., Meyer, E., Bushnell, M. C., and Duncan, G. H. 1991: Multiple representations of pain in human cerebral cortex. *Science*, 251, 1355–8.

Tuke, D. H. 1884: *Illustrations of the Influence of the Mind upon the Body in Health and Disease Designed to Elucidate the Imagination, Second Edition*. Philadelphia: Henry C. Lee's.

Vogt, B. A., Derbyshire, S., and Jones, A. K. 1996: Pain processing in four regions of human cingulate cortex localized with co-registered PET and MR imaging. *European Journal of Neuroscience*, 8, 1461–73.

Wall, P. D. 1989: Introduction. In P. D. Wall and R. Melsack (eds), *Textbook of Pain*, 2nd edn. New York: Churchill Livingotone.

Wittgenstein, L. 1953: *Philosophical Investigations*. Trans. by G. E. M. Anscombe. New York: Macmillan.

17

Points of View from the Brain's Eye View: Subjectivity and Neural Representation

Pete Mandik

Conscious experiences are supposed by many to be subjective in the sense of being perspectival or from a point of view (see, for example, Nagel, 1974, 1986; Tye, 1995). Allegedly, the subjectivity of consciousness is beyond the grasp of science, which is objective (Nagel, 1986). If you wanted a scientific understanding of consciousness, how would you solve the problem of subjectivity? One strategy would be to deny that consciousness is really subjective. Another might be to deny that science is really objective. Yet a third, and the one I favor, conserves both the subjectivity of experience and the objectivity of science. I present a way of looking at subjectivity and objectivity that allows for an objective, scientific – indeed, neuroscientific – understanding of the subjectivity of conscious experience.

My general strategy will be to assimilate the subjectivity of experience to the subjectivity of certain mental representations and give an account of how the subjectivity of representation can be understood within an objective naturalistic framework. My account is naturalistic in that I sketch a neuroscientifically informed philosophical account of what it means for mental representations to be subjective (that is, perspectival or imbued with a point of view). Previous accounts – such as those advocated by Georges Rey (1997), Michael Tye (1995), and William Lycan (1996) – understand subjective representations as akin to indexical linguistic expressions. My alternative account does not make indexicality a requirement of subjectivity. Instead, I analyze subjectivity in terms of subject relativity: subjective representations have the function of carrying information about the representing subject. I discuss the neural bases of egocentric representations of space. I also discuss ways in which my account may generalize to non-visual and non-spatial sensory modalities. Finally, I relate my account of perspective to discussions of the so-called *knowledge argument* that the subjectivity of consciousness is a bar to the success of physicalism.

What does it mean for something to be subjective in the way that experiences are said to be subjective, that is, perspectival or embodying a point of view? To get a handle on what it means for experience to be perspectival, we must first note that experience is representational. When we experience the world as being a certain way, we thereby mentally represent the world as being a certain way. When I have a conscious experience of a fire engine being bright red and six feet away from me, the experience itself is neither bright red nor six feet away from me. The experience itself is a state of my nervous system that represents the fire engine as being bright red and six feet away from me. To understand what it means for experience to be perspectival I propose, then, to begin by answering the following question: What does it mean for a representation to have or be from a point of view?

A philosophically popular kind of answer to the above question begins by latching on to the literary usage of the phrases "point of view" and "perspective" (see, for example, Lycan, 1996). Prose written from the so-called first-person point of view employs indexical terms to refer to the author or alleged author. One famous example would be the opening line of Melville's *Moby Dick:* "Call me Ishmael." The second-person point of view uses indexical/demonstrative terms to address the reader, as in the sentence "You may be wondering what will happen next." Typical instances of prose written in the third-person point of view are devoid of indexical and demonstrative terms, as in "On November 19, 1999, Pete Mandik purchased a portrait of a platypus."

There is a less literary and more literal way of understanding perspective and point of view. This is the rather literal sense that pictorial representations embody a point of view. Consider two photographs taken of a person's face: the first may be head-on, the other may show the head in profile. The camera that produced the photos occupied two different points of view with respect to the person's head. The representational contents of the photographs produced include content about these points of view. This is why we can tell by looking at the photographs whether the camera occupied a point of view in front of or to the side of the subject.

Note that the pictorial sense of "point of view" is rather literal but not totally literal. Few imagistic representations contain enough information to specify a particular *point* of view. For example, a typical map of Chicago presents a "bird's eye view" of the city, but abstracts away from any information that would specify a point of view positioned over the Sears Tower as opposed to the Hancock Building. This is not to say that the map abstracts entirely away from point of view: it does contain enough information to specify that one is viewing the city from above rather than below and that one is viewing the city from a location over a part of the United States as opposed to a location over Japan. Another way of seeing how pictorial perspective abstracts away from literal, geometric, *points* of view is in the notion of what is known as aerial or atmospheric perspective. Pictures employing atmospheric perspective depict things in the far distance as being fainter, hazier, and bluer than things that are closer to the viewer. The device of atmospheric perspective exploits facts about the behavior of light in the atmosphere to depict relations of distance between the viewer and what is pictured, and is a frequently employed device in landscape painting. The sort of relations captured by atmospheric perspective abstract away from the precise location of the viewer. The pictorial sense of perspective can be understood as not necessarily implicating a precise *point* of view. Thus the pictorial notion of perspective is just the notion that part of the representational content of the picture includes relations between what is pictured and the viewer. Such relations would include being in front of or to the side of a face or a house. Other relations would include being closer to one pictured object than another.

Pictorial perspective is not only a property of photographs, drawings, and paintings. Mental representations also exhibit pictorial perspective. In describing pictures as involving pictorial perspective I said that the representational contents of the pictures include relations between the things pictured and the viewer. Extending this account of perspective to mental representations yields the thesis that some mental representations include in their contents relations between the representing subject and that which is represented. Thus such representations are egocentric (self-centered) since they represent relations that things bear to the representer. I turn now to the case that there are such mental representations.

The point that mental representations exhibit pictorial perspective can be bolstered by both phenomenological and empirical scientific considerations. Consider first the phenomenological support. Just as two different photographs of the same house may be from two different points of view, so may two different visual experiences be from two different points of view. What it is like to look at the house from a point to the north of it may be quite different from what it is like to look at the house from a point to the west of it. Thus the perceptual representations involved in the two different experiences exhibit pictorial perspective. We need not actually perceive a house to have mental representations that exhibit pictorial perspective. One may dream of seeing a house from one point and then another and notice the differences in point of view. One may also imagine looking at the house from the north and then the west and in each instance a difference in point of view is introspectible in imagination. These points based on introspection may be enhanced by evidence from psychological studies, to which I now turn.

Much psychological research in the past several decades has concerned the nature of mental images and speaks to the issue of the existence of perspectival representations. A classic example is due to Roger N. Shepard and his colleagues (Shepard and Cooper, 1982). These researchers had subjects look at simultaneously presented pairs of objects. The second member of each pair was either the same as the first or a mirror image. Further, pair members could differ from each other in their rotations in depth and in the picture plane. The researchers found that the time it took for subjects to make "same-different" judgments increased monotonically with increases of rotational displacement between pair members. Shepard took these reaction time data as evidence that subjects were rotating mental images to see if they would match the stimulus. The evidence that Shepard collected also serves as evidence for the existence of pictorially perspectival mental representations. A mental image at any given stage of a rotation constitutes a perspectival representation because at each point in rotation, the image represents what the object would look like from a particular point of view.

Some theorists have postulated that mechanisms similar to those postulated for image rotation may be at work in visual object recognition. Humans recognize visually presented three-dimensional objects with only two-dimensional projections on the retina as a guide. Somehow, we are able to recognize objects seen from unfamiliar viewpoints, that is, based on unfamiliar projections on to our retinas. Certain studies of the accuracy and reaction times in visual recognition tasks implicate perspectival representations. Such studies typically examine the reaction times and accuracy of recognition judgments of objects seen from unfamiliar viewpoints. In such studies, average length of reaction time and judgment accuracy vary monotonically with the degree of rotational deviation (in depth or on the picture plane) from familiar views of an object. These correlations are taken as evidence for the hypothesis that visual object recognition is mediated by a normalization mechanism. The stored representation of an object is one or more encoded "views" that encode only two-dimensional information based on previous retinal projections. Recognition of familiar objects seen from unfamiliar viewpoints involves a match between a stored view and the perceptual view via a normalization mechanism which compares the views (e.g. Bülthoff and Edelman, 1992; Shepard and Cooper, 1982; Ullman, 1989). For example, this might involve mentally rotating an image (Shepard and Cooper, 1982). Object recognition, as well as imagery, may involve perspectival representations.

Perspectival representations also surface in accounts of navigation. Just as there is evidence that an object may be recognized better from one point of view than another, so may a destination be better arrived at from one starting point than another. There is a rich body of research on navigation by rats. One kind of experiment examines the performance of lesioned rats in the Morris water maze, an apparatus filled with water in which rats can swim. Objects such as small platforms can be placed in this area. Milk powder can be added to the water to make it opaque, and the level of the water can be adjusted so that when a platform is submerged it is not visible to rats swimming in the maze. Eichenbaum et al. (1990) employed a water maze in which rats had to swim to a platform visible during training trials, but occluded by the opaque water in the testing trials. Varied visual stimuli were positioned around the maze to serve as orientation cues. The experimenters trained intact and hippocampal-system damaged rats to swim to the platform from a given start location. During test trials, both the intact and damaged rats were able to swim to the platform if they were started from the same location as in the learning trials. However, the performances of the intact and damaged rats diverged widely when they were started from novel locations in the water maze: intact rats were still able to navigate to the platform, whereas the hippocampal damaged rats either required much longer to find the platform or never found it.

Much discussion of these sorts of investigations of the neural bases of rat navigational abilities has concerned the proposal that the hippocampus is a locus of allocentric ("other-centered") representations of spatial locations; when it is damaged rats must rely on merely egocentric ("self-centered") representations of spatial locations (see, e.g., O'Keefe and Nadel, 1978).

Evidence of egocentric and allocentric representations of spatial locations is not confined to studies of rats. For example, Feigenbaum and Rolls (1991) recorded the electrical activity of individual neurons in the hippocampus of macaque monkeys. In their study, they looked for neurons that were maximally responsive to the spatial location of a visual stimulus. They then changed the spatial relation of the monkey with respect to the stimulus so that, although the stimulus had not moved, it projected to a different part of the monkey's retina. Neurons which were maximally responsive to stimuli in that location regardless of what part of the retina the stimulus projected to were regarded as encoding allocentric representations of that spatial location. In contrast, neural activity maximally responsive to a spatial location defined relative to the site of retinal projection is regarded as an egocentric representation of that location. Feigenbaum and Rolls (1991) report that the majority (but not all) of the cells in the hippocampus that they investigated were allocentric representations of spatial location. Where the hippocampus has been widely implicated as the locus of allocentric representations of spatial locations, many kinds of egocentric representations of space (including, for instance, head–centered and shoulder–centered representations) have been localized in regions of the posterior parietal cortex (see Stein, 1992, and Milner and Goodale, 1995, for general discussions of egocentric representations in parietal cortex).

However, questions of *where* egocentric representations are located are not as interesting as questions of *what* egocentric representations are. I propose to explore such questions against the backdrop of naturalistic theories of representational content. Here I will assume an informational psychosemantics, the crux of which is the view that a brain state represents X by virtue of having the function of carrying information about (being caused by) X (Dretske, 1995). See Bechtel (chapter 18, this volume) for a discussion of this kind of view and its significance in neuroscience. While informational psychosemantics offers these conditions as both necessary and sufficient for representations, here I need only regard them as sufficient. Elsewhere I argue against their necessity (Mandik, 1999) and offer an additional set of sufficient conditions: X may represent Y if X has the function of causing Y. As discussed there, this (procedural instead of informational) account is most appropriate to accommodating the view, widespread in neuroscience, that activation of areas in motor cortex represents efferent events.

We have here then two kinds of condition sufficient for X to represent Y: one expressed in terms of X's causing Y, one in terms of Y's causing X. I will combine the two and say that it is sufficient for X to represent Y that X has the function of being causally related to Y (alternately: causally covarying with Y), thus covering both the informational (affector) and procedural (effector) cases. I make no claims on these sufficient conditions also being necessary. I have neither the resources nor the need here to offer a full–blown account of representation, an account that can supply conditions for all kinds of mental representations. My main concern here is with accounts adequate for representation as it figures in sensory experience. I will assume without further argument that causal covariational accounts are adequate.

The neurobiological paradigm for causal covariational semantics is the *feature detector*: one or more neurons that are (1) maximally responsive to a particular type of stimulus, and (2) have the function of indicating the presence of a stimulus of that type. Examples of such stimulus-types for visual feature detectors include high-contrast edges, motion direction, and colors. A favorite feature detector among philosophers is the alleged fly detector in the frog. Lettvin et al. (1959) identified cells in the frog retina that responded maximally to small shapes moving across the visual field. The inference that such cells have the function of detecting flies and not just any small moving thing is based on certain assumptions about the diet and environment of frogs.

Using experimental techniques ranging from single-cell recording to sophisticated functional imaging, neuroscientists have recently discovered a host of neurons that are maximally responsive to a variety of stimuli. Among these are neurons that are particularly sensitive to the spatial locations of stimuli. In some cases the neurons are responsive to locations relative to the subject, thus giving rise to perspectival, or egocentric, representations of spatial locations. In other cases, the neurons are responsive to locations independent of the relations between the location and the subject, thus giving rise to non-perspectival, or allocentric, representations of spatial locations.

Now we are in a position to see how the notion of perspectival representations such as egocentric representations of locations may be accommodated by a casual covariational psychosemantics. A subject S has a perspectival representation R of X if and only if the representational content of R includes relations S bears to X. Cashing out the notion of representation in terms of the teleological and causal covariational account described above yields the following formulation. A subject S has a perspectival representation R of X if (but maybe not only if) R has the function of causally covarying with X and relations Z1–Zn S bears to X. In the case of spatial representations, on which I will focus for now, the relations in question will be spatial relations. Later I will generalize this definition of perspectival representation to non-spatial sensory modalities.

One class of spatial perspectival representations is provided by neurons with retinocentric receptive fields. Such a neuron, whether in cortex or in the retina itself, demonstrates a pattern of activity maximally responsive to the occurrence of a specific kind of electromagnetic radiation in a certain spatial location defined relative to the retina. It is a plausible and widespread assumption that activity in neurons with retinocentric receptive fields represents (or "encodes" or "codes for") luminance increments in retina-relative spatial locations. If this assumption is correct, then we can see how such neural representations conform to the account of perspectival representations. In this example R is a certain kind of activity in a certain neuron in S's nervous system, X is a luminance increment and Z1–Zn include the spatial relations X bears to S (especially spatial relations to S's retina).

For another example, consider neural activity that represents goal locations for saccades. Analogous to the receptive fields of sensory neurons, motor neurons have what we may call *effective* fields. The effective field of a neuron may be a region in space that an organism may move or reach toward in response to activity in a particular neuron. There are neurons that control saccades that have as effective fields head-relative spatial locations. If it is correct to speak of activity in such neurons as *representing, encoding,* or *coding for* head-relative spatial locations, then we have another instance of perspectival representations. Activity in such motor neurons represents the movement of the eye toward a location in space defined relative to the subject's head. Such neural activations do not simply causally covary with the movement of the eyes to a certain location, but the movement of eyes to a certain location defined relative to the subject, and in this instance relative to the subject's head. And if these activations have the function of causally covarying with these subject-relative locations then they constitute perspectival representations of spatial locations.

A brief word needs to be said about the compatibility of mental imagery and the causal covariational account of representation. Some researchers favor an account of imagery whereby images represent in virtue of resembling that which is represented (see Kosslyn, 1994). There is much literature on this issue, and suffice it to say, few agree that resemblance is necessary for representation, even in cases where the representations are images. To illustrate the point, consider finding a creature inside of which we found something that looked like this: _/\/_ . Suppose that we wondered whether this constituted a representation of something. It resembles a mountain range. Might it be a representation of a mountain range? It also represents saw-teeth, a row of evergreen trees, and abandoning visual resemblance, we may say that it resembles a noise with a certain waveform. Which does it represent? Resemblance, as many have pointed out, underdetermines representation, and even in cases in which representations do resemble what they represent, functional causal covariation may be called in to do the job of disambiguation. For instance, a photograph of Joe equally resembles Joe and Joe's identical twin Moe. But the photograph is a photograph of Joe and not Moe in virtue of Joe's position in the causal chain that led up to the creation of that photograph.

These points about the role of causal covariation in determining the contents of imagistic representations are consistent with a common analysis of mental imagery. According to this analysis, imagery is the off-line utilization of perceptual (and perhaps motor) processes that are typically used on-line (see, for example, Chapter 19, this volume). Since that is all there is to imagery, imagery need not resemble what it is an image of. Such a view allows for imagery in non-visual modalities such as olfaction, where the possibility of resemblance between the representation and the represented seems obscure. I mention these points not to settle any ongoing controversies regarding mental imagery, but only to show that the existence of imagistic mental representations is neither necessarily nor obviously incompatible with accounts of representation that make having the function of causally covarying with X sufficient for representing X.

Another alleged incompatibility of imagery and causal psychosemantics can be shown to be merely apparent. Recall that part of the account on offer is that a subject S's representation R of X is perspectival if it has the function of carrying information about relations between S and X. Rick Grush (personal communication) objects that this kind of account is inadequate for the perspective embodied in imagery on the grounds that:

> Taking S to be the actual real S then entails that my imagining seeing the Eiffel Tower right now from the north is not perspectival, because it carries no information about the relations between S (the actual me sitting here in the café) and the Eiffel Tower.

Grush may be correct that the event of imagining does not carry information about the relations between S and the Eiffel Tower. On a particular occasion one may be caused to imagine the Eiffel Tower by something other than the Eiffel Tower, and thus that particular imagining would not carry information about the Eiffel Tower. However, on the account of imagery sketched in the previous paragraph, the event of imagining involves a state that has the function of carrying information about the relations between S and the Eiffel Tower. Imagining involves running off-line what is run on-line during perception: states that are supposed to carry information in the perceptual case may also be pressed into service for off-line imaginings. Thus the off-line states employed in imagining the Eiffel Tower owe their representational content to the information they are supposed to carry in the on-line perceptual case.

I call the analysis of perspective I offer *pictorial* in order to contrast it with the analysis of perspective offered by Lycan (1996) and others in terms of indexicals and the literary convention of first-person point of view. Pictures are the prototypical instances of representations with pictorial perspective, but it is important to empha-size that they are not the only instances. Pictures of a car from two different points of view may be encoded as bitmaps, which may themselves be translated into strings of ones and zeros or sentences describing the occupants of every cell in the bitmap's two-dimensional array. One may find it natural to suppose that such resulting strings of numerals and sentences may retain the representational content of the pictures from whence they came without themselves being pictures. This is not to say, however, that bitmaps have all and only the content of the images they encode. There may be some differences in the representational contents of the images and the bitmaps. However, despite possible differences of representational content, there are also considerable similarities. If a bitmapped photograph represents a car, then the corresponding numerical string does too. After all, the picture of a car is recoverable from the bitstring. And if the bitstring retains the representational content of being about a car, then there seems no reason to deny that the bitstring also retains the representational content of being about a car *as seen from a particular point of view*. Thus, if a picture is perspectival, then its corresponding bitstrings are perspectival, even though the corresponding bitstrings are not pictures.

Bitstrings are not the only instances of representations with pictorial perspective that are not pictures. Activations in neurons with retinocentric receptive fields are another example. Such neural activations represent the occurrence of stimuli at spatial locations defined relative to the retina. I take it as obvious that the activation of a single neuron is not a picture even in cases in which such activation may be a spatial representation. Below I will discuss the possibility of perspectival representations of temperature, thus giving yet another example of representations with pictorial perspective that are not pictures.

Representations with pictorial perspective are not necessarily pictures. Nor are they necessarily indexical. Above I mentioned that I intend the pictorial analysis to contrast with indexical analysis, but have not shown that this is indeed the case. The reason why representations with pictorial perspective are not necessarily indexical has to do with *particularity*. Indexicals necessarily involve particularity in a way that non-indexicals and egocentric representations do not. The representational content of the utterance "I am here now" picks out a particular individual at a particular location at a particular time. Even when indexicals and demonstratives are used to pick out universals, as in saying "this shade of red" while holding up a chip of paint, reference to the universal (the shade of red) piggy-backs on the particulars that secure the indexical content: the particular paint chip held by the particular individual at a particular time. In contrast, a picture can exhibit perspectival content without picking out any particular. A drawing of the 1991 Chevy Cavalier may be used not to represent any particular 1991 Chevy Cavalier. It may instead be used to represent a corresponding universal, say, the general category that all and only 1991 Chevy Cavaliers belong to. Nonetheless, the picture, in being a two-dimensional representation of a three-dimensional object, represents the car from one or another point of view. For instance, the picture may show the front of the car but not the back. Of course, the conventions of photography may have an indexical element: a photo represents me and not my twin in virtue of being appropriately caused by me, not my twin. But a drawing of the general body style of the 1991 Chevy Cavalier need not have its representational content determined in the way a photo of a particular 1991 Chevy Cavalier would. But both would be perspectival – both would implicate points of view (types of points of view, not tokens). Thus, there is a kind of perspective – pictorial perspective – that is underdetermined by indexical content. Something may exhibit pictorial perspective without being or containing an indexical.

This completes my sketch of pictorial perspective and its neural underpinnings. All the examples I employed involved spatial representations. One obstacle to claiming that pictorial perspective is an appropriate way of thinking of the perspectivality of conscious experience will depend on extending the account to non-spatial cases. While we experience spatial locations and relations, we also experience non-spatial features of the world, including colors, odors, and temperatures. In order to show that the account of perspective I favor can be extended to non-spatial cases, I next discuss how experiences of temperature may involve perspectival representations of temperature.

In a detailed examination of thermoperception (drawing on the work of H. Hensel, 1982), Akins (chapter 20, this volume) finds grounds for questioning whether thermoreceptors allow the brain to represent features of the objective world (in this case, temperatures). Akins argues that in order for thermoperception to be in the business of representing temperature, three conditions must obtain. First, there must be constant correlation between receptor activity and temperature stimuli. Second, the activity of the receptors must preserve relevant structure of the stimuli (e.g. greater and lesser activity in the receptor must reflect greater and lesser temperature). Third, the sensory system must be servile in the sense that it does not embroider upon the information extracted from the environment. Akins argues that thermoperception fails all three criteria.

While Akins's critique provides an obstacle to viewing thermoreceptors as representing objective properties, her analysis leaves entirely open the possibility that these receptors serve to represent temperature in an egocentric way. Thus, on the account I favor, the contents of the deliverance of these sensory systems are less like reporting that the water is 5 degrees Celsius and more like reporting that the water is too cold for me. I agree with Akins that thermoperception is what she calls a *narcissistic* system. I depart from Akins in that I view thermoperception as representing narcissistic properties whereas Akins holds that thermoreception does not represent at all.

I focus on two aspects of Akins's discussion of thermoreceptors (cold receptors and warmth receptors) that lead to thinking of the human thermoperceptive system as producing perspectival representations of temperature. The first concerns the fact that thermoreceptors are not distributed across the skin in a uniform concentration, and that different concentrations of thermoreceptors give rise to different sensations given a particular skin temperature. The ratio of cold to warm receptors varies across different parts of the body. One result of this is that different parts of the body may have varying degrees of comfort for water at a given temperature. Water that feels comfortably tepid on the hands may feel shockingly cold when dumped over the top of the head. The second aspect of thermoreceptors that lend them to a perspectival view is that they have dynamic response functions. A thermoreceptor's response to a given temperature at a given time is in part a function of what its response at a previous instant was. This is evident in a well-known example of the context sensitivity of temperature perception. Prior to submerging your hands in a bucket of tepid water, hold one hand in a bucket of ice and the other hand in hot water. The tepid water in the bucket will feel hot to the previously chilled hand and much cooler to the previously heated hand.

Both the differing concentrations of receptors and the dynamic response functions give rise to a many-to-one mapping of temperature sensations and temperatures. A sample of water of a given temperature will give rise to many different sensations depending on the concentrations of receptors and the level of their previous activity. These many-to-one mappings are arguably and plausibly part and parcel of the proper functioning of thermoreceptors. A given temperature may be more hazardous to tissue in one part of the body than another, and thus, a more sensitive alarm system may be accomplished by varying receptor concentrations. Dynamic responses may be adaptive since a rapid change of temperature can be damaging to tissue even if it occurs in a range of temperatures that would otherwise be harmless.

Akins sums up these aspects of thermoreceptors by describing them as *narcissistic*: they are less concerned with how things are independently of the organism and more concerned with how things relate to the organism, thus echoing the human narcissist's favorite question: "so what does this have to do with me?" She takes the narcissism of thermoperception to count against the claim that thermoreceptive sensory systems represent at all. I favor the alternative interpretation that thermoperception represents temperature, albeit in a perspectival way. A given temperature sensation does not just represent a temperature of a region on or near the skin but represents temperatures as being of varying degrees of hazard or harmlessness to the subject's tissues. The output of a thermoreceptor in response to a given temperature does not represent a given temperature *per se* but instead whether the given temperature is, for example, too hot, too cold, or just right. The property of being too hot cannot be defined independently of answering the question "too hot for whom?" and the subject relativity of such a property is what makes it narcissistic. Thermoreceptors include in the representational contents of their outputs relations that the temperatures bear to the representing subject, much in the way that retinocentric representations of spatial locations represent locations defined relative to the subject.

Arguably, perspectival representation may be found in examples beyond spatial and thermal perception. Our detection of chemicals in olfactory and gustatory senses may not be in the job of simply representing the presence of a certain chemical but also representing the chemical as noxious or poisonous or nutritious. But these are properties that can only be defined in relation to the organism: one man's meat is another man's poison and all that. Thus any system that has the function of causally covarying with such properties thereby produces perspectival representations of chemical concentrations: representations of chemical-involving narcissistic properties.

Akins objects to the proposal that thermoreception represents narcissistic properties on the grounds that the proposal depends on accepting the Detection Thesis: the claim that "*each and every sensory system functions to detect properties*" (p. 383 below; emphasis in original). Akins then gives two reasons for doubting the Detection Thesis: the first is that it is overly strong given the relatively small amount of evidence regarding sensory function collected to date. The second is that in at least one case – the case of proprioception – interpreting sensory activity as having the function of detecting is unhelpful and unenlightening.

I propose to grant Akins the falsity of the Detection Thesis, since such a concession leaves unscathed the proposal I favor. For convenience I will call the proposal *the narcissistic representation proposal* – the proposal that thermoperception represents narcissistic properties. Akins misconstrues the logic of the situation in asserting that the narcissistic representation proposal depends on the Detection Thesis. Contra Akins, the Detection Thesis is not a necessary condition on the truth of the narcissistic representation proposal. The narcissistic representation proposal plausibly has as a necessary condition the truth of the thesis that *at least one* sensory system functions to detect properties. But it is not at all obvious how it could have as a necessary condition the claim that *"each and every"* sensory system functions to detect properties. And for my immediate purpose, that of establishing the plausibility of non-spatial representations that nonetheless have pictorial perspective, the Detection Thesis may be disregarded as irrelevant.

I have proposed that some mental representations exhibit pictorial perspective. We need to tie this into consciousness. Do states of consciousness possess this kind of perspective? And what about the so-called *knowledge argument* that has figured heavily in discussions of the subjectivity of consciousness?

Regarding whether conscious states exhibit this kind of perspective, the answer seems a resounding "yes." The thermoperception examples are all examples of conscious sensations that vary independently of actual temperature: what enters into sensation includes relations of the temperature to states of the subject. Water of a given temperature may feel colder on the head than on the hands. Likewise, the remarks about the phenomenology of visual experience lead naturally to finding pictorial perspective in conscious representation. My percept of seeing a house differs depending on where I am standing. It depends on my literal point of view.

I turn now to the famous "knowledge argument" against physicalism (Nagel, 1974; Jackson, 1982). The main alternative account of subjectivity is the indexical account, which has been developed largely in response to the knowledge argument. In offering an alternative account of subjectivity, the question arises of whether I am abandoning an especially powerful response to the knowledge argument. Before answering this question, I briefly describe the knowledge argument and the indexical response to it.

The gist of the knowledge argument is as follows. A person that has never had any experiences of seeing a red thing may nonetheless have exhaustive knowledge of the physical goings-on in the nervous system of an individual seeing red. Such a knowledgeable person may know all the physical facts about seeing red without having a red experience. But suppose this knowledgeable individual were to finally have a red experience. Many find it intuitive to suppose that such an individual would learn something new, namely, they would learn what it is like to see red. Anti-physicalistic conclusions are supposed to follow on the supposition that learning what it is like to see red involves learning some new fact. Prior to having the experience, the subject knew all the physical facts; thus in learning a new fact upon having a red experience, the subject learns a non-physical fact. Thus, allegedly, physical facts do not exhaust all the facts, since they do not include certain facts about experience. Knowledge of physical facts leaves out knowledge of subjective facts.

The indexical physicalistic response to Nagel is due to philosophers such as Lycan (1996), Tye (1995), and Rey (1997). To take one instance as representative: Lycan's account of subjectivity is as follows. Experiences are representations. My visual experience of my blue coffee mug is a mental representation of the mug as being blue. When I introspect my experience, I form a second-order representation of the first-order representation of the coffee mug. Other people may form syntactically similar second-order representations, but those representations will be about their first-order states, not my own. The crucial analogy here is to the use of indexicals in speech. When I say "my leg hurts" I am referring to my leg, and only I can refer to my leg by using that utterance. You may use a syntactically, morphologically, and phonologically similar construction: you may utter the words "my leg hurts," but in doing so, you would be representing your leg, not mine. Analogously, only I can represent my first-order states by the introspective application of self-referential indexical concepts. And this, according to Lycan, is the ultimate explication of subjectivity.

At least in the case of Lycan (1996) and Tye (1995), explicating the subjectivity of consciousness in terms of indexicality grows directly out of a certain kind of physicalistic response to the knowledge argument. This kind of response seeks to maintain physicalism while granting that there is some sense in which Mary learns something new upon having her first experience of red. After having the red experience for the first time, Mary is able to correctly think to herself the thought she would express by saying "*this* is what it is like to experience red." Prior to having the experience, she was not in a position to correctly employ the indexical/ demonstrative thought. But the state of affairs she picks out with the demonstrative is one she was able to pick out by other means prior to having the experience. This is analogous to how, prior to visiting my house, you are incapable of correctly thinking to yourself the thought you would express by saying: "Pete Mandik lives *here*." Though you cannot refer to my home with *here* until you get there, you can nonetheless refer to it by other means prior to your arrival, e.g. under the description "Pete Mandik lives at *such-and-such* address."

It may be thought that my account of subjectivity, in not explicating subjectivity in terms of indexicality, deprives the physicalist of a powerful response to the knowledge argument. This worry may be assuaged in two general ways, one negative and one positive. The first is to point out that the indexical response to the knowledge argument is not particularly powerful. The second involves indicating how the pictorial account of subjectivity offers a superior physicalist response to the knowledge argument.

The indexical response to the knowledge argument is one of the many responses to the knowledge argument that hinge on notions of *modes of presentation*. This kind of response grants that the person in question learns something new but only in the sense of learning to apply a new description to an old fact. What differs in the old and new knowledge is only *the mode of presentation* of the facts known – there is no difference in the facts themselves. This mode of presentation defense of physicalism falls prey to the objection that in learning to apply a new mode of presentation to an old fact, the subject learns a new fact, namely, that the new mode of presentation applies to the old fact. Thus, given the presupposition that the subject already knew all of the physical facts, this new fact must be non-physical (Alter, 1998).

On the alternate response to the knowledge argument that I favor, the first premise of the argument is false: it is false that the subject could know all the physical facts without having an experience of red. I favor the view that there are both objective and subjective physical facts. What a subject can learn only by having an experience of red is a subjective, yet nonetheless wholly physical fact. (A similar view is defended by Deutsch, unpublished.)

The account of pictorial perspective described above renders intelligible the compatibility of physicalism and subjectivity. It allows us to see how a physicalistic framework can tolerate, first, physical properties that depend for their existence on representations and, second, physical properties that can be represented only by the representations that they depend on. Thus, such physical facts are subjective in the classical senses of being, first, mind–dependent, and, second, knowable only by a restricted mode of access. These notions may be briefly characterized by reference to imagistic representations.

The subject-dependence involved may be sketched as follows. What an image of X represents is the way X would look like from some particular location. What the image represents depends for its existence on the process of its representation. Precisely what is represented cannot be characterized independently of specifying the point of view of the representing subject. For example, part of what is represented is what the object looks like from one location as opposed to another. Consider a pictorial representation, say a photograph, of a complex object like the Statue of Liberty. Imagine that portions of the surface of the statue are painted black so that from a particular point of view only black regions of the statue could be seen, but from any other point of view, much of the statue's non–black surfaces can be seen. Consider the set of black regions of the statue. What is it that unifies those regions as a set? What is common to all and only those regions? The point of view occupied by a viewer – a generator of pictorial representations – is the unifying essence of those particular regions. It is in this sense, then, that the things that are represented depend on being represented. Of course, there is a sense in which they would exist even if no one were to represent them. But in specifying the set consisting of all and only the spatial regions captured in the image, one does not carve nature at the joints, but instead carves nature into a gerrymandered collection of items that would be of no interest apart from their involvement in a particular representation. That much of neural representation is concerned with similarly gerrymandered properties should not come as an enormous surprise. For instance, it makes sense that an animal's chemoreceptors would be less interested in carving nature into the periodic table of elements and more interested in carving nature into the nutrients and the poisons – categories that make no sense apart from the needs of a particular organism.

The restricted epistemic access involved may be sketched in terms of imagistic representations as follows. What is represented by an image can only be represented, without addition or deletion, by an image. Even a string of numerals coding a bitmap for an image does not have all and only the representational content of the image. The numeral string constitutes, in part, a recipe for constructing an image, and in doing so, it has content that the image itself lacks. The old saw about a picture being worth a thousand words is false: a picture is worth no number of words. This point cannot be adequately argued here, but suffice it to suggest that it is not obviously incompatible with physicalism that there are properties represented in sensory experience that may only be represented in sensory experience. *Part* of what is represented in, say, olfaction may be conveyed in some other mode of representation like spoken language: "your perfume smells like vanilla and roses." But the suggestion that *all and only* what is represented in olfactory experience can only be represented in olfactory experience is an entirely physical possibility, if not a physical actuality. That experience is perspectival in this sense, allows us to conceive of physical facts that may be knowable only by a restricted mode of access, that is, physical facts that may only be represented by specific sensory experiences.

These remarks are far too brief to *establish* that thinking of subjectivity in terms of pictorial perspective renders the knowledge argument against physicalism ineffectual. However, these remarks do show, at a minimum, that it is not obvious that abandoning the indexical account of subjectivity leaves the physicalist defenseless in the face of the knowledge argument.

In this chapter I have articulated an account of what it is for neural representations to be perspectival, that is, to represent the world from a particular point of view. On this account, such representations have the function of causally covarying with subjective properties: properties defined in terms of the representing subject. Imagistic mental representations serve as a prototype for such representations, and afford us a sketch of why a neuroscientific view of the brain need not exclude subjectivity.

References

Alter, T. 1998: A limited defense of the knowledge argument. *Philosophical Studies*, 90, 35–56.

Bülthoff, H. H., and Edelman, S. 1992: Psychophysical support for a two-dimensional view interpolation theory of object recognition. *Proceedings of the National Academy of Science*, 89, 60–4.

Deutsch, M. (unpublished). Subjective physical facts.

Dretske, F. 1995: *Naturalizing the Mind.* Cambridge, MA: MIT Press.

Eichenbaum, H., Stewart, C., and Morris, R. G. M. 1990: Hippocampal representation in spatial learning. *Journal of Neuroscience*, 10, 331–9.

Feigenbaum, J., and Rolls, E. 1991: Allocentric and egocentric spatial information processing in the hippocampal formation of the behaving primate. *Psychobiology*, 19 (1), 21–40.

Hensel, H. 1982: *Thermal Sensations and Thermoreceptors in Man.* Springfield, IL: Charles Thomas.

Jackson, F. 1982: Epiphenomenal qualia, *Philosophical Quarterly*, 32, 127–36.

Kosslyn, S. M. 1994: Image and Brain. Cambridge, MA: MIT Press.

Lettvin, J. Y., Maturana, H. R., McCulloch, W. S., and Pitts, W. H. 1959: What the frog's eye tells the frog's brain. *Proceedings of the IRF*, 47, 1940–51.

Lycan, W. 1996: *Consciousness and Experience.* Cambridge, MA: MIT Press.

Mandik, P. 1999: Qualia, space, and control. *Philosophical Psychology*, 12, 47–60.

Milner, A. D., and Goodale, M. A. 1995: *The Visual Brain in Action.* Oxford: Oxford University Press.

Nagel, T. 1974: What is it like to be a bat? *Philosophical Review*, 83, 435–50.

Nagel, T. 1986: *The View from Nowhere*. Oxford: Oxford University Press.

O'Keefe, J., and Nadel, L. 1978: *The Hippocampus as a Cognitive Map*. Oxford: Clarendon Press.

Rey, G. 1997: *Contemporary Philosophy of Mind*. Oxford: Blackwell.

Shepard, R. N., and Cooper, L. A. 1982: *Mental Images and their Transformations*. Cambridge, MA: MIT Press.

Stein, J. F. 1992: The representation of egocentric space in the posterior parietal cortex. *Behavioral and Brain Sciences*, 15 (4), 691–700.

Tye, M. 1995: *Ten Problems of Consciousness: A Representational Theory of the Phenomenal Mind*. Cambridge, MA: MIT Press.

Ullman, S. 1989: Aligning pictorial descriptions: An approach to object recognition. *Cognition*, 32, 193–254.

Questions for Further Study and Reflection

1   According to Crick and Koch, whenever some information is represented in the neural correlate of consciousness it is represented in consciousness itself. What neuroscientific methods allow one to know what information (if any) is represented in consciousness or its neural correlates?

2   Why might very short-term memory be important to visual consciousness? Consider, in your answer, what consciousness is alleged to be useful for.

3   Are you conscious of things that you are not paying attention to? Can you pay attention to things without being conscious of them? What would count as evidence – neuroscientific or otherwise – that would be relevant to answering either question?

4   Prinz claims that in introspective consciousness one has a second-order representation of a first-order representational state. What, more specifically, is the second-order state a representation of? Does it represent the first-order state's content, the vehicle, or both?

5   Besides pain, are there any other aspects of the nervous system for which it is an open question whether or not they are sensory systems?

6   Hardcastle speaks of "pain information." What is pain information about?

7   In what situations are egocentric representations more useful for an organism than allocentric representations? When would allocentric representations be more useful?

8   If pain subsystems employ representations, would they be egocentric or allocentric? Does your answer render pain subjective or objective?

Part V

Representation

Introduction

Pete Mandik

Topics of representation have long vexed philosophers. Historically, philosophers boggled at the question of how one can know the external world if one's access to it is always mediated by representations. A topic that has had special urgency in contemporary discussions is the topic of the very nature of representation: how is it that something inside of a person can come to represent, designate, or stand for something external? Meanwhile, on the scientific front, notions of representation enjoyed wide and frequent commerce in both cognitive sciences and neurosciences. A foundational idea in cognitive psychology and cognitive neuroscience is that mental operations involve information processing: the manipulation of representations in computational processes. There have been two main points of interface between scientists and philosophers regarding representation. The first is in philosophy of science where the scientific use and status of concepts of representation are topics of study (exemplified by chapters 18 and 21). The second is in philosophy of mind where philosophers turn to science to bolster philosophical claims about representation (exemplified by chapters 20 and 21). An issue pertinent to both points of interface is the question of whether there are any representations after all. Pro-representational positions are defended in chapters 18 and 19. Anti-representational positions are defended in chapters 20 and 21.

In chapter 18, Bechtel defends the existence of representation as defined in terms of having the function to carry information. In chapter 19 Grush defends representation, but with a different spin on the definition. For Grush, the key notion is one of emulation: the use of certain neural processes as models of distal systems. In chapter 20 Akins challenges traditional philosophical accounts of representation that see sensory function as the basic case of representational activity. Akins examines thermoperception and argues that at least some basic sensory functions are devoid of representations. In chapter 21 Stufflebeam takes a broad view of the use of notions of representation in multiple disciplines and proposes an account of representation that is neutral among them. He then questions whether contemporary accounts of brain function benefit from positing representations.

18

Representations: From Neural Systems to Cognitive Systems

William Bechtel

One of the hallmarks of *cognitive* explanations of behavior is that they appeal to mental representations and operations over them. For example, in explanations of problem-solving behavior, cognitivists posit representations of the current situation, the goal state, and of possible operations that might be performed, and construe problem solving as involving such things as comparisons of the representations of the current state and goal state and alterations in the representations of the current state so as to determine the consequence of various operations. Many of the most vociferous debates in the cognitive science literature have focused on the format of representations – whether they take the form of language-like symbols (or even natural language expressions themselves), pictures or images, or the sorts of distributed representations identified in connectionist models.

Given the propensity of scientists and philosophers to challenge even the most basic assumptions of their inquiry, it should not be surprising that some 30 years into the cognitive revolution some investigators – in particular, those advocating dynamical explanations of cognitive phenomena – would begin to challenge the very need for representations. However, although appeals to internal representations are rampant in cognitive science, they are not limited to that field. Biologists in a variety of fields refer to molecular structures as representations. Of special importance to our purposes, neuroscientists, especially behavioral and cognitive neuroscientists, routinely refer to brain activities as representations. This raises a question: should the attacks on representations that have emerged recently in cognitive science apply equally to appeals to representations in neuroscience?[1] I will argue, rather, that there is a viable notion of representation that is being employed in neuroscience that is not subject to the dynamicist's challenge. This answer, however, raises a further question: is this notion of representation adequate for the purposes for which cognitivists have appealed to representations. In the last section I will sketch how one might build up from the conception of representation employed in neuroscience to one adequate for cognitive accounts.

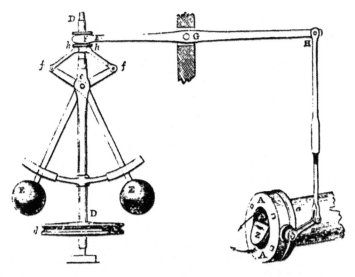

Figure 18.1 Watt's centrifugal governor for a steam engine. Drawing from J. Farley, *A Treatise on the Steam Engine: Historical, Practical, and Descriptive* (London: Longman, Rees, Orme, Brown and Green, 1927).

1 The Dynamicist's Critique of Representations

The basic complaint of the dynamicist critics of representations is that physical systems lacking representations could accomplish many or all of the various tasks for which cognitivists had postulated internal representations (and operations involving them). One strategy the critics used to mount their case was to present examples of systems that accomplish tasks for which we might be tempted to posit representations (e.g. problem-solving tasks) and to demonstrate ways of accomplishing such tasks without internal representations. Van Gelder's (1995) example of the governor James Watt introduced for his steam engine is perhaps the best known.

The task facing Watt was to regulate the output of steam from a steam engine so that the flywheel would rotate at a constant speed regardless of the resistance being generated by the appliances connected to it. Watt's governor was ingeniously simple (figure 18.1). He attached a spindle on a flywheel driven by the steam generated by the steam engine, and attached arms to the spindle which would, as a result of centrifugal force, open out in proportion to the speed at which the flywheel turned. A mechanical linkage between the arms connected the arms to the steam valve so that, when the wheel turned too fast, the valve would close, releasing less steam, thereby slowing the flywheel, but when the flywheel turned too slowly, the valve would open, releasing more steam and speeding up the flywheel. It is in part the simplicity of this device that led van Gelder to reject a representationalist interpretation in which the angle of the arms represented the speed of the flywheel. But, van Gelder argued, the Watt governor solves the very kind of problem for which a cognitivist might be tempted to develop a representationalist solution – one, for example, in which the present and desired speed of the flywheel as well as the operations of opening and closing the steam valve are represented, and rules that apply to the representations are invoked to determine how much to open or close the valve (think of how a computer might be programmed to perform this task). Thus, the Watt governor, for van Gelder, demonstrates how one might perform cognitive tasks that seemingly require representations without them, motivating the program of trying to account for cognitive activities more broadly without positing representations.

One response to van Gelder's strategy is to grant that some simple tasks for which one might be tempted to posit representations do not in fact require them, but to contend that there are other tasks that are "representation-hungry" and do require them (Clark and Toribio, 1994). In this vein, Grush (chapter 19, this volume) argues that we find representations only in contexts where the system does not have immediate access to what is represented, and so representations stand in for what is not physically or temporarily present. In contrast, I have adopted an extreme position and argued that there are representations even in the Watt governor (Bechtel, 1998). Many philosophers have found this to be a mission of madness. They maintain that the notion of representation loses its interest if we can indeed find representations in such simple systems. In this chapter I argue that this is not the case by showing that this is precisely the kind of system for which cognitive neuroscience finds the notion of representation indispensable. When techniques such as single-cell recording and neuroimaging are employed to understand how the brain performs cognitive tasks, implicitly the notion of representation on which they are relying is the same as I invoked in identifying representations in the Watt governor. Thus, rather than being too simple to have representations, the Watt governor provides a simple, illustrative case where we can readily observe the importance of appeals to representations.

2   The Watt Governor as an Exemplar Representational System

Central to demonstrating the presence of representations in the Watt governor is the articulation of an analysis of what it is for something to be a representation. In its most basic sense, a representation is something (an event or process) that stands in for and carries information about what it represents, enabling the system in which it occurs to use that information in directing its behavior. In the simplest case, the system actually acts upon that which is represented (see figure 18.2). For example, we employ maps to stand in for the actual geography, and use those maps to coordinate our behavior in a way that is responsive to that geography – finding a route to a target location or circumventing an obstacle. The angle arms of the Watt governor play such a standing-in role: they stand in for the speed of the steam engine

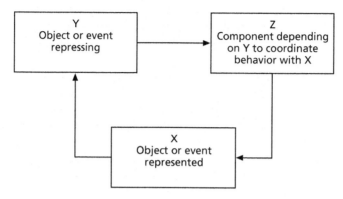

Figure 18.2   Three components in an analysis of a representation: the representation Y carries information about X for Z, which uses Y in order to act or think about X.

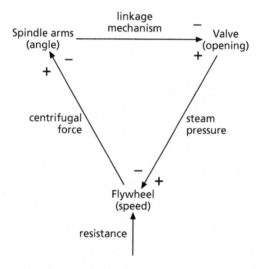

Figure 18.3   Application of this scheme to the Watt governor. The spindle arms carry information about the speed of the flywheel for the valve opening system, which uses the angle to determine the opening, thereby regulating the speed of the flywheel.

in a mechanism that controls the opening and closing of the steam valve and it is because they so stand in that the governor is able to regulate the flow of steam appropriately for the current speed of the flywheel. If someone did not immediately perceive how the governor performed its function, what one would do is explain how the angle of the arms responds to the speed with which the flywheel is turning (i.e. it carries information about the speed), thereby making this information available in a format that can be used by the linkage mechanism to control the flow of steam and thus the speed of the flywheel at a future time (figure 18.3).

This characterization of what it is to be a representation has two components. One is the idea of an information-bearing state. As Dretske (1981), among others, has argued, information can be analyzed causally – an effect of a process carries information about its cause (specifically, that an event of the kind that could cause a particular effect had indeed occurred). Thus, thunder carries information about an electrical discharge. But Millikan (1984) stresses that being a causal effect is not sufficient to turn something into a representation. Something becomes a representation because of how it is used by a system in controlling its behavior. It is because a system is *designed* so as to use a state to inform it about something distal that makes the state a representation.

In this analysis, emphasis is placed on *design*, on how the system is designed to use information-bearing states in coordinating its behavior. One does not just look to how it in fact uses a particular information-bearing state. Otherwise, we end up with an account in which the meaning of a representation depends on whatever ways it is used. An account which appeals to actual use is often referred to a conceptual role semantics, and such accounts have been subject to powerful criticisms (Cummins, 1996). But in my analysis, it is only the designed use that matters. An appeal to the notion of design, of course, requires an account of what it is for something to be designed to do one thing rather than another. In biological systems the usual route is to argue that what something was designed to do was that which its predecessors did more successfully than their competitors and, thus, contributed to their differential reproductive success. It can be relatively easy to tell a story about how a particular trait of an organism contributed to its differential reproductive success (these are often criticized as "just-so stories"), but difficult (though not impossible) to actually demonstrate that the trait really did contribute in this way (Griffiths, 1997). In the case of artifacts that we have designed, we sometimes have privileged access to information about what they were designed to do.

According to this analysis of representations, only some of the things that cause a representation to appear are what it represents. It is in this way that one can account for misrepresentation, and show how states or events can represent things that are not their immediate cause. In biological evolution, a mechanism may be selected in which a state is only rarely actually produced by that which it represents – if it is important that the system respond to the distal state when it occurs, a system may evolve that is subject to many false alarms. In systems designed by humans, we can take this to the extreme – we can design a system in which a particular internal state would occur (or at least be more likely to occur) in response to a distal state, but where the distal state has never occurred and all occurrences of the response have been false alarms (e.g. a system designed to detect a nuclear missile attack).

Behind at least some of the opposition to finding representations in the Watt governor is the fear that if they occur there, representations are everywhere and there is no theoretical bite in calling something a representation (see Haselager and de Groot, forthcoming, for such an argument). Although I will soon argue that there is a point to some promiscuousness about representations, let me first show that under this analysis, representations are not as promiscuous as feared. States or events that carry information are indeed ubiquitous. Any causal effect carries information about what caused it. In emphasizing the consumer, however, we emphasize the kind of system in which the representation occurs. The Watt governor is a particular kind of physical system – a system that is designed to employ states bearing information to control its behavior. While there are many such systems within evolved organisms and human artifacts, not all natural systems employ internal states in such a manner. Representations are found only in those systems in which there is another process (a consumer) designed to use the representation in generating its behavior (in the simplest cases, to coordinate its behavior with respect to what is represented) and where the fact that the representation carried that information was the reason the down-line process was designed to rely on the representation.

Even though this analysis does not make representations as ubiquitous as might be feared, it does entail that representations will appear in many non-cognitive systems. Many physiological systems (e.g. simple biochemical systems such as fermentation where reactions are controlled by feedback mechanisms in which the availability of the product of the reaction determines whether it will continue) will, on this construal, employ representations since processes in them as well as the responses to them were selected because of the process's information-bearing role. But this, I would contend, is as it should be. Representational or intentional vocabulary is in fact regularly used in the sciences dealing with such systems. Without it, scientists would be hard pressed to explain how these systems perform the tasks for which they appear to have evolved. In this context, the Watt governor provides a simple model of the role representational discourse is playing in allowing us to explain such systems.

3   Representational States in Brains

As I noted above, representational talk is widely used in behavioral and cognitive neuroscience. A particularly clear example is Penfield and Rasmussen's (1950) homuncular maps of how the sensory and motor cortex represents different parts of the body (in which they emphasize in terms of the distorted sizes of the homunculus's organs how the maps devote more area to regions from which more sensory input is received or over which there is more motor control – see figure 4.2, above).[2] This is extremely natural insofar as neuroscientists are attempting to analyze how various components of the brain gather and use information in the course of producing behavior. In this section I will draw upon neuroscience research into visual processing in the brain, described more fully in Part III, to show that the appeals of neuroscientists to representations do indeed conform to the analysis in the previous section.

Although there are various choices as to what sorts of neural states or processes serve a representational function, a reasonable place to begin is to assume that it is the firing rate or firing pattern of individual neurons that serves as representations.

If that is the case, what one needs in order to understand the representational system of the brain is means to record the electrical activity of individual neurons and then to determine their representational content. Techniques for implanting electrodes in or next to individual neurons and then amplifying the signal and playing it through an audiometer were developed in the early decades of the twentieth century. The challenge was then to make sense of the firing patterns that could be detected. Although I am arguing that the focus in construing a state or event as a representation is on the consumer of the representation, neuroscientists typically begin by trying to correlate neural activity with external processes that they might represent (as with the method of single-cell recording – see chapter 4, this volume). This is, indeed, quite sensible and does not undercut my claim. An extremely useful first step in determining what the system takes a state or event to represent is to ascertain what information a state or event might carry. Then one can ask the question of how the system was designed to use the information.

Penfield and Rasmussen's homuncular representation of sensory cortex was arrived at through single-cell recording (the homuncular representation of motor cortex resulting from related experiments in which weak electrical stimuli were supplied through electrodes to see what motor organs the stimulated cells would affect). Talbot and Marshall (1941) were the first to apply single-cell recording to cells in the occipital lobe, which was already recognized as figuring in visual processing on the basis of lesion studies. They recorded from cells in Brodmann's area 17 (which later came to be known as V1) in anesthetized cats and monkeys, and correlated activations there with the location of stimuli in the animal's visual field. They showed that each cell responded to a stimulus in a particular part of the visual field (the cell's *receptive field*) and confirmed that cells were so organized in the cortex that they constituted a map that preserved the topography of the visual field (see figure 18.4). To determine what the firing of a cell represented, however, required a further step – determining what kinds of stimuli, when present in a cell's receptive field, caused it to fire. Stephen Kuffler (1953) provided the model for this type of investigation. He recorded from cells in the retina and the lateral geniculate nucleus (LGN) and determined that they were most activated either by a light stimulus in the center of the cell's receptive field surrounded by a dark area (an *on-center* cell), or by a light surround of a dark center (an *off-center* cell; see figures 13.3 and 13.4, above).

As recounted in chapters 10 and 13, Hubel and Wiesel (1962, 1968) extended Kuffler's approach to primary visual cortex, discovering that cells there responded to bars of light. They found that some cells, which they identified as *simple cells* (figure 13.3, C–G), responded to bars with specific orientation only in specific areas of the cell's receptive field, whereas others, which they referred to as *complex cells*, responded to bars with a specific orientation anywhere within the receptive field. The result of Hubel and Wiesel's research was to suggest that firing of cells in V1 represented bars or lines at specific orientations in parts of the visual field.

Chapter 13, above, relates how, in the years after Hubel and Wiesel's pioneering research, discovery of cells in different prestriate, temporal, and parietal lobe areas which responded to specific visual features (color in V4, motion in MT/V5, shapes

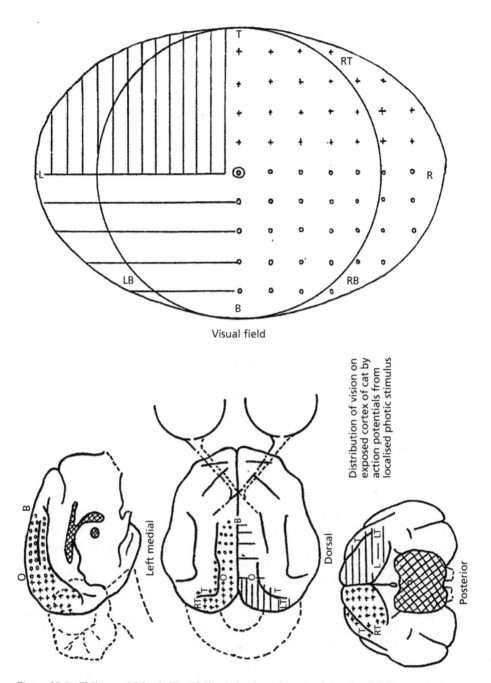

Figure 18.4 Talbot and Marshall's (1941) projection of areas of the visual field on to primary visual cortex in the cat, based on single-cell recording.

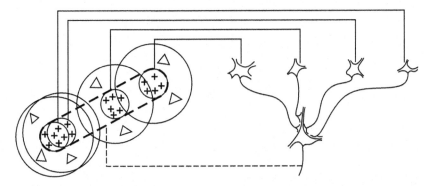

Figure 18.5 Hubel and Wiesel's proposal for a wiring diagram linking with excitatory connections from lateral geniculate cells with *on* centers to a simple cell in primary visual cortex that is excited by an oriented bar of light. From Hubel and Weisel, 1962, p. 142. Reprinted with permission of the Physiological Society.

in inferior temporal cortex, and location information in posterior parietal cortex) continued with vigor. Van Essen and Gallant (chapter 12, this volume) indicate that 33 brain areas have been identified as principally involved in visual processing. Many of these have been shown to represent different specific types of visual information and currently much effort is being extended to developing accounts of how these representations figure in an overall processing scheme.

As I noted, most of this work has relied on establishing what information might be carried by different neural firings. The account of representation I endorsed, however, placed primary emphasis on the consumer of the representations. But a concern with the consumer of the representation is at least tacit in such research. Investigators would have had little interest in figuring out the information relation between brain activity and distal stimuli unless they assumed the brain was using this information in further processing. To determine how the brain actually consumes this information requires developing processing models which show how activations in later areas in visual pathways utilize the information encoded in earlier stages of processing. The early work of Hubel and Wiesel again provided an exemplar. They proposed simple wiring diagrams of how LGN cells might be wired to simple V1 cells and simple V1 cells in turn to complex V1 cells which were designed to show how downstream cells could compute their representations from more upstream representations (see figure 18.5). By showing that processes at each stage respond to those upstream to arrive at their characteristic response, they show that the upstream processes were representations.[3] Even in this example, the processing is more complex than in the Watt governor, where the consumer employed a simple mechanical linkage. Often complex computational models are required to characterize the processing mechanisms through which representations are used by later stages in processing (this is especially true when it is recognized that the feedback or recurrent processing plays a major role). Much current work, especially in computational neuroscience, is devoted to developing processing models showing how later visual areas can generate their representations from what is represented in earlier visual areas.

Part of the challenge in understanding the consumption of visual representations is that the visual areas are often many steps removed from any behavior, for the performance of which the organism relies on visual information. This is especially true in the ventral pathway from primary visual areas into temporal areas, through which object recognition is assumed to occur (see chapter 11, this volume). Object representations developed in inferior temporal cortex are thought to be employed in higher cognitive processes such as reasoning and problem solving, the detailed neural mechanisms of which have not yet been identified. In the dorsal pathway proceeding to the posterior parietal cortex, on the other hand, it has been easier for researchers to focus on the question of how putative representational states figure in guiding motor action. The basic strategy was developed by Goldman-Rakic (1987) in her research on prefrontal brain areas. She found cells, for example, that were active when a monkey had to temporarily remember a specific direction in which it was directed to move its eyes after a delay interval. She was able to establish that certain cells were active only when the monkey would correctly remember the particular direction of eye movement.

Even when researchers discover sufficient cues to develop an interpretation of what a neural pattern represents, the process of interpretation is quite indirect. This is nicely illustrated in a puzzle Larry Snyder has attempted to address. Focusing on neurons in the parietal cortex which fire when the animal is required to move its arm to a specific location, he asked whether these neurons represented the location to which the animal was *attending* or the location to which the animal *intended* to move. To do this, Snyder developed experimental protocols in which monkeys had to attend to one location while holding an intention to move to another location. This revealed that different areas in the parietal cortex were involved in representing the intention than were involved in simply attending to a location (Snyder et al., 1997; Batista et al., 1999).

A word of caution is needed. Since the brain is much more complex than the Watt governor and we lack independent access to the design process, hypotheses about what is represented by specific neural activity must be treated as extremely tentative. The project of single-cell recording is limited by the stimuli one thinks to test. It was through serendipity that Hubel and Wiesel thought to test bar stimuli in V1 and that Gross thought to test hand-shaped stimuli for an area in the inferior temporal cortex. It would be easy for a researcher simply to fail to test whether a particular stimulus would drive a cell. In this light it is important to note that Van Essen and Gallant (chapter 12, this volume) found that esoteric stimuli, such as expanding or rotating stimuli, would cause specific MSTd cells, which fired weakly in response to straight line movements, to fire vigorously. Moreover, one should not assume that the cell is only carrying information about the stimulus that causes it to fire more vigorously. As Van Essen and Gallant stress, less than full responses may still carry important information that can be used by downstream consumers. Thus, cells may not be feature detectors, but may be better construed as filters with a representational profile. Finally, as Akins (chapter 20, this volume) emphasizes, neurons do not respond to objective features of the world that we assume they would represent, when we think about how we might design a system from scratch. For example, they may not respond to absolute properties, such as temperature, but rather their response may be relative to the current state of the organism (e.g. whether the stimulus is warmer or colder than background stimulation). (Although Akins rejects a representational analysis, see chapter 16, this volume, for a representational account compatible with these findings.)

In this section I have been focusing on how neuroscientists investigate the representational content of neural firings. A skeptic about representations might question what the scientist gains by identifying particular neural activities as representations. Why not settle for a simple causal model identifying the dynamics of activity in the brain? To answer this question, we need to bear in mind what the goal of inquiry is. Neuroscientists generally assume that the brain is a complex machine whose activity allows the organism to coordinate its behavior with features of its environment. As with the Watt governor, once one has identified components in the system, one wants to know how they facilitate the organism as it extracts information about its environment and deploys that information in determining its behavior. Someone examining Watt's invention might ask: What do the spindle arms do in the governor? The answer would be: They represent information as to whether the flywheel is moving too fast or too slow and pass this information to the valve that controls the steam flow. Similarly, one might ask: What does area V4 do in the brain? The kind of answer a neuroscientist would offer is that it represents information about the color of stimuli and provides this information to other areas which use it to determine the identity of the stimulus. Accordingly, construing internal states as representing various aspects of the environment is critical to this endeavor.

4   From Brain Representations to Cognitive Representations

One response to the line of argument in the previous section is to acknowledge that there may be a role for positing representations that fit the above analysis in neuroscience, but to deny that such representations are sufficient for the business of cognitive science. Indeed, the representations that have figured in many cognitive models employ a much more complex format, often one drawn from formal logic and natural languages. Some have argued that language-like or propositional representations are required if we are to account for the cognitive abilities exhibited at least by humans, since humans must encode complex relational information and be able to extract representations of components as needed. Fodor (1975, 1987) has argued that only representations that share critical properties with language, especially that of having a compositional syntax and semantics, are adequate for modeling thought. When compositional rules are invoked, lexical items are put together according to syntactic rules in such a way that the meaning of the composed structure is built up from the meaning of the component lexical items.

The key to Fodor's argument is the observation that human cognition (and perhaps that of other species) exhibits a number of special properties, especially productivity and systematicity. Productivity and systematicity are properties manifest in natural languages, and Fodor argues that they are exhibited in thought as well. Productivity with respect to language refers to the capacity to indefinitely extend the corpus of sentences in a language; applied to thought, it refers to the fact that the range of possible thoughts is not bounded. Systematicity with respect to language refers to the fact that there are relations between the sentences of a language such that if one string is well formed, so is another that results from appropriate substitutions. For example, if *the florist loves Mary* is a sentence of English, so is *Mary loves the florist*. Applied to thought, it designates the fact that a cognitive system that can think one such thought automatically has the capacity to think the other. In a linguistic system in which sentences are composed employing syntactic rules, these properties arise automatically, and would accrue equally to a cognitive system if it employed representations that are language-like in relying on a compositional syntax. Just as he has faulted the representations found in connectionist networks as incapable of accounting for these properties (Fodor and Pylyshyn, 1988), Fodor would find the sort of representation found in the Watt governor or identified in the activities of individual neurons to lack the requisite compositionality and thus to be incapable of exhibiting these properties.

One unfortunate consequence of grounding explanations of cognitive capacities in language-like representations is that it leaves unanswered the question of how such representations might be embodied in the brain. It is clear that the brain is a mechanism that can comprehend and produce linguistic structures, and so must have tools for representing such structures, but it is far less clear that it uses language-like structures for its own internal representations. So there is motivation for starting with representations of the sort discussed in the previous section – ones that seem to figure in the brain itself. The analysis of representations cannot end there, however. Rather, one must show how to build up from the sorts of representations found in the brain to those that exhibit the requisite compositionality.

While filling in the gap between the sort of neural representations I have been discussing and ones that exhibit productivity and systematicity may seem like a tall order, Larry Barsalou's recent work on concepts suggests how it might be done (Barsalou, 1999). Attacking amodal language-like symbols (symbols not tied to a particular sensory modality), Barsalou has argued that "perceptual representations can play *all* of the critical symbolic functions that amodal symbols play in traditional systems, such that amodal symbols become redundant." Barsalou is clear that the perceptual representations he is considering are neural – he describes perceptual symbols as "records of the neural states that underlie perception." (Although much of his discussion focuses on visual perception, he intends his account to include perception in other modalities, including perception of emotion and introspection.)

The attempt to ground cognition in perception goes back at least to the seventeenth-century Empiricists in philosophy, such as Locke. Their program has been much ridiculed, but the target in most attacks is the view that perception gives rise to static pictures or images (images of which we are consciously aware) that are holistic recordings of the input. Perceptual representations for Barsalou, however, are not (despite his reference to them as "records of neural states") pictures or images – they are not recordings. In particular, they are interpreted in such a way that "specific tokens in perception (i.e. individuals) [are bound] to knowledge for general types of things in memory (i.e. concepts)." The key to this move is a proper understanding of neural processing in vision – the brain is not constructing a picture of the world (if it did, it would then need another perceiver to view the picture), but an analysis of the visual input geared to action. This is already suggested by the way the brain decomposes visual processing, with different brain areas analyzing distinct features of a scene as color, shape, or location. Neural activity in different brain areas represents categorization and conceptualization of the visual input – it contains *this* shape, *this* color, or occurs at *this* location.

Barsalou refers to perceptual representations as schematic representations in that only certain features of the perceptual input are represented. He appeals to psychological research on attention to show how a schematic representation is constructed – selective attention isolates and emphasizes pieces of information that is given in perception and facilitates storage of these features in long-term memory. Recent neural research on attention could support the same analysis. Relying on the evidence that different features of stimuli are analyzed in different brain areas, Corbetta et al. (1993) have shown that when subjects are required to differentially attend to different properties of stimuli, brain areas responsible for processing those features are activated, indicating that particular features are being processed. The fact that perceptual symbols are schematic in this manner allows them to be indeterminate in ways that pictures cannot – representing a tiger, for example, as having stripes, but not a determinate number of stripes.

In addition to emphasizing the schematic character of perceptual representations, Barsalou also emphasizes their dynamic character. Different neural records are related temporally in experience, and they give rise to simulations of the way we can attend to different parts of an object over time or the way it itself changes over time. (Like a perceptual representation itself, a simulation is not just a repetition of previous experiences, but a composed structure in which individual components can be put together differently on different occasions. Barsalou refers to the organizing information specifying how different perceptual representations can be related as *frames*, thereby invoking previous cognitive science research on the type of complex information structures that seem to figure in cognition.) For Barsalou, this allows individual perceptual representations to be integrated into what he terms "simulation competences," a capacity that is expanded as humans learn languages which allow them to index and control features in a simulation.

For Barsalou, linguistic representations extend the capacities of the conceptual system built on perceptual representations. He proposes that:

> As people hear or read a text, they use productively formulated sentences to construct a productively formulated representation that constitutes a semantic interpretation.

Conversely, during language production, the construction of a simulation activates associated words and syntactic patterns, which become candidates for spoken sentences designed to produce a similar simulation in a listener.

But it is clear that while linguistic indexing supplements the cognitive capacities provided by perceptual symbols, it is the perceptual symbols themselves that do the cognitive work for Barsalou. In fact, linguistic symbols are, for him, acquired as simply additional perceptual symbols. Thus, it is important for him to show that they can have the sorts of properties Fodor argued were needed for cognition – productivity and systematicity – without appealing to language-like representations. Barsalou maintains that the very features of perceptual symbols that I have already reviewed provide him the resources to do this.[4] The key is that perceptual symbols and simulations are built up componentially, and thus, just as with linguistic representations, they can be continually put together in new ways, thereby accounting for productivity. They also permit substitutions of different component representations, thereby accounting for systematicity. Barsalou illustrates this potential by employing diagrams much like those used by cognitive linguists (Langacker, 1987). Figure 18.6 is an example. It illustrates how perceptual symbols for object categories (A) and spatial relations (B) can be (C) combined, even (D) recursively, to productively generate new representations. The symbols in this diagram (e.g. the balloon and airplane in A) are not intended as pictures, but to stand for perceptual representations, that is, configurations of neurons that would be activated in representing these objects. The boxes with thin solid lines are intended to represent simulation competences that have developed over many experiences with the object or relation and represent it schematically. The boxes with thick slashed lines then represent particular simulations that might be generated from the simulation competences by combining them, sometimes recursively.

The preceding is only a partial sketch of Barsalou's account of perceptual symbols (he goes on to suggest how even abstract concepts such as *truth* can be constructed from perceptual representations), but it does indicate that there are ways of building up from the sorts of representations found in the brain. The key ingredient in his account is the construal of the kind of analysis the visual system performs, by having different neurons represent such things as shape and color of stimuli, as involving categorization and conceptualization. The separately analyzed features afford composition, thereby providing a resource similar to that Fodor identified for language-like representations. (Perceptual symbols, however, do not thereby become implementations for Fodorian language-like symbols – perceptual symbols are modality-specific and the particular features of the symbols themselves generally specify definite features in what they represent. Unlike amodal language-like symbols, the particular embodiment of the symbols as patterns of neural firing in particular brain regions is important to the information that they carry. One consequence of this, which Barsalou happily endorses, is that different individuals, with different learning histories, are likely to have somewhat different representations.)

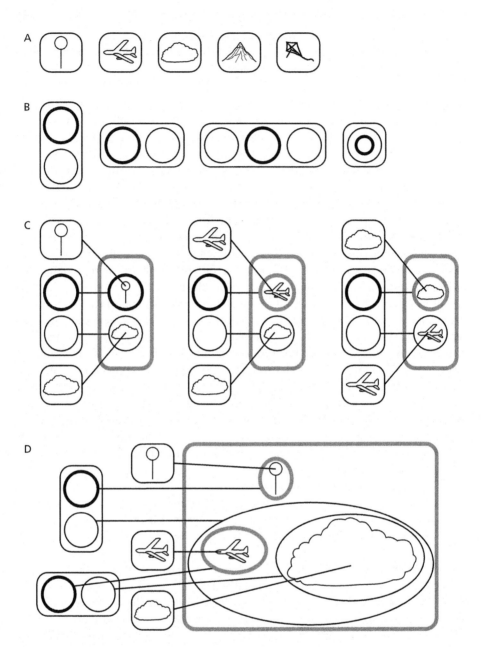

Figure 18.6 Barsalou's representation of how perceptual symbols for object categories (A) and spatial relations (B) implement productivity through combinatorial (C) and recursive (D) processing. Boxes with thin solid lines represent simulation competences; boxes with thick dashed lines represent simulations. From Barsalou, 1999, p. 593. Reprinted with permission of Cambridge University Press.

5 Conclusions

I began this chapter articulating a minimal notion of representation, wherein a representation is an information-bearing state or event which stands in for what it represents and enables the system in which it operates to utilize this information in coordinating its behavior. I argued that representations of this minimal sort are found even in the Watt governor, a simple mechanical device that has figured in some of the opposition to the invocation of representations in cognitive science. I also contended that this minimal notion is what is required for most neuroscientists' references to representations in the brain. In the final section, I tried to show how one might build these neural representations up into representations that exhibit properties such as productivity and systematicity, which have been argued to be characteristics of thought.

Notes

1   For arguments that answer this question in the affirmative, see chapter 21, this volume.

2   Although I am focusing primarily on representations whose content is fixed by what causes them, the accounts can be generalized to representations in the motor system in which the content is specified by what the representations are designed to cause (Mandik, 1999). Although the story gets more complex, one can even conceive of representations for which we need to appeal to both their sensory causes and motor effects in specifying their content.

3   Eventually, researchers hope to discover complete pathways through the system that result in behavioral responses that are appropriate to the information represented at each stage in the system. Until this stage is reached, each imputation of a representation to processes in the system involves taking out a promissory note that can only be repaid by future research.

4   In his own discussion, Barsalou uses the term *productivity* somewhat differently, referring to the ability of subjects to supply instantiations by filling in schemas that were created by filtering out features of the initial perceptual situation. In his treatment of this filling-in Barsalou allows for supplying features that were not part of the initial perception, thus allowing for novelty, including novel representations that violate physical principles. Thus, what he terms productivity is one way of generating new representations, but clearly not the only one present in his account of perceptual symbols.

References

Barsalou, L. 1999: Perceptual symbol systems. *Behavioral and Brain Sciences*, 22, 577–609.

Batista, A. P., Buneo, C. A., Snyder, L. H., and Andersen, R. A. 1999: Reach plans in eye-centered coordinates. *Science*, 285, 257–60.

Bechtel, W. 1998: Representations and cognitive explanations: Assessing the dynamicist's challenge in cognitive science. *Cognitive Science*, 22, 295–318.

Clark, A., and Toribio, J. 1994: Doing without representing? *Synthese*, 101, 401–31.

Corbetta, M., Miezin, F. M., Shulman, G. L., and Petersen, S. E. 1993: A PET study of visuospatial attention. *Journal of Neuroscience*, 13 (3), 1202–26.

Cummins, R. 1996: *Representations, Targets, and Attitudes*. Cambridge, MA: MIT Press.

Dretske, F. I. 1981: *Knowledge and the Flow of Information*. Cambridge, MA: MIT Press/Bradford Books.

Fodor, J. A. 1975: *The Language of Thought*. New York: Crowell.

Fodor, J. A. 1987: *Psychosemantics: The Problem of Meaning in the Philosophy of Mind*. Cambridge, MA: MIT Press.

Fodor, J. A., and Pylyshyn, Z. W. 1988: Connectionism and cognitive architecture: A critical analysis. *Cognition*, 28, 3–71.

Goldman-Rakic, P. S. 1987: Circuitry of primate prefrontal cortex and regulation of behavior by representational memory. In J. M. Brookhart, V. B. Mountcastle, and S. R. Geiger (eds), *Handbook of Physiology: The Nervous System*, vol. 5. Bethesda, MD: American Physiological Society, 373–417.

Griffiths, P. E. 1997: *What Emotions Really Are*. Chicago: University of Chicago Press.

Haselager, W. F. G., and de Groot, A. D. (forthcoming): Representationalism versus anti-representationalism: a debate for the sake of appearance?

Hubel, D. H., and Wiesel, T. N. 1962: Receptive fields, binocular interaction and functional architecture in the cat's visual cortex. *Journal of Physiology (London)*, 160, 106–54.

Hubel, D. H., and Wiesel, T. N. 1968: Receptive fields and functional architecture of monkey striate cortex. *Journal of Physiology (London)*, 195, 215–43.

Kuffler, S. W. 1953: Discharge patterns and functional organization of mammalian retina. *Journal of Neurophysiology*, 16, 37–68.

Langacker, R. 1987: *Foundations Of Cognitive Grammar*, Vol. 1. Stanford, CA: Stanford University Press.

Mandik, P. 1999: Qualia, space, and control. *Philosophical Psychology*, 12, 47–60.

Millikan, R. G. 1984: *Language, Thought, and Other Biological Categories*. Cambridge, MA: MIT Press.

Penfield, W., and Rasmussen, T. 1950: *The Cerebral Cortex in Man: A Clinical Study of Localization of Function*. New York: Macmillan.

Snyder, L. H., Batista, A. P., and Andersen, R. A. 1997: Coding of intention in the posterior parietal cortex. *Nature*, 386, 167–70.

Talbot, S. A., and Marshall, W. H. 1941: Physiological studies on neural mechanisms of visual localization and discrimination. *American Journal of Ophthalmology*, 24, 1255–63.

van Gelder, T. 1995: What might cognition be, if not computation. *Journal of Philosophy*, 92, 345–81.

19

The Architecture of Representation

Rick Grush

Introduction

There is no notion more crucial to the study of thought and cognition than *representation*. It is the fact that cognitive systems traffic in representations that sets them apart from merely complex and interesting, but non-cognitive, systems like elms, oceans, and microwave ovens. As crucial as this notion is, and as much theoretical attention and industry as has been devoted to it, it remains a frustrating enigma. Frustrating because it seems to be very simple – a representation is something that stands for something else – and yet it has proven quite resistant to any systematic, plausible, and revealing analyses.

In this paper I outline, apply, and defend a theory of natural representation which is, in the first instance, a theory of how representations are constructed and used by natural systems such as nervous systems. The main consequences of this theory are: (1) representational status is a matter of how physical entities are used, and specifically is *not* a matter of causation, nomic relations with the intentional object, or information; (2) there are genuine (brain-)internal representations, contra theorists who maintain that only *external* symbols can be representations; (3) such representations are really *representations*, and not just farcical pseudo-representations, such as attractors, principal components, state-space partitions, or what-have-you;[1] and (4) the theory allows us to sharply distinguish those complex behaviors which are genuinely cognitive from those which are merely complex and adaptive, contra dynamical systems-theoretic and related views which treat cognitive phenomena as just complex adaptive behavior on the same continuum with "simple" sensorimotor integration.

Section 1 will briefly introduce some terminological distinctions and develop an example of representational activity which motivates these distinctions. Section 2 attempts to provide fairly precise characterizations of these concepts in terms of control theory and some associated mathematical apparatus. We will then be in a position to describe an architecture, which I call the emulation theory of representation (ETR), that is a necessary condition on a system's representational status. Section 3 provides an example from robotics research which both instantiates the architecture I describe, and is a clear example of representational activity. Section 4 turns to the brain and provides evidence to the effect that the brain does instantiate this architecture for maintaining and using internal representations for use in motor control and imagery. Section 5 extends ETR in order to account for certain crucial features of representation. Section 6 answers some common objections.

1 Representation and Presentation

Representations are entities which stand for something else – or better, they are entities which are used to stand for something else. This second characterization brings out something not explicit in the first, that of a user. If this second definition, and my gloss on it, are correct, then a representation is a part of a three-way relationship which also includes a user and a target. So far so good. Some may quibble over the need for a user, but that is not where the real problem lies. The real problem has been, and continues to be, the choice of states for which theorists attempt to give a representational analysis. Specifically, sensory states have been used as a model for representational states, the idea presumably being that sensory states represent the world to the subject. The thought seems to be that an analysis of the representational status of sensory states, once adequately done, will then be able to be generalized to other sorts of internal representational states. This choice has the consequence that the theoretical role of a user is optional (e.g. the retinotopic projections in various parts of the central nervous system (CNS) represent the visual scene even if these projections are not being attended to or used by anything else in the brain, one might argue).

But this is exactly wrong. A useful theory of representation must not treat sensory input as representational. Such information will be better treated as *presentational*. To see the distinction consider the following analogy. I am playing a game of chess with a friend; however, I am not in the immediate vicinity of my friend or the chess board. Rather, I am at some other location where I learn about my opponent's moves, and issue my own moves, via the telephone. Now I cannot keep track of everything in my head, so I keep with me a board which I use to keep track of what the "official" game board looks like. But, and this will mark the crucial distinction, I also keep a *second* board with me. This second board I use to try out moves, perhaps long sequences of moves, and to assess the possible consequences of those moves and countermoves. These two uses for chess boards are quite distinct. The first board's use is to accurately mirror the state of the official board, accurate information about this board being crucial to my chances of success in the game. The second board is emphatically *not* used to accurately mirror the state of the official board. Its usefulness for assessing possible positions depends on its not having to carry information about the actual state of the official board.[2] It might be thought that one can get by with one board – one can try out moves and then put the pieces back before making the official move. This is entirely correct. But note that what one is doing in this case is putting the same board to the two different uses I described. One uses the board now to present the real game position, and now takes it "off-line" to try out moves.[3]

According to the theory I will advance, only the second board is a representation, the first can perhaps be described as a presentation. Most of the recent philosophical literature on internal representation and content has been tying itself in knots because it has not distinguished internal representations from internal presentations, and has been trying futilely to give a theory of content for internal presentations, and then subjecting these theories to constraints and intuitions which properly belong to representations.[4] The pitfalls of this conflation should be nowhere more evident than where analyses of inaccurate perception (mistaking a horse on a dark night for a cow, for example) are taken to be relevant to the question why something can be a representation even when it is not an accurate mirror.

As I have remarked, what distinguishes presentations from representations is the use they are put to. A presentation is used to provide information about some other, probably external in some sense, state of affairs. It can be used in this way because the presentation is typically causally or informationally linked to the target in some way.[5] The representation's use is quite different: it is used as a counterfactual presentation. It is, in very rough terms, a model of the target which is used off-line to try out possible actions, so that their likely consequences can be assessed without having to actually try those actions or suffer those consequences.[6] Second, and this is implicit in the first point, the ability to use an entity as an off-line stand-in depends crucially on its not being causally linked to, and its not necessarily carrying information about, the entity it represents.

This issue is actually quite touchy. Given my claim that a representation need carry no information about the target, two sorts of objections have been raised. First, to stick with the chess board analogy, the structure of the representational board must mirror the structure of the real board if it is to be of any use. As one person put it, "The second board, of course, carries many sorts of information about the first board that are necessary for it to be a representation of the first board, including the board's structure and the kinds of pieces that can be involved."[7] This is correct. If I might be allowed to distinguish two sorts of information, say *state* information and *structure* information, I want to say that representations need carry none of the first, but I admit they must have at least some of the second. The distinction would be roughly this: structure information tells one about the laws and generalizations which govern the operation of a system as well as the system's gross and relatively permanent features, while state information tells one about specific contingent features of a system, and hence which laws and generalizations are in effect. I will say a bit more about this in section 5, note 17. But for now I will note that when I discuss information, I mean state information. I feel justified in doing this because this is the way that the term is used by proponents of informational, covariational, and teleological theories of representation and content, and it is with these theorists that I am quarreling.

Second, it might be thought, even if we restrict the discussion to state information, that the representation will carry information about the target. The fact that my black bishop is currently on the second board entails that it has not already been captured on the official board. Thus the second board does in fact carry information about the official board, in that knowledge of the state of the second board reduces the uncertainty about the state of the official board. But this is not because of any constraints on representation, but is rather because in this case I only have need to represent certain classes of situation, namely those which I might encounter on the real board. Thus, given my purposes, and the fact that I am not completely daft, the state of the second board will reduce the uncertainty about the state of the official board. This information is more properly a result of my purposes and not a result of requirements for representation, however. I can, after all, put my black bishop back on the second board after it has been captured to determine where I went wrong, or if there were other moves I might have made to avoid the capture.

I will be arguing in due course that the proper way to understand the representational brain is to understand the interplay of three distinct sorts of entity: controllers (these correspond to the players which actually make and try out moves in the chess game analogy), presentations (these will typically be sensory or perceptual states), and internal representations (or emulators). With these distinctions in hand, many of the problems which have frustrated attempts to understand natural representation dissolve. But before we can apply these distinctions with any clarity, they must be refined.

2 Emulation

The place to look for illumination on this topic is control theory. It will be useful to start with open-loop, or feedforward, control. Such a system is shown schematically in figure 19.1. The system includes a target system (often called the "plant" in the control literature, thus I will use "plant" and "target system" interchangeably) that can be described as a set of system variables, some of which will be control variables, and some of which will be output variables, and perhaps others which are neither. For instance, one might describe a car with a number of variables (amount of fuel, engine rpm, acceleration, mass, etc.), some of which are control variables (pressure on the gas pedal, torque on the steering wheel), and some of which are output variables (the position of the speedometer needle). The goal is to get some of the target's system variables to certain goal values – one wants the car's speed to be between x and y mph, and its orientation to be exactly n degrees.

The controller is a system whose purpose is to get the crucial variables of the target system to within their goal parameters – the driver would usually be the controller of a car, for instance. Controllers typically need two pieces of information to do their job, the current state of the target, and a specification of the goals. (In order to get the car to 55 mph, one needs to know not only the goal speed, but the car's current speed as well. What action is appropriate will depend on both these factors.)

Figure 19.1   Open-loop control.

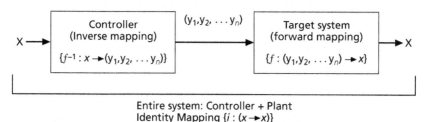

Entire system: Controller + Plant
Identity Mapping $\{i : (x \rightarrow x)\}$

Figure 19.2   Forward and inverse mappings.

So in the simplest case, we can imagine the following sort of process. A controller is given a goal state and the current state of the target. From this information it determines an appropriate set of actions, perhaps an action sequence, which will get the target within the goal parameters. It issues these commands to the target. The target system, which starts in its initial state, then undergoes state changes as a function of the commands sent to it, and if everything is working correctly, the target ends up in the appropriate goal state. Once we notice that the entire system, controller plus target, implements an identity mapping (from goal states to goal states), we can characterize the controller as the inverse of the target (see figure 19.2). That is (modulo the initial state specification), the controller implements a mapping from goal states (x in the figure) to command sequences (the y_i's in the figure), and the target implements a mapping from command sequences to goal states. The mapping performed by the target is called the forward mapping, while that performed by the controller is the inverse mapping.

Open-loop control has the virtue of being conceptually simple, and allowing us to introduce useful distinctions between types of mappings. Its usefulness as a control architecture is more questionable. Closed–loop control (also known as feed-back control) is often more effective, flexible, and efficient than open–loop control. In a closed-loop control system, there are sensors that are sensitive to various parameters of the plant, and these sensors feed this information back to the controller, which can then effectively change or continue its ongoing command sequence in real time as needed (see figure 19.3).

So, for instance, a thermostat is a simple sort of closed-loop control system. The thermostat itself is the controller, and the plant in this case consists of everything else: the heater, cooler, thermometer, and the room. The thermostat sends commands to the plant (specifically, the heater and the cooler) in order to influence the tem-

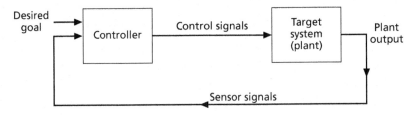

Figure 19.3 Closed-loop control.

perature of the room, and it gets constant information (feedback) about the current state of the room, particularly its ambient temperature, from a thermometer. This frees the thermostat from having to determine the entire command sequence at once, as an open-loop controller would. Such an open-loop scheme would be brittle anyway, because the system would not be able to adjust the command sequence on the basis of unexpected perturbances. Feedback control allows the controller to produce and update the command sequence in real time as a function of its information about the progress of the plant. Because of this, in most cases, closed-loop control allows for much simpler controller design than open-loop systems do. For instance, the closed-loop thermostat can simply compare the current temperature with the desired temperature at each time step, and on the basis of that comparison, do one of three things; turn (keep) heater on, turn (keep) air conditioner on, or turn (keep) both off.

Closed-loop control is often revered as the state of the art in control technology. Indeed, when the only competitor countenanced is open-loop control, this conclusion is understandable. But closed-loop control has a number of problems as a control architecture, and it has an additional problem if one attempts to use it as a model of cognitive activity. One general problem with closed-loop systems is that they can be quite sensitive to feedback delays. To illustrate: suppose a thermostat is situated such that the information it receives concerning the temperature of the room it controls is, say, four hours old (perhaps the thermostat is controlling, via radio signals, the temperature of a room on a space station some four light-hours away). If the thermostat is trying to get the room to 70 degrees, and is getting information to the effect that the room is 60 degrees, it will turn on the heat. In fact, because the feedback is delayed four hours, the thermostat will keep the heater cranked for four hours longer than it needs to – by which time the room can be expected to be nice and toasty.[8] At that point, the thermostat gets information to the effect that the room has reached 70 degrees. It turns off the heater, but then it continues to be told that the heat is rising. So it turns on the air conditioner. For the next four hours, the thermostat continues to be told that the temperature in the room is rising, so it keeps the air conditioner on. Of course, the room will be below 70 degrees for four hours before the thermostat turns the air conditioner off. You get the idea. In general, depending on various parameters of the system, such as how much delay there is in the feedback, and how responsive the plant is to control signals, the plant may go into oscillations or instabilities as a result of such delays.

In addition to problems with closed-loop control itself, there is an additional problem arising from the attempt to use closed-loop control systems as a model for cognition.[9] This is the problem of *obligatory action* – a controller is only a controller when in closed-loop contact with the target system. When decoupled, like a detached thermostat or a Watt governor lying on the shop floor, controllers do *nothing*, and in particular they do nothing *cognitive*. Humans, on the other hand, seem able to do all sorts of cognitive things even when not actively engaged with a target system or interactive environment, contra some of the stronger claims issued from the embedded cognition camp.[10] I can close my eyes and plan the quickest route home after work when I learn about a traffic jam on my normal route. Additional examples of this sort of thing are, I imagine, easy to construct.

I do not intend to get into a discussion of the merits or demerits of looking at matters this way, but rather to render the entire debate obsolete by explaining exactly how systems represent targets even when not coupled to them; that is, how cognitive systems avoid the problem of obligatory action by being able to act on representations in lieu of acting on external environments. The key to this is to recognize that a *cognitive system* is not just a controller, but a controller together with a forward model (or as I shall call it, an *emulator*).

To see what a forward model is, consider the following solution to the feedback delay problem faced by the thermostat. Suppose that before the space station was sent away, that is, before there was a feedback delay problem, researchers decided to train a neural network to mimic the forward mapping of the plant. In this case, the plant is the room, its heating and cooling systems, and its associated sensors (the thermometer). Inputs to the plant are commands to the heater and cooler, and outputs are thermometer readings. The training proceeds by letting the thermostat control the temperature of the room, and at every time interval (preferably very short), the neural network is given as input the current temperature and current command, and the training signal is the temperature which is produced at the next time step. After training, the neural network will exactly implement the forward mapping. It is thus called a forward model, or emulator.

Once one has an emulator, it is possible to implement pseudo–closed-loop control (see figure 19.4). Here, the controller's output is split into two copies. One is sent to the plant, as usual, and the other is sent to the emulator. Because the emulator implements the same input–output function as the plant, the controller can use feedback supplied by the emulator just as well as it could use the real feedback from the plant itself. Once this happens, it will no longer matter if the real feedback signal is delayed – it is not being used anyway. Provided that the emulator is accurate, and that *its* output is not delayed, all will proceed without a hitch.

Of course doing this introduces other problems. If the emulator is not perfect, for instance, and there is no means of keeping the state of the emulator close to the state the plant is in, it will eventually wander, and then the control signals are likely to be inappropriate. But for the present conceptual point we can safely ignore these

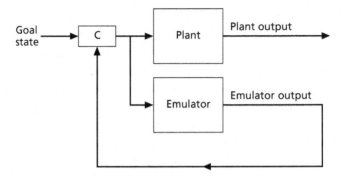

Figure 19.4   Pseudo-closed-loop control.

issues. The important point is the notion of an emulator, an entity which implements the forward mapping, which is plug-compatible with the plant, and can thus be run in parallel with the plant (as in pseudo-closed-loop control) or can be run instead of the plant (examples of this will be provided shortly).

This brief exposition of a few control architectures should allow us to gain clarity on the distinction between presentations and representations. A presentation, as I shall use the term, can be thought of as information about the current state of whatever actual system one is dealing with. This will often be sensory information. A representation, by contrast, is an emulator of some system, which can be run off-line to provide mock sensory information about what the real target system would do under various conditions. It is, to belabor the point, something which is used to stand for something else. If this is correct, then we can say that a necessary condition for a system being a representing cognitive system is that it has the capacity to internally emulate some external system with which it can also interact overtly.

3   An Example from Robotics

I have distinguished representations from presentations, and attempted to spell this difference out in terms of control loops – a presentation is perceptual information from a target, while a representation is an internally maintained forward model of the target, which can supply mock information for various purposes, including counterfactual reasoning. But an example can be worth a thousand definitions.

Murphy (Mel, 1990) is a robot whose task is to maneuver its arm around obstacles in order to get its hand to an object which it will then be in a position to grasp. Murphy's arm has three joints, a shoulder, elbow, and wrist, all of which have their range of motion limited to a single plane. Above this plane sits a video camera whose signal is used to drive a 64×64 grid of units which act as a sort of visual display – a 64×64 grid of pixels. This grid of units is what Murphy uses to guide its arm movements around the workspace without contacting obstacles.

Each of the units of the grid is actually a connectionist unit whose activity is normally driven by the video camera. But each of these units also receives inputs from the robot's motor output.[11] That is, the motor output is split into two copies, one of which normally gets sent to the arm, and a second which gets sent to the grid units. Each unit then gets information from two sources, visual information from the video camera, and information concerning the motor commands which Murphy issues.

During the first phase of Murphy's operation, it simply moves its arm through a sample of arm configurations. During this time, the grid units monitor both the motor commands sent to the arm, as well as the resultant video input to the grid which drives the units. Because these units see both the input and the output of the forward mapping which the arm/camera system implements – it gets a copy of the motor command and is fed with a visual image which that command leads to – they are thus able to learn the forward mapping of the plant. In this case, the forward mapping is a function from joint angle commands to visual grid displays. The *plant* implements this mapping by actually sending the joint angle command to the arm, which moves accordingly, and the video camera sends an image of the resultant configuration to the grid. But once the units have learned the forward mapping, they can implement this function without going though the real arm/camera system. Thus, with the real arm and the video camera off-line, the grid is able to process a copy of the motor command to create a mock visual display which is similar to the display which would have been produced by the video camera pointed at the arm as it carried out those same motor commands.

And this is exactly what Murphy does after learning. It gets an input of a visual scene including its arm's initial location, the object location, and obstacle locations. It then takes the real system (its arm and the video camera) off-line, and drives the visual grid, which is an emulator, directly with an efferent copy of the motor command. Murphy is then able to create an image of its arm moving around the workspace. When it sees its arm (now an image of its arm) run into an image of an obstacle, it backs up and tries again, just as it would when actually operating its arm. It continues this way until it discovers a route from the initial arm configuration to the goal location. It then puts the arm and camera system back on-line, and implements this solution.

According to the terminology I introduced earlier, the visual grid, when on-line and being driven by the actual arm/camera system, is a *presentation* of the workspace. When the arm/camera system is taken off-line and the grid is being used as a forward model driven by an efferent copy, the visual grid is a *representation* of the workspace.

But we can go further than this. What the forward model does is allow Murphy to engage in *counterfactual reasoning*. It can determine that if it were to issue joint-angle command sequence R, its arm would impact an obstacle, but if it were to implement joint-angle command sequence S, its arm would move to the goal without impact. At the end of this reasoning process, it implements the correct sequence. And even though the system is a relatively simple one whose operation could be described in terms of the flow of activation in connectionist units, it is also easy to see how we can provide a representational explanation of its operation. Why did Murphy *not* move its arm to location X? Because it found (believed) that if it did, its arm would hit an obstacle. What is Murphy thinking about now? It is contemplating the use of command sequence R. No doubt my use of the terms "believe" and "contemplate" are a bit strained with such a simple system. But I hope that does not divert attention from the fact that some sort of representational description is almost forced by Murphy's use of the emulator to stand for the real system.

So this, in brief, is the emulation theory of representation (ETR). A representation is something which is used to stand for something else. And this amounts to a controller which normally operates some system being able to decouple from that system, and couple instead to a plug-compatible emulator. When it does this, the emulator is standing for the real system. It represents it, not because it is causally linked to the real system, not because it carries information about the state of the real system, but because it is used to stand for it.

4   The Human CNS

We have before us a starting point for a theory of internal representation and an example of a robot which implements the architecture described in the theory. Now I will turn to the human nervous system and provide evidence that it uses emulators to represent entities external to it. I will discuss two sorts of case; emulators of the musculoskeletal system (MSS) and emulators of the motor-visual loop.[12]

We saw in section 2 how feedback delays can cause problems for closed-loop control systems. There is considerable evidence to the effect that, in part because of relatively slow axonal conduction velocities[13] proprioceptive feedback from the limbs is too slow to be used to guide fast voluntary movements which are executed in less than about 200–450 ms.[14] This might lead one to suspect that such movements are executed open-loop. But this seems not to be correct either, as there is evidence that there are adjustments made to the motor program, as quickly as 70 ms after movement initiation, which have the effect of correcting initial inconsistencies in the first part of the trajectory (cf. van der Meulen et al., 1990).

This apparent paradox – that it looks as though there are corrections made to the motor program on the basis of peripheral feedback before peripheral feedback can be used effectively – is dissolved when one realizes that the feedback used to make adjustments to the motor program could be supplied by an internal emulator. One way to do this would be via the pseudo-closed-loop architecture described in section 2. In other words, when the movement must be executed too quickly for peripheral feedback to be of use, an internally generated "mock" proprioceptive signal, generated with a musculoskeletal emulator, can be used to adjust the motor program instead.

A number of researchers (e.g. Gerdes and Happee, 1994; Kawato, 1990; Kawato et al., 1987; Wolpert et al., 1995) have developed models based on this insight.

According to Kawato (1990), there is a neural circuit involving the cerebellum (especially the dentate nucleus) and the red nucleus which acts as a model of the MSS. The models of Wolpert et al. (1995), and Gerdes and Happee (1994) are more sophisticated, positing not a simple pseudo-closed-loop architecture, but rather variants of Kalman filters which use forward models. In the model of Wolpert et al., to a first approximation, this filter combines the deliverances of the real target and the forward model in a time-varying manner, so that the motor centers rely exclusively on the output of the forward model during the initial phases, and as time progresses rely more and more on the "real" proprioceptive information. Whatever the details, there is converging experimental and theoretical evidence from human motor performance which indicates that the human CNS in fact uses internal forward models of the body.

It should be noticed that this solution to a motor control problem has an unexpected benefit: the tools the nervous system evolved to solve this problem can also be used to generate motor imagery if the MSS is simply taken off-line. Motor imagery is no more than an internally generated proprioceptive image (like visual imagery, only not visual), and this is exactly what the forward model generates in order to solve the feedback delay problem. Consistent with this hypothesis is the finding that many of the same motor areas which are used to drive overt motor behaviors are also active during motor imagery (Decety et al., 1990; Fox et al., 1987; Ingvar and Philipsson, 1977; Roland et al., 1980).[15]

It is possible to address visual imagery with the same architecture. The account of motor imagery I sketched exploited the fact that there are learnable regularities in the mapping from initial proprioceptive states and motor commands to future proprioceptive states, and thus that an emulator driven off-line by efferent copies can produce proprioceptive imagery. The same is true for visual imagery. Given a visual input (retinal projection, primary visual cortical projection, whatever) and a motor command (such as a saccade, or a step forward), the next visual input is at least in part predictable (e.g. a translation in the direction opposite the saccade, or an enlargement of the projection of the object one is walking towards). Mel (1986) has developed a model of mental rotation, zoom, and pan using this insight. The model is a robot that can move around, towards, and away from objects while looking at them, has the ability to learn this forward mapping, and can then take itself off-line, and "mentally" rotate, zoom, and pan images of objects.

In the human case, there is increasing evidence that many of the same areas that subserve vision are also implicated in visual imagery, that is, evidence that these areas are driven not only by sensory organs, but can also be driven by internal efferent copies, much like Murphy. To take but one example of dozens, Farah et al. (1992) conducted a series of imagery experiments on a patient before and after unilateral occipital lobectomy. The subject was asked to imagine a number of common objects (ruler, car, bicycle, etc.), and to imagine them getting closer and closer until they just began to overflow the edges of the visual image boundary. The fascinating result was that after the unilateral occipital lobectomy, the distance at which imagined objects began to overflow the boundary increased dramatically for objects which were primarily horizontally oriented, but did not significantly alter for vertically oriented objects. One plausible explanation for this finding is that visual imagery is the result of the use of a forward model which drives (at least some of) the same visual areas driven by overt vision. After the removal of one occipital lobe, the forward model has a screen, so to speak, only half as wide to drive. Imagined horizontal objects will accordingly be farther away when they begin to push the edge of the mind's eye.

5   Articulation

Even though the theory of representation I have outlined here may be a promising starting point for understanding natural cognition, there are a number of serious inadequacies as it stands. While addressing all of them is beyond the scope of this paper, in this section I will address one which is crucial.

It will have been noticed that when I have talked about this ETR, I have claimed that the emulator represents the target system (e.g. the musculoskeletal emulator [MSE] represents the MSS). But even though this is representation, it is a weak variety. While this might perhaps work for imagery (which is what I have primarily been discussing), it is not clear that it will work for representation in general. This is because we have one entire system which represents, in some holistic way, some different system. What might be preferable in some cases is to have entities which represent, individually, components of the target system. So for instance, given that specific physical parameters of joints and muscles are relevant components of the MSS for purposes of its movements, one would like to be able to point to entities in the CNS which represent, say, specific physical parameters of the forearm, or the index finger. A similar problem is manifest with Murphy – it has a representation of its workspace (the visual grid when run off-line), but there is no distinguishable entity which has the dedicated task of representing Murphy's hand. As Murphy imagines a movement, first some units will be active, then those will spin down as others become active (this is what movement across the grid amounts to), and this is about as close as one is able to get to a representation of the hand.

But suppose that the emulator not only implements the forward mapping, but does so because it is composed of parts, or *articulants*, each of which represents some aspect of the target system. A target system will typically be describable as a dynamical system with N parameters, $e_1, e_2, \ldots, e_n$. Some of these, let us say $e_1, e_2, \ldots, e_i$ will be input parameters, that is, parameters whose equations of evolution include one or more parameters of the controller. Others, say $e_{i+1}, e_{i+2}, \ldots, e_k$ will be output parameters, that is, parameters of the target which directly influence the evolution of one or more of the parameters of the controller. The rest of the target system's $N - k$ parameters we can call internal parameters – these will neither be directly influenced by, nor will directly influence, any parameters of the controller. Now let us suppose that we have an emulator which implements the forward mapping of the target because it also is an N-parameter dynamical system, with $e^*_1, e^*_2, \ldots, e^*_i$, as inputs, $e^*_{i+1}, e^*_{i+2}, \ldots, e^*_k$ as outputs, and the rest of the $N - k$ e^*s as internal parameters, just like the real target (the first two conditions guarantee that the emulator and target are "plug-compatible"). Suppose further that the dynamic of the emulator is formally equivalent to the dynamic of the target.[16] In such a case, not only does the emulator represent the target system, but it will also be true that the emulator articulant e^*_h represents target system parameter e_h. In such a case we can say that the emulator is articulated into components which represent components of the target system.[17]

This has several advantages. First, it makes semantics much easier. To the question "What represents this aspect of the target system?" there will be a clear answer. Second, and more importantly, this gets us out of the regress problem.[18] A cognitive system (a system which not only interacts with its environment, but consists of a controller and an articulated environmental emulator to which it can couple to represent its environment) can have an internal representation of some feature of its environment F. What makes the emulator articulant F^* represent F is the fact that the controller interacts with the emulator, and the emulator's articulant F^*, in a way analogous to the way it interacts with the environment, and the environmental feature F. Furthermore, the controller does not need to *interpret* anything as anything else, except in the innocuous and non-question-begging sense that it interacts with something as it interacts with something else. It simply enters into dynamical interaction with one system rather than the other. Representation is cashed out in terms of use, and use is cashed out in terms of selective dynamical coupling. We are thus not forced into a representational regress, antirepresentationalism, mere instrumentalism about representational descriptions; nor does the problem of obligatory action arise.

Furthermore, we are in a position to see one clear criterion that distinguishes genuinely cognitive systems from merely complex and adaptive, but non-cognitive, systems. It is a necessary condition for a system to be a cognitive system that it consists not only of a controller (which may, by itself, be able to interact adaptively with the system's environment), but also that the system be able to internally emulate aspects of its environment, via use of a forward model, in such a way as to allow the controller to selectively decouple with the real environment and couple instead with the internal emulator.

6   Objections and Replies

1   Why do we need to describe such systems as representing anything? Rather, can't we just explain the behavior of Murphy, for example, as the operation of a complex dynamical system?

This objection seems to presuppose that only in cases where one is *forced* to describe a system in representational terms is one licensed to legitimately do so at all. But I cannot see why anybody should be moved by this. Indeed, any materialist will agree that there are, in principle, physical explanations for all behaviors – but this is not seen as prima facie incompatible with legitimate representational explanation. The best criteria for the legitimacy of a representational explanation are whether or not such explanations make sense of the behavior of the system, and whether or not it is useful to interpret aspects of the system's internal antics as cognition about something else. In Murphy's case, to take the example at hand, both of these criteria are met. A representational account of what Murphy is doing when it goes off-line is natural and, in some sense, just plain right.

2   It has not been established that the motor control systems of the human CNS use motor emulators, as your theory suggests they do.

This objection misunderstands my rhetorical strategy. The existence of MSEs is not a premise on which the larger argument is based, but merely an example of an application of the architecture. It may in fact turn out that motor control does not use such emulators. It would not follow from this that the emulation theory of representation is wrong, but only that fast voluntary motor control does not employ representations. Nonetheless, there is very compelling evidence that the human CNS does use such models (a small sample of which is cited in section 4). And the speculation that nervous systems do use such models for motor control purposes may shed light on the evolution of cognition (see section 7).

3   Even if ETR works for imagery, it is not at all clear how it is meant to work for other aspects of cognition.

It is not known to what extent cognition is in fact based on imagery, but it could be a much greater extent than is often supposed. To point out but a few examples: much current work in cognitive linguistics (cf. Lakoff, 1987; Langacker, 1987, 1990, 1991) seeks to explain many key aspects of language in terms, in part, of imagery and image schemas. There are likewise rich connections between imagery and memory, suggesting that memory may not store information in an amodal propositional form, but rather as modality-specific images (see Paivio, 1995, for review). Furthermore, given that emulators are just neurally implemented models, the door is open to accommodating many of the insights of Johnson-Laird's (1983) mental models framework for reasoning and inference.

How much of cognition can be accounted for in terms of imagery or image schemas is an open empirical question. Until such questions have been answered, the fact that emulators need not traffic in propositions or first-order predicate calculus is no strike against them.

4   It does not appear to be the case that the representations posited by ETR will be such as to exhibit well-known properties of representational content such as failures of substitution.

Actually, one of the more exciting precipitates of the current account is that it does in fact address this issue, but again, I can only sketch this roughly.[19] Emulators do not represent the target system *per se*, but represent it *as interacted with*. Any given physical object or system actually implements an infinite number of dynamical systems, and it is possible to interact with a physical system *as* any number of these systems. So for instance a computer can implement a dynamical system which describes a Turing machine, as well as implementing a dynamical system describable as a mass-spring (imagine here the computer attached to a wall with a spring, and oscillating back and forth). I could then interact with the computer either as a Turing machine, or as a mass-spring system (I might program it, I might try to predict when it will reverse direction). The articulated emulator, *qua* emulator, explicitly implements only one dynamical system,[20] and so it is only that aspect which is represented. My musculoskeletal emulator represents my arm, to a rough approximation, as *the body part with such-and-such physical/dynamical properties, and not as the last item seen by Smith before he went unconscious or as the twentieth heaviest object in the room.* Emulation is necessarily emulation of some aspect of the target, where aspects are, in the first instance, individuated by use and means of interaction and prediction.

5 *"Standing for" is not a relation distinctive of representations, because signals in the sensory/presentational systems also stand for something else, namely the current state of the target.*

But this is exactly what I am trying to get away from. The distinction between representation and presentation is a distinction between *standing for* and *providing information about*. Perhaps it is easier to grasp this distinction if one distinguishes *standing IN for* and *providing information about*. If I am on a football team and get injured, another player can take my place – can, so to speak, stand in for me. There is all the difference in the world between this, and simply providing information about me, e.g. via a video camera, to the undermanned team. So much difference that I am frankly surprised at how often this distinction is not grasped. In order for X to stand for Y, X must do something that Y would do, and be doing it instead of Y. Healthy replacement players can do this, information about injured players cannot. Similarly, sensory states do not stand in for the objects in any sense. A certain pattern of activity in my primary visual cortex, say pattern T, when I look at a tree is not at all standing in for a tree, because pattern T is not doing anything which is normally done by a tree.

7 Conclusion

In *Neurophilosophy*, Patricia Churchland (1986) writes:

> To follow evolution's footsteps in discovering how basic principles of motor control are refined and upgraded to yield more complex systems is a productive strategy. Additionally, it may be a shift in focus that allows us a breakthrough in the attempt to understand the higher functions. . . . If we can see how the complexity in the behavior that we call cognition evolved from solutions to basic problems in sensorimotor control, this

can provide the framework for determining the nature and dynamics of cognition.

(p. 451)

This is exactly what I hope to have done. If the theory I have very tentatively sketched here is correct, then representation and cognition are dependent on emulation, and the strategy of emulation plausibly may have arisen as a solution to a clear problem in motor control. Given fixed and relatively slow axonal conduction velocities, the twin evolutionary pressures of greater size and greater speed work at cross purposes. Pseudo-closed-loop control (or some other similar strategy employing a forward model, such as Wolpert et al.'s (1995) Kalman filter) provides a straightforward solution to this problem, and requires only relatively humble ingredients: an efferent copy of the motor command, and some way to do associative learning. And once this simple solution is in place, nervous systems have a powerful new tool. A tool which makes possible imagery, representation, and cognition itself.

Perhaps we should be grateful that axons are as slow as they are.

Notes

1 "... while dynamical models are not based on transformations of representational structures, they allow plenty of room for representation. A wide variety of aspects of dynamical models can be regarded as having a representational status: these include states, attractors, trajectories, bifurcations and parameter settings" (van Gelder and Port, 1995). This is just one of an increasingly large number of cases where the notion of representation is being bleached to the point where theorists feel comfortable calling any bit of theoretical exotica a representation if it is some state or process which allows the system to behave appropriately, even if there is no hint of an account of what the content of such a representation might be. Of course, such things may be genuine representations, but this will be because they have some specifiable content, and not just because some mathematically minded theorist appeals to it as part of a non-psychological behavioral explanation.

2 I have been informed that Cummins (1996) makes a similar distinction. I thank Pete Mandik for pointing this out to me.

3 As we shall see, there is reason to believe that certain cortical areas have exactly this character – during perception they are driven largely by peripheral sensory organs, but they can be taken off-line to support imagery.

4 A few famous examples include Dretske (1981), Fodor (1987), and Millikan (1984).

5 It is not my purpose in this paper to examine presentations in any detail. I will continue to gloss their function as providing accurate information about the environment, even though I think that this is not quite correct. See Akins (1996) for a critique, with which I am sympathetic, of the view that the role of sensory systems is to provide accurate or veridical information about the environment.

6 This statement is reminiscent of Craik's (1943) theory of representation and cognition. Indeed, major portions of my project, as expressed in Grush (1995; in preparation) can be seen as spelling out Craik's insights in more precision and detail.

7   This point was made by an anonymous referee for this journal [*Philosophical Psychology*].

8   I am here ignoring the fact that in this example the total delay in the control loop would actually be eight hours, because the command signal will take four hours to get to the room.

9   Van Gelder (1995) urges us to think of the Watt governor, a closed-loop control system, as a model for cognition.

10 ". . . we must learn to think of an agent as containing only a latent potential to engage in appropriate patterns of interaction. It is only when coupled with a suitable environment that this potential is actually realized through the agent's behavior in that environment" (Beer, 1995). Amazingly, Beer (and many others) sees the problem of obligatory action not as a problem, but as a premise on which to base a call for a reconception of cognitive activity.

11 In Murphy's case, the motor output is a joint angle configuration, and not an effector command. The forward mapping Murphy learns is thus the forward kinematics, and not the forward dynamics. This is not of consequence to the present point, however.

12 Much more detail on these types of emulators and others, and more evidence for them, can be found in Grush (1995; in preparation).

13 The muscle spindle and Golgi tendon organ signals, which constitute the major proprioceptive mechanoreceptors, are Ia afferents, which are large mylinated fibers having the fastest conduction velocities of all afferent axons. Such fibers are fast *compared with other types of axons*. But the delay is not exclusively a product of the axonal conduction velocity, but of synaptic relays, and central processing of the signal, as well as delay introduced from the efferent side of the process. Thus the feedback delay is the total delay in the loop. I will continue to gloss this total delay, somewhat inaccurately, as due to "slow axonal conduction velocities."

14 Cf. Denier van der Gon (1988) and Ito (1984). There is considerable debate as to the exact delay involved in the proprioceptive loop; most proposals seem to indicate it as being between 200 and 450 ms. The fastest latency for human arm motions I have ever seen defended in print is 125 ms, and even this is longer than the 70 ms which is when adjustments seem to be made to the motor sequence (see below). The usual paradigm for testing feedback delays is the tendon vibration technique (see Redon et al., 1991), where a vibrator is placed on a tendon, causing the muscle spindles to misjudge the joint angle. Even with proprioceptive information being distorted, there is no change in the details of a given movement, as compared with a non–vibration control, in less than about 200–450 ms, which implies that proprioceptive information cannot influence the motor program in less than that amount of time.

15 To pick out one of a hundred examples of converging evidence: Vilayanur Ramachandran (personal communication) has found that most phantom limb patients fall into one of two groups: those who can voluntarily control their phantom limbs and those who cannot. It turns out that in almost all cases, those who cannot move their phantoms experienced a period of pre-amputation paralysis, while those who can move their limbs did not. On the present theory, this is to be expected. If the musculoskeletal emulator learns and updates the forward mapping by monitoring the operation of the musculoskeletal system, then when there is a period of paralysis, the mapping learned is that no matter what the motor command is, the proprioceptive result is "no movement." When the amputation occurs without a pre-operative paralysis period, there is no information to contradict the operation of the emulator, and hence no reason for the forward model to

change (keep in mind the crucial difference between (1) proprioceptive information to the effect that there was no movement, and (2) no proprioceptive information about any movements).

16   The same equations of evolution are obtained by replacing the *e*s with *e**s, the only difference being what the parameters are physically parameters of, but that need not affect the dynamic. E.g. $de_n/d_t = -(k/m)e^*_m$ might be the dynamic for a mass-spring system (where e_m is displacement and e_n is velocity), while $de^*_n/dt = -(k/m)e^*_m$ might be the dynamic of a neural oscillator, where e^*_m and e^*_n are firing frequencies of appropriately coupled neurons. For a more detailed example of this, as well as evidence to the effect that the human motor emulator is articulated in this way, see Grush (1995, chapters 3 and 4).

17   Moreover, it is arguably the case that the articulated emulator is best thought of as a *theory of the target domain*. Doing so allows me to make contact with some of Paul Churchland's views, especially as expressed in Churchland (1989). I address the comparison of my views and those of Churchland to some degree in Grush (1995, chapter 4). Finally, I can now make clear the distinction, mentioned in section 2, between state information and structure information. The articulated emulator carries structure information about the target because it implements the same dynamical system as the target. The emulator's structure in such cases then carries information about the target's structure. But the emulator carries no state information about the target, because, in general, one can set the emulator's state (the specific values of the *e**s) to any value whatsoever to see what the target system would do if its *e*s had those values, and there need be no correlation whatsoever between the values of the *e**s and the *e*s.

18   By "regress problem" I mean a common objection to theories of representation that maintain that a representation requires a user. The objection is that it seems that the user must interpret the representation to be a representation of something, which is tantamount to saying that the user must be able to represent in order to treat other entities as representations. The user then becomes a homunculus, whose capacities for interpretation are unexplained.

19   I discuss opacity phenomena in more, though still greatly inadequate, detail in Grush (1995, chapter 7). Initial appearances notwithstanding, the treatment there is compatible with, and in some sense presupposes, the remarks of this section.

20   The *qua* clause is important, because of course the emulator itself also implements an infinite number of dynamical systems – my brain could be attached to a wall with a spring, for instance. But this mass-spring dynamical system is not an emulator used by my brain because it does not have the appropriate coupling parameters with my motor centers.

References

Akins, K. 1996: Of sensory systems and the "aboutness" of mental states. *Journal of Philosophy*, 93, 337–72.

Beer, R. 1995: A dynamical systems perspective on agent–environment interaction. *Artificial Intelligence*, 72, 173–215.

Churchland, P. 1986: *Neurophilosophy*. Cambridge, MA: MIT Press.

Churchland, P. 1989: *A Neurocomputational Perspective*. Cambridge, MA: MIT Press.

Craik, K. 1943: *The Nature of Explanation*. Cambridge, UK: Cambridge University Press.

Cummins, R. 1996: *Representations, Targets and Attitudes*. Cambridge, MA: MIT Press.

Decety, J., Sjoholm, H., Ryding, E., Stenberg, G., and Ingvar, D. 1990: The cerebellum participates in cognitive activity: Tomographic measurements of regional cerebral blood flow. *Brain Research*, 535, 313–17.

Denier van der Gon, J. J. 1988: Motor control: Aspects of its organization, control signals and properties. In *Proceedings of the 7th Congress of the International Electrophysiological Society*. Amsterdam: Elsevier.

Dretske, F. 1981: *Knowledge and the Flow of Information*. Cambridge, MA: MIT Press.

Farah, M., Soso, M. J., and Dasheiff, R. M. 1992: Visual angle of the mind's eye before and after unilateral occipital lobectomy. *Journal of Experimental Psychology: Human Perception and Performance*, 18, 241–6.

Fodor, J. 1987: *Psychosemantics*. Cambridge, MA: MIT Press.

Fox, P. T., Pardo, J. V., Petersen, J. V., and Raichle, M. E. 1987: Supplementary motor and premotor responses to actual and imagined hand movements with positron emission tomography. *Neuroscience Abstracts*, 398, 1433.

Gerdes, V. G. J., and Happee, R. 1994: The use of an internal representation in fast goal-directed movements: A modeling approach. *Biological Cybernetics*, 70, 513–24.

Grush, R. 1995: *Emulation and Cognition*. PhD dissertation, University of California, San Diego. Available at URL http://www.artsci.wustl.edu/~philos/rgrush.html.

Grush, R. in preparation: *The Machinery of Mindedness*.

Ingvar, D., and Phillipsson, L. 1977: Distribution of the cerebral blood flow in the dominant hemisphere during motor ideation and motor performance. *Annals of Neurology*, 2, 230–7.

Ito, M. 1984: *The Cerebellum and Neural Control*. New York: Raven Press.

Johnson-Laird, P. 1983: *Mental Models*. Cambridge, MA: Harvard University Press.

Kawato, M. 1990: Computational schemes and neural network models for formation and control of multijoint arm trajectories. In W. T. Miller, R. S. Sutton, and P. J. Werbos (eds), *Neural Networks for Control*, Cambridge, MA: MIT Press.

Kawato, M., Furukawa, K., and Suzuki, R. 1987: A hierarchical neural network model for control and learning of voluntary movement. *Biological Cybernetics*, 57, 447–54.

Lakoff, G. 1987: *Women, Fire and Dangerous Things*. Chicago: University of Chicago Press.

Langacker, R. 1987: *Foundations of Cognitive Grammar*, vol. 1. Palo Alto, CA: Stanford University Press.

Langacker, R. 1990: *Concept, Image and Symbol*. New York: Mouton de Gruyter.

Langacker R. 1991: *Foundations of Cognitive Grammar*, vol. 2. Palo Alto, CA: Stanford University Press.

Mel, B. 1986: A connectionist learning model for 3-d mental rotation, zoom, and pan. In *Proceedings of the Eighth Annual Conference of the Cognitive Science Society*, Hillsdale, NJ: Lawrence Erlbaum Associates, 562–71.

Mel, B. 1990: Vision-based robot motion planning. In W. T. Miller III, R. S. Sutton, and P. Werbos (eds), *Neural Networks for Control*, Cambridge, MA: MIT Press.

Millikan, R. 1984: *Language, Thought, and Other Biological Categories*. Cambridge, MA: MIT Press.

Paivio, A. 1995: Imagery and memory. In M. Gazzaniga (ed.), *The Cognitive Neurosciences*, Cambridge, MA: MIT Press.

Redon, C., Hay, L., and Velay, J. L. 1991: Proprioceptive control of goal-directed movements in man, studied by means of vibratory muscle tendon stimulation. *Journal of Motor Behavior*, 23, 101–8.

Roland, P. E., Larsen, B., Lassen, N. A. and Skinhoj, E. 1980: Supplementary motor area and other cortical areas in organization of voluntary movements in man. *Journal of Neurophysiology*, 43, 118–36.

van der Meulen, J. H. P., Gooskens, R. H. J. M., Dennierr Van Der Gon, J. J., Gielen, C. C. A. M., and Wilhelm, K. 1990: Mechanisms underlying accuracy in fast goal-directed arm movements in man. *Journal of Motor Behavior*, 22, 67–84.

van Gelder, T. 1995: What might cognition be, if not computation? *Journal of Philosophy*, 91, 345–81.

van Gelder, T., and Port, R. 1995: It's about time. Editors' introduction to R. Port and T. Van Gelder (eds), *Mind as Motion: Exploration in the Dynamics of Cognition*, Cambridge, MA: MIT Press.

Wolpert, D., Ghahraramni, Z., and Jordan, M. 1995: An internal model for sensorimotor integration. *Science*, 269, 1880–2.

20

Of Sensory Systems and the "Aboutness" of Mental States

Kathleen Akins

1 Introduction

Our thoughts, we believe, are "about" things. When I look at a picture of the Eiffel Tower, I am thinking about an object, a particular structure that exists in Paris; when I remember the fragrance of gardenias, I recall a property that certain flowers have; and when I wonder how George Smith is getting along, I have in mind a particular person, a philosopher I met at Tufts University. Somehow or other, my thoughts are linked to properties and objects. The question, of course, is "how?" How are mental events tied to the objects they represent – what is "aboutness" and how does it come into being?

[In this paper] I want to discuss recent naturalistic theories of "aboutness," naturalistic theories of, roughly, how certain psychological or neural states come to represent, stand for, or be about properties and objects. The theories at issue are, among others, the theories of the Churchlands, Dennett, Dretske, Fodor, Papineau, Stampe, and Sterelny.[1] What I wish to show is that the naturalists' project, as commonly conceived, rests upon an intuitive and seemingly banal view of what the senses do – that the senses function to inform the brain of what is going on "out there," in the external world and in one's own body – and that this "banal" view about sensory function is false or, more moderately, unlikely to be true in the strong guise that the naturalists' project requires. Hence I suspect that the naturalists' project, at least in its present form, is unlikely to work.

. . . Like most naturalists, I too believe that our thoughts/neural states are about the world – about the scent of gardenias, about the Eiffel Tower, about the tickles and itches in one's feet – and that this fact of "aboutness" is fundamentally a *biological* fact about persons. Broadly construed, I share the naturalists' goal. It is the trodden path that has me worried. . . .

2   The Naturalist Camp and the Traditional
View of the Senses

What, then, is the naturalists' project? Part of the problem with characterizing current naturalistic theories of representation is simply that, as a "camp," they comprise quite a diverse lot. That is, there is little agreement among the naturalists about what the project *is* – what exactly requires explanation – much less how the naturalistic theory should go. . . .

[That said], all of the above naturalists share a common project. Each is trying to provide a naturalistic explanation of psychological representation and each is concerned (wholly or in part) with how, in the natural order of things, representational content comes into being – how a psychological state comes to be about a property or object. In an ecumenical spirit, then, I will call "the" property at issue *aboutness* and take the naturalists' project as one of explaining aboutness.

More specifically . . . four broad assumptions . . . that naturalistic theories of aboutness usually espouse are these: that to give a theory of aboutness is to explain a relation between a representational state and some property or object; that this relation is to be explained in the terms of the natural sciences without recourse to semantic predicates; that this psychological relation will be ultimately explained by some relation from the natural sciences; and that the simple perceptual case is in some sense "basic," that it constitutes the most likely starting point for such a theory. . . . [Let me explain this fourth point.] . . . virtually, all naturalistic theories agree [that] . . . if we want to understand how our mental states come to be about the world, we should begin with those occasions on which our connection to the world is most clear-cut – when, say, the subject sits, with his eyes open, staring at the world before him and has visual experiences of the forest beyond him. There is general agreement, in other words, that if there are any simple cases of aboutness to be found, it is static perceptual events that will furnish them. To put this another way, we all recognize that the relations of our ordinary mental states to their objects can be very complex. We have thoughts about things that are presently inaccessible to us by perception; we can ponder the nature of objects that are fictitious or abstract; we can think about things with which we are only vaguely acquainted, or that we are unable to identify, or about which we may have false beliefs. I can think about the Eiffel Tower even if I have never seen it; I can wonder how a certain George Smith is doing even though there are a number of philosophers named "George Smith"; I can ponder the nature of causality or irrational numbers. These, we agree, are the difficult cases, the ones we ought best leave aside, at least until there is a story to tell about the simple case – when the object of perception is physically present to the subject and when he or she has some thoughts or representations that must surely be about that object.[2] It is the simple perceptual case that will, in the end, ground the aboutness of complex representational states. . . . This agreement, I think, is a surprising fact given the diverse explanatory goals of naturalistic theories. If the goals of the authors are so heterogeneous, why do their theories have essentially the same broad theoretic *form*?

Enter the "Traditional view" of the senses, our intuitive understanding of sensory function. To present the traditional view in its strongest guise, think for a moment about the solipsistic plight of the brain. There floats the brain, the "I," encased and protected by the skull. It is the control center of the body, the mover of limbs, the planner of actions, the origin of all the body's thoughts and feelings. This central function notwithstanding, the brain resides in a kind of intracranial isolation. It has only mediated access, through its sensory and motor attachments, to its central concerns, to the body and the external world. But for a series of outgoing and incoming "wires," the brain is alone with its thoughts. It is because of the brain's solipsistic existence, then, that the senses have a clear role to play. They are the brain's window on the world. The senses show the brain, otherwise blind, how things stand, "out there," both in the external world and in its own distal body.

To flesh out this picture, consider one example of external perception, peripheral thermoreception, the system that reacts to surface skin temperature. Think about the sensations that we have of temperature – the bitter icy cold of a January wind, the pleasant coolness of a chilled drink in hand, the warmth of the spring sun, the singeing pain of a hot iron. From this internal evidence, one might guess that our thermal sensations are the products of skin receptors that are finely tuned to surface temperature. Like miniature biological thermometers, the receptors record the temperature of their immediate surround, its ups and downs. If, say, the skin temperature is 89 degrees Fahrenheit, then the receptors must send a signal, some firing pattern, that "means" 89 degrees Fahrenheit; if the skin is cooler, say 83 degrees, the receptors must give an appropriate response (perhaps they fire less rapidly). The receptors, we think, must react with a unique signal, one that correlates with a particular temperature state. Thus the brain gains information about what it is like "outside."

Although we realize that human thermoreception probably does not function exactly as described above (for we all know about the various temperature illusions), the example captures a certain feature of the traditional view: sensory systems must be veridical in some sense of the word. If the senses are the brain's window on the world, then any system worth its salt (and functioning correctly) ought to provide an accurate account of just how things are: the brain must be able to tell, from the signals it receives, how things stand in the world.

First, it would not do to have a sensory system that is fickle or unreliable in its reports. No one would want a thermoreceptive system that suddenly started to send a signal, usually reserved for a skin temperature of 89 degrees Fahrenheit, at arbitrary intervals; nor should the system be such that, when the skin *is* 89 degrees Fahrenheit, the thermoreceptors fail to register that fact, forget to send the signal at all. What good, *qua* informant of the brain, would such a system be? In non-metaphorical terms, this aspect of sensory veridicality is usually expressed as that of *constant correlation*: if a signal is to be informative ("tell the truth"), it must be produced when and only when a particular stimulus (or stimulus set) is present.

Second, if the brain is to have an accurate understanding of how the world lies, then constant correlation is not enough. The relevant *structure* of the external events or properties must also be preserved by the sensory signals: the representational relations among the sensory signals must mirror the relevant relations in the sensed domain. In the case of thermoreception, the brain must be able to discern a one-dimensional relation between temperatures – whether this temperature is greater or less than some other one. So a thermoreceptive system must be, well, thermometer-like: the relations among the temperature states in the world ought to be at least roughly discernible from (if not strictly isomorphic to) the relations among the individual thermoreceptive signals. Constant correlation alone is not very helpful, especially if the question before the brain is whether or not to pull a hand away from the warming fire (that is, is the hand getting too warm now?).[3]

Third, a veridical sensory system is also a *servile* system (or if you prefer, a system that acts as the brain's loyal retainer.) By "servile," I do not mean a system that reacts in only strictly prescribed (law-like) ways to the world's sensory impingements. Much of sensory processing, we now believe, involves active or "top-down" processing: the brain uses stored information, default assumptions or hypothesis generation and testing in order to solve the computational problem at hand ("what is that object?") or to increase its efficiency. Quite obviously, this kind of activity does not challenge our intuition of veridicality: such sensory systems are still trying to represent the world, in the face of limited information, as accurately as possible. What would not jibe with our notions of a proper (veridical) sensory system is one that actively embroidered upon the nature of the impinging sensory stimuli or that simply made things up. What we expect from sensory systems, in other words, is that they are the brain's "ontological drones." In the service of the brain, they toil tirelessly to report the "what, when, and where" of the world's events. They do not interject their own opinions into their reports; they do not slyly skew the information to reflect their own interests or prejudices. Their job is to state the facts. To say that a veridical sensory system must be "servile" then, is to say that it represents the world as accurately as possible, without embroidery or fiction, given the information available. (I take servility to be a slightly different notion than that of reliability as given above, although perhaps it is merely a subtype. By analogy, the problem with the town gossip is not that, for any event in town, he might fail to have something to say about it. What we doubt is the *accuracy* of what he says. One worries that it will incorporate his own likes and dislikes, prejudices and interests.)

On the traditional picture, then, the senses, using a system of signals that captures the structure of a domain of external properties, tell the brain, without exaggeration or omission, "what is where." It is this view of the senses that dovetails with the philosophical problem at hand. The naturalists' project, stated in its most general form, is to explain the psychological relation of aboutness in some way that fits neatly with our scientific picture of the world. On the traditional view of the senses, there is also a relation of aboutness, a relation between any state of a veridical sensory system and its object. Indeed, fulfilling this relation is the *raison d'être* of the senses – that without which the brain would not know how things stood, either in its own body or in the external world beyond it. So what the traditional view of the senses provides, from the natural sciences, is exactly the sort of relation a theory of aboutness requires – hence an obvious starting place for any naturalistic theory of aboutness. Not all intentional states, the naturalist freely admits, are sensory states; nor need the subject have ever had sensory contact with a given object of thought; nor indeed are the intentional objects of such states necessarily physical objects or properties, the sort of entities with which the subject could have sensory contact. Still, the path to choose is clear, the naturalist will contend. Indeed, at first glance, it is the only path visible at all – the only path out from the solipsistic existence of the brain.

3   Narcissistic Sensory Systems

The problem with the traditional theory of sensory processing, I think, is that it is not universally or even generally true. In some cases, sensory mechanisms do behave as one intuitively expects, in accordance with the traditional view. More often than not, however, sensory systems fail to be "veridical" in the sense given above. Indeed, if one had to pick a single predicate to describe all sensory systems, it is that each and every sensory system is, well, "narcissistic."

As a first pass at explaining this metaphor, think of the usual case, the human narcissist, a person whose worldview is informed by exactly one question: "But how does this all relate to ME?" . . .

Needless to say, a narcissist's world picture, being informed by but one question, is strangely askew. In this regard, it is important to realize that the problem with the narcissistic worldview is not just that her range of interests is idiosyncratic, that self-interest directs her attention toward a limited portion of the world, a small part of what other people see. Rather, by asking the narcissistic question, the *form* of the answer is compromised: it always has a self-centered glow. . . . The narcissist cannot see herself and her relation to the world in an objective light, as one life among others. Hence her understanding of the emotions and actions of other people, and of events in the world in general, will necessarily incorporate her own particular interests. For the most part, it is not possible for the narcissist to stand back and remove herself from the picture – and that, of course, is exactly the property which gives the narcissist away. I do not mean to imply by this that the narcissist never gets things right, never sees the world for how it is. Sometimes the question "How does all this relate to me?" is just the right question. "What does that person want from me now?" is the appropriate thing to wonder in a dark alley. Hence, it yields a legitimate – "veridical" – answer. It is nonetheless true that the narcissist's question shapes her worldview at all times, and this is so regardless of whether the question is "appropriate" or not.

Turn again to a simple sensory system, that of thermoreception. As I said above, our first guess about thermoreceptors is that they act like small thermometers, signaling the brain about the location and the temperature state of the skin. In fact, however, the nature of thermoreception is quite different (the following draws largely upon the work of Hensel, 1982). First, the apparently continuous temperature gradient that we feel is not the result of the continuous response of a single thermomechanism. Instead, our sensations are the result of the action of four different types of receptors: two thermoreceptors, "warm spots" and "cold spots," and two pain receptors (nociceptors), that fire only in the extreme conditions of very high or very low temperature. At very high and very low temperatures, we feel only pain, sensations that are qualitatively indistinguishable from one another. In the middle zone, we rely upon one or the other kind of thermoreceptors for our sensations of warmth and cold. (The temperature at which warm spots begin to respond is roughly the same temperature at which cold spots leave off.) Second, there are far more cold receptors than warmth receptors, the exact ratio differing from location to location. So, for example, on the face, the nose has a ratio of 8:1 cold to warm spots, the cheeks and the chin are somewhat more sensitive to warmth with a ratio of 4:1, and the lips (entirely counterintuitively) are sensitive to only cold, with almost no warm spots whatsoever. One result of this variability is that different parts of the body are more sensitive to heat or cold than others. Because conscious sensation is the result of cumulative neural response, the more receptors there are, the more cumulative neural activity. This is a fact that will strike you as immediately plausible if you imagine wading into a cold lake. As a matter of fact, some steps *are* harder to take than others. Another result of this variability is that the "temperature of neutrality" – the "comfort zone" as thermostat makers call it – differs from one area of the body to another. Whether a temperature feels comfortable is as much a function of location on the body as it is of temperature.

What do the individual receptors do? Each kind of receptor, warm spots and cold spots, have both a static and a dynamic function (figure 20.1). The receptor's response to a constant temperature is its static function, represented by a curve that plots the rate at which the neuron fires against the stimulus temperature. For both the warm spots and cold spots this response is nonlinear; the static functions for the two receptors also differ, one from the other. The warm receptor responds over a narrow range of temperatures with a steep rise in firing rate at the high end; then the firing abruptly halts at a maximum temperature. The cold receptor has a less intuitive response pattern. It has a wider window of response, a gentle curve with a maximum response at the midpoint, tapering off thereafter. The static functions of neither the warm spots nor the cold spots are thermometer-like, with a certain set increase in firing rate per degree of temperature change.

Both kinds of thermoreceptors also have a dynamic function, a response to temperature change. When the temperature of a warm spot is increased, a burst of activity occurs; then the firing rate gradually slows and settles into a new higher base rate, a rate determined by the static function. For example, after shoveling a snowy walk, putting your hands under only tepid water initially feels very warm. The dense liquid causes sudden energy transfer, a sudden increase in temperature, and the dynamic functions burst of neural activity. When the temperature of your hands stabilizes, the neural activity also decreases and the water feels as it normally would, cool. What

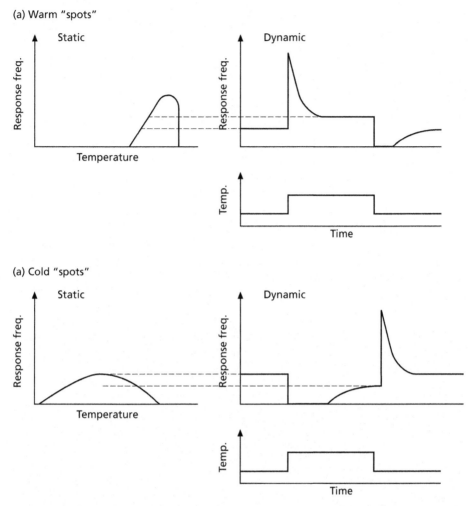

Figure 20.1 The static and dynamic functions of the warm and cold receptors. To illustrate dynamic function, both receptors were subjected to a sudden temperature increase and then a sudden temperature decrease. The warm receptor (a) shows a dynamic response to temperature increase alone; the cold receptor (b) shows a dynamic response to temperature decrease alone. (Adapted from Hensel, 1982.)

is important here is that the size of the initial activity burst is *variable*: it depends upon the starting temperature. So, for example, if you transfer your hand from tepid to warm water, say a 40-degree Fahrenheit change in temperature, there will be a sudden burst of activity in the warm receptors. But if you then transfer your hand from warm to hot water, again a difference of 40 degrees Fahrenheit, the dynamic burst will be much greater. The higher the starting temperature, the larger the initial burst of the warm receptor. Thus, the transition from warm to hot water is felt more keenly than the change from tepid to warm water. Note also that when a warm spot is cooled there is no neural burst at all, only a gradual decrease in firing rate until the new lower base rate (determined by the static function) is reached. These same principles apply to cold spots as well, except in reverse. A *cold* stimulus evokes a dramatic over-compensatory response; a *warm* stimulus causes a gradual decrease in activity; and the *lower* the initial skin temperature, the greater the initial burst of activity in response to a cold stimulus. (These opposing characteristics are, in fact, the defining functional qualities of warm and cold receptors.)

As one can see, the response properties of the thermoreceptors are complex. There is no single thermometer-like receptor or even two "abutting" thermometer-like receptors; the warm spots and cold spots exhibit two different nonlinear (static) functions; the transducers exhibit adaptation and behave in a context-dependent fashion (the dynamic response depends upon the current base-line skin temperature). What is more, our temperature sensations do not result directly from the functional properties of individual receptors but are determined by the characteristics of the neural population as a whole. Where there are more cold receptors, the area seems colder; if a location has few warm spots, heat is hardly noticed at all.

All of this seems somewhat strange on the traditional view of sensory processing, of thermoreception as a system that disinterestedly records temperature facts. Just how inept could this system be? Viewed as narcissistic, however, the system makes perfect sense. What the organism is worried about, in the best of narcissistic traditions, is its own comfort. The system is not asking "what is it like out there?" – a question about the objective temperature states of the body's skin. Rather it is *doing* something – informing the brain about the presence of any relevant thermal events. Relevant, of course, to itself.

In this light, reconsider the old illusion created by placing one hand in cold water, the other in hot and then, and after a few minutes, placing both hands simultaneously in some tepid water. Stupid sensors. They tell you that the tepid water is two different temperatures. But the sensors seem less dull-witted if you think of them as telling you how a stimulus is affecting your skin – that one hand is suddenly cooling while the other is rapidly warming. Skin damage can occur from extremes of temperature (being burnt or frozen) but also from rapid temperature changes alone, even if the changes occur within the temperature range for healthy skin. So the information provided by the dynamic function – whether there has been a change in temperature, in which direction and how rapidly that change is occurring – is crucial to the organism for avoiding skin damage. It allows you to pull away your hand before it is burnt, to remove your hand when the change in temperature will be fatal to the skin.

From this perspective, one can also see why it makes good sense to have a dynamic function that is context-dependent – why the burst rate depends upon both the amount of change and the starting temperature. If the skin temperature is already high, any increase is likely to cross the upper limits of skin tolerance. So the extreme bursts of the dynamic function, produced by stimuli at the upper end of warm spot function, warn that tissue damage is about to occur. (If the temperature-sensitive nociceptors fire, then damage has already been done.) In other words, the system produces strong signals whenever potentially dangerous temperature changes occur. From this perspective, the utility of the dynamic function is clear.

As another example, think again about a cold mountain lake – this time, about putting your head into a cold mountain lake. Most of us have had the experience of going for a hike and then, after a strenuous day, coming upon an inviting-looking lake. First you put your hands in the water to test the temperature; then, after the initial shock is over (after the dynamic burst of the cold spots has subsided), you decide to risk a swim. Note, however, that making your brain very cold is not a good idea. Heat loss through the scalp is one of the many ways that human life comes to an end. So, from an evolutionary perspective, it is not surprising that the scalp has an abundance of cold receptors – that when you dive in, the water feels colder to your head than it did to your hands earlier. Given how dangerous heat loss can be, it makes sense to receive a sharp warning signal whenever the scalp is cooled, before you lose the sense to react rationally. More generally, if one looks at the complexity of the human body and at the large range of behaviors we are capable of performing, it is clear that the importance of thermal stimuli can vary, depending upon where the stimuli occur. You need to keep the top of your head warm; you need to make sure that your nose is not frozen in weather where your cheeks will still be fine; you need to be able to sense both cold and warm objects with your hands. Hence, the evolutionary "solution" of a variable ratio of cold to warm spots. Through a difference in the number of cold and warm spots, the cumulative signals are tailor-made to reflect the relevance of temperature stimuli for a particular place on the body.

Looking back at thermoreceptive function, then, one realizes that this system is not merely inept, a defective indicator of surface temperature. Rather, the system as a whole constitutes one solution to man's various thermal needs – that he be warned when thermal damage is occurring or before it is likely to occur, when temperature changes are likely to have specific consequences, and so on. These are the thermal problems which have been "posed" over time to the evolving organism and which have resulted in the idiosyncratic functions described above. In other words, this system, *qua* system, reflects a constellation of interacting facts about one species – the environmental conditions encountered by *Homo sapiens* (for example, the thermal properties of the environment), the developed physiology of the species (for example, the mortality conditions of the skin, the details of internal and external thermoregulation), and the range of behaviors of which the species is capable.[4]

Let me now make explicit how the above characterization of thermoreception fails to fit the traditional view. Recall from the ways in which, according to the traditional view, sensory systems are veridical: each signal must correlate with some property (or range of properties) in the world; the structure of the relevant relations between the external properties must be preserved in a systematic encoding of those relations, and; third, a sensory system must be servile in that it must be seen to be reconstructing, without fiction or embellishment, the properties, objects, and events of the world external to the brain. In this case, the property of interest is, obviously enough, skin temperature at a specific part of the body. So the question that one would expect the thermoreceptive system to answer is "what is my skin temperature at x?" But this is not what it does.

First, particular thermal sensations do not necessarily correlate with any particular temperature or any particular temperature change. Because thermal sensations are a function of the firing rates of a neural population and because the absolute number and ratio of the two different receptors differ from one part of the body to another, exactly the same skin temperature can give rise to a variety of sensations. This is one reason. For another, the elliptical static response properties of cold receptors ensures ambiguity in its signals. As the skin becomes colder, a cold receptor fires more and more rapidly until the receptor's maximal response level is reached; then, as the temperature continues to drop, the firing rate starts to decrease. So a single cold receptor will fire in exactly the same way for two different stimuli, a temperature at the low end of its response range and a temperature at the high end. Necessarily, its signals (alone) are ambiguous. Nor do thermal sensations reflect temperature change. Here, it is the context dependency of the dynamic function of both warm spots and cold spots that presents the problem. The felt change in temperature for a specific temperature change will depend upon the starting temperature of the skin. If the temperature of a warm spot is increased at the bottom of its response range, the dynamic burst will be small; if it is warmed at the top of its response range, the burst will be very large. So neither absolute temperature nor temperature change are recorded by thermal sensations.

Second, for almost the same reasons, thermal sensations as a whole do not reflect the structure of thermal stimuli, whether stimulus T1 is greater than, less than, or equal to stimulus T2. The water, as we wade into it, initially lessens the skin temperature of each body part about equally; but it certainly does not feel that way. Some parts feel much colder. Similarly, cold receptors, with their elliptical static response function, represent disparate pairs of temperature stimuli as being equal – very low ones and relatively high (tepid) ones. Even warm receptors, with their nonlinear static responses (which abruptly stop at a certain maximal rate), do not quite get the picture right. Lower temperatures do elicit lower firing rates but the differences in temperature between the warmer and hotter stimuli are not uniformly recorded. Linear temperature increases are encoded as nonlinear changes.

Third, the thermoreceptive system embroiders its account of the temperature states of the world. Unlike many other sensory systems, it is probably unfair to say that this particular system actually manufactures fictions in the course of its ordinary function, but it does appear prone to chronic exaggeration. At the lower and upper limits of response for the cold spots and warm spots, respectively, a small temperature change elicits a hysterical response. Moreover, given the differing ratios of cold spots to warm spots, no body part simply reports its surface temperature. Each body part exaggerates its own state in accordance with its own interests and sensitivities. So the thermoreceptive system is hardly the ontological drone that the traditional view had imagined, the tireless reporter of the "what, where, and when" of surface temperature.

[More generally] in imagining the function of any sensory system, we slide very easily from the question "to what are the receptors responding?" – a question about the nature of the proximal stimulus which will evoke a receptor response (for example, mechanical deformation or light in a certain range of wavelengths) – to the question "what do the signals of the system detect?" In the grip of the traditional view, we treat these as one and the same question. But despite the entirely rule-governed nature of sensory systems, the question "what is the system detecting?" may not be apt. Rather, . . . [w]hen we examine a sensory system, . . . we are looking at an evolved solution to a specific informational problem. Perhaps it is a system that is able to increase or decrease the sensitivity of its receptors in response to present needs (without recording the change in sensitivity), that applies filters for the creation of specific features, that adjusts an outgoing signal in order to ensure certain characteristics in the signal return – or that encodes the facts about skin temperature in some way designed to aid creature comfort. The relevant question to ask of such a system is "what is it doing?" This is a question to which the answer might be "it is measuring, with variable discrimination, the animal's tilt away from the vertical in order to maintain its upright posture," "it is providing a visual signal that can be easily processed for movement information," "it is monitoring the stretch of the flexor muscle in order to adjust the length of the tensor by an equal amount," or "it is indicating an edible insect"– answers that may or may not make reference to processes of veridical perception.

4   Defending the Traditional View: Three Objections and Replies

In the face of the above kind of neurophysiological example (for human thermore-ception is merely one among many, chosen for its familiarity), there are a number of responses one might give in order to vindicate the traditional view. I shall give three of the most common replies. The first response is, in effect (but not in intention) an a priori defense of the traditional view; the second response is a straightforward denial of the thesis that sensory systems are, at bottom, narcissistic; the third response, the most promising of the three, grants that sensory systems are narcissistic – but it denies any conflict with the traditional view.

The a priori defense

Given the above example of thermoreception, one might take away quite a different lesson, not a lesson about sensory processing *per se*, but a metaphysical view about the nature of *properties*. Daniel Dennett, for example, in talking about our conscious perception of secondary properties, takes the clearly self-interested nature of sensory systems to show that their normal function can *define* properties in the world. It does not matter that our sensations of, say, redness fail to correlate with any easily delineable set of surface reflectance properties, a single property that is sanctioned by the scientific image. There is redness in the world because we have a disposition to respond in a "discriminating" fashion to "red" stimuli.[5] That is, if a sensory system responds in one and the same way to a rather diverse set of properties and conditions, then that disorderly set simply *is* a property of the world. I am not sure that Dennett himself would wish to frame the view in this general form but let us assume this strong view for the sake of argument.

Setting aside the merits of this view *qua* metaphysical thesis about properties, note that as an answer to the question "are sensory systems veridical?" it is empty. If properties are defined by the ordinary causes of sensory signals, then, by definition, all sensory systems are veridical: the system "captures" the structure of the domain of external properties just because it *defines* that domain and, of course, any signal reliably records, without exaggeration or omission, whatever stimuli ordinarily cause it. It is only against a firm prior ontology of properties, one that is specified and justified independently of actual sensory response, that questions about veridicality have any bite at all. Thus, on this reply, what seemed like a prototypically empirical question – "are the senses veridical?" – is recast as a question of definition, the answer to which has the status of an analytic truth. This alone, of course, is not a reason to dismiss the response – perhaps the traditional view ought to be granted the status of analytic truth – but this seems to me unlikely. (Below, in answering the third objection, a more decisive objection to this response will emerge.)

The appeal to signal information

Whether or not a sensory state appears to correlate with a particular external property, this response begins, sensory states nonetheless *carry information* about their causes. Take the case of thermal sensations. Although any given thermal sensation might be caused by a large variety of temperature states, a given thermal sensation still carries information about its cause. It just does not wear this information on its sleeve. After all, thermoreceptors act in orderly ways – the response functions of both warm spots and cold spots are shown in figure 20.1 and there is some function that sums the responses of individual receptors. So starting at the beginning of the process, with the initial state of the thermoreceptive system (the current static response rate of the receptors), plus the neural population characteristics (the ratio of warm spots to cold spots and their numbers), and stimulus temperature of the skin, one could predict the sensations of the subject. Conversely, starting with the resultant thermal sensation (or population response), plus the initial state of the system and the neural population characteristics, one could compute the value of the stimulus temperature. In other words, it is possible, using the appropriate calculations, to deduce skin temperature from the thermal sensation under standard conditions – and this is just what it means to say that a signal (thermal sensation) carries information about a source (skin temperature). Hence, in this sense, an informational sense, thermal sensations do reliably indicate temperature states.

There is certainly some truth to this response. Given the causal regularities that govern sensory systems (for there is nothing supernatural about them), the sensations or signals they give rise to will often carry information about their causes. That is, whenever there is a computable function that describes the input–output relation, the response of a sensory system will carry information about its causes. In the case of thermal sensations, sometimes our sensations carry information about skin temperature and temperature change, sometimes not. This is because the static response of the cold receptors does not define a computable function, hence there is no algorithm that could determine, given the population response of the cold spots (a sensation of cold), the stimulus temperature.[6] Indeed, a bell-shaped response curve for sensory receptors is more the norm than the exception, but let us put aside this fact for the moment. The question that concerns us here is whether, given that information about the stimulus is often carried in the sensory signal, this will be of any practical use in constructing a theory of aboutness.

The problem here is that it makes little sense to identify the contents of sensory states with whatever information they carry. Return to the example of wading into the lake. There you stand in water up to your thighs, debating the relative advantages and disadvantages of total submersion. (Would not a bit of splashing and swishing be just as effective, you wonder?) Eventually you wade in, gasping and sputtering at various intervals. Now, on the informational view, all this silliness is for naught. The dynamic bursts of the various cold-spot populations, no matter where they may lie, all signal exactly the same thing – a single temperature for the water or, if you like, a single temperature change. This is the information carried objectively by each population of cold spots – information that you already had when you waded in up to your thighs. Of course, *you* may behave a little oddly, as you immerse yourself, but there is nothing odd about the behavior of your thermoreceptive system. It reliably indicates a single change in skin temperature, no matter what parts are presently getting wet.

This, of course, is a parody of the information response, but it makes a serious point. Information that is carried by, but not encoded in, a signal is information that is available only *in theory*. To say that the information is *present* is to say only that there exists a computable function which, if used, would yield the correct result – in this case, the actual temperature change. It is present, as it were, from the point of view of the universe. But no creature has ever acted upon information that is available only in principle. Thus to posit a theory of aboutness based upon signal information alone is to a posit a theory that, by hypothesis, makes no connection with an organism's action, behavior, or thought. I take it that this is not what the objector had in mind.

Rather, when an objector points out that even a narcissistic response contains information about its causes, the hope is that somewhere down the line, this information is *extracted*. Despite our conscious narcissistic sensations or the narcissistic responses of sensory systems, somewhere higher up in each sensory system the appropriate calculations are made and the true state of the world is inferred. (Thus, for example, in the case of human color vision, the ambiguity inherent in the bell-shaped response curve of each cone – ambiguity between intensity and frequency information – is partially resolved by the overlapping range of responses of the three cone types. Through comparison, ambiguity is resolved.) How else could a sensory system be of use? This suggestion, however, amounts to little more than an expression of one's faith in the traditional view. Empirically, there is little reason to think that all sensory systems carry within them the means to "decode" their own responses.

Once again, in the case of thermoreception, there is no anatomical or physiological evidence to suggest that the thermoreceptive system has the additional information needed to compute stimulus temperature – that it has knowledge of the number of receptors at each location, or the nature of the static and dynamic functions of the receptors, or that it keeps track of the system's initial state. Nor is there any further psychological or behavioral evidence that our thermoreceptive systems have this information or make these calculations. Given a thermal sensation alone, we have no capacity to consciously "see through" the narcissistic thermal sensation to the objective skin temperature (or temperature change) nor, behaviorally, do we act as if (unbeknownst to us) our brains make these kinds of calculations. This information is simply not available to us through sensory means. (Of course, there are many other ways that one might use thermal sensations to infer temperature. After reading Dr Spock, you will know what temperature of milk correlates with a certain warm feeling on your wrist. Knowing this, the sensation can be used as an indicator of a particular milk temperature. But to make this inference requires exactly those cognitive capacities that the naturalistic theory hopes to explain – intentional thoughts about Dr Spock, baby books, wrists, sensations *qua* sensations, and so on. The thermoreceptive system alone does not provide this information.) The point, here, is a general one: when a sensory system uses a narcissistic strategy to encode information, there need not be any counteracting system which has the task of decoding the output state. If the very point of a narcissistic encoding is to match the incoming information to an organism's behavioral needs, then for the most part, there is no reason to re-encode the sensory information into a veridical format.

The appeal to detector cells and biologically salient properties

Of the three objections, the third is the most promising as well as the most obvious: granted that sensory systems utilize narcissistic encodings, why not say that the function of a sensory system is to detect *narcissistic properties*, properties that are defined relative to an organism's interests? The shocking cold of the lake that you feel on your scalp, for example, is a property, one that *that* sensation (population response) detects – namely the property of being-too-cold-for-my-head. The same will hold for the properties detected by similarly narcissistic mechanisms. To put this another way, when the question "what temperature states do thermal sensations indicate?" was asked, the answer was restricted to the temperature readings of some scientific scale, degrees Fahrenheit or Celsius or Kelvin. On this division of the world, the answer was "no": one thermal sensation does not indicate any single skin temperature (or temperature change). But if we look to the neuroethological or neurophysiological literature on sensory function, we find descriptions that make use of a variety of biologically salient properties, properties that are not described by the predicates of physics and chemistry. We find, that is, descriptions of "detectors" in simple organisms, some of which respond selectively to "legitimate" properties such as magnetic north, being a complex sugar or a certain amino acid, or having a certain wavelength, others of which respond to "messy" properties such as vertical symmetry, small flying insects, movement in the left of the visual field, or being a poisonous substance. Moreover, referring to these messy properties is essential to characterizing the function of these detectors; they are an ineliminable part of the neurobiological description. Thus if we recognize "narcissistic" properties, along with "legitimate" and "messy" properties, as biologically salient, we can recast sensory systems in a way that conforms to the traditional view – as detectors or reliable indicators of external properties.

What makes this a particularly good objection, of course, is that much of it is right. I take it as incontrovertible that there are neurons that act as detectors, and that these detectors are tuned to properties such as predator and prey. Biologically salient properties are exactly those properties that neuroethology and neurophysiology appeal to in describing sensory function and those without which sensory function *qua* function could not be characterized. Moreover, prima facie, there is no reason to disbar narcissistic properties from the club, if neurobiological description demands them. Admitting all of this, however, does not give the critic what he needs.

Notice, first, just how strong the suggestion is *qua* neurobiological thesis. The claim is that *each and every sensory system functions to detect properties*, be they narcissistic properties (defined relative the organism's needs), biologically salient "messy" properties (for example, the property of vertical symmetry) or "legitimate" properties (those recognized by the other physical sciences, say, the property of containing NaCl). Call this the *detection thesis*. This is an extremely strong universal claim about sensory function and hence about the form of all our future biological explanations of the senses. Moreover, it is a universal claim made in advance of the lion's share of empirical research on, and biological theorizing about, a vast topic – a claim, the only present empirical basis for which could be a few notorious examples from the comparative neurobiological literature, for example the famous fly detectors of the frog. It is a testimony to the strong intuitive pull of the traditional view, I think, that the prematurity and all-encompassing nature of the detection thesis seems to have escaped most of its backers.

Note that even if one were to take all of the present literature on sensory processing and think up narcissistic or messy properties for those systems to detect, this would not provide any evidence for the universal claim. There are, after all, many devices, natural and human-made, for which one could claim a rough function from input to output states, for example, for vacuum cleaners and stereo amplifiers, as well as lungs, livers, and intestines. With a bit of imagination and a good sense of humor, one could think up some "messy" or "narcissistic" properties for these devices to detect. What the objector is claiming, however, is not merely that we could conjure up properties for each system to detect. He is making the stronger claim that neurobiologists will usefully describe the systems as such – that characterizing sensory systems as detecting properties will *always* provide us with some further insight into how or why the mechanisms function as they do. In other words, the detection thesis is committed to the view that our best biological explanations will characterize sensory systems as detecting. As I said above, this is a very broad empirical claim, especially when one realizes the infant state of neurobiological research on the senses. On what grounds might someone believe it to be true?

Above, in giving the example of the human thermoreceptive system, I argued that the neurobiological research does not support the traditional view nor, for that reason, the detection thesis – that we must ask the broader question "what is the system doing?" not merely "what is the system detecting?" In the literature of that nascent science, one finds, along with references to the now-famous detector cells, a variety of other descriptions. There, the neurobiologist begins his research by trying to pinpoint the informational problem(s) that the system must solve (a good alternative to poking blindly about) and then looks for mechanisms that solve them. Here, because the problems and answers are often framed in standard engineering terms, function is explained using standard engineering concepts. Thus one reads about mechanisms that "turn up the gain," "act as a resistor," "apply a cut-off filter," "use a step-function," "shift a spectral sensitivity," and so on. Surely, the objector says (or must be saying), we could profitably recast this "engineering talk" in terms of detection and indeed surely we will, once the full explanations are in hand. I think not. Let me give one short example of a sensory process for which this sort of "re-casting," far from illuminating function, would positively obscure it.

Consider our proprioceptive system, the sense through which we know where our limbs are in space (without looking, that is).[7] In this system, the receptors are not external transducers, ones affected by events beyond or on the skin.[8] Rather, our sense of proprioception relies upon muscle spindles, internal receptors that are stimulated by the mechanical stretching of the muscles at our joints. In this system, the "engineering" dilemma that arises is that the range of activity to be encoded exceeds the ability of the neurons to respond. The elbow joint, for example, has a wide range of angular extension; as you extend your arm from its fully bent to the straight position, the flexor and tensor muscles are shortened and lengthened considerably. Sensory neurons, however, have a fairly restricted range of response – there is a limit on just how quickly a neuron can fire (i.e. four times a second). One has, then, what looks like a dilemma: if the muscle spindles are set to respond over the full range of joint movement they will give only coarse-grained information about muscle stretch; but if spindles are adjusted to provide fine-grained information about muscle extension, they will respond over only a limited range of joint movement. What to do? The neural solution to this dilemma is a certain kind of feedback process: central control alters the sensitivity of the receptors, the response of the muscle spindles to change in muscle length. As the limb extends, and as the muscle becomes longer and longer, central control "lowers the gain" on the spindle response. This allows the spindle to maintain a continuing, fine-grained response throughout the entire range of muscle movement – although it does so at the "expense" of determining muscle length or extension.

This kind of problem, encountered by the proprioceptive system, is a very common one in sensory processing. Quite often the range of stimulation over which the system must respond far exceeds the ability of a sensory neuron to signal those events. (For example, the visual system of the cat, it is thought, responds over an illumination range of about 16 log units!) Sometimes the solution to the problem is to build a variety of receptors, in order to cover the full range of stimuli; just as often, the solution is to use a mechanism that adjusts sensitivity. What is important to realize, here, is that there need not be any further device that records the "position" of the gain mechanism. If the purpose of a stretch receptor is to keep the flexor and tensor muscles balanced (as one stretches, the other contracts), the absolute muscle length is not what is important (at least not for this task). One could say, of course, that the stretch receptors detect muscle length if one wanted to give a rough characterization of what they do (after all, they do react to mechanical stimulation – the muscle stretching). But this way of talking quickly becomes misleading, say in the face of motor dysfunction, if one wants to explain why a person has tensor rigidity (or spasticity). For a careful characterization, one that will explain both function and misfunction, talk of detection only confuses the matter. Thus while the proximal stimulus is the stretching of the muscle, the function of the muscle spindles is characterized more broadly, in terms of what the system is doing.

As I said above, the third objection depends upon a very strong and broad empirical conjecture about sensory function. Quite obviously, given that neurobiology is just getting off the ground and that, as a result, there are almost no complete accounts of any single sensory system, this question, about the correct principle of sensory processing, remains open. Still, the objector's claim is very strong, and (my guess is) not well grounded in any neurobiological evidence. Is there any other science, then, apart from neurobiology, that might prove the detection thesis?

A different defense of the detection thesis is based upon general evolutionary considerations. After all, the defense goes, organisms simply could not have survived if their senses did not function as reliable detectors of salient properties. If animals are going to find prey, avoid predators and attract mates, then their sensory systems must detect these properties of the world. As Patricia Churchland once said, in the course of defending an early theory of representational content called *correlational content*: "I take it as obvious that if there were no systematic relations between the external world and states of the brain, the animals could not survive or if they did it would be a miracle in the strict sense of the word. An owl would not know where the mouse is and he would not intercept it as it runs" (1986b, p. 260). Stating this view in an even stronger form, Fred Dretske says:

> Without such internal indicators, an organism has no way to negotiate its way through its environment, no way to avoid predators, find food, locate mates and do the things

it has to do to survive and propagate. *This, indeed, is what perception is all about* (Akins's italics). An animal's senses . . . are merely the diverse ways nature has devised for making what happens inside an animal depend, in some indicator-relevant way, on what happens outside. If the firing of a particular neuron in a female cricket's brain did not indicate the distinctive chirp of a conspecific male, there would be nothing to guide the female in its efforts to find a mate.

(1988, p. 62)

Moreover, because evolutionary theory explains just those factors that affect a species' survival or demise, evolutionary biology will make reference to exactly those environmental properties to which the creature reacts (or fails to react). To illustrate, a typical explanation in evolutionary biology might go something like this: "at the beginning of the egg-laying season, the parasites chose with care their egg-laying sites, rejecting less than optimal places; when the days grew shorter, signaling the end of the season, the parasites became far less discriminating, choosing exactly those kinds of sites which were rejected earlier." In the standard literature of behavioral ecology and evolutionary biology, even parasites are seen to recognize "optimal and suboptimal egg-laying sites," to detect the "shortening of the days" and "the end of the breeding season." What allows us to advance the detection thesis, then, is the simple fact that, in order to survive, all animals must react to salient properties of the environment – and evolutionary biologists cannot explain the factors affecting survival without characterizing the organism's reactions as such.

Again, there is much truth to the view. Of course it is true that, as a whole, an animal's behavior must be directed toward certain salient objects or properties of its environment. Objects (and their properties) are important to the survival of all creatures. But from this fact alone, one cannot infer that the system of sensory encoding, used to produce that behavior, uses a veridical encoding. That is, it does not follow from the fact that the owl's behavior is *directed toward* the mouse or that the brain states bear systematic relations to stimuli, that there are any states of the owl's sensory system that are *about* or serve to detect that property of being a mouse. What is required for survival of the owl, for example, is that it finds its way to the mouse in a timely fashion and that, once it gets there, the owl can grab the mouse. To do so, the owl's visual system need not encode external space veridically, with anything like, say, a cartographer's topographic map, with the spatial relations of the world exactly mirrored by the spatial relations of the sensory map. Nor need the owl intercept the mouse by literally computing its trajectory based upon its present motion, nor need it use the deliverances of a mouse detector as part of the calculation. It is an open question what kinds of systems are actually used. In the case of the cricket, detector cells, for the chirp of the male conspecific, are used. But this is only one neural solution to one behavioral problem. Nothing about the directedness of an organism's *behavior* yields a firm conclusion about the directedness of the *internal states* of its sensory system.

Setting aside the question of whether there is any empirical evidence for the detection thesis, let me turn to a second response to it. Even if we granted the objection – even if we admitted that narcissistic properties will invariably figure in our neurobiological descriptions – this would not give the naturalist what he wants, namely the beginnings of a theory of aboutness. The suggestion made is that states of sensory systems are about narcissistic properties, properties defined by relation to the subject's interests – *relational properties*. However, what the naturalist wishes to explain, in the end, are representations which are about *objective properties* and events of the world. Whatever the neurophysiological facts about the thermoreceptive system, that is, we do not merely *feel* temperature sensations *qua* sensations in our skin. We have thoughts about the independent temperatures of objects. I put my hand out, grasp the coffee cup, and feel its warmth. I see it *as* a coffee cup and feel the cup *as* having a particular property, warmth. Similarly, when I put my hands into the dishwater, I feel *the water* as being warm, as having an objective property that is independent of my perceptions. Consider again the illusion produced by cooling one hand and warming the other, then placing both hands in a single bucket of tepid water. Here the illusion, as you place both hands in the tepid water, is that *the water* is both hot and cold. That is what makes the exercise so surprising. There would be no illusion of a contradiction, however, if one assigned content as the objector suggests. Let one sensation indicate the property "precipitous drop in right-hand temperature" and the other sensation indicate "precipitous increase in left-hand temperature" and there is no explicit contradiction to be found. So the problem for the naturalist lies in the rather large gap between what our ordinary perceptions are about and the representational contents which the detection of narcissistic properties would assign, namely mental states that are about "what is good for me," subject-dependent properties, as it were. To put this another way, the naturalist's hope was to find a relation between sensory states and external properties, a relation that would ground (in some undefined way) a theory of aboutness. But in order to save the traditional view, the objector introduces narcissistic properties and thereby gives up the link to the objective properties of the world. Prima facie, this is not a promising starting point for the naturalist theory. I will have more to say about this later.

This same argument against the third objection, the detection thesis, also works against the first objection, the a priori defense. On the a priori view, recall, sensory mechanisms define properties of the world – large disjunctive sets of *objective properties*. Thus, on the a priori view, an individual temperature sensation defines a property, the disjunctive set of exactly those thermal stimuli which would evoke a particular sensation. The "hot" feeling will define one "property," a disjunctive set of thermal properties; the "cold" feeling will define another different disjunctive set. Does not this way of describing the properties "detected," as disjunctive sets of non-relational properties, circumvent the problem? No. Even given this re-description of the properties, the a priori view fails to capture the content of ordinary thermal perceptions. When I put both hands into tepid water, the properties detected by the left hand and the right hand comprise two different disjunctive sets. *Ex hypothesi*, one temperature of water can bring about these two different properties in the course of normal function. Hence, there is no contradiction between what my right hand says and what my left hand says (as there would be if there were no overlap between the disjuncts of both sets). In retrospect, this is exactly the problem one would expect given the form of the a priori defense: it simply takes a set of relational, narcissistic properties and redefines them in nonrelational terms. Defining the very same "properties" by means of a different device, however, does not alter the very nature of those properties. Thus, even though the a priori view picks out sets of objective properties with which to match our temperature perceptions, it, too, picks out the *wrong* sets.

5 Philosophical Implications

So far, nothing that has been said casts any doubt upon our commonsense view about our perceptions *qua* intentional representations of the world. We have thoughts about the temperature of objects. We have thoughts about the Eiffel tower (real or fictive), about George Smith (the very one I met at Tufts), and the fragrance of gardenias (however vague or unreliable). However our sensory systems work, we are creatures who represent a world of objects, properties, and events. Granting all that has been said above, there is no reason, here, to question the existence of mental representations writ large. The philosophical problem of aboutness still stands.

On the other hand, if the above arguments are to be believed, sensory systems do not seem to help us understand this fact. Let us take stock of the situation. First, sensory systems exemplify (usually quite elegant) solutions to specific processing problems. Sometimes reliable correlations are used (as in the case of the frog's proverbial fly detector) and sometimes they are not (as in the case of the thermoreceptors). The wiring fits the problem. Second, these kinds of nonrepresentational systems are characteristic of not only simple creatures, for example the ones that swim around. These are the kinds of systems that connect *us* to the world. At our sensory and motor peripheries, our systems have narcissistic properties. Third, much of a simple organism's behavioral repertoire can be accounted for without the use of anything other than such narcissistic systems, without anything that looks like an internal representation of objects and properties. If one wants to build a simple organism that swims around and eats anchovies, it is not clear that *any* representational states must be used at all. For many classes of behavior, some of which are quite complex, "making it work" is all that counts. This is how sensory function looks from the point of view of the neurophysiologist.[9]

What lessons about representation can be wrought from these two viewpoints? First, insofar as the neurophysiologist's view conflicts with our firm intuitions about how a sensory system should work, we ought to be suspicious of our intuitions. These are, after all, intuitions about a quintessentially empirical question – "how do sensory systems work?" – a question to which a scientific answer is only now forming. Perhaps we have simply painted sensory systems in our own first-person image. Because our own conscious perceptions are genuinely representational (we have thoughts about the world, its objects, properties, and events), we have expected that both the sensory systems of simple creatures and the parts of our own complex systems will have this property as well, or at least some rough facsimile of it. We have assumed, without warrant, that all sensory systems (or parts thereof) are, in some sense or other, about the world as well.

More importantly, however, the distance between the neurophysiologist's view of sensory systems and our first-person perspective on conscious perception should raise a genuine puzzle (or rather, yet another genuine puzzle) about representation. We ought to wonder why and how we came to represent the world as we do, given the way in which our neural systems anchor us to the world, given how our sensory and motor systems function. The initial assumption of naturalistic theories has been that even simple perceptual systems must have states that have, at least in a limited sense, aboutness. If an organism did not know, at least roughly, what was out there, when, and where, how could it possibly survive? This is why even simple systems must represent the objects and properties of the world. But if the function of sensory systems is not to inform the brain (the ganglia?) about properties of the world *per se*, then genuinely representational systems are not merely the next bells and whistles added on to an established evolutionary trend. (First there was representation, and then there was *more*.) They are not mere refinements of or embellishments on an ongoing representational strategy. Our ability to represent the external world as containing objects, properties, and events constitutes a distinct – different – capacity of an organism. What exactly is this capacity, and for what reasons did it come about? If an organism can get about, feed itself, and reproduce all without representational mechanisms, what neural/environmental "problem" was answered by the evolution of intentional states? How does this representational capacity work and more specifically, how do our conscious intentional perceptions seem to form an apparently seamless union with our narcissistic sensory systems?

Let me try to set out this puzzle in more detail. On this view of things, aboutness constitutes something like an ontological "capacity," an ability to impose stability, order, and uniformity upon a conception of the world (and sometimes the world itself) on the basis of stimuli that do not themselves exhibit these properties. Walk around your kitchen and imagine the images that thereby are reflected on the back of your retinae (pretend, that is, that you are Bishop Berkeley making a cup of tea). The image of your refrigerator looms larger and then smaller as you walk towards it then back away; despite uniform paint, its three sides have different spectral reflectances (given the location of the light source and each side's proximity to other colored surfaces); as you turn your head, the image shape changes dramatically; when you turn around and walk out the kitchen door, the image disappears completely. Add to this the fact that the images on both retinae are somewhat different, for each eye views the world from a slightly different vantage point. And on and on. That none of these changes in the stimuli matter to your perception of the world is a remarkable fact. You see the fridge as having a constant shape, with a uniform surface color, standing in a single place in the world. Despite the differences between this fridge and the hundreds of others you have seen, you regard it as one instance of a type, as a refrigerator; despite the dents and scratches it has gathered over the past years, you know it is the very same refrigerator that you purchased six years ago, the same one that sat in a warehouse several miles away. That you come to glean this stable ontology, of particulars that instantiate types, of particulars that occupy stable places in the world, is an astounding capacity. It requires that you (your brain) find stability despite stimulus change and uniformity, despite real difference, despite the dissimilarities between objects of the same type and the changes in objects over time. It does not matter whether, in the world, these stabilities and uniformities actually exist (or whether, in some cases, we merely impose stability via ontological categories). To conceive of types and tokens, places and objects as existing at all, given our sensory access to the world, is a fantastically difficult task. Call this the *ontological project*.

On the other hand, there is that small task, assigned to sensory systems, of getting the job done, of directing motor behavior. For the most part, this is not an ontological project, a task for which it is an advantage to see the world according to stable categories or as containing re-identifiable places and particulars. For one, because sensory systems encode information symbiotically with motor needs, the similarities and differences, uniformities and discontinuities that the senses "record" need not exist in the world. What matters is whether the system of encoding is effective, whether the encoded similarities and differences are useful. . . .

There is another reason, more difficult to explain, why sensory systems might be considered "pre-ontological," or unconcerned with a delineated world. Take the philosopher's favorite reptile, the frog. When the fly-detector cells react, they indicate (in normal circumstances) the presence of a fly, but not the presence of any particular fly (call him "Herbert"). Just a fly, whichever one happens to present itself. For a frog, at least, it does not matter whether this fly is Herbert or Harold; it does not need to keep track of or identify Herbert *as a particular*, whatever that could mean ("You've flown your last flight, Herbert!"). Nor, for the purposes of eating the fly, does it matter that Herbert is in any particular place. The coordinates of the moving spot on the retina encode the fly's position relative to the frog and this is exactly the information which the tongue-swiper needs – the direction of swipe for the tongue relative to the frog. Nor, for the purposes of consuming the fly, does the frog need to know where, in the world, Herbert is located ("The last time I saw Herbert, he was sitting in the Savoy. . . ."). All of these are familiar points from the philosophical literature. The same lesson, however, can be applied to our own case. When push comes to shove, so to speak, much of the information needed to make a movement is of the very same sort. Although you see the fly as a particular (and may even call him "Herbert"), reaching out to grab the fly will require information, not about Herbert *qua* particular fly (here in the lounge of the Savoy), but about his velocity relative to your arm motion, his position relative to your hand, and so on.[10] *That* information, without which you cannot catch the fly, is no different in kind than the information required by the frog. It is extremely precise information about a particular but not *qua* a particular. Call the narcissistic encoding of this type of information the *sensory-motor project*.

It is the gap between the needs of the sensory-motor project and the demands of the ontological project, I want to claim, that calls out for explanation. What were the behavioral/environmental conditions in virtue of which the development of an ontology – and hence things and properties to have thoughts about – came to have survival value to our predecessors?[11] Exactly what kinds of abilities or capacities are required in order to represent a stable ontology, of types and tokens, objects and places? And how exactly does the information provided by our sensory systems co-exist with, form a whole with, the ontology imposed by a representational system? (Given that we *feel* thermal sensations as a function of a receptor population response, how does that fit with our conception of temperature as an objective property of objects?) These are the questions about representational directedness that immediately arise.

To grant that there is this sort of puzzle – even to ask the above questions about directedness without answering them – is to admit, *pace* Wilfrid Sellars, that in an important sense we do not really know what "aboutness" is. Certainly, at the outset, a vague realism about the directedness of mental/neural events is adopted: representations are "tied" to objects and properties and hence (there being no good reason to suppose otherwise) bear some kind of relation to them. But if we do not know exactly what it means to regard a particular as a particular, to see this thing as being of a certain type, this place as the same place, and so on – hence what kinds of capacities or abilities are involved in having representations that are about those things – then we do not know, in any substantive sense, in what that relationship consists. We only trust that it *is*.

To admit the above, I think, is to take on a different view about what kind of intertheoretic explanations are appropriate to intentional phenomena. On the naturalistic scheme, recall, the idea is to explain the relation of aboutness, in the first instance, by appeal to some other relation postulated by the natural sciences. Exactly how this natural relation figures in the larger theoretical picture varies from theory to theory, given the diversity of the authors' explanatory goals. Sometimes the natural relation is thought to confer a sort of "proto-intentionality" upon specific sensory states, states which, are hypothesized to form a "ground" for the genuinely intentional states of folk theory (see, for example, Dretske's *Knowledge and the Flow of Information*, 1981); in other theories, the natural relation is simply identified with the "aboutness" relation of, say, a computational state (for example, the Churchlands' notion of "calibrational content," developed in Paul Churchland's *Scientific Realism and the Plasticity of Mind*, 1979). Once one realizes that the demands of the sensory-motor project are, for the most part, distinct from the demands of the ontological project, however, one realizes that sensory systems need not be veridical reporters as portrayed by the traditional view. Hence sensory states need not bear the expected natural relation to external properties – the natural relation that was to mirror or ground the aboutness of mental or computational states. To ask the above questions is to admit, then, that the directedness of mental events constitutes a distinct set of representational abilities and capacities, which at this point we can only roughly define. If this is so, then there is no longer any reason to think that a relational property at one level of theoretic explanation (the psychological/computational level) will have a clear mapping on to, or grounding in, any relational property at another (the neurophysiological or biological). The relation of aboutness need not be explained primarily in terms of some other natural relation at all.

Most importantly, questions of the above kind serve to shift the focus of theoretic attention away from static perceptual states. Recall the naturalists' hope that by closely scrutinizing simple sensory events and their relations to external causes, we would gain a toehold on the phenomenon of directedness – by understanding the most straightforward cases of mental directedness, we would have a route into more complex intentional phenomena. Because all of the senses, on the traditional view, are veridical, it will be the static perceptual case *qua* correlational state that will provide the essential key to aboutness. Once the assumptions of the traditional view are set aside, however, there is no assumption that sensory states will, in general, be about external properties of the world. More importantly, one can no longer assume that for those fully intentional perceptual states which indeed are about the world, it is the static perceptual case – where we sit with our eyes open staring at the object of perception – that will provide us with insight into that relation. Trace out the causal path between the object of perception, the stimulation of the receptors, and whatever neural events that thereafter eventuate, and this alone will not *explain*, in the required sense, how genuine representation arises. The explanation of any particular perception can take place only against a background theory of the representational capacities at work. So, it is a theoretic understanding of those capacities *qua* capacities that will give the explanation of aboutness bite – and these are not capacities that we will understand simply by scrutinizing with care the static perceptual case. (This is not to say that the study of sensory systems is of no utility to the intentionality theorist – on the contrary – but that looking for correlations between sensory states and external properties is not what the study of perception will be all about.)

Acknowledgments

This paper began as the first chapter of my doctoral dissertation, "On Piranhas, Narcissism and Mental Representation: An Essay on Intentionality and Naturalism", Ann Arbor, Michigan, 1989, and has gone through a number of incarnations as both papers and talks. The comments of many people have been very helpful – but I owe special thanks to Joseph Malpeli, Daniel C. Dennett, Kim Sterelny, Jaegwon Kim, Martin Hahn, the members of the Embedded Computation Project at Xerox PARC and the members of the Spatial Representation Project at King's College, Cambridge. My largest debt, however, is to Brian C. Smith, whose views on intentionality have profoundly influenced my own philosophical views.

This paper was written, in part, with the financial support of the Center for the Study of Language and Information (under a grant from the Systems Development Foundation) and the Xerox Palo Alto Research Center.

Notes

1   See, for example, P. M. Churchland (1979), Churchland and Churchland (1981), P. S. Churchland (1986a), Dennett (1987), Dretske (1981, 1988), Fodor (1987, 1990), Papineau (1984), Stampe (1975), and Sterelny (1990).

2   Of course, simplicity is not the only reason why the static perceptual case has seemed such an obvious starting point. Nor need this assumption hinge entirely on any view of the senses. For philosophers with an empiricist bent, who believe that the content of *all* of our mental/computational states is somehow wrought from the content of perceptual states, perception is not merely the most simple cases of mental directedness. Perceptual states are also the most *basic*. If it is our perceptions that somehow provide the raw materials for our more complex intentional states (about unperceived, fictive, or abstract properties), then perceptual states *ground* all other intentional relations. For an empiricist, this methodological point of entry is dictated by his philosophical program. Even a rationalist, however, will find the study of the simple perceptual case a perfectly obvious starting point. All one need believe is that, in the absence of all sensory input, there would not be (could not be) any thought at all, that sensory input is a necessary condition for intentional states. On this view, perceptual states are also "basic," although not in the strong sense endorsed by the empiricist, given above.

3   For a well-worked-out example of the notion of isomorphic representations, see Gallistel (1990).

4   To make this point more tangible, try to imagine the thermoreceptive system of some other creature, say the penguin.

5   In *Consciousness Explained* (1991, p. 382), Dennett says:

> What property does Otto judge something to have when he judges it to be pink? The property he calls "pink." And what property is that? It's hard to say but that should not embarrass us, because we can say why it is hard to say. The best we can do, practically, when asked what surface properties we detect with color vision, is to say, uninformatively, that we detect the properties we detect. If someone wants a more informative story about these properties there is a large and incompressible literature in biology, neuroscience, and psychophysics to consult.

6   My thanks here to José Antonio Diez Calzada for clarifying this point.

7   For an extremely interesting account of what it is like to live *without* proprioception, that is, with only visual knowledge of body position, see Cole (1995).

8   The following information on stretch receptors comes from Carew and Ghez (1984).

9   And to many computer scientists. For one defense of this view, see Brooks (1991).

10   For some interesting research on cortical visual maps that are "arm-centered," see Granziano, Yap, and Gross (1994).

11   To put this another way, there are certain circumstances when particulars (one's offspring or mate) impinge upon the lives of simple creatures and for which having detector cells can be just what is needed. Detectors are reliable (indeed, they are often much more reliable than, say, the mechanisms of human facial recognition) and they are usually economical as well, requiring few neural resources (an ant, one must realize, does not have many neurons to spare). But while detectors may signal the presence of a particular, they do not represent them *as* particulars. So one can rephrase the above question

as follows: what kinds of informational problems about particulars require, or are better served by, something other than detectors?

References

Brooks, R. 1991: Intelligence without representations. *Artificial Intelligence*, 47, 139–60.

Carew, T., and Ghez, C. 1984: Muscles and muscle receptors. In E. Kandel and J. Schwartz (eds), *Principles of Neural Science*, 2nd edn. New York: Elsevier Science, 454ff.

Churchland, P. 1979: *Scientific Realism and the Plasticity of Mind*. Cambridge: Cambridge University Press.

Churchland, P. M., and Churchland, P. S. 1981: Functionalism, qualia and intentionality. *Philosophical Topics*, 1, 121–45.

Churchland, P. S. 1986a: *Neurophilosophy: Toward a Unified Science of the Mind-Brain*. Cambridge, MA: Bradford Books.

Churchland, P. S. 1986b: Replies to comments from "Symposium on Patricia Smith Churchland's *Neurophilosophy*." *Inquiry*, 29, 139–273.

Cole, J. 1995: *Pride and a Daily Marathon*. Cambridge, MA: MIT Press.

Dennett, D. 1987: *The Intentional Stance*. Cambridge, MA: MIT Press.

Dennett, D. C. 1991: *Consciousness Explained*. New York: Little John.

Dretske, F. 1981: *Knowledge and the Flow of Information*. Cambridge, MA: MIT Press.

Dretske, F. 1988: *Explaining Behaviour: Reasons in a World of Causes*. Cambridge, MA: MIT Press.

Fodor, J. A. 1987: *Psychosemantics: The Problem of Meaning in the Philosophy of Mind*. Cambridge, MA: MIT Press.

Fodor, J. A. 1990: *A Theory of Content and Other Essays*. Cambridge, MA: MIT Press.

Gallistel, C. R. 1990: *The Organization of Learning*. Cambridge, MA: Bradford/MIT Press.

Granziano, M., Yap, G., and Gross, C. 1994: Coding of visual space by premotor neurons. *Science*, 266, 1054–7.

Hensel, H. 1982: *Thermal Sensations and Thermoreceptors in Man*. Springfield, IL: Charles Thomas.

Papineau, D. 1984: Representation and explanation. *Philosophy of Science*, 51 (4), 550–72.

Stampe, D. 1975: Show and tell. In B. Freeman et al. (eds), *Forms of Representation*, Amsterdam: North-Holland, 221–45.

Sterelny, K. 1990: *The Representational Theory of Mind*. Cambridge, MA: MIT Press.

21

Brain Matters: A Case Against Representations in the Brain

Robert S. Stufflebeam

1 Introduction

Obviously enough, the *raison d'être* of the neurosciences is to model the brain and to explain how it works. Although neuroscience is a broad church, nothing so binds the methodologically diverse research fields named by *neuroscience* as does the conviction that without internal representations over which to operate (e.g. to compute, to use, to manipulate, or to store), the brain could not do what it does. This view is called *representationalism*. Neuroscientists are by no means alone in their commitment to representationalism. Indeed! Cognitive scientists of nearly all theoretical persuasions tend to posit internal representations to explain how intelligent systems work.

Given the cognitive and neural sciences' commitment to representationalism, does it follow that internal representations are "real entities" to be found among the causal components of the brain's systems? Alternatively, do internal representations "really" exist?

In relation to the constraints naturalism places on both what exists and how we can know what exists, such ontological questions are, at best, problematic. After all, a commitment to naturalism entails that there is no privileged standpoint outside science from which to settle ontological questions. Yet because philosophical labor contributes as much to the construction and evaluation of theories as does the collection of data, a commitment to naturalism does *not* entail that nonexperimentalists should simply defer to experimentalists on matters of ontology. Nevertheless, to accept naturalism is to accept the view that the existence of an entity depends on its status within an empirical theory. As Quine says, "[e]verything to which we concede existence is a posit from the standpoint of a description or the theory-building process, and simultaneously real from the standpoint of the theory that is being built" (1960, p. 22). Hence, if one's theory requires positing internal representations to explain how the brain (or whatever) works, then in relation to *that* theory, internal representations are "real."

But not all theories are equal. And as there are several theories currently jockeying for position as the paradigm within which to explain how brains and other intelligent systems work, the "revolutionary period" in which we now find ourselves, while exciting, makes it difficult to evaluate the representation-related claims (and ontologies) of competing theories.

The problem is not that the meaning of the term *representation* varies considerably from theory to theory. In fact, most cognitive scientists would assent to the following loose analysis of representation: (1) a representation is an "entity" that bears content; (2) an "entity" bears content just in case it *conveys information about, symbolizes, depicts,* or otherwise "stands for" something else; and (3) an "entity" that bears content *within* a system is an *internal representation*. And with respect to the relation between human brains and representations, almost everyone would assent to these views as well: (4) human brains produce, store, and recall *internal* representations of places, faces, facts, fears, and fancies for use in at least *some* cognitive and perceptual tasks; (5) human brains produce *external* representations (e.g. maps, drawings, meaningful linguistic speech or writing, etc.); and (6) human brains produce machines that are themselves representation producers (e.g. cameras, camcorders, tape recorders, etc.). Clearly, there *is* much concerning representations about which almost everyone agrees.

Unfortunately, there is much less agreement concerning how to identify "entities" that qualify as representations and whether anything bears content intrinsically. These are matters to be settled by a general theory of representation. There are several in service. Not all of them are computation-friendly.

For instance, according to the theory of representation championed by Searle (1980, 1990), the intentionality of creatures with minds makes *mental states* the only "entities" that qualify as *internal* representations. Because minds do not emerge from the manipulation of symbols, brains are not "computers" – at least not in any interesting sense. Computationalists disagree. According to the theory of representation championed by proponents of symbolic-digital computation, *mental states* and *symbols* are the only "entities" that qualify as representations – internal or otherwise (Newell and Simon, 1976; Fodor, 1987; Newell, 1990). After all, intelligent systems process internal information in the service of their goals, aims, or purposes, internal information processing occurs only through computation, and computation requires operations over a sophisticated "medium" of discrete internal symbolic representations. Proponents of this view then reinterpret claims from connectionists and neuroscientists in order to describe processing within neural networks as symbol manipulation. Connectionists and neuroscientists disagree. According to the theory of representation championed by connectionists, internal representations within a neural network architecture are not symbols, but *distributed patterns of activation*. But not everyone considers nonsymbolic-analog processing to be a type of computation, including many neuroscientists. According to the theory of representation that seems to be championed by most neuroscientists, internal representations occur whenever an entity bears information, processes information, consumes information, or otherwise "uses" information. Because information processing occurs at *every* structural level of organization in the brain, internal representations are ubiquitous.

To be sure, everyone is free to interpret competing theories and computational frameworks relative to the vocabulary and ontological commitments of one's own. But such practices often strain the vocabulary beyond recognition, beg the question against the possibility of nonsymbolic computation, and commit us to an ontology of internal representations so robust that representation-talk is rendered vacuous and we lose whatever explanatory gain such talk is supposed to contribute to our understanding of how brains and other intelligent systems work. My purpose for this chapter is to defend a general theory of representation that avoids these problems.

2 Disambiguating Representation

The term *representation* means either (1) the *process* of representing – standing for, symbolizing, or depicting some other thing or event; or (2) the *entity* or state that stands for, symbolizes, or depicts some other thing or event. To avoid the ambiguities that attend the unconsidered use of the term "representation," I individuate representations as follows:

R = X is a representation if and only if
(i)   there exists a Y such that X stands for Y;
(ii) $X \neq Y$; and
(iii) X is not a representation-producing process.

Let me make just a few quick comments in defense of each of these conditions. First, one should not read too much into the existence of some Y being necessary, for it does not follow that pictures of, say, unicorns, could not be representations. The point, simply, is that for a representation to *be* a representation, it must stand for *something*, and that something has to *be* in some sense, even if only as an uninstantiated concept. Second, what I am after here is a theory of representation that offers a high degree of generality. As such, though "stands-for" is admittedly vague, for now I shall make no attempt to specify the precise nature by which something may stand for something else.

The following reasons make **R** a compelling basis for the individuation of representations. First, **R** is neutral as to the kinds of "things" that can be representations, so it does not circumscribe representations too narrowly. Thus, **R** permits "things" that are not mental, internal, or computation-dependent to be representations. Second, **R** is neutral as to the processes by which representations are produced, so it does not beg the question against any computationalist camp. Third, **R** does not require the existence of any representation user or consumer.

Explaining these merits of **R** will entail identifying the demerits of competing conceptions; that can get a bit messy. Still, I endeavor (below) to defend each of **R**'s virtues in more or less the order presented above.

Not all representations are mental entities or states of cognitive systems

Representations are posited in explanations by theorists in many domains, not just those pertaining to brains and cognitive processing. For example, *Mona Lisa* the painting, a *mental image* of the Mona Lisa, and the *concept* "MONA LISA" are each, according to some theory or other, a *representation*. More importantly, with **R** as the standard by which to individuate representations, they would remain so. This would *not* be the case if – following Fodor (1987) – **R** restricted the things that may be representations to only "symbols and mental states." Hence, **R** avoids circumscribing representations too narrowly. Such is not the case with symbolic-computation-dependent conceptions of representation.

Moreover, although what constitutes an intelligent system is a matter of some dispute, it is fairly uncontentious to assert that nonmental representations figure in at least *some* of *our* cognitive processing. For example, without the aid of a calculator, suppose you were required to divide 96.769 by 3.837. With pen and paper in hand, you would first write "3.837" and "96.769." Then in accordance with the rules for division, you would manipulate those *external* symbols to derive the answer. Since nonmental tokens of numerals stand for numbers, they too ought to count as representations.

Finally, that there are "witless" systems oblivious to the representations they either manipulate or produce (Haugeland, 1991, pp. 69–70) is a fact that in no way affects the status of the representations themselves: if something meets **R**'s conditions, then regardless of the cognitive status of the system that produced it, it is a legitimate representation.

Not all representations are symbolic structures

R also avoids circularity in another way. Namely, it does not equate representations with states (or products) of an internal symbol system. But defining *representation* in terms of symbols or symbolic processing begs the issue against connectionists, my appeal to "things" and a "stands-for relation" does not obviously beg the question against either computational camp.

But while classicists may beg the question against connectionists, it should be noted that they (often) define *symbol* using the same vocabulary I have used to define *representation*. For instance,

> the standard view [is] that symbols *stand for* something and that the token of a symbol occurring in some place in a structure carries the interpretation that the symbol stands for something within the context that is specified because of the symbol token's location.
>
> (Newell, 1990)

And when fleshing out the components of physical symbol systems, Newell and Simon (1976) draw a fundamental distinction between, on the one hand, the "processes" or operations over "symbols" and "expressions" (or strings of symbols), and on the other, the "entities" being operated over (the symbols and expressions):

> A physical symbol system consists of [1] *a set of entities*, called symbols, which are phys-
> ical patterns that can occur as components of [2] another type of entity called an
> expression (or symbol structure). Thus a symbol structure is composed of a number
> of instances (or tokens) of symbols related in some physical way (such as one token
> being next to another). At any instant of time the system will contain a collection of
> these symbol structures. Besides these structures, the system also contains [3] *a collec-
> tion of processes* that operate on expressions to produce other expressions: processes of
> creation, modification, reproduction, and destruction. A physical symbol system is a
> machine that produces through time an evolving collection of symbol structures.
>
> (Newell and Simon, 1976)

Hence, although the processes are in some sense "entities," *the processes are not them-
selves symbols*. Symbols are representations – or at least most of them are. Indeed, there is not anything a classicist would want to call a symbol that would not be a representation according to **R**. Consequently, my proscription against "processes" being representations is not arbitrary. In any event, we shall revisit this issue in due course.

Not all representations are states of consumption/use

The representational status of *some* representations is clearly dependent on their being part of a triadic relation between the content bearer, the content, and the user of the content. For example, strings of letters would not be words if some language community did not deem them meaningful. Kosslyn (1994) convincingly argues that subjectively generated mental images, like words, are also dependent on a representation-user. My point here is that such representations do not exhaust what count as representations. Some representations (e.g. photos) do not depend on any representation-user. And if something meets **R**'s conditions, then regardless of whether someone or something can be said to use its content or information, it is a legitimate representation. But lest I get too far ahead of myself, I shall wait to defend this claim until section 4.

Examples

Thus, **R** is neutral not just among the kinds of things that can be representations; it is neutral as to their internal or external status. It is also neutral regarding the processes – whether classicist or connectionist, natural or artificial, etc. – by which representations are produced. As such, **R** seems to capture not only what is necessary and sufficient for something to be a representation, it seems to do so without begging the sorts of questions that are at issue here. This has positive as well as negative consequences.

On the up side, **R** avoids some of the problems that plague theories that limit internal representations to mental entities, symbols, etc. As such, because each of the following "entities" meet all of **R**'s conditions, they are representations: (1) the RETINAL IMAGES caused by light bouncing off my American-made, two-wheeled means of transportation; (2) PHOTOGRAPHS of my Harley; (3) DRAWINGS of brains, photographs of brains, and the PRODUCTS OF NEURAL IMAGING are all representations; (4) PERCEPTS (e.g. the image I see when I look at Sophie, my cat); (5) NAMES; e.g. *Sophie, Normal, IL*, etc.; (6) WORDS or (some) SYMBOLIC EXPRESSIONS; (7) TOPO-GRAPHIC MAPS (and each of their feature symbols), etc.

On the down side, "almost anything may stand for almost anything else" (Goodman, 1976). Indeed, because almost anything can satisfy **R**'s conditions, it follows that almost anything can *be* a representation. For example, we can append to the above list of arguably *intrinsic* representational entities, the following *extrinsic*, though equally legitimate ones: (8) my HARLEY is a representation because it stands for the blueprint from which it was constructed; (9) ROCKS: take three rocks. Arrange them on the floor in such a way as to depict the location Washington, DC, relative to, say, Chicago and LA. Because each of the rocks now stands for a particular city, according to **R**, each rock is a representation, as is the arrangement of rocks itself; (10) GOVERNORS: suppose I built a device à la Watt (see chapter 18, this volume) that regulated the speed of an engine. At any time t, let us suppose, the angle of the arms of this governor correlates to the speed of the engine it is regulating. As such, the state of the arms of the governor at t could be said to stand for the engine speed, and as such, be a representation. Of which more anon.

However "legitimate" such representations might be, there is something *ad hoc* about the representational status of Harleys, rocks, etc. This is so because some representations (e.g. photos, names, and so on) seem to have content intrinsically, whereas others (e.g. Harleys, rocks, and states of Watt governors) can merely be treated as if they bore content. After all, **R** entails that so long as X "causally correlates" with Y, spuriously or otherwise, because causal correlation is one way something can be said to "stand for" something else, causal correlation is sufficient to fulfill the stands-for condition. Indeed, causal correlation has been taken by some to be the hallmark of the stands-for relation (e.g. Tye, 1995). Thus, not only are **R**-style representations ubiquitous, **R** does not capture what one ordinarily means by "real" or "genuine" representations: "Real representations," as one might say, are things which would not exist save as bearers of content. While the representational status of governors may ultimately be as "genuine" as that of photos, surely ordinary rocks do not count as "genuine" or "intrinsic" representations.

The tension here should be obvious: because anything can be treated *as if* it were a representation, even what one might call an "extrinsic" representation would not only *be* a representation, it would be a "real" one as well. Indeed, if we dropped the proscription against processes being content bearers, they too could be treated as though they were in a stands-for relation to something else. As such, practically anything in the universe could *be* a representation.

But this does not worry me. What does is this: Through a computational or information-processing description or interpretation, internal computational states become imbued with content, thereby becoming internal representations. Through such a practice, there is no limit as to what become representations in "explanations" of computational processing, even processes. But if intelligent systems truly trafficked in internal representations, they should bear their content on their sleeves, as do, say, photos or words. Otherwise, there would not be any nonintentional entities or states upon which to ground the intentional output of cognitive processing. Moreover, if internal states became "genuine" representations just in virtue of the interpretation *we* put on cognitive processing, then representation-talk would be "so unconstrained as to be without content" (Hatfield, 1991, p. 167). Aside from avoiding ambiguity, such is the reason why I have adopted the proscription against processes being representations.

Thus, the *problem of representation* is not that of determining what a representation is. Rather, it is discovering a principled way of distinguishing between *intrinsic* and *extrinsic* ones. In other words, given that almost anything can be a representation, how can one tell whether a given **R**-style representation is a genuine representation or merely an as-if one? If none can be found, then the ontological status of *internal* representations in the brain would be problematic indeed. So too would the efficacy of representation-talk in explanations of how brains and other intelligent systems do what they do.

3   Intrinsic vs. Extrinsic Representations

Because the content-bearing status of *some* representations *is* parasitic upon our descriptions or interpretations, no one questions whether *extrinsic representations* exist. As such, what is at stake is not whether *some* "things" bear content in virtue of the interpretation or description *we* (or other agents) confer upon them. Rather, what is at stake is whether *all* representations are extrinsic representations.

If the contrast class is not empty, it includes "entities" that must bear content *independent* of our (or some other agent's) descriptions or interpretations, their ontological status as content bearers must not waver over time, and they must be the sorts of things that would not exist save as content bearers. In short, an *intrinsic representation* must be ontologically dependent on being a bearer of content. That much is clear. Put a bit more formally,

> **I** = X is an intrinsic representation just in case
> (i) X meets **R**'s conditions, and
> (ii) X is ontologically dependent on being a content bearer.

Now comes the hard part. Does anything meet **I**'s conditions? If intrinsic representations exist, what is the feature they possess that satisfies the condition of being ontologically dependent on bearing content? Let us call this feature "ϕ." Several candidates suggest themselves (see figure 21.1); each needs to be evaluated. But first we need to clarify what intrinsicness involves – and what it does not.

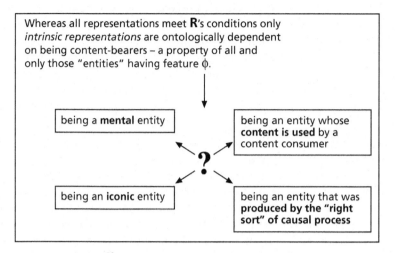

Figure 21.1   Candidates for the feature φ that makes a representation intrinsic.

What *intrinsic* does not *mean*

According to Dennett (1987, p. 288), there is a "great divide" among philosophers of mind regarding whether anything possesses "intrinsic" or "original" intentionality. If so, what are these things?

Those who accept what he calls the *doctrine of original intentionality* believe that whereas some of our artifacts may have intentionality derived from us, we have "original (or intrinsic) intentionality, utterly underived" (Dennett, 1987, p. 288). Among those Dennett cites as subscribing to this doctrine are Fodor, Dretske, Burge, and Putnam. For *Fodor and company*, since mental states exhibit intrinsic intentionality, mental states are *essentially* representations. Thus, mental states are *intrinsic* representations.

Among those cited as denying this doctrine are Dennett, Millikan, the Churchlands, Quine, and almost everyone in Artificial Intelligence (AI). For *Dennett and company*, representations, mental or otherwise, have "derived intentionality" or "as-if intentionality": though all representations have content, what makes a representation a representation is that it gets *used or interpreted as a representation*. Thus, whether something bears content is *derivative* on or *dependent* upon its manner of use.

I mention all this for two reasons. First, I wish to make it clear that the intrinsic vs. extrinsic distinction is not one of my own making. Second, and more importantly, coupling talk of ontological dependence with talk of intrinsicness seems to engender the following problem for those who champion intrinsic representations. Namely, if the intrinsic status of a representation is parasitic upon ontological dependence, and if ontological dependence is in turn parasitic upon having feature φ, then whether something is an intrinsic representation would be parasitic upon having feature ϕ. But if what makes something an intrinsic representation is its having feature ϕ, Dennett is correct to point out that *nothing* would be an intrinsic representation.

Apparently, for Dennett, the only candidates for intrinsic representations would be all and only those content-bearing things whose status as such *is not* ontologically dependent on anything else. The problem, as Dennett well knows, is that nothing emerges *ex nihilo*. So, the chances are not just slim of any content-bearing thing being produced noncausally and hence nonderivatively – the chances are nil. Thus, in Dennett's sense of *intrinsic*, he is right: nothing is intrinsic in *that* sense. But since no one who uses *intrinsic* does so in Dennett's sense, surely Dennett's standards for intrinsicness are too high.

Here is the moral: since everyone in Dennett and company save Dennett endorses one manner or other for distinguishing between genuine and as-if representations, one should not treat *intrinsic* as synonymous with *innate* or *wholly underivative*. While it is true that our ontological posits are always made against the backdrop of our theory, it is *also* true that if the intrinsic vs. extrinsic distinction fails, it should do so for reasons *other* than our having set the standard for intrinsicness impossibly high.

Although Fodor and company consider mental states to be intrinsic representations, not one of them claims that *being a mental entity* or *being a natural entity* is what *makes* a mental state an intrinsic representation. Rather, they hold that mental states are intrinsic representations because mental processing exhibits a privileged sort of causal processing (e.g. information processing). Thus, even for those who would consider mental states to be intrinsic representations, they do not do so just because such states are either mental or natural. That is a good thing. After all, whereas privileging "the mental" would beg the question concerning whether *non-mental* intelligent systems traffic in internal representations, privileging "the natural" would beg the question concerning whether *man-made* intelligent systems do (or could ever) traffic in such entities. Besides, as candidates for ϕ, *being a mental entity* and *being a natural entity* beg the question concerning whether *brains* traffic in intrinsic representations.

The hunt for ϕ via photos

Given the diverse collection of *token* representations identified earlier, we know of a host of representation-*types* whose tokens are sure to meet **R**'s conditions (see figure 21.2). For those tokens to be *intrinsic* representations, meeting **R**'s conditions is only half the battle. The other half is more interesting. It concerns whether any of them meet **I**'s other condition. Some of them do.

For instance, Harleys will continue to exist even after they lose their content-bearing status, *photos of Harleys* would not. This is not due to any special property of Harleys. On the contrary, regardless of what a photo depicts, *all photos* have the property of bearing content for as long as they exist. Not having one's content-bearing status vary over time is what ontological dependence is all about. That photos

$$\left\{\begin{array}{c}
\text{Symbols} \\
\text{Photographic} \\
\text{Numerals} \\
\text{Names} \\
\text{World} \\
\text{Maps} \\
\text{Retinal images} \\
\text{Percepts} \\
\text{States of governors} \\
\text{Harleys} \\
\text{Rocks}
\end{array}\right\}$$

Figure 21.2   Some of the "types" of representation that correspond to the representation "tokens" identified in the text. Although it would be much easier to deal just with representation tokens, I am after a level of generality that requires type-talk.

have it explains why, for example, a photo of a brain in a neuroscience textbook does not cease to depict something even if no one ever opens the book. Such is the case because the content-bearing status of photos is not dependent on the continued existence of the camera, me, you, or anything else save the photo. Regardless of how one individuates the content of photos, they simply would not be photos if they were not ontologically dependent on being in a stands-for relation to whatever they depict.

Now comes the hard part: *Why* is it that photos remain representations for as long as they exist? What is the feature all photos share that matters? Finding it is what the search for ϕ is all about. Here is the short list of candidate properties shared by all photos: (1) each is an *iconic* content bearer; (2) each is produced by a mechanistic process; and (3) each photo conveys information. The answer lies somewhere among these alternatives. And because meaningful linguistic speech and writing are not iconic content bearers, let us ignore whether being an iconic content bearer matters.

4   Why Having Been Produced by the "Right Sort" of Processing Matters

Photos, drawings, and certain other representation-tokens play a large role in our cognitive lives. For instance, whereas a photo (or a drawing) is indeed worth a thousand words when one's aim is to explain how the brain is structurally organized, a thousand words is better than ten thousand gestures should no photo be available. Photos and words are both representation-types. Iconic representations are of great pedagogical value for their ability to capture a great deal of theoretical labor succinctly. Why do we rely on such external representations? The answer is, to paraphrase Clark (1989), that biological intelligent creatures such as ourselves tend not to store or process information in costly *internal* ways when we can use the structure of the environment or our artifacts as "a convenient stand-in" for *internal* information-processing operations (Clark, 1989, p. 64). Hence, there are good cognitive scientific grounds for being interested in the representational properties of photos and stand-ins whose function is to ease our internal cognitive labor. Nevertheless, this is not the reason I have belabored the discussion of photos here.

What we have been after is determining *why* it is that photos remain representations for as long as they exist. Although everything that we *could* call representation was produced by *some* sort of causal process, not all representation-producing processes are equal. Therein lies the value of having belabored the discussion of the content-bearing ability of photos: *photos remain content bearers for as long as they exist because they were produced by a process whose designed function is to produce content bearers.* That is what ϕ is: namely, having been produced by the "right sort" of processing. Put in such a way as not to beg the question of whether *natural* mechanistic processes are those that produce intrinsic representations: something is ontologically dependent on being a content bearer just in case it was produced by a mechanistic process whose function (by design or nature) is to produce content bearers. I call such processes *first-order representation-producing processes.*

Here is the story so far. All representations are "things" that stand for something; that is, all representations are content bearers. Representations exist. Some representations (e.g. rocks) *seem* to bear content only under such-and-such interpretation. Other representations (e.g. photos) *seem* to bear content independent of our interpretations. Thus, it is plausible to believe that there are two general classes of representations: *extrinsic representations* and *intrinsic representations.* The former are content-bearing "things" whose status as representations depends just on *our* (or some other agent's) description or interpretation. The latter are content-bearing "things" whose status as representations is parasitic upon being ontologically dependent on bearing content; that is, they must bear content *independent* of our (or some other agent's) descriptions or interpretations. I take it that *having been produced by a first-order representation-producing process* is necessary for something to be ontologically dependent on bearing content. Intrinsic representations exist. And if cognitive processing in the brain is mediated by internal representations, they too should be intrinsic ones.

Anything meeting **R**'s theory-neutral identity conditions would be a legitimate representation, but anything meeting **I**'s conditions would be an *intrinsic* one. And if internal representations need to figure in explanations of computational processing, those representations need to be intrinsic representations. Otherwise, the representations that are doing the explanatory work are not *internal* representations, but internal *states* that become representations just in virtue of our (or some other agent's) interpretations.

Thus, at the heart of my theory of representation is the *assumption* that all intrinsic representations are produced by first-order representation-producing processes. Although one can defend how one arrives at one's assumptions, as I have done above, one cannot *prove* that one's assumptions are right. What one *can* do is to test them.

Insofar as no representations, intrinsic or otherwise, ever emerge *ex nihilo*, it is good that representation-producing processes are causal processes. Not all causal processes are representation-producing processes. At least they are not according to *this* theory – and for good reason. After all, I am defending what I hope is an *interesting* theory about what counts as a representation. Such is not the case for theories wedded to causal covariance or information processing.

5   Does "Use of Information" Matter?

Cognition is the flexible coupling of perception and action. Whether direct or complex, this coupling depends on representing information and operating upon it. Thus, representation and its partner, processing, are the most fundamental of ideas in cognitive science. *Representations* are bundles of information on which processes operate. Cognitive processes such as perception and attention encode information from the world, thus creating or changing our representations. Processes of reasoning and decision making operate on representations to form new beliefs and to specify particular actions. *Process* refers to the dynamic use of information. *Representation* refers to the information available for use.

(Billman, 1998, p. 649)

Besides re-emphasizing the distinction between representations and the processes which operate over representations, I have selected this passage because it captures what many researchers in the cognitive and neurosciences take to be the hallmark of representation. Namely, a content bearer's content (or information) must be used, consumed, interpreted, manipulated, or otherwise operated over by an "agent" for it to be a representation. Use-dependent notions of representation have been championed by Millikan (1989a, 1989b), Bechtel (chapter 18, this volume), Grush (chapter 19, this volume), and scores of cognitive scientists. According to this view, representation *per se* is always a *triadic* relation. On this view, something is a representation just in case it is a content bearer whose content gets "used" by a content consumer.

Insofar as it is reasonable to evaluate the assumptions of competing theories, is this view right? Must the content of X be used for it to be a representation? If so, we need not worry about whether **I**'s conditions succeed in picking out *intrinsic* representations; rather, we need to worry about whether **R**'s conditions succeed in picking out even *extrinsic* ones. Hence,

r = X is a representation if and only if
- (i)   there exists a Y such that X stands for Y (where *stands for* means *conveys information about*);
- (ii)   there exists a Z such that Z uses, consumes, interprets, manipulates, or otherwise functions in a manner dependent on the information X bears;
- (iii) $(X \neq Y)$ and $(X \neq Z)$ and $(Y \neq Z)$; and
- (iv) X is not a representation-producing process;

Let us ignore worries about whether representation-producing process can be representations. Because Millikan, Grush, and Bechtel individuate representations in a use of information–dependent way, I think **r** is consistent with their views on the nature of representation. According to Millikan:

> What *makes* a thing into an inner representation is, near enough, that its function is to represent. But, I shall argue, the way to unpack this insight is to focus on representation *consumption*, rather than representation *production*. It is the devices that *use* representations which determine these to be representations and, at the same time, determine their content. . . . The part of the system which consumes representations must *understand* the representations proffered to it. . . . [I]nformation could not serve the system as information unless the signs were understood by the system, and, furthermore, understood as bearers of whatever specific information they, in fact, do bear.
>
> (Millikan, 1989a)

Hence, for a representation to *be* a representation, it must "function as a sign or representation *for the system itself*" (1989a, p. 245).

For Grush,

> Representations are entities which stand for something else – or better, they are entities which are used to stand for something else. This second characterization brings out something not explicit in the first, that of a user. If this second definition, and my gloss on it, are correct, then a representation is a part of a three-way relationship which also includes a user and a target. So far so good. Some may quibble over the need for a user, but that is not where the real problem lies. The real problem has been, and continues to be, the choice of states for which theorists attempt to give a representational analysis.

For Bechtel,

> In its most basic sense, a representation is something (an event or process) that stands in for what it represents and is used as a stand in by a system in which it occurs to coordinate its behavior in a manner that depends upon what is represented.

Given that many representation-tokens *are* dependent on a representation user (or users), and because "users" of representations are not limited to things with minds, it is reasonable to suppose that the condition (ii) above captures something essential about the nature of representations.

However, all use of information-dependent conceptions of representation have the following flaws. I think the flaws are fatal.

First, information processing and the use of information occur not just at every structural level of organization in the brain, they occur throughout the body. Bechtel is commendably explicit about this:

> Many physiological systems (e.g. simple biochemical systems such as fermentation where reactions are controlled by feedback mechanisms in which the availability of the

product of the reaction determines whether it will continue) will, on this construal, employ representations since processes in them as well as the responses to them were selected because of the process's information bearing role.

Consequently, if representations should be individuated in a use of information–dependent fashion, then practically every state of every structure in every system in every organ throughout the body becomes an internal representation. This not only renders representation-talk vacuous, we lose whatever explanatory gain such talk is supposed to contribute to our understanding of how brains and other intelligent systems work.

Second, is it reasonable to individuate representations in such a way that an "entity" could be a representation at time-slice$_1$, not a representation at time-slice$_2$, then a representation (or not) at time-slice$_n$? Insofar as rocks and scores of extrinsic representations are concerned, yes. Their content-bearing status over time may waver. Photos are another matter. A buried photo does not cease to be a representation when its content is not be used by a content consumer. *Mutatis mutandis*, neither do TV images that no one sees, words in books that no one reads, etc., etc. Why? Well, it is not *always* the case that what makes something a content bearer is that the bearer's content gets used. Rather, some things bear content simply because they were produced via a (designed or evolved) process whose function is to produce content bearers. But if some cognitively interesting things can be content bearers regardless of whether anything or anyone can be said to use them, then condition (ii) is not necessary. Although we *could*, as a matter of convention, hold that r identifies the necessary and sufficient conditions for a "thing" to be a representation, doing so would exclude a host of intrinsically representational entities from being representations.

Third, use-dependent notions of representation beg one of the main questions at stake here; namely, "*Are* there internal representations in the brain?" And if the use-condition does not hold for a *general* account of what is necessary for something to be a representation, then it stands no chance of being that in virtue of which something is an *intrinsic* representation. Such is the lesson to be drawn about buried photos.

Hence, or so it would seem, the status of something *being* a representation is separate from the issue of whether anything uses the content it bears. As such, let us retain **R** as our standard for the identity conditions. If something meets **R**'s conditions, then regardless of the cognitive status of the system that produced it, regardless of whether anything can be said to use its content, it is a legitimate representation. The question, however, is whether it is an intrinsic or an extrinsic one.

6 Computation Matters

Computationalism would . . . be false if it could be shown both that the brain was an analog processor and that *this analog nature mattered crucially for cognition.* This is because computationalism is the hypothesis that cognition is computation, not analog processing.

(Dietrich, 1994, p. 15)

Computational processes are always implemented within a computational framework – classicist, connectionist, Bayesian, etc. Since cognitive processes are assumed to be computational processes, explaining how an intelligent system works requires positing some computational framework. The link between computationalism and representationalism appears to be direct, for without a "medium" of internal representations, computational systems could not compute. Thus, it is claimed, all computational systems require internal representations as a "medium" of computation.

I disagree. Elsewhere I have defended the position that one can be a full-blooded computationalist without committing oneself to the view that computation requires a medium of internal representations (Stufflebeam, 1997, 1998a, 1998b). At the core of this position are the following claims: analog computers are full-blooded computers, the nonsymbolic-analog processing found in neural networks is a type of computation, and distributed patterns of activation among the processing units make poor candidates for being intrinsic representations. In relation to the received view on computation, these claims are heretical. In any event, I will not attempt to defend them now. Space will not permit it. What I will do is to sketch out a computational framework within which to explain how the brain works, but one that does not render representation-talk vacuous.

Computational systems

We do not find computational systems "in the wild" because whether something is a computational system is not *just* an empirical matter. Rather, something is a computational system always relative to a *computational interpretation* – a description of a system's behavior in terms of some function *f*.

For example, all the major bodies in our solar system move in predictable orbits. The orbit or outward behavior of each of these bodies has been interpreted in terms of differential equations. Because differential equations are mathematical functions *par excellence*, our solar system is a computational system. This does not mean that computational systems are ubiquitous. After all, even if the behavior of every physical system in the universe *can* receive a computational interpretation, not every one of them *does*. Hence, not everything *is* a computational system.

Of course, our solar system is not an *intelligent* system. Nevertheless, the solar system and human beings both share some important features: (1) their behavior emerges from the tightly coupled interaction with simpler systems; and (2) their behavior can be interpreted in terms of differential equations. As these are the features of *dynamical systems*, it follows that some intelligent systems are dynamical systems. Although it might seem obvious that persons, like planets, are embedded within an environment rich in other dynamical systems that shape their behavior, explaining *intelligent* systems within the framework of dynamic systems theory

(DST) is *very* controversial. This is so, in part, because DST forces cognitive scientists to re-examine the practice of wedding intelligent systems to *internal* computational processing. Aside from being motivated to dispel problematic ontological commitments, DST (among other approaches) aims to reconnect cognitive processing with the world. The issue of *re*-connection arises because in *behaviorism*, which was the former received scientific view of cognition, internal processing paled in causal significance when compared to the environment. But in the *current* received scientific view, *computationalism*, the environment is all but ignored in favor of internal processing. *Cognitivism*, which is the view that the mind is to the brain as a program is to a symbolic-digital computer, carries this computational solipsism to its logical extreme. But for most proponents of DST, cognition is seen as the product of an agent who is closely coupled with her environment. On this view of intelligent systems, not only does cognitive processing get *extended* out into the environment, the boundaries for the intelligent system do so as well.

Here is the lesson: Not all computational interpretations license a commitment to internal representations.

Actual computation and the brain

According to contemporary orthodoxy, actual computation requires "the rule-governed manipulation of internal symbolic representations" (van Gelder, 1995, p. 345). This explains: (1) why many cognitive scientists treat *computation* as shorthand for *symbolic-digital processing*; (2) why *computer* is treated as synonymous with *digital computer*; and (3) why Searle and other anticomputationalists focus their critical attention on computational "explanations" based either on the *architecture* of digital computers or *simulations* of digital computers.

Nevertheless, there seem to be at least two paradigms of actual computational processing:

In *symbolic-digital processing*, computation involves simple mechanical operations applied to discrete and explicit symbol-tokens – determinable quasi-linguistic structures that mediate the combinatorial, rule-following production of a system's output. It is the sort of processing occurring in *classicist* systems.

In *nonsymbolic-analog processing*, computational operations are defined over continuous sets involving direct, quantitative input–output relations (i.e. as the input varies continuously, so do the outputs). It is the sort of processing occurring in systems implementing parallel distributed processing [PDP]. The computational primitives here are analog quantities or "distributed representations" – i.e. distributed patterns of activation among the processing units.

Let us cut to chase. To explain how a given system does what it does, positing internal representations would be required just in case the system trafficked in "entities" whose content-bearing status does *not* depend just on *our* descriptions or interpretations. Adopting the less-than ideal vocabulary found in the literature on intentionality, I call such content bearers *intrinsic representations*. Again, take your average photograph. Not only would it bear content even if no one were to see it, it will bear content for as long as it exists. Such is the case because photos are ontologically dependent on being content bearers. They have this feature because unlike, say, rocks, photos are produced by a process designed to produce content bearers. Not everything has this feature. The contrast class is *extrinsic representations* – content-bearing "entities" whose status as representations *does* depend just on *our* descriptions or interpretations. Anything can be described *as if* it bore content, so anything can be an extrinsic representation. Not everything is an *intrinsic* representation.

Do brains produce internal intrinsic representations? I think they do. Whereas photos are produced by a mechanistic process *designed* to produce "entities" that are ontologically dependent on being content bearers, a plausible *evolutionary* analog would be the products of mental imagery. Surely "mental images" of one's past experiences are intrinsic "representations" if anything is. Linguistic tokens are another candidate. After all, once the content of a linguistic item is "fixed," at least for a particular linguistic community during such-and-such time, tokens of that type will always bear content. So, if either mental images or linguistic utterances are intrinsic representations, *some* intrinsic representations are products of biological cognitive processing. What is at issue is this: Do internal representations mediate the computational processes underlying the production of such representations?

I think not, because brains are analog computers. After all, brains possess all the features of a system that implements a nontrivial type of nonsymbolic-analog computation: (1) information processing does not involve simple, mechanical operations applied to determinable symbol-tokens; (2) the materials of the brain matter (as does its structural and spatial organization); (3) biological information processing occurs in time; (4) biological computational operations are defined over continuous sets; and (5) the behavior of "real" neural networks can be described as satisfying differential equations or statistical functions, neither of which are "algorithms" or "programs."

But are brains the same sort of analog computer as "artificial" neural networks? Not quite. In fact, one of the chief differences between the two is that "artificial" neural networks (like PCs) "are essentially passive input-output devices" (Perkel, 1990, p. 38). The inputs and outputs in such networks are representations only because they become meaningful at the hand of the one running the simulation. Consequently, in every *simulated* neural network, there is a "gap" between the computational system and the environment. With brains, this gap is not present. Such is the case in vision, where the environment (along with its regularities) remains stable and available for continuous resampling. But this is presentation, not *re*-presentation. The analog nature of this processing matters. So too does the analog spatial organization of the materials by which the processing occurs.

There are other important differences between "artificial" neural networks and "real" ones – populations of neurons in the brain. Above all, whereas analog quantities do the computational work in brains, in artificial neural nets it is the patterns of activation among the processing units, so-called "distributed representations." Both of these "mediums" are poor candidates for being representations. Let us ignore worries about causal covariance, multiple ends, and whether nondiscrete "entities" or processes make intrinsic content bearers. Instead, note that neither analog quantities nor distributed patterns of activation *mediate* nonsymbolic-analog processing; rather, they *are* the processing. Thus, *distributed patterns of activation in the production of representations are not themselves representations.*

7 Conclusion

What I hope to have accomplished here is to defend a general theory of representation that avoids straining the vocabulary and committing us to an ontology of internal representations so robust that representation-talk is rendered vacuous. When coupled with a nonsymbolic-analog computational framework, it turns out that there is a great deal of processing within the brain that can be explained without positing internal representations.

However, given the right sort of interpretation, analog quantities or distributed patterns of activation, like anything else, can *be* representations. Yet since it is the interpretational process alone that makes them representations, at best, they are *extrinsic* ones. While such constructs are descriptively useful, trying to pass them off as *internal* representations trivializes whatever gain representation-talk is supposed to contribute to our understanding of how brains work. It also immunizes representationalism from being falsified. So much the worse for representationalism. As I have said elsewhere, with respect to the ontology of brains, representation-related conservatism is a small price to pay for a commitment to naturalism, hallowed be its name.

References

Billman, D. 1998: Representations. In W. Bechtel and G. Graham (eds), *A Companion to Cognitive Science*, Oxford: Blackwell, 649–59.

Clark, A. 1989: *Microcognition*. Cambridge, MA: MIT Press.

Dennett, D. C. 1987: *The Intentional Stance*. Cambridge, MA: MIT Press.

Dietrich, E. 1994: Thinking computers and the problem of intentionality. In E. Dietrich (ed.), *Thinking Computers and Virtual Persons: Essays on the Intentionality of Machines*, San Diego, CA: Academic Press, 3–34.

Fodor, J. A. 1987: *Psychosemantics: The Problem of Meaning in the Philosophy of Mind*. Cambridge, MA: MIT Press.

Goodman, N. 1976: *Languages of Art: An Approach to a Theory of Symbols*, 2nd edn. Indianapolis, IN: Hackett.

Hatfield, G. 1991: Representation in perception and cognition: Connectionist affordances. In W. Ramsey, S. P. Stich, and D. E. Rumelhart (eds), *Philosophy and Connectionist Theory*, Hillsdale, NJ: Lawerence Erlbaum Associates, 163–95.

Haugeland, J. 1991: Representational genera. In W. Ramsey, S. P. Stich, and D. E. Rumelhart (eds), *Philosophy and Connectionist Theory*, Hillsdale, NJ: Lawrence Erlbaum Associates, 61–89.

Kosslyn, S. M. 1994: *Image and Brain: The Resolution of the Imagery Debate*. Cambridge, MA: MIT Press.

Millikan, R. G. 1989a: Biosemantics. In S. P. Stich and T. A. Warfield (eds), *Mental Representation: A Reader*, Cambridge, MA: Blackwell, 1994, 243–58.

Millikan, R. G. 1989b: In defense of proper functions. *Philosophy of Science*, 56, 288–302.

Newell, A. 1990: *Unified Theories of Cognition*. Cambridge, MA: Harvard University Press.

Newell, A., and Simon, H. A. 1976: Computer science as empirical enquiry: Symbols and search. In J. Haugeland (ed.), *Mind Design*, Cambridge, MA: MIT Press, 1981, 35–66.

Perkel, D. H. 1990: Computational neuroscience: Scope and structure. In E. L. Schwartz (ed.), *Computational neuroscience*, Cambridge, MA: MIT Press, 38–45.

Quine, W. V. O. 1960: *Word and Object*. Cambridge, MA: MIT Press.

Searle, J. R. 1980: Minds, brains and programs. *Behavioral and Brain Sciences*, 3, 417–24.

Searle, J. R. 1990: Is the brain a digital computer? *APA Proceedings*, 64 (3), 21–37.

Stufflebeam, R. S. 1997: *Whither Internal Representations? In Defense of Antirepresentationalism and Other Heresies*. Doctoral dissertation. St Louis, MO: Washington University.

Stufflebeam, R. S. 1998a: Computation without representation: Nonsymbolic-analog processing. In M. Gams, M. Paprzycki, and X. Wu (eds), *Mind Versus Computer: Were Dreyfus and Winograd Right?*, Amsterdam: IOS Press, 171–89.

Stufflebeam, R. S. 1998b: Representation and computation. In W. Bechtel and G. Graham (eds), *A Companion to Cognitive Science*, Oxford: Blackwell, 636–48.

Tye, M. 1995: *Ten Problems of Consciousness: A Representational Theory of the Phenomenal Mind*. Cambridge, MA: MIT Press.

van Gelder, T. 1995: What might cognition be if not computation? *Journal of Philosophy*, 92 (7), 345–81.

Questions for Further Study and Reflection

1   Compare the views of all four authors with respect to the following questions:
    How (if at all), does one distinguish between a mere information-bearing state
    and a representation?
    How (if at all), does the notion of a user, or consumer, of represented informa-
    tion serve to define what is to count as a representation?
    When (if at all), do sensory states count as representational states?

2   What does it mean to "stand in for," in the context of representation? What sorts
    of entities can "stand in for," and what sorts of entities can be "stood in for"?

3   Akins makes the surprising claim that sensory systems need not, and generally
    do not, convey veridical information about the world. How does she meet the
    evolutionary objection to this claim; namely, that organisms could not survive
    unless their senses provided them with an accurate picture of the surrounding
    environment?

4   What sensory systems, if any, seem more likely than thermoperception to involve
    representations? Besides sensory systems, what other mental processes seem
    especially involved in representation? What mental processes seem especially
    devoid of representation?

5   What are some major things that mental and nonmental representations have in
    common? What are some major differences?

6   If it is possible to describe the behavior of a system without positing internal
    representations, does this have any bearing on the question of whether or not a
    system actually possesses internal representations? If we granted, for the sake of
    argument, that at least some systems do possess internal representations, would
    it follow that an explanation of that system's behavior would have to take these
    internal representations into account?

7   What is the relation between representation and intentionality? Are all inten-
    tional states representational states, and vice versa?

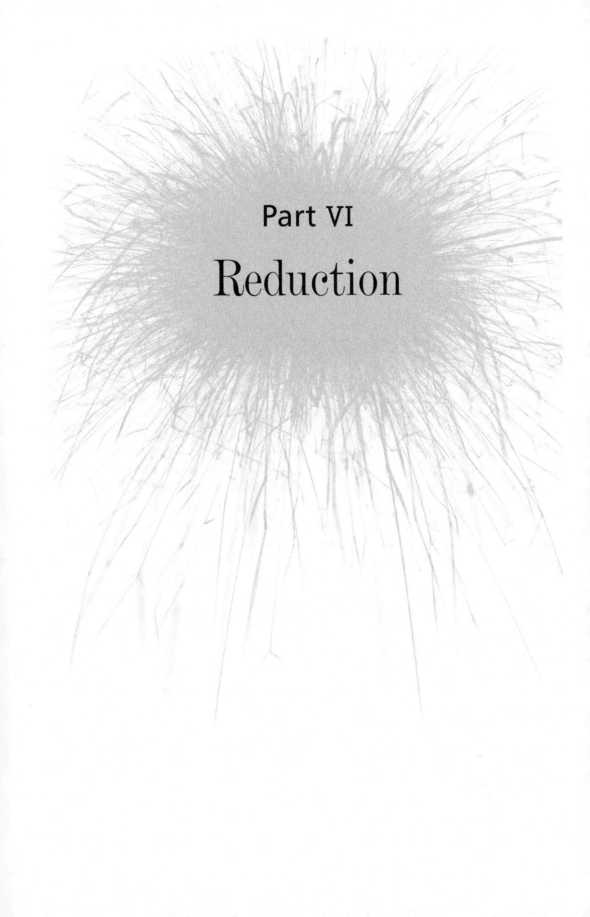

Part VI
Reduction

Introduction

Jennifer Mundale

We began with readings concerning the foundations of neuroscience, then explored several of the most absorbing and challenging topics at the intersection of philosophy and neuroscience. We now take a metascientific turn and, in this final selection of readings, address the intertheoretic relation between neuroscience and psychology.

In the following readings, although McCauley and the Churchlands disagree about the precise nature of the relation between psychology and neuroscience, they at least agree that there is a relation, and so share in their rejection of the autonomy thesis (a view held by Fodor, for example, who denies that there are any useful or interesting theoretical connections to be made between the two disciplines). They also share a commitment to some form of reductionism, or the view that psychology is reducible to neuroscience. Of course, there are several different kinds of reductionism, and what McCauley and the Churchlands disagree about is just what *sort* of reduction might usefully be constructed between these two disciplines. Roughly, though, to say that psychology is reducible to neuroscience means that some or all of the theories and/or entities of psychology can be correlated with (and possibly replaced by) the theories and/or entities of neuroscience.

In the first essay, the Churchlands defend reductionism against such problems as determinism, the irreducibility of intentionality, the inexplicability of high-level or emergent properties in materialistic terms, and the multiple instantiation of mental states. They support the feasibility of reducing the theories of psychology to the theories of neuroscience by adducing parallel cases of reduction in other sciences. The prototypical examples they provide differ widely with respect to the final status of the reduced theory. The three different outcomes for the reduced, or higher-level theory are: (1) it is eliminated in favor of the more general, or lower-level theory (as in the elimination of the phlogiston theory of combustion); (2) it is subsumed under the more general theory, perhaps as a special case of it (as in the reduction of Kepler's laws of planetary motion to Newton's more general laws of motion); or (3) it neither disappears nor is subsumed, but is enriched, influenced, and shaped by the more general theory (as in the relation between chemistry and physics). The Churchlands set out to defend reductionism in general, but do not specify which of these three will most closely describe the final shape of a reduced psychology.

Drawing from the larger corpus of the Churchlands' work, McCauley argues that they have not been clear or consistent about which model of intertheoretic relations they are promoting. He identifies three conflicting models in their work, which he terms co-evolution$_M$, co-evolution$_S$, and co-evolution$_P$. Co-evolution$_M$ is a form of reductionism in which psychological theories may be revised or even replaced in the process of reducing psychology to neuroscience, but in principle, something of the discipline of psychology remains. If carried a step further, co-evolution$_M$ becomes co-evolution$_S$, and the reduced psychology is eliminated in favor of neuroscience. This is the model for eliminative materialism, a position long associated with the Churchlands. As McCauley admits, these two need not be seen as conflicting alternatives, but could be viewed as two sequential stages of a larger process: first reduce psychological theories to neuroscientific ones, then eliminate the psychological theories altogether. Finally, there is co-evolution$_P$, which McCauley finds to be most prevalent in the Churchlands' later work. In this case, the primacy of the lower-level science is weakened in the interests of explanatory pluralism. Psychology and neuroscience mutually inform and enrich each other, and each is subject to revision in light of compelling results in the other. Obviously, this position is incompatible with the eliminative co-evolution$_S$.

This pluralistic form of co-evolution is the position which McCauley himself advocates, and the rest of his essay is devoted to arguments against co-evolution$_S$, and for co-evolution$_P$. One of the claims McCauley makes here is that cases of eliminative co-evolution rarely, if ever, take place between two disparate theoretical levels (interlevel elimination), but only occur within theoretical levels (intralevel elimination). Interlevel connections provide enormous heuristic value and lead to the enhancement of both lower-level and higher-level theories. In particular, theories at the psychological level are indispensable to lower-level neuroscientific theories, and vice versa.

In the last essay, the Churchlands respond to McCauley by challenging his claims about the rarity of interlevel elimination of theories. They also argue that, even if we accept the importance of higher-level theory, there is nothing to prevent us from replacing psychology with another theory at that same level (hence, a "co-level competitor"). The co-level competitor the Churchlands have in mind is contemporary connectionist theory, and the latter part of their essay provides us with a preliminary view of what intertheoretic and intratheoretic relations might look like with connectionist theory occupying the upper bunk.

22

Intertheoretic Reduction: A Neuroscientist's Field Guide

Paul M. Churchland and Patricia S. Churchland

"Reductionism" is a term of contention in academic circles. For some, it connotes a right-headed approach to any genuinely scientific field, an approach that seeks intertheoretic unity and real systematicity in the phenomena. It is an approach to be vigorously pursued and defended.

For others, it connotes a wrong-headed approach that is narrow-minded and blind to the richness of the phenomena. It is a bullish instance of "nothing-butery," insensitive to emergent complexity and higher-level organization. It is an approach to be resisted.

This latter reaction is most often found within the various social sciences, such as anthropology, sociology, and psychology. The former attitude is most often found within the physical sciences, such as physics, chemistry, and molecular biology. Predictably, then, the issue of reductionism is especially turbulent at the point where these two intellectual rivers meet: in the discipline of modern neuroscience.

The question at issue is whether it is reasonable to expect, and to work toward, a reduction of all psychological phenomena to neurobiological and neurocomputational phenomena. A large and still respectable contingent within the academic community remains inclined to say no. Their resistance is principled. Some point to the existence of what philosophers call *qualia*, the various subjective qualitative characters displayed in our sensations: think of pain, the smell of a rose, the sensation of redness, and so forth. These qualia, it is held, are beyond the possibility of any materialist explanation or reduction (Jackson, 1982; Nagel, 1974). Others point to the semantic content or *intentionality* of our thoughts, and make a similar claim about its irreducibility (Popper and Eccles, 1978; Searle, 1980, 1990). Others claim that the most important aspects of human behavior are explicable only in terms of high-level *emergent properties* and their correlative regularities, properties that irreducibly encompass the social level, properties such as loyalty to a moral ideal, perception of a political fact, or the recognition of a personal betrayal (Taylor, 1970, 1987). Yet others see a conflict with the important and deeply entrenched idea of *human freedom* (Popper and Eccles, 1978). Finally, some materialists raise what is called the problem of *multiple instantiation*. They point to the presumed fact that conscious intelligence could be sustained by physical systems other than the biochemistry peculiar to humans – by a system of transistors, for example – just as a nation's financial economy can be sustained by tokens other than silver coins and paper bills. But no one thinks that macroeconomics can be reduced to the chemistry of metals and paper. So why think that psychology should be reducible to the neurobiology of terrestrial humans? (Fodor, 1975).

Our aim here is threefold. First, we will try to provide a useful overview of the general nature of intertheoretic reduction, as it appears in the many examples to be found in the history of science. Expanding our horizons here is important, since little is to be learned from simply staring long and hard at the problematic case at issue, namely, the potential reduction of psychological phenomena to neural phenomena. Instead, we need to look at cases where the dust has already settled and where the issues are already clear. Second, we will identify the very real virtues that such cases display, and the correlative vices to be avoided. And finally, we will attempt to apply these historical lessons to the case here at issue – cognitive neuroscience – and we will try to meet the salient objections listed above.

1 Intertheoretic Reduction: Some Prototypical Cases

Since nothing instructs like examples, let us briefly examine some. One of the earliest cases of intertheoretic reduction on a grand scale was the reduction of Kepler's three laws of astronomical motion by the newly minted mechanics of Isaac Newton. Kepler's theory was specific to the motions of the solar planets, but Newton's theory at least purported to be the correct account of bodily motions in general. It was therefore a great triumph when Newton showed that one could deduce all three of Kepler's laws from his own theory, given only the background assumption that the mass of any planet is tiny compared to the great mass of the sun.

Kepler's three planetary laws are:

1 All planets move on ellipses with the sun at one locus.
2 A given planet always sweeps out equal areas in equal times.
3 The square of a planet's period is proportional to the cube of its mean orbital radius.

Newton's three laws of motion are:

1 Inertial motion is constant and rectilinear.
2 Acceleration = force/mass.
3 For any change in momentum something suffers an equal and opposite change in momentum.

To these laws we must add his gravitation law: $F = Gm_1m_2/R^2$.

Kepler's account thus turned out to be just a special case or a special application of Newton's more encompassing account. And astronomical motions turned out to be just a special instance of the inertial and force-governed motions of massive bodies in general. The divine or supernatural character of the heavens was thereby lost forever. The sublunary and the superlunary realms were thereby united as a single domain in which the same kinds of objects were governed by one and the same set of laws.

Newton's mechanics also provides a second great example of intertheoretic reduction, one that did not emerge until the nineteenth century. If his mechanics successfully comprehends motion at both the astronomical and the human-sized scales, then what, it was asked, about motions at the microscopic scale? Might these be accounted for in the same way?

The attempts to construct such an account produced another unification, one with an unexpected bonus concerning the theory of heat. If we assume that any confined body of gas consists in a swarm of submicroscopic corpuscles bouncing around inside the container according to Newton's three laws, then we can deduce a law describing the pressure they will collectively exert on the container's walls by repeatedly bouncing off them. This "kinetic" law has the form

$$PV = 2n/3 \cdot mv^2/2$$

This law had the same form as the then already familiar "ideal gas law,"

$$PV = \mu R \cdot T$$

(Here P is pressure and V is volume.) Although they are notationally different, the expressions "$2n/3$" and "μR" both denote the amount of gas present in the container (n denotes the number of molecules in the container; μ denotes the fraction of a mole). The only remaining difference, then, is that the former law has an expression for the *kinetic energy of an average corpuscle* ($mv^2/2$) in the place where the latter has an expression for *temperature* (T). Might the phenomenon we call "temperature" thus *be* mean kinetic energy at the molecular level? This striking convergence of principle, and many others like it, invited Bernoulli, Joule, Kelvin, and Boltzmann to say yes. As matters were further pursued, mean molecular kinetic energy turned out to have *all* of the causal properties that the classical theory had been ascribing to temperature. In short, temperature turned out to *be* mean molecular kinetic energy. Newtonian mechanics had another reductive triumph in hand. Motion at all three scales was subsumed under the same theory, and a familiar phenomenal property, *temperature*, was reconceived in a new and unexpected way.

It is worth emphasizing that this reduction involved identifying a familiar *phenomenal* property of common objects with a highly unfamiliar microphysical property. (By "phenomenal," we mean a property one can reliably discriminate in experience, but where one is unable to articulate, by reference to yet simpler discriminable elements, just how one discriminates that property.) Evidently, reduction is not limited to conceptual frameworks hidden away in the theoretical stratosphere. Sometimes the conceptual framework that gets subsumed by a deeper vision turns out to be a familiar piece of our commonsense framework, a piece whose concepts are regularly applied in casual observation on the basis of our native sensory systems. Other examples are close at hand: before Newton, *sound* had already been identified with compression waves in the atmosphere, and *pitch* with wavelength, as part of the larger reduction of commonsense sound and musical theory to mechanical acoustics. A century and a half after Newton, *light* and its various *colors* were identified with electromagnetic waves and their various wavelengths, within the larger reduction of geometrical optics by electromagnetic theory, as outlined by Maxwell in 1864. *Radiant heat*, another commonsense observable, was similarly reconceived as long-wavelength electromagnetic waves in a later articulation of the same theory. Evidently, the fact that a property or state is at the prime focus of one of our native discriminatory faculties does not mean that it is exempt from possible reconception within the conceptual framework of some deeper explanatory theory.

This fact will loom larger later in the chapter. For now, let us explore some further examples of intertheoretic reduction. The twentieth-century reduction of classical (valence) chemistry by atomic and subatomic (quantum) physics is another impressive case of conceptual unification. Here the structure of an atom's successive electron shells, and the character of stable regimes of electron-sharing between atoms, allowed us to reconstruct, in a systematic and thus illuminating way, the electronic structure of the many atomic elements, the classical laws of valence-bonding, and the gross structure of the periodic table. As often happens in intertheoretic reductions, the newer theory also allowed us to explain much that the old theory had been unable to explain, such as the specific heat capacities of various substances and the interactions of chemical compounds with light.

This reduction of chemistry to physics is notable for the further reason that it is not yet complete, and probably never will be. For one thing, given the combinatorial possibilities here, the variety of chemical compounds is effectively endless, as are their idiosyncratic chemical, mechanical, optical, and thermal properties. And for another, the calculation of these diverse properties from basic quantum principles is computationally daunting, even when we restrict ourselves to merely approximate results, which for purely mathematical reasons we generally must. Accordingly, it is not true that all chemical knowledge has been successfully reconstructed in quantum-mechanical terms. Only the basics have, and then only in approximation. But our experience here firmly suggests that quantum physics has indeed managed to grasp the underlying elements of chemical reality. We thus expect that any particular part of chemistry can be approximately reconstructed in quantum-mechanical terms, when and if the specific need arises.

The preceding examples make it evident that intertheoretic reduction is at bottom a relation between two distinct conceptual frameworks for describing the phenomena, rather than a relation between two distinct domains of phenomena. The whole point of a reduction, after all, is to show that what we thought to be two domains is actually one domain, though it may have been described in two (or more) different vocabularies.

Perhaps the most famous reduction of all is Einstein's twentieth-century reduction of Newton's three laws of motion by the quite different mechanics of the Special Theory of Relativity (STR). STR subsumed Newton's laws in the following sense. If we make the (false) assumption that all bodies move with velocities much less than the velocity of light, then STR entails a set of laws for the motion of such bodies, a set that is experimentally indistinguishable from Newton's old set. It is thus no mystery that those old Newtonian laws seemed to be true, given the relatively parochial human experience they were asked to account for.

But while those special-case STR laws may be experimentally indistinguishable from Newton's laws, they are logically and semantically quite different from Newton's laws: they ascribe an importantly different family of features to the world. Specifically, in every situation where Newton ascribed an intrinsic property to a body (e.g. mass, or length, or momentum, and so forth), STR ascribes a *relation*, a two-place property (e.g. *x* has a mass-relative-to-an-inertial-frame-F, and so on), because its portrait of the universe and what it contains (an unitary 4-D space-time continuum with 4-D world-lines) is profoundly different from Newton's.

Here we have an example where the special-case resources and deductive consequences of the new and more general theory are not identical, but merely similar, to the old and more narrow theory it purports to reduce. That is to say, the special-case reconstruction achieved within the new theory parallels the old theory with sufficient systematicity to explain why the old theory worked as well as it did in a certain domain, and to demonstrate that the old theory could be displaced by the new without predictive or explanatory loss within the old theory's domain; and yet the new reconstruction is not perfectly isomorphic to the old theory. The old theory turns out not just to be narrow, but to be false in certain important respects. Space and time are not distinct, as Newton assumed, and there simply are no intrinsic properties such as mass and length that are invariant over all inertial frames.

The trend of this example leads us toward cases where the new and more general theory does not sustain the portrait of reality painted by the old theory at all, even as a limiting special case or even in its roughest outlines. An example would be the outright displacement, without reduction, of the old phlogiston theory of combustion by Lavoisier's oxygen theory of combustion. The older theory held that the combustion of any body involved the *loss* of a spirit-like substance, phlogiston, whose precombustion function it was to provide a noble wood-like or metal-like character to the baser ash or calx that is left behind after the process of combustion is complete. It was the "ghost" that gave metal its form. With the acceptance of Lavoisier's contrary claim that a purely material substance, oxygen, was being somehow *absorbed* during combustion, phlogiston was simply eliminated from our overall account of the world.

Other examples of theoretical entities that have been eliminated from serious science include caloric fluid, the rotating crystal spheres of Ptolemaic astronomy, the four humors of medieval medicine, the vital spirit of pre-modern biology, and the luminiferous ether of pre-Einsteinian mechanics. In all of these cases, the newer theory did not have the resources adequate to reconstruct the furniture of the older theory or the laws that supposedly governed their behavior; but the newer theory was so clearly superior to the old as to displace it regardless.

At one end of the spectrum, then, we have pairs of theories where the old is smoothly reduced by the new, and the ontology of the old theory (that is, the set of things and properties that it postulates) survives, although redescribed, perhaps, in a new and more penetrating vocabulary. Here we typically find claims of cross-theoretic identity, such as "Heat is identical with mean molecular kinetic energy" and "Light is identical with electromagnetic waves." In the middle of the spectrum, we find pairs of theories where the old ontology is only poorly mirrored within the vision of the new, and it "survives" only in a significantly modulated form. Finally, at the other end of the spectrum we find pairs where the older theory, and its old ontology with it, is eliminated entirely in favor of the more useful ontology and the more successful laws of the new.

Before closing this quick survey, it is instructive to note some cases where the older theory is neither subsumed under nor eliminated by the aspirant and allegedly more general theory. Rather, it successfully resists the takeover attempt, and proves not to be just a special case of the general theory at issue. A clear example is Maxwell's electromagnetic theory (hereafter, EM theory). From 1864 to 1905 it was widely expected that EM theory would surely find a definitive reduction in terms of the mechanical properties of an all-pervading ether, the elastic medium in which EM waves were supposedly propagated. Though never satisfactorily completed, some significant attempts at reconstructing EM phenomena in mechanical terms had already been launched. Unexpectedly, the existence of such an absolute medium of luminous propagation turned out to be flatly inconsistent with the character of space and time as described in Einstein's 1905 Special Theory of Relativity. EM theory thus emerged as a fundamental theory in its own right, and not just as a special case of mechanics. The attempt at subsumption was a failure.

A second example concerns the theory of stellar behavior accumulated by classical astronomy in the late nineteenth century It was widely believed that the pattern of radiative behavior displayed by a star would be adequately explained in mechanical or in chemical terms. It became increasingly plain, however, that the possible sources of chemical and mechanical energy available to any star would sustain their enormous outpourings of thermal and luminous energy for only a few tens of millions of years. This limited time scale was at odds with the emerging geological evidence of a history numbered, in the *billions* of years. Geology notwithstanding, Lord Kelvin himself was prepared to bite the bullet and declare the stars to be no more than a few tens of millions of years old. The conflict was finally resolved when the enormous energies in the atomic nucleus were discovered. Stellar astronomy was eventually reduced all right, and very beautifully, but by quantum physics rather than by mere chemistry or mechanics. Another reductive attempt had failed, though it was followed by one that succeeded.

2 The Lessons for Neuroscience

Having seen these examples and the spectrum of cases they define, what lessons should a neuroscientist draw? One lesson is that intertheoretic reduction is a normal and fairly commonplace event in the history of science. Another lesson is that genuine reduction, when you can get it, is clearly a good thing. It is a good thing for many reasons, reasons made more powerful by their conjunction.

First, by being displayed as a special case of the (presumably true) new theory, the old theory is thereby *vindicated*, at least in its general outlines, or at least in some suitably restricted domain. Second, the old theory is typically *corrected* in some of its important details, since the reconstructed image is seldom a perfect mirror image of the old theory, and the differences reflect improvements in our knowledge. Third, the reduction provides us with a much *deeper insight* into, and thus a *more effective control* over, the phenomena within the old theory's domain. Fourth, the reduction provides us with a *simpler* overall account of nature, since apparently diverse phenomena are brought under a single explanatory umbrella. And fifth, the new and more general theory immediately *inherits all the evidence* that had accumulated in favor of the older theory it reduces, because it explains all of the same data.

It is of course a bad thing to try to force a well-functioning old theory into a procrustean bed, to try to effect a reduction where the aspirant reducing theory lacks the resources to do reconstructive justice to the target old theory. But whether or not the resources are adequate is seldom clear beforehand, despite people's intuitive convictions. And even if a reduction is impossible, this may reflect the old theory's radical falsity instead of its fundamental accuracy. The new theory may simply eliminate the old, rather than smoothly reduce it. Perhaps folk notions such as "beliefs" and "the will," for example, will be eliminated in favor of some quite different story of information storage and behavior initiation.

The fact is, in the neuroscience and psychology case there are conflicting indications. On the one side, we should note that the presumption in favor of an eventual reduction (or elimination) is far stronger than it was in the historical cases just examined. For unlike the earlier cases of light, or heat, or heavenly motions, in general terms we already know how psychological phenomena arise: they arise from the evolutionary and ontogenetic articulation of matter, more specifically, from the articulation of biological organization. We therefore *expect* to understand the former in terms of the latter. The former is produced by the relevant articulation of the latter.

But there are counterindications as well, and this returns us at last to the five objections with which we opened this chapter. From the historical perspective outlined above, can we say anything useful about those objections to reduction? Let us take them in sequence.

The first concerns the possibility of explaining the character of our subjective sensory qualia. The negative arguments here all exploit the very same theme, namely our inability to imagine how any possible story about the objective nuts and bolts of neurons could ever explain the inarticulable subjective phenomena at issue. Plainly this objection places a great deal of weight on what we can and cannot imagine, as a measure of what is and is not possible. It places more, clearly, than the test should bear. For who would have imagined, before James Clark Maxwell, that the theory of charged pith balls and wobbling compass needles could prove adequate to explain all the phenomena of light? Who would have thought, before Descartes, Bernoulli, and Joule, that the mechanics of billiard balls would prove adequate to explain the prima facie very different phenomenon of heat? Who would have found it remotely plausible that the pitch of a sound is a frequency, in advance of a general appreciation that sound itself consists in a train of compression waves in the atmosphere?

We must remember that a successful intertheoretic reduction is typically a complex affair, as it involves the systematic reconstruction of all or most of the old conception within the resources of the new conception. And not only is it complex, often the reconstruction is highly surprising. It is not something that we can reasonably expect anyone's imagination to think up or comprehend on rhetorical demand, as in the question, "How could *As possibly* be nothing but *Bs*?"

Besides, this rhetorical question need not stump us if our imagination is informed by recent theories of sensory coding. The idea that taste sensations are coded as a four-dimensional vector of spiking frequencies (corresponding to the four types of receptor on the tongue) yields a representation of the space of humanly possible tastes which unites the familiar tastes according to their various similarities, differences, and other relations such as betweenness (Bartoshuk, 1978). Land's retinex theory of color vision (Land, 1977) suggests a similar arrangement for our color sensations, with similar virtues. Such a theory also predicts the principal forms of color blindness, as when one's three-dimensional color space is reduced to two dimensions by the loss of one of the three classes of retinal cones.

Here we are already reconstructing some of the features of the target phenomena in terms of the new theory. We need only carry such a reconstruction through, as in the historical precedents of the objective phenomenal properties noted earlier (heat, light, pitch). Some things may indeed be inarticulably phenomenal in character, because they are the target of one of our basic discriminatory modalities. But this in no way makes them immune to an illuminating intertheoretic reduction. History already teaches us the contrary.

The second objection concerned the meaning, or semantic content, or intentionality of our thoughts and other mental states. The anti-reductionist arguments in this area are very similar to those found in the case of qualia. They appeal to our inability to imagine how meaning could be just a matter of how signals interact or how inert symbols are processed (Searle, 1980, 1990; for a rebuttal, see Churchland and Churchland, 1990. Searle, strictly speaking, objects only to a purely computational reduction, but that is an important option for neuroscience, so we shall include him with the other anti-reductionists.) Such appeals, as before, are really arguments from ignorance. They have the form, "I can't *imagine* how a neurocomputational account of meaningful representations could possibly work; therefore, it can't possibly work." To counter such appeals in the short term, we need only point out this failing.

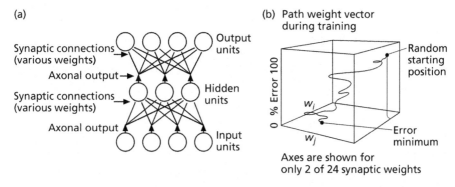

(a)

Synaptic connections
(various weights)

Axonal output→

Synaptic connections
(various weights)

Axonal output

Output
units

Hidden
units

Input
units

(b) Path weight vector
during training

Random
starting
position

% Error 100

0

w_i

w_j

Error
minimum

Axes are shown for
only 2 of 24 synaptic weights

Figure 22.1 (a) A simple feedforward artificial "neural network." Inputs are coded as a pattern or vector of activation levels across the input units. This pattern is conveyed toward the hidden units, but is transformed as it passes through the bank of intervening synaptic connections of various weights. Each hidden unit then sums the weighted inputs it receives and assumes an activation level appropriate to that sum. Thus results a second pattern or vector of activation levels across the hidden units. The story is repeated for the final layer of output units, which assumes a third pattern of activations. The network is thus a vector-to-vector transformer. Precisely what transformation it embodies is dictated by the specific configuration of its synaptic weights. With sigmoid output functions at the first two layers, networks of this kind can approximate any computable transformation whatever. (b) A schematic portrayal of an abstract "weight space" for the simple network. Learning in such a network consists in the successive modification of its weight configuration in order to incrementally reduce its performance error. This process continues until the network finally performs the input–output transformation implicit in the many examples on which it was trained (from P. M. Churchland, 1989).

To counter them in the long term requires more. It requires that we actually produce an account of how the brain represents the external world and the regularities it displays. But that is precisely what current theories of neural network function address. According to them, real-time information about the world is coded in high-dimensional activation vectors, and general information about the world is coded in the background configuration of the network's synaptic weights. Activation vectors are processed by the weight configurations through which they pass, and learning consists in the adjustment of one's global weight configuration (see figure 22.1). These accounts already provide the resources to explain a variety of things, such as the recognition of complex objects despite partial or degraded sensory inputs, the swift retrieval of relevant information from a vast content-addressable memory, the appreciation of diffuse and inarticulable similarities, and the administration of complex sensorimotor coordination (P. M. Churchland, 1989). We are still too ignorant to insist that hypotheses of this sort will prove adequate to explain all of the representational capacities of mind. But neither can we insist that they are doomed to prove inadequate. It is an empirical question, and the jury is still out.

The third objection complains that what constitutes a human consciousness is not just the intrinsic character of the creature itself, but also the rich matrix of relations it bears to the other humans, practices, and institutions of its embedding culture. A reductionistic account of human consciousness and behavior, insofar as it is limited to the microscopic activities in an individual's brain, cannot hope to capture more than a small part of what is explanatorily important.

The proper response to this objection is to embrace it. Human behavior is indeed a function of the factors cited. And the character of any individual human consciousness will be profoundly shaped by the culture in which it develops. What this means is that any adequate neurocomputational account of human consciousness must take into account the manner in which a brain comes to represent, not just the gross features of the physical world, but also the character of the other cognitive creatures with which it interacts, and the details of the social, moral, and political world in which they all live. The brains of social animals, after all, learn to be interactive elements in a community of brains, much to their cognitive advantage. We need to know how they do it.

This is a major challenge, one that neuroscientists have not yet addressed with any seriousness, nor even much acknowledged. This is not surprising. Accounting for a creature's knowledge of the spatial location of a fly is difficult enough. Accounting for its knowledge of a loved one's embarrassment, a politician's character, or a bargaining opponent's hidden agenda, represents a much higher level of difficulty. And yet we already know that artificial neural networks, trained by examples, can come to recognize and respond to the most astonishingly subtle patterns and similarities in nature. If physical patterns, why not social patterns? We confront no problem in principle here. Only a major challenge.

It may indeed be unrealistic to expect an exhaustive global account of the neural and behavioral trajectory of a specific person over any period of time. The complexity of the neural systems we are dealing with may forever preclude anything more than useful approximations to the desired ideal account. The case of chemistry and its relation to quantum physics comes to mind. There also, the mathematics of complex dynamical systems imposes limits on how easily and accurately we can reconstruct the chemical facts from the physical principles. This means that our reduction will never be truly complete, but we rightly remain confident that chemical phenomena are nothing but the macro-level reflection of the underlying quantum physical phenomena even so. As with chemical phenomena, so with psychological phenomena.

This brings us to the fourth objection, concerning the threat that a reduction would pose to human freedom. Here we shall be brief. Whether and in what sense there is any human freedom, beyond the relative autonomy that attaches to any complex dynamical system that is partially isolated from the world, is an entirely empirical question. Accordingly, rather than struggle to show that a completed neuroscience will be consistent with this, that, or the other preconceived notion of human freedom, we recommend that we let scientific investigation *teach us* in what ways and to what degrees human creatures are "free." No doubt this will entail modifications for some people's current conceptions of human freedom, and the complete elimination of some others. But that is preferable to making our current confusions into a standard that future theories must struggle to be consistent with.

The fifth and final objection claims an irreducibly abstract status for psychology, on grounds that a variety of quite different physical systems could realize equally well the abstract organization that constitutes a cognitive economy. How can we reduce psychological phenomena to neurobiology, if other physical substrates might serve just as well?

The premise of this objection will likely be conceded by all of us. But the conclusion against reduction does not follow. We can see this clearly by examining a case from our own scientific history. Temperature, we claimed earlier, is identical with mean molecular kinetic energy. But strictly speaking, this is true only for a gas, where the molecules are free to move in a ballistic fashion. In a solid, where the particles oscillate back and forth, their energy is constantly switching between a kinetic and a potential mode. In a high-temperature plasma, there are no molecules at all to consider, since everything has been ripped into sub-atomic parts. Here temperature is a complex mix of various energies. And in a vacuum, where there is no mass at all, temperature consists in the wavelength distribution – the "black-body curve" – of the EM waves passing through it.

What these examples show us is that reductions can be domain-specific: in a gas, temperature is one thing; in a solid, temperature is another thing; in a plasma, it is a third; in a vacuum, a fourth; and so on. (They all count as "temperature," since they interact, and they all obey the same laws of equilibrium and disequilibrium.) None of this moves us to say that classical thermodynamics is an autonomous, irreducible science, forever safe from the ambitions of the underlying microphysical story. On the contrary, it just teaches us that there is more than one way in which energy can be manifested at the microphysical level.

Similarly, visual experience may be one thing in a mammal, and a slightly different thing in an octopus, and a substantially different thing in some possible metal-and-semiconductor android. But they will all count as visual experiences because they share some set of abstract features at a higher level of description. That neurobiology should prove capable of explaining all psychological phenomena in humans is not threatened by the possibility that some *other* theory, say semiconductor electronics, should serve to explain psychological phenomena in *robots*. The two reductions would not conflict. They would complement each other.

We have elsewhere provided more comprehensive accounts of how recent work in neuroscience illuminates issues in psychology and cognitive theory (P. S. Churchland, 1986; P. M. Churchland, 1989). We conclude here with two cautionary remarks. First, while we have here been very upbeat about the possibility of reducing psychology to neuroscience, producing such a reduction will surely be a long and difficult business. We have here been concerned only to rebut the counsel of impossibility, and to locate the reductive aspirations of neuroscience in a proper historical context.

Second, it should not be assumed that the science of psychology will somehow disappear in the process, nor that its role will be limited to that of a passive target of neural explanation. On the contrary, chemistry has not disappeared through the quantum-mechanical explication of its basics; nor has the science of biology disappeared, despite the chemical explication of its basics. Moreover, each of these higher-level sciences has helped to shape profoundly the development and articulation of its underlying science. It will surely be the same with psychology and neuroscience. At this level of complexity, intertheoretic reduction does not appear as the sudden takeover of one discipline by another; it more closely resembles a long and slowly maturing marriage.

References

Bartoshuk, L. M. 1978: Gustatory system. In R. B. Masterton (ed.), *Handbook of Behavioral Neurobiology*, vol. 1, *Sensory Integration*, New York: Plenum Press, 503–67.

Churchland, P. M. 1989: *A Neurocomputational Perspective: The Nature of Mind and the Structure of Science*. Cambridge, MA: MIT Press.

Churchland, P. M., and Churchland, P. S. 1990: Could a machine think? *Scientific American*, 262, 32–7.

Churchland, P. S. 1986: *Neurophilosophy: Toward a Unified Understanding of the Mind/Brain*. Cambridge, MA: MIT Press.

Fodor, J. A. 1975: *The Language of Thought*. New York: Crowell.

Jackson, F. 1982: Epiphenomenal qualia. *Philosophical Quarterly*, 32, 127–36.

Land, E. 1977: The retinex theory of color vision. *Scientific American*, 237, 108–28.

Nagel, T. 1974: What is it like to be a bat? *Philosophical Review*, 83, 435–50.

Popper, K., and Eccles, J. 1978: *The Self and its Brain*. New York: Springer.

Searle, J. 1980: Minds, brains, and programs. *Behavioral and Brain Sciences*, 3, 417–57.

Searle, J. 1990: Is the brain's mind a computer program? *Scientific American*, 262, 26–31.

Taylor, C. 1970: Mind-body identity: A side issue? In C. V. Borst (ed.), *The Mind/Brain Identity Theory*, Toronto: Macmillan, 231–41.

Taylor, C. 1987: Overcoming epistemology. In K. Baynes, J. Uohman, and T. McCarthy (eds), *After Philosophy: End or Transformation?*, Cambridge, MA: MIT Press, 464–88.

23

Explanatory Pluralism and the Co-evolution of Theories in Science

Robert N. McCauley*

1 Introduction

Over the 1990s Patricia and Paul Churchland have made major contributions to philosophical treatments of intertheoretic reduction in science. The historic importance of this issue in the philosophy of science is patent and so, therefore, is the importance of the Churchlands' contributions. Their insistence on the centrality of this issue to discussions in the philosophy of mind may, however, be even more praiseworthy in an era when many in that field (even among those who claim the mantle of naturalism) make repeated declarations about the status of the pertinent sciences and the mind–body problem generally in what often appears to be blithe ignorance of both those sciences and the relevant literature in the philosophy of science since 1975.

In a recent, joint paper the Churchlands (1990) discuss and largely defuse five well-worn objections (concerning qualia, intentionality, complexity, freedom, and multiple instantiation) to the reduction of psychology to neurobiology. My concerns with that putative reduction and with the Churchlands' account of the overall process are of a very different sort.

Two models have traditionally dominated discussions of intertheoretic relations. After briefly surveying the contrasts between them, section 2 examines how the Churchlands' account of these relations in terms of a continuum of intertheoretic commensurability captures those models' respective advantages in a single proposal. That section ends by examining how Patricia Churchland's subsequent discussions of the co-evolution of theories enhances this account by exploring some of its underlying dynamics. In short, the co-evolution of theories concerns cross-scientific interactions that change the position of a particular intertheoretic relationship on the Churchlands' continuum.

In section 3 I locate some revealing equivocations in the Churchlands' discussions of "the co-evolution of theories" by distinguishing three possible interpretations of that notion that wind their ways through the Churchlands' work and through *Neurophilosophy* in particular. With the aid of a distinction concerning levels of analysis that I have developed elsewhere, I argue, in effect, that the Churchlands' account of the co-evolution of theories and their model of intertheoretic reduction obscure critical distinctions between three quite different types of intertheoretic relations. Section 4 positions these three types within a more fine-grained account of intertheoretic relations that will offer a basis for evaluating their relative merits as analyses of the interface of psychology and neuroscience.

One of these three, the picture of co-evolution modeled on the dynamics of scientific revolutions, has attracted the most attention. This interpretation has encouraged the recurring eliminativist inclinations concerning folk psychology for which the Churchlands are renowned, but, of the three, it is also the interpretation that is least plausible as an analysis of the relations between psychology and neuroscience. Psychology (folk or otherwise) may well undergo substantial revision, and future scientific progress may well lead to the elimination of some psychological theories, but the Churchlands have offered an unhelpfully oversimplified account of the intertheoretic dynamics in question.

In section 5 I shall support and elaborate upon another of the interpretations of co-evolution that emerges from *Neurophilosophy* by, among other things, examining a case (concerning the connectionist network, NETtalk) that the Churchlands and their collaborators have highlighted. This third interpretation recognizes not merely the value of integrating scientific disciplines but of preserving a plurality of semi-autonomous explanatory perspectives. Although the Churchlands now often seem to favor this third interpretation too, some of their comments continue to conflate the three distinct types of intertheoretic relations.

2   Three Philosophical Models of Intertheoretic Relations in Science

Until the late 1970s (at least) two models of intertheoretic relations in science dominated philosophers' attention. The first, a general-purpose model of intertheoretic relations, was deeply rooted in logical empiricism; the second, in effect a model of theory change, emerged largely in reaction to the first (Bechtel, 1986). I shall briefly discuss them in order.

Although Ernest Nagel's *The Structure of Science* (1961) contains the most time-honored treatment of theory reduction, Robert Causey's *Unity of Science* (1977) probably provides the most comprehensive discussion of the topic. Their general approach to theory reduction proceeds within the constellation of commitments that characterize logical empiricism, including the assumptions that a satisfactory account of scientific rationality requires heed to justificatory considerations only, that scientific theories are best understood as complex propositional structures and best represented via formal reconstructions, that scientific explanation results from the deduction of explananda from scientific laws, that scientific progress results from the subsumption of reigning theories by theories of even greater generality, and that science ultimately enjoys an underlying unity of theory and ontology.

This model conceives theory reduction as a special case of deductive-nomological explanation. It is a special case because the explanandum is not a statement describing some event but rather a law of the reduced theory. In order to carry out such reductions, the premises in the most complex cases of heterogeneous reductive explanations must include

1 at least one law from the reducing theory;
2 statements indicating the satisfaction of the requisite initial conditions specified in that law;
3 bridge laws which systematically relate – *within a particular domain* delineated by appropriate boundary conditions – the terms from the pertinent law(s) of the reducing theory to those from the law of the reduced theory;
4 statements indicating the satisfaction of those boundary conditions (under which the events described in the law of the reducing theory realize the events described in the law of the reduced theory that is to be explained).

Such premises permit a straightforward deduction of the law of the reduced theory.

Because the boundary conditions included in the bridge laws are cast in terms of predicates characteristic of the reducing theory, the reduction reflects an asymmetry between the two theories. The reducing theory explains the reduced theory, finally, because the reducing theory encompasses a wider array of events within its explanatory purview. This set of events, presumably, includes all of the events the reduced theory explains and more, so that the principles of the reducing theory are both more general and more fundamental. The most popular showcase illustration is the reduction of the laws of classical thermodynamics to the principles of statistical mechanics.

When the reducing theory operates at a lower level of analysis than the reduced theory, the added generality of its principles is a direct function of this fact. These are cases of *microreductions* where a lower-level theory and its ontology reduce a higher-level theory and its ontology (Oppenheim and Putnam, 1958; Causey, 1977). Microreductionists hold that if we can exhaustively describe and predict upper-level (or macro) entities, properties, and principles in terms of lower-level (or micro) entities, properties, and principles, then we can reduce the former to the latter and replace, at least in principle, the upper-level theory.

Virtually all discussions of intertheoretic relations presuppose this arrangement among (and within) the sciences in terms of levels of analysis (see, for example, Churchland and Sejnowski, 1992, pp. 10–11). Numerous considerations contribute to the depiction of the architecture of science as a layered edifice of analytical levels (Wimsatt, 1976). Ideally, moving toward lower levels involves moving toward the study of increasingly simple systems and entities that are ubiquitous, enduring, and small. Conversely, moving from lower- to higher-level sciences involves moving toward studies of larger, rarer systems of greater complexity and (often) less stability and whose history is less ancient. Because the altitude of a level of analysis is directly proportional to the complexity of the systems it treats, higher-level sciences deal with increasingly restricted ranges of events having to do with increasingly organized physical systems.[1]

As a simple matter of fact, often more than one configuration of lower-level entities can realize various higher-level kinds (especially when functionally characterized). The resulting multiple instantiations highlight both the importance and the complexity of the boundary conditions in the bridge laws of heterogeneous microreductions. Critics of the microreductionist program (e.g. Fodor, 1975) see that complexity as sufficient grounds for questioning the program's feasibility in the case of the special sciences, while more sympathetic participants in these discussions such as Robert Richardson (1979) and the Churchlands (1990) suggest that when scientists trace out such connections between higher- and lower-level entities in specific domains they vindicate the overall strategy, while recognizing the *domain specificity* of its results.

Reductionists differ among themselves as to the precise connections between entities at different levels that are required for successful reductive explanation. They all agree, however, that the theories which are parties to the reduction should map on to one another well enough to support systematic connections, usually contingent identities, between some, if not all, of the entities that populate them. The test of the resulting contingent identities is met, ultimately, by the explanatory successes the reductions accomplish (McCauley, 1981; Enc, 1983).

Feyerabend (1962) and Kuhn (1970) are the most prominent proponents of the second major account of intertheoretic relations. They forged their early discussions largely in response to both the logical empiricist program and its reductionist blueprint for scientific progress. Feyerabend emphasized how scrutiny of many of the showcase illustrations of intertheoretic reductions revealed the *failure* of these cases to conform to the logical empiricists' model. Kuhn discussed numerous examples in the history of science where successive theories were not even remotely plausible candidates for the sort of smooth transitions the standard reductive model envisions. Instead, Kuhn proposed that progress in science consists of extended intervals of relative theoretical stability punctuated by periodic revolutionary upheavals. Both hold that the cases in question involve conflicts between *incommensurable* theories.

Although the subsequent literature is rife with assessments of this claim (Thagard, 1992, offers the most suggestive of recent treatments), the critical point for now is that, whatever incommensurability amounts to, it stands in stark opposition to any model of intertheoretic relations that requires neat mappings between theories' principles and ontologies capable of supporting strict deductive-nomological explanations. The history of science provides ample evidence that where such incompatibility is sufficiently severe the theory and its ontology that are eventually deemed deficient undergo elimination. Stahl's system of chemistry is the preferred illustration, but Darwin's theory of inheritance could serve just as well.

The unmistakable sense that both of these models of intertheoretic relations describe some actual cases fairly accurately and that they each capture important insights about the issues at stake, their profound conflicts notwithstanding, could induce puzzlement. An account of intertheoretic relations in terms of a continuum of commensurability that Paul Churchland (1979) initially sketched and which the Churchlands have subsequently developed (P. S. Churchland, 1986, pp. 281f; Churchland and Churchland, 1990) substantially resolves that perplexity by reconciling those conflicts and allotting to each model a measure of descriptive force.

The Churchlands point out that, in fact, different cases of intertheoretic relations vary considerably with respect to the commensurability of the theories involved. So, they propose that such cases fall along a continuum of relative intertheoretic commensurability, where, in effect, the two models sketched above constitute that continuum's end points.

One end of the continuum represents cases where intertheoretic mapping is extremely low or even absent. These are cases of *radical* incommensurability where revolutionary science and the complete elimination of inferior theories ensue. Whatever vagueness may surround the notion of "incommensurability," the Churchlands are clearly confident that the developments which brought about the elimination of the bodily humors, the luminiferous ether, caloric fluid, and the like, involve sufficiently drastic changes to justify the sort of extreme departures from the traditional model of reduction that Kuhn and Feyerabend advocated.

At the other end of this continuum, where the mapping of one theory on another is nearly exhaustive and the former theory's ontology is composed from the entities the latter theory countenances, the most rigorous models of theory reduction most nearly apply (e.g. Causey, 1972). The constraints that proponents have imposed on theory reduction are so demanding that it is a fair question whether *any* actual scientific case qualifies. The Churchlands have urged considerable relaxation of the conditions necessary for intertheoretic reduction. Instead of conformity to the rigorous logical and ontological constraints traditional models impose, Paul Churchland (1979; see also Hooker, 1981; Bickle, 1992) suggests that the reducing theory need only preserve an "equipotent image" of the reduced theory's most central explanatory principles. The reduction involves an *image*, since the reducing theory need not duplicate every feature of the reduced theory's principles, but only enough of their salient ones to suggest their general character and to indicate their systematic import (see Schaffner, 1967). That image is *equipotent*, though, since the reducing theory's principles will possess all of the explanatory and predictive power of the reduced theory's principles – and more. From the standpoint of traditional models, Churchland proposes a form of *approximate* reduction, which falls well short of the logical empiricists' standards, but which also suggests how true theories (e.g. the mechanics of relativity) can correct and even approximately reduce theories that are false (e.g. classical mechanics). Switching to the metaphor of imagery is appropriate, since, as William Wimsatt, (1976, p. 218) noted over a decade ago, if the standard models of reduction allege that a false theory follows from a true one, the putative deduction had better involve an equivocation somewhere!

In recent years the Churchlands have each enlarged on this continuum model. For example, within his neurocomputational program Paul Churchland has advanced a prototype activation model of explanatory understanding that, presumably, includes the understanding that arises from *reductive* explanations.

Churchland holds that the neurocomputational basis of explanatory understanding resides in the activation of a prototype vector within a neural network in response to impinging circumstances. A distributed representation of the prototype in the neural network constitutes the brain's current best stab at detecting an underlying pattern in the blooming, buzzing confusion. For Churchland explanatory understanding is an array of inputs leading to the activation of one of these existing prototypes as opposed to another.

Churchland insists that the activation of a prototype vector increases, rather than diminishes, available information. It involves a "speculative *gain*" in information (1989, p. 212). Thus, contrary to anti-reductionist caricature, this account of explanatory understanding implies that reductive explanations *amplify* our knowledge. The originality of the insights that a reductive explanation offers depends upon the novel application of existing cognitive resources, that is, of an individual's repertoire of prototype vectors. Consequently, reductive explanation involves neither the generation of new schemes nor the destruction of old ones. The approximate character of intertheoretic reductions is a function of this "conceptual redeployment" on which they turn (1989, p. 237). In conceptual redeployment a developed conceptual framework from one domain is enlisted for understanding another. In short, successful reductive explanation rests on an analogical inference by virtue of which we deem an image of a theory equipotent to the original. Having established the initial applicability of an existing, alternative prototype vector, it inevitably undergoes a reshaping as a consequence of exposure to the newly adopted training set. This reshaping of activation space is the neurocomputational process that drives the remaining co-evolution of the reductively related theories.

In her discussions of the co-evolution of theories, Patricia Churchland has introduced a dynamic element into the continuum model. She suggests that the position of two theories' relations on this continuum can change over time as they each undergo adjustments in the light of one another's progress.

The suggestion that scientific theories co-evolve arises from an analogy with the co-evolution of species and from the picture of the sciences briefly outlined above. On the co-evolutionary picture the sciences exert selection pressures on one another in virtue of a general concern for supplying as much coherence as possible among our explanatory schemes. If the various sciences are arranged in tiers of analytical levels, then each will stand at varying distances from the others in this structure. Typically, proximity is a central consideration in assessing the force of selection pressures. Thus, the pivotal relationships are those between a science and those sciences at immediately adjacent levels. For example, the presumption is that the neurosciences below and the socio-cultural sciences above are more likely to influence psychology than are the physical sciences, since they are located below the neurosciences and, therefore, at an even greater distance.

It is this process of the co-evolution of theories and the Churchlands' account of it that will dominate the remainder of this chapter. I shall attend to its implications for the relationship of cognitive psychology to the sort of neurocomputational modeling that the Churchlands endorse.

3   Three Ways Theories Might Co-evolve

Patricia Churchland's *Neurophilosophy* (1986) contains the most extensive available discussion of reduction in terms of the co-evolution of theories.[2] Churchland focuses on the relation between neuroscience and psychology, but her discussion clearly aspires to morals that are general. Her comments at various points seem to support three different co-evolutionary scenarios, though two of them are, quite clearly, closely related. The three are distinguished by the locations on the Churchlands' continuum to which they predict co-evolving theories will incline.

On some occasions Churchland suggests that psychology and the neurosciences will co-evolve in the direction of approximate reduction. She states, for example, that "the co-evolutionary development of neuroscience and psychology means that establishing points of reductive contact is more or less inevitable. . . . The heart of the matter is that if there is theoretical give and take, then the two sciences will knit themselves into one another" (1986, p. 374). The metaphor of two sciences knitted into one another implies an integration that is tight, orderly, and detailed. Although Churchland, presumably, does not think that integration will satisfy the traditional microreductionists' stringent demands on intertheoretic mapping, talk of knitting two sciences into one another, the ongoing pursuit of a unified model of *reduction* (Churchland and Churchland, 1990), and a new interest in establishing psy-chophysical identities echo commitments of traditional microreductionism, where the sort of reductive contact in question led to talk of an "in-principle replaceabil-ity" of the reduced theory in which the lower-level theory enjoys both explanatory and metaphysical priority. More recently, the Churchlands have been clear about the futility of attempts to replace upper-level theories, but they still generally subscribe to the explanatory and metaphysical priority of the lower-level theory – especially in the case of psychology and neuroscience (see, for example, P. S. Churchland, 1986, pp. 277, 294, 382). Certainly, a co-evolutionary account of intertheoretic relations has no problem translating the general microreductive *impulse*. Within this frame-work it amounts to the claim that the selection pressures exerted by the science at the level of analysis *below* that of the theory in question will have an overwhelmingly greater effect on that theory's eventual shape and fate than will the sciences above (see section 5 below).

On the Churchlands' account, such intertheoretic integration would enable the neurosciences to supply an equipotent image of psychological principles. Paul Churchland's speculations about the neural representation of the sensory qualia associated with color vision might constitute an appropriate illustration. The fit between our commonsense notions about our experiences of colors and the system of neural representation he proposes is quite neat (1989, pp. 102–8). Hereafter I shall refer to this sense of the co-evolution of theories as "co-evolution$_M$," that is, co-evolution in the direction of approximate microreduction. In the Churchlands' joint discussion (1990, chapter 6.1), where it plays both a predictive and normative role, this notion of reduction receives considerable attention. The Churchlands clearly hold that "it is reasonable to expect, and to work toward, a reduction of all psychological phenomena to neurobiological and neurocomputational phenomena" (1990, p. 249).

Co-evolution$_M$ is not the only account of co-evolution in *Neurophilosophy*, for, as the Churchlands have subsequently asserted, in the case of psychology and neuroscience, "there are conflicting indications" about the direction in which conjectures at these two levels of analysis will likely co-evolve (1990, p. 253). If integration is the fate of psychology and neuroscience, Patricia Churchland repeatedly hints that this will only occur *after* psychology's initial demolition and subsequent reconstruction in accord with the mandates of the neurosciences. She claims, for example, that "the possibility that psychological categories will not map one to one on to neurobiological categories . . . does not look like an obstacle to reduction so much as it predicts a fragmentation and reconfiguration of the psychological categories" (1986, p. 365). With this second view, as with the first, no question arises about where the blame lies, if the theories of psychology and neuroscience fail to map on to one another neatly (see Wimsatt, 1976). *At least for the short term*, Churchland seems to expect that this intertheoretic relation will migrate in just the *opposite* direction on the continuum of intertheoretic commensurability from what co-evolution$_M$ predicts, that is, toward a growing *in*commensurability that predicts a fragmentation of *psychological* categories.

If the "fragmentation and reconfiguration" of psychological categories involved only the elaboration or adjustment (or even the in-principle replaceability) of psychological theories by discoveries in the neurosciences, co-evolution$_M$ might suffice. On this second view, though, this process can lead to the eventual eradication of major parts of psychology. So, for example, Churchland remarks that "there is a tendency to assume that the capacities at the cognitive level are well defined . . . in the case of memory and learning, however, the categorial definition is far from optimal, and *remembering stands to go the way of impetus*" (1986, p. 373; emphasis added[3]). Here Churchland anticipates that just as the new physics of Galileo and his successors ousted the late medieval theory of impetus, so too shall advances in neuroscience dispose of psychologists' speculations about memory. This, then, is co-evolution$_S$ (co-evolution producing the eliminations of theories characteristic of scientific revolutions) in which the theoretical perspectives of two neighboring sciences are so disparate that eventually the theoretical commitments of one must go – in the face of the other's success.

Co-evolution$_S$ underlies the position for which the Churchlands' advocacy has been famous, namely, eliminative materialism.[4] They have contended that progress in the neurosciences will probably bring about the elimination of folk psychology as well as any other psychological theories that involve commitments to the propositional attitudes (presumably including much of mainstream cognitive and social psychology). Just as scientists banished phlogiston and caloric fluid, so too will the propositional attitudes be expelled as neuroscience progresses. The psychological conjectures in question (will) fail to match the descriptive, explanatory, and predictive successes of their neuroscientific competitors. Moreover, their substantial dissimilarities to those alleged competitors preclude any sort of reconciliation. Consequently, numerous theoretical notions in psychology stand to go the way of impetus. This is the predicted result when the Churchlands emphasize, among those "conflicting indications," the *uncongenial* relations between psychology and neuroscience.

Revising their extreme eliminativism, the Churchlands sometimes seem to intend these two interpretations to address different stages in the co-evolutionary process (as I suggested above): first, the demolition of much current psychology via co-evolution$_S$ followed by, second, the reconstruction of a neuroscientifically inspired psychology via co-evolution$_M$. The crucial point for now is that these two interpretations of co-evolution hold that the relationship between two theories will, over time, shift in one direction (as opposed to the other) on the Churchlands' continuum.

An obvious question arises, though. If either direction is possible, then what are the variables that determine the direction of any shift? (This question presses the revised version of eliminativism no less than the original.) The Churchlands have not addressed this question directly, because they have recognized that the complexities of the intertheoretic relations in question and of the relationship of psychology to neuroscience, in particular, require more. Enter the third interpretation.

One of Patricia Churchland's extended comments about her general model of reduction (1986, pp. 296–7) is especially revealing, since it reflects at various points the influence of all three interpretations.

> . . . some misgivings may linger about the possibility of reduction should it be assumed that a reductive strategy means an exclusively bottom-up strategy. . . . These misgivings are really just bugbears, and they have no place in my framework for reduction.
>   . . . if the reduction is smooth, its reduction gives it [the reduced theory] – and its phenomena – a firmer place in the larger scheme. . . . If the reduction involves a major correction, the corrected, reduced theory continues to play a role in prediction and explanation. . . . Only if one theory is eliminated by another does it fall by the wayside.
>   . . . coevolution . . . is certain to be more productive than an isolated bottom-up strategy.

The second paragraph traces points on the continuum. It alludes initially to co-evolution$_M$ – its final sentence to co-evolution$_S$. It is the first and third paragraphs, though, where shadows of a third interpretation appear.

Closely related to co-evolution$_M$ is co-evolution$_P$ (co-evolution as explanatory pluralism). Their many similarities notwithstanding, it is worth teasing them apart. As a first pass, where co-evolution$_M$ anticipates increasing intertheoretic integration largely guided by and with a default preference for the lower level, co-evolution$_P$ construes the process as preserving a diverse set of partially integrated yet semiautonomous explanatory perspectives – where that non-negligible measure of analytical independence rests at each analytical level on the explanatory success and the epistemic integrity of the theories and on the suggestiveness of the empirical findings. Co-evolution$_M$, in effect, holds that selection pressures are exerted exclusively from the bottom up, whereas co-evolution$_P$ attends to the constraints imposed by the needs and demands of theories operating at higher levels.

These apparently small differences are but the fringe skirmishes of some of the most basic epistemological and metaphysical battles in the philosophy of science. Space limitations preclude extensive development, but broadly, if they are not persuaded by co-evolution$_S$, physicalists prefer co-evolution$_M$, since it suggests a science unified in both theory and ontology that accords priority to the lower (i.e. physical) levels. More pragmatically minded philosophers opt for co-evolution$_P$, forgoing assurances of and worries about a unified science and metaphysical purity in favor of enhanced explanatory resources. For nearly a decade now the Churchlands have been negotiating their interests in unified science and metaphysical purity on the one hand with their interests in enhanced explanatory resources and internalism on the other (see McCauley, 1993, and note 10 below). The relaxation of their eliminativism and their emerging preference for co-evolution$_P$ indicate the influence of pragmatic currents in their thought.

Co-evolution$_P$ is prominent in *Neurophilosophy* and even more so in subsequent work.[5] Patricia Churchland claims that "the history of science reveals that co-evolution of theories has typically been mutually enriching," that "[r]esearch influences go up and down and all over the map," that "co-evolution typically is . . . interactive . . . and involves one theory's being susceptible to correction and reconceptualization at the behest of the cohort theory," and that "psychology and neuroscience should each be vulnerable to disconfirmation and revision at any level by the discoveries of the other" (1986, pp. 363, 368, 373, 376).

Figure 23.1 seems the most plausible interpretation of the relationship between these three notions of co-evolution and the earlier continuum model; it roughly indicates the regions of that continuum where the cases covered by the three types of co-evolution end up (see Churchland and Churchland, 1990, p. 252).

Section 4 below will suggest that in figure 23.1 the picture of intertheoretic relations and of co-evolution, in particular, is oversimplified to the point of distortion. The intertheoretic dynamics of scientific revolutions are quite different from those of approximate microreduction and explanatory pluralism. Crucially, co-evolution$_P$ is incompatible with co-evolution$_S$. The *mutual* intertheoretic enrichment co-evolution$_P$ envisions will not arise, if neuroscience is radically reconfiguring (let alone eliminating) psychology. Neither the history of science nor pragmatic accounts of scientific practice offer much reason to think that co-evolution$_S$ provides either an accurate description or a useful norm for the relationship between psychology and neuroscience or for any such relationship between theories in sciences operating at *different* analytical levels.

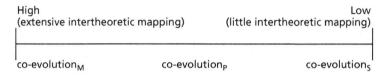

Figure 23.1   Three notions of co-evolution situated on the Churchlands' continuum.

The differences between co-evolution$_P$ and co-evolution$_M$ are also important. At stake is the question of the relative priority of neuroscientific (lower-level) and psychological (upper-level) contributions to the science of the mind/brain. This topic will dominate section 5. In criticizing co-evolution$_S$ and curtailing co-evolution$_M$, the aim of the next two sections is, ultimately, to endorse and develop the notion of explanatory pluralism.

4   Exploring Explanatory Pluralism: Debunking Co-evolution$_S$

Enlisting a distinction Wimsatt (1976) introduced between *intralevel* and *interlevel* contexts, I have previously developed a model of intertheoretic relations that discloses why we should not expect advances in neuroscience to eliminate much psychology directly (McCauley, 1986). More generally, it suggests that co-evolution$_S$ does not very happily model the co-evolving relations of theories at different levels.

The sorts of unequivocal eliminations of theories and ontologies that co-evolution$_S$ countenances arise in *intralevel* contexts involving considerable incommensurability. These contexts concern changes *within* a particular science over time. They include the classic cases that philosophers group under the rubric of "scientific revolutions" – impetus, phlogiston, caloric fluid, and the like. Within a particular level of analysis some newly proposed theory proves superior to its immediate predecessor with which it is substantially discontinuous. When the scientific community opts for this new theory, most traces of its predecessor rapidly disappear. Since they offer incompatible accounts of many of the same phenomena, the new theory *explains* the old theory *away*.

By contrast, intralevel situations where the mappings between theories are reasonably good fall near the other end of the Churchlands' continuum. Here the new theory *explains* its predecessor, which it also typically corrects. Scientists regard the earlier theory's domain as a special case to which the new theory applies and for which the old theory continues to suffice as a useful calculating heuristic. Although corrected and incorporated as a special case into a more general theory, Newton's laws of motion work well for most practical purposes.

A new theory disrupts science less to the extent it preserves (rather than overthrows) the cherished insights and conceptual apparatus of its predecessors. It may require reinterpretation of established notions ("planets," "genes," "grammar acquisition," etc.), but changes are evolutionary only when they preserve a fair measure of intensional and extensional overlap with their predecessors. When succeeding theories in some science are largely continuous, no one speaks of elimination. The change is evolutionary, not revolutionary. Consequently, the new theory is perfectly capable of providing an equipotent image of the old. These are the cases where the new theory overwhelmingly *inherits* the evidence for the old.

Revolutionary or evolutionary, progress within some science eventually eliminates features of earlier theories. In revolutionary settings the changes are abrupt and the elimination is (relatively) immediate. In evolutionary contexts incompatibilities accrue over time. Although the transition from one theory to its immediate successor may be more or less smooth, over a series of such transitions all traces of ancestral theories may completely disappear. Consider the fate of "natural motions" from Aristotelian through Newtonian mechanics (McCauley, 1986, pp. 192–3). Similarly, over the past hundred years the "memory trace" has undergone considerable evolutionary transformation. Some theorists would argue that the reinterpretations have been so substantial that the original notion (and what it allegedly referred to) has virtually vanished.[6]

Interlevel relations concern theories at *different* (typically neighboring) levels of analysis at a particular point in time (in contrast to intralevel cases concerned with successive theories at the *same* level of analysis). The Churchlands' continuum maps on to interlevel cases too.

When sciences at adjoining levels enjoy substantial intertheoretic mapping (in situations approximating classic microreductions) they heavily constrain one another's form – otherwise, why would anyone have attempted to characterize their relations in terms of deductive logic and identity statements? This is the effect of the knitting of two sciences into one another that co-evolution$_M$ envisions. A well-integrated lower-level theory has resources sufficient to reproduce the explanatory and predictive accomplishments of the corresponding upper-level theory; however, this often comes at considerable computational expense. As the Churchlands have emphasized, this does not disgrace the higher-level theory or lead to the evaporation of the phenomena it seeks to explain.

When considering interlevel cases with relatively unproblematic intertheoretic relations, the Churchlands, like the traditional reductionists before them,[7] have focused exclusively on their *resemblances* to the intralevel settings described above (see, for example, P. S. Churchland, 1986, p. 294). After all, here too the elaborations of the upper-level theory's central concepts that the lower-level, *reducing* theory offers often correct the less fine-grained, upper-level theory's pronouncements. However, because the theories are tightly knit, the upper-level theory still provides a useful and efficient approximation of the lower-level theory's results. This sounds quite like the cases of scientific evolution described above.

Beneath these resemblances, though, lie small but revealing differences. First, unlike the evolutionary intralevel cases, the reduced theory in interlevel situations does not stand in need of technical correction in every case. For a few situations at least, its results will conform precisely with those of the lower-level theory, because for these cases it adequately summarizes the effects of all relevant lower-level variables.[8] This contrasts with the *inescapable*, if often negligible (from a practical standpoint), divergence of the calculations of some theory and its successor, such as classical mechanics and the mechanics of relativity (see Churchland and Churchland, 1990, p. 251). In interlevel cases corrections can arise because the upper-level theory is insufficiently fine-grained to handle certain problems. By contrast, in intralevel cases corrections always arise because the earlier theory is wrong – by a little in evolutionary cases, by a lot in revolutionary ones. It follows that the upper-level theory is not always a *mere* calculating heuristic (as the replaced predecessor is in cases of scientific evolution). Moreover, the upper-level theory's heuristic advantages in well-integrated interlevel contexts are typically enormous, compared with intralevel cases. The divergence of computational effort between the classical and statistical solutions for simple problems about gases (an interlevel case) dwarfs that between classical mechanics and the mechanics of relativity for simple problems about motion (an intralevel case). Of a piece with this observation, the Churchlands quite accurately describe the quantum calculations of various chemical properties (another interlevel case) as "daunting" (1990, p. 251). Finally, the upper-level theory lays out regularities about a subset of the phenomena that the lower-level theory encompasses but for which it has neither the resources nor the motivation to highlight. That is the price of the lower-level theory's generality and finer grain.

If these considerations are not compelling, scrutiny of interlevel circumstances that support relatively little intertheoretic mapping reveals far more important grounds for stressing the distinction between interlevel and intralevel settings. Here two sciences at adjacent levels address some common explananda under different descriptions, but their explanatory stories are largely (though not wholly) incompatible. On the Churchlands' view, this is just the relationship between neuroscience and most of folk psychology, and if remembering is to go the way of impetus, the relationship between neuroscience and some important parts of scientific psychology as well.

If all of these intertheoretic relations should receive a *unified* treatment, as traditional reductionists, the Churchlands (e.g. Churchland and Sejnowski, 1990, p. 229), and figure 23.1 suggest, then it is perfectly reasonable to expect elimination in those interlevel situations involving significant incommensurability. The problem, though, is that neither the history of science, nor current scientific practice, nor the scientific research the Churchlands champion, nor a concern for explanatory pluralism offers much reason to expect theory elimination in such settings.

Incommensurability in interlevel contexts neither requires the elimination of theories on principled grounds nor results in such eliminations in fact. Admittedly, in the early stages of a science's history it is not always easy to distinguish levels of analysis and, consequently, to distinguish what would count as an interlevel, as opposed to an intralevel, elimination. Crucially, though, the history of science and especially the history of late nineteenth- and twentieth-century science offer no examples of large-scale interlevel theory elimination (particularly of the wholesale variety standard eliminativism and co-evolutions envision) once the upper-level science achieves sufficient historical momentum to enjoy the accoutrements of other recognized sciences (such as characteristic research techniques and instruments, journals, university departments, professional societies, and funding agencies). The reason is simple enough. Mature sciences are largely defined by their theories and, more generally, by their research traditions (Laudan, 1977); hence, elimination of an upper-level theory by a lower-level theory may risk the elimination of the upper-level scientific enterprise! (Presumably, this is why Nagel always spoke of the reduction of a *science*, rather than of a theory, when addressing interlevel cases.)

A motive for undertaking interlevel investigation (especially when the intertheoretic connections are not plentiful) is to explore one science's successful problem-solving strategies as a means to inspire research, provoke discoveries, and solve recalcitrant problems at another level. (Bechtel and Richardson, 1993, focus in particular on the problem of understanding the operation of mechanisms.) Monitoring developments in theories at neighboring levels is often a fruitful heuristic of discovery. The strategy's fruitfulness depends precisely on the two sciences maintaining a measure of independence from one another.

This is the mark of explanatory pluralism and co-evolution$_p$. A paucity of interlevel connections only enhances the (relative) integrity and autonomy of the upper-level science. As Wimsatt notes: "in interlevel reduction, the more difficult the translation becomes, the more *irreplaceable* the upper-level theory is! It becomes the only practical way of handling the regularities it describes" (1976, p. 222). The theories at the two levels possess different conceptual and explanatory resources, which underscore different features of their common explanandum. They provide multiple explanatory perspectives that should be judged on the basis of their empirical success – not on hopes about their putative promise for the theoretical (or ontological) unification of science. For the pragmatically inclined, explanatory success is both sufficiently valuable and rare that it would be imprudent to encourage the elimination of any potentially promising avenue of research. As Churchland and Sejnowski remark, "the co-evolutionary advice regarding methodological efficiency is 'let many flowers bloom'" (1992, p. 13).

The Churchlands have argued famously, though, that folk psychology is barren (P. S. Churchland, 1986, pp. 288–312; P. M. Churchland, 1989, pp. 2–11). Those arguments have provoked an entire literature in response (see Greenwood, 1991, and Christensen and Turner, 1993). I am sympathetic with the Churchlands' arguments, at least when they wield them against positions in the philosophy of mind that deny the explanatory goals and the conjectural and fallible character of folk psychology. That folk psychology offers explanations and that it is conjectural and fallible are both correct. That is just not the whole story, though.

The pivotal question for a pragmatist is whether folk psychology can contribute to the progress of our knowledge, or, better, whether folk psychology contains resources that may aid subsequent, more systematic psychological theorizing. Attribution theory, the theory of cognitive dissonance, and other proposals within social psychology employ as rich versions of the propositional attitudes as does folk psychology (Bechtel and Abrahamsen, 1993). Moreover, as Dennett (1987) has emphasized, employing the intentional stance aids theorizing about operative subsystems in sub-personal cognitive psychology.[9] These are just two fronts where *psychological* science seems to be simultaneously employing and, ever so gradually, *transforming* familiar folk psychological notions. Arguably, then, the Churchlands may have underestimated the possible contribution of the resources of folk psychology, because they have been insufficiently attentive to their role in social psychological and cognitive theorizing (McCauley, 1987, 1989). Indeed, they *sometimes* disregard the psychological altogether.[10] (See, however, note 14 below.)

I suspect that such neglect is born of insisting on a unified account of inter-theoretic relations and of entertaining images of co-evolution$_S$, in particular. The Churchlands are correct to emphasize the salient role of theory elimination in sci-entific progress, but these eliminations are *intralevel* processes and most univocally so (1) when the levels in question concern scientific pursuits as well established as neuroscience and psychology, and (2) when those levels are construed as thickly, i.e. as inclusively, as the distinction between those two sciences implies. The theo-ries and characteristic ontologies informing Stahl's account of combustion and Young's account of the propagation of light were replaced by theories (with new ontologies) that operated at the same levels of analysis and that were identified, both now *and then*, as continuations of the research traditions associated with those levels. Elimination in science is principally an intralevel process.

That is not to assert that interlevel considerations play no role. Even with levels of analysis so thickly construed, I do *not* mean to deny that scientists' decisions at lower and higher levels influence theoretical developments at a given level. Nor do I wish to deny that at that targeted level such developments can involve eliminations. Rather, the critical point is that these influences are reliably *mediated* by develop-ments in the conceptual apparatus and research practices that are associated with the research tradition of the targeted level (see Bechtel and Richardson, 1993, especially chapter 8; and Bechtel, 1996).

If it is *construed as an explanatory construct*, I agree with the Churchlands that much of folk psychology may well undergo substantial revision and, perhaps, even elimination eventually.[11] What I am suggesting, though, is:

1 that those changes will occur primarily as a result of progress within social and cognitive psychology; that is, that they will arise as the consequence of intralevel processes within the psychological level of analysis;
2 that, by virtue of the role of intentional attributions in the theories of social and cognitive psychology, this displacement will probably be quite gradual; that is, that, so far, the changes are proving evolutionary, not revolutionary;
3 that theoretical developments within those subdisciplines of psychology will mediate whatever co-evolutionary influence neuroscience has in this outcome.

Mapping the Churchlands' continuum on to the intralevel–interlevel distinction yields the arrangement in figure 23.2. It readily accommodates co-evolution$_M$ and co-evolution$_P$, but co-evolution$_S$ finds no obvious home. The point is that the inter-

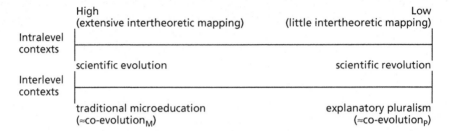

Figure 23.2 Mapping degrees of intertheoretic continuity (the Churchlands' continuum) on to intralevel and interlevel contexts.

action of psychology and neuroscience, like all co-evolutionary situations, is a case of *interlevel* relations. In short, co-evolution$_S$ embodies a category mistake. It conflates the dynamics of the co-evolution of theories at different levels of analysis with those of scientific revolutions, which are intralevel processes.[12]

What follows on this revised picture in figure 23.2 about interlevel cases that reflect substantial incommensurability? In fact, I think such cases are extremely rare, especially if the sciences in question are well established, since part of becoming a well-established science is precisely to possess theories that recognizably cohere with at least some features of theories at contiguous levels. Arguably, the distinctions between levels of analysis already presume the extreme improbability of such radical incompatibility between theories operating at adjoining *scientific* levels. (Of course, not all explanatory theories are scientific theories.) If analyses diverge in nearly all respects, then it may no longer be clear that they share a common explanandum – tempting some researchers to adopt obscurantist strategies of metaphysical extravagance.[13]

The problems surrounding co-evolution$_S$ notwithstanding, in elaborating Wimsatt's metaphor of the co-evolution of theories the Churchlands have fundamentally reinvigorated the study of change in interlevel relations over time (arguably initiated in Schaffner, 1967). As with this section, the next will say more about co-evolution$_P$ by opening with more about what it is not.

5   Exploring Explanatory Pluralism: Beyond Co-evolution$_M$

The demand in science for coherence of theories at adjacent levels of analysis is an additional motive, beyond the promise of new discoveries, for probing possible interlevel connections. The motive is to ascertain whether or not research at nearby levels coheres with and supports scientists' findings and, if it does not, to explore possible adjustments to increase the probability of such mutual support. This can, among other things, clarify respects in which the two sciences share a common explanandum.

In the long term scientists' concern for coherence among their results inevitably tends to encourage better intertheoretic mapping in interlevel settings. Forging such connections produces new discoveries in the respective sciences. One strategy, though certainly not the only one, is to advance hypothetical identities between theoretical ontologies in order to power an engine of discovery. The relationship between Mendelian genetics and biochemical genetics over the first half of the twentieth century is an especially apt illustration of two related research programs at neighboring levels of analysis aiding one another through the investigation of a series of proposals about which structures were, in fact, the genes. Scientists' two primary motives for inquiries into research at neighboring levels, then, are finally one and the same. This might seem to suggest that co-evolution$_M$ predominates; however, a number of countervailing considerations (some of which are briefly examined in this section) favor an explanatory pluralism where the sciences maintain some independence of theory, method, and practice. So, even approximate microreduction need not be inevitable.

Two issues especially distinguish co-evolution$_M$ and co-evolution$_P$. The first concerns the relative metaphysical, epistemic, and/or explanatory priority of upper- and lower-level theories in the co-evolutionary process. The second concerns the grounds offered for any disparate assignments of these priorities.

The default assumption adopted in an analysis of co-evolution$_M$ that accords with the traditional microreductionistic rationale for physicalism attributes comprehensive priority to lower levels. Classical microreduction would forecast a co-evolutionary process where the overwhelming majority of the selection pressures are exerted from the bottom up. The upper-level theory may contribute in the process of discovery, providing an initial vocabulary and problems for research, but sooner or later it must conform to the lower-level theory's expectations. Here the grounds for this priority rest not merely on the theoretical maturity and superior precision lower-level theories typically enjoy (with which pragmatism has no complaint) but also on presumptions about those theories' metaphysical pre-eminence (see note 10).

Occasionally,[14] the Churchlands seem to subscribe to a version of co-evolution$_M$ that resembles this position. For example, Churchland and Sejnowski emphasize "the importance of the single neuron models [among the various sublevels of analysis within neuroscience] as *the bedrock and fundament* into which network models *must* eventually fit" (1992, p. 13; emphasis added).[15] Although the Churchlands have avoided the traditional microreductionists' fervor about the replaceability of the reduced theory at the upper level (e.g. Churchland and Churchland, 1990, p. 256), their repeated emphasis on lower-level theories' corrections of upper-level theories also suggests that selection pressures are largely unidirectional, especially when they treat these lower-level elaborations as of a piece with corrections in intralevel contexts where substantial ontological modification *is* sometimes part of the package.

Co-evolution$_M$ will prove relevant to but a small percentage of cases, at best. On the one hand, if co-evolution$_M$ is supposed to issue in the classical microreductionist program (presumably it is not), then all of the familiar objections and caveats apply – plus at least one important additional one. The sort of tight integration with a dominant lower-level theory to which classical microreduction aspires must inevitably restrict research at the higher level. If there ever was a microreduction that conformed to all of the logical and ontological constraints imposed by the classical model, for example Causey's (1977) version, it would endow the lower level with an explanatory and metaphysical priority that would discourage all motives for theoretical novelty at the higher level. It would encourage only those paths of research at the higher level that promised to preserve its tight fit with the theory at the lower level. Its effect, in short, would be to check imaginative scientific proposals.

On the other hand, if co-evolution$_M$ is supposed to result only in the weaker analogical relation to which the Churchlands' model of approximate reduction looks, then the points of reductive contact may prove less extensive than the knitting metaphor suggests, and the microreductionist case for the explanatory, epistemic, and metaphysical priority of lower levels ends up seeming somewhat less compelling, especially once we have teased apart the differences in the "corrections" that occur in interlevel and intralevel contexts.

The case for co-evolution$_P$, however, does not turn exclusively on the problems the two competing conceptions face. Scrutiny of actual cases, including those in cognitive neuroscience to which the Churchlands have devoted particular attention, strongly suggests that the outcome of the co-evolution of theories is usually as co-evolution$_P$ describes. Instead of driving inexorably toward comprehensive theoretical and practical integration where the lower-level theory governs, scientific opportunism is usually closer to the truth in most interlevel forays. At least initially, scientists periodically monitor developments at nearby levels searching for either interlevel support, tantalizing findings, or both.

Churchland and Sejnowski's survey of proposals concerning the neural basis of working memory is a fitting illustration (1992, pp. 297–305). Not only did the concept of "working memory" emerge out of theoretical developments in experimental psychology, but so did many of the findings that guide neural modeling. For example, Churchland and Sejnowski point explicitly to the discovery of a short-term memory deficit for verbal materials in some subjects. They also highlight the ability of various interference effects both to dissociate working memory from long-term memory in normal subjects and to dissociate subsystems of working memory (linked with auditory, visuospatial, and verbal materials) from one another. These discoveries in experimental psychology provided both inspiration and direction for neural modeling. They also constitute a set of findings that any relevant neuroscientific proposal should make sense of.

On even the most exacting philosophical standards, this last consideration is *epistemically* significant. Theoretical proposals and the research they spawn at the higher level do not merely contribute to the process of discovery at the lower level. The upper-level science provides a body of *evidence* against which the science at the lower level can evaluate competing models. This evidence is particularly useful, precisely because it frequently arises *independently* of the formulation of the specific lower-level models to whose assessment it contributes. It helps to ensure the independent testability of the models in question.

It has been widely conceded that upper-level theories can play a catalytic role in the process of *discovery* at the lower level. Indeed, sometimes the conceptual resources and research techniques of a lower-level science are basically insufficient to enable practitioners even to recognize some of that level's fundamental phenomena without aid and direction from an upper-level science (Lykken et al., 1992, constitutes a particularly intriguing recent illustration). In the previous section we also saw how microreductionistic proposals to subordinate upper-level explanations to lower-level explanations risk needlessly downplaying valuable resources for dealing with the often huge computational burdens lower-level theories entail. Upper-level theorizing (e.g. in transmission genetics) contributes usefully to everyday scientific *problem solving*, even after lower-level research (e.g. in molecular genetics) indicates the microlevel story is far more complicated. Scientific endeavors at different levels regularly display what Robert Burton (1993) has called a "strategic interdependence." Now we can see that upper-level theorizing also initiates research that can contribute to lower-level developments pertaining directly to *justification*. Microreductionistic proposals to subordinate upper-level sciences to lower-level sciences, either epistemically or metaphysically, risk needless evidentiary impoverishment.

The value of this evidence turns precisely on the fact that the research arose within a context of scientific theorizing and investigation sufficiently removed and sufficiently autonomous of the lower-level research to insure an honest check. These psychological findings do not occur in isolation. They arise in the course of ongoing theorizing and research at the psychological level. Their value to neuroscience rests in part on the fact that they emanate from a tradition of psychological theorizing and experimentation that neuroscience has not dominated. This is why it is worth while for each level of analysis to maintain a measure of independence.

As Churchland and Sejnowski note, experimental psychology has a century of findings (and theorizing) from which neuroscientists and neurocomputational modelers may draw (1992, p. 27; see also p. 240). Nothing more clearly illustrates the sort of scientific opportunism explanatory pluralism envisions than one of Sejnowski and Rosenberg's papers (1988) in defense of the claim that NETtalk plausibly models operative processes in human learning and cognition. (It is a fair question at what level of analysis connectionist modeling should be located. On the criteria I identified in section 2 above, it seems to occur at a level below that of social and cognitive psychology. Churchland and Sejnowski clearly regard it as a form of neurocomputational modeling. It is worth noting that Michael Gazzaniga places Sejnowski and Rosenberg's chapter (1988) in the first half of his book, which concerns "*Neurobiologic* Considerations in Memory Function" rather than in the second half, which concerns "Psychological Dimensions of Memory Function in Humans.")

NETtalk is a connectionist system that converts English text into strings of phonemes (Sejnowski and Rosenberg, 1987). It is a three-layer, feedforward network that employs the standard backpropagation learning algorithm. On any given trial

NETtalk receives seven inputs corresponding to a window of seven letters (including punctuation or spaces between words, if they happen to arise). The desired output is the correct phoneme associated with the fourth item in the window. The three places on either side of the fourth item provide the network with information about how context affects pronunciation.

NETtalk's performance is nothing short of remarkable. It captures most of the regularities in English pronunciation and many of the irregularities as well. After 50,000 training trials with words, its accuracy with phonemes approaches 95 percent and it is virtually perfect with stresses and syllable boundaries.

The critical question for now, however, is what evidence Sejnowski and Rosenberg might cite to support the claim that NETtalk models processes that resemble those involved in human learning and cognition. A model of co-evolution as explanatory pluralism suggests that attention to the findings of experimental psychology might prove just as helpful here as attention to research on neural structure, and, in fact, not only do Sejnowski and Rosenberg look to psychology, they look to one of those century-old findings about *remembering*, the spacing effect.

The spacing effect is the finding that distributed practice with items enhances the probability of their long-term retention more than massed practice does. If occasions for rehearsal are spaced out over time, the probability is high that memory performance will exceed that from employing some small number of massed practice sessions of comparable duration at the outset. Massed repetition facilitates memory when retention intervals are extremely short. In practical terms, the spacing effect makes cramming for an exam not nearly so helpful as regular, daily preparation, whereas retention of two new telephone numbers supplied by Directory Assistance requires immediate, massed rehearsal, if they cannot be written down.

In the course of investigating the various hypotheses that psychologists have offered for explaining the spacing effect, researchers have demonstrated its robustness across a huge variety of experimental settings, materials, and tasks. Thus, Sejnowski and Rosenberg suspect that it reflects "something of central importance in memory" (1988, p. 163). Consequently, if NETtalk can be induced to exhibit the spacing effect, this is by no means trivial. It would be even more striking if its exhibition of the effect was similar in form to documented human performance.

Because of NETtalk's architecture the obvious comparison is with studies of cued recall. Sejnowski and Rosenberg chose a design after Glenberg (1976). The design called for training NETtalk up in the standard fashion, and then presenting it with the cues from a list of 20 paired associates where those cues were strings of six random letters and their associated responses were randomly generated phoneme and stress strings six characters long. (This ensured that NETtalk's performance at the test could not be a function of any information it had acquired about English pronunciation.) During both the spacing interval between training opportunities and the retention interval before the test, NETtalk was presented with English distractor words that were part of its original training corpus. Both training on the paired associates and distractor episodes included feedback via backpropagation. The order of the presentations to NETtalk in the experiment was as follows:

1   2, 10, or 20 presentations of each of the 20 paired associate cues;
2   a spacing interval of 0, 1, 4, 8, 20, or 40 distractors;
3   2, 10, or 20 re-presentations of each of the 20 paired associate cues;
4   a retention interval of 2, 8, 32, or 64 distractors;
5   a test of NETtalk's accuracy in cued recall of the 20 paired associates.

In short, NETtalk displayed the spacing effect: "A significant spacing effect was observed in NETtalk: retention of nonwords after a 64–item retention interval was significantly better when presented at the longer spacings (distributed presentation) than at the shorter spacings. In addition, a significant advantage for massed presentations was found for short-term retention of the items" (Sejnowski and Rosenberg, 1988, p. 167). Moreover, although direct comparison was impossible, NETtalk's overall response profile resembled that of Glenberg's human subjects.

The interlevel interaction here benefits both cognitive psychology and neurocomputational modeling. Sejnowski and Rosenberg briefly review the two major theoretical proposals for explaining the spacing effect in cognitive psychology, pointing out that neither the encoding variability hypothesis (e.g. Bower, 1972) nor the processing effort hypothesis (e.g. Jacoby, 1978) can account for all of the available data. They then suggest a further hypothesis focusing on the form in which information is encoded in a connectionist network, that is, on the form of the memory representation. They propose that the short-term advantage of massed practice and, particularly, the longer-term advantage of distributed practice are at least partially explicable in terms of the dynamics of connectionist nets.

Crucially, Sejnowski and Rosenberg do *not* construe their hypothesis as competing with (let alone correcting or eliminating) the two psychological proposals. (They have, after all, explored but one set of findings concerning cued recall.) Instead, they emphasize its compatibility with each. They claim correctly that it offers "a different type of explanation" at "a different level of explanation" (1988, p. 170). They explicitly discuss ways in which the notions of "encoding variability" and "processing effort" could map on to the dynamics of connectionist networks. These finer-grained accounts of these processes in terms of a network's operations suggest bases for *elaborating* the two hypotheses.

If the co-evolution of research in interlevel contexts yields the explanatory pluralism for which I have been plumping, then it is not only the lower level that offers the aid and comfort, nor is it only the higher level that receives it. As the neural modeling of working memory illustrates, here too psychological findings provide both evidentiary support and strategic guidance to lower-level modeling of brain functioning. Sejnowski and Rosenberg remark that "those aspects of the network's performance that are similar to human performance are good candidates for general properties of network models" (1988, p. 171). Their project reflects a general strategy for the testing and refinement of neurocomputational models that relies on the relative independence of work in experimental psychology. Features of particular networks that enable them to mimic aspects of the human performance that psychology documents themselves deserve mimicry in subsequent modeling of human cognition.

What is especially clear about the contribution of higher levels in this example is Sejnowski and Rosenberg's explicit acknowledgment of just how far "guidance" can go. "When NETtalk deviates from human performance, there is good reason to believe that a more detailed account of brain circuitry may be necessary" (1988, p. 172). Their comment accords nicely with the account of explanatory pluralism I have been developing. Unlike the picture of co-evolution inspired by the tradition of microreductionism, a pragmatically inspired explanatory pluralism permits no a priori presumptions about lower-level priority. Sejnowski and Rosenberg readily allow that our psychological knowledge enjoys sufficient integrity to forcefully urge further *elaboration* of analyses of brain systems formulated at lower levels.[16] This would be no less (and no more) a correction of the lower-level theory (and its ontology) than are the lower-level "corrections" of upper-level theories (and their ontologies) that the Churchlands have sometimes been wont to stress.

Such divergences, then, are not grounds for dismissal. They are, rather, opportunities for advance. The co-evolution of sciences (not just theories) at contiguous levels of analysis preserves the plurality of explanatory perspectives that the distinctions between levels imply, because leaving these research traditions to their own devices is an effective means of insuring scientific progress.

Notes

*   I wish to express my gratitude to William Bechtel and Donald Rutherford for their many helpful comments on an earlier draft of this paper.

1   The story is even more complex, since each level of analysis has both a synchronic and a diachronic moment for which separate theories have been developed. See McCauley (forthcoming). At the biological level, for example, cell biology is one of the synchronic subdisciplines focusing on the structures within the cell whereas evolutionary biology is devoted to the study of change in forms of life over time. The Churchlands have confined their discussions almost exclusively to synchronic examples.

2   One of the first, if not *the* first, is Wimsatt's (1976) classic discussion.

3   Although they concur with Churchland's judgment that the folk psychological notion of a unitary faculty of memory is probably wrong, Hirst and Gazzaniga (1988, pp. 276, 294, and 304–05) seem to adopt a far more sanguine view about the contributions of psychology (both folk and experimental) to our understanding of memory. They recognize that the fragmentation of "memory" need not lead to its elimination. (See section 5 below.)

4   . . . and the position from which they have generally (though not unequivocally) retreated over the past few years.

5   See Churchland and Sejnowski (1990, p. 229), Churchland, Koch, and Sejnowski (1990, pp. 51 and 54), and Churchland and Sejnowski (1992, pp. 10–13).

6   Consider the discussion in Neisser (1967).

7   Interestingly, Ernest Nagel's *The Structure of Science* (1961), the *locus classicus* of traditional research on reduction, implicitly recognizes the importance of distinguishing between intralevel and interlevel contexts. Nagel consistently describes intralevel cases

as involving the reduction of *theories* and interlevel cases as involving the reduction of *sciences*.

8   This is, in part, the result of the same considerations that motivate the Churchlands and Richardson's (1979) arguments that alleged reductions that conform to traditional microreductionistic standards can only be domain-specific.

9   – or in neuropsychology, as Churchland and Sejnowski's (1992, p. 282) discussion of the role of the hippocampus in short-term memory illustrates. See P. S. Churchland (1986, p. 361).

10 An interesting illustration arises in Churchland and Sejnowski's discussion of the major levels of organization in the nervous system (1992, pp. 10–11). Their diagram of the relevant levels tops out at the central nervous system with no mention of psychology. The obvious defense is to note that the diagram addresses *anatomical* structures of the nervous system only. Fair enough. What is telling, though, is a footnote (1992, p. 11, footnote 5) to this discussion. Churchland and Sejnowski concede that a more compre-hensive account would include a *social* level above the central nervous system. At least for the purposes of this discussion, they seem not even to countenance the possibility that cognitive research may capture organizational structure of explanatory significance not immediately reducible to the neurophysiological. (See, too, Sejnowski and Church-land, 1989, p. 343.)

A meta-level comment: the physicalist holds that metaphysical manifestness (which, remember, is *physical* manifestness for the physicalist) constrains what will count as *satisfactory* explanation, whereas the pragmatist proposes that explanatory success *should* constrain metaphysical commitment. If that diagnosis is correct, the ongoing negotiation in the Churchlands' work I described in section 3 is, at its root, one about competing norms.

11 I should emphasize that I am speaking of the elimination of folk psychology as an explanatory construct *within* scientific psychology. The elimination of the principles of folk physics centuries ago in physics has had little effect on its persistence among the folk.

12 The illustrations the Churchlands (1990) offer in support of their "overview of the general nature of intertheoretic reduction" (p. 249) proceed in the following order:

1   the reduction of Kepler's laws to Newton's (intralevel);
2   the reduction of the ideal gas law to the kinetic theory – emphasizing (p. 250, some emphasis added) that "this reduction involved *identifying* a familiar *phenomenal* property of common objects with a highly unfamiliar *micro*-physical property" (interlevel);
3   the reduction of classical (valence) chemistry by atomic and subatomic (quantum) physics (interlevel);
4   the reduction of Newtonian mechanics to the mechanics of Special Relativity (intralevel);
5   the elimination of phlogiston by Lavoisier's oxygen theory of combustion (intralevel).

13 But just as progress in tracing the relevant biological systems preserved the vitality of organisms without vitalism, so too is progress at tracing the relevant psychological systems slowly revealing how we can preserve the cleverness and wondrous experiences

of intelligent creatures without dualism. The interlevel influences of neuroscience will no more co-opt or eliminate psychological theorizing than the interlevel influences of chemistry co-opted or eliminated physiological theorizing.

14 As noted near the end of section 3, the Churchlands more often seem to endorse an account of co-evolution resembling co-evolution$_P$. In Churchland and Sejnowski 1990 (p. 250) and 1992 (p. 240), they not only advocate a form of explanatory pluralism, but they explicitly include the psychological sciences.

15 Conceding that it will not involve a single model nor direct explanations of higher levels in terms of events at the molecular level, Churchland and Sejnowski nonetheless aspire to a "unified account" of the nervous system, where "the integration [will] consist of a chain of theories and models that links adjacent levels" (Sejnowski and Churchland, 1989, p. 343).

16 If neurocomputational modeling of networks constitutes a higher level of analysis than does the study of particular neurons (and it certainly seems to on Churchland and Sejnowski's view – 1992, p. 11), then Churchland and Sejnowski's (1992, pp. 183–8) take on recordings of single cells' response profiles in the visual cortex is an illustration of just the sort of circumstances that the Sejnowski and Rosenberg citation allows for – one in which higher-level research impels a re-evaluation of lower-level doctrines.

    Churchland and Sejnowski (following Lehky and Sejnowski, 1988) argue that neurocomputational research on the visual system's ability to extract shapes exclusively from information about shading reveals that the conventional interpretation of the function of receptive fields of neurons in the visual cortex may well be wrong. That interpretation, which arose from single-cell studies, holds that these neurons function as edge and bar detectors. Churchland and Sejnowski maintain that this interpretation ignores the cells' projective fields. Hidden units in Lehky and Sejnowski's model developed receptive fields with similar response profiles; however, these orientations were the result of training the network on the shape from shading task. "In a trained-up network, the hidden units represent an intermediate transformation for a computational task quite different from the one that has been customarily ascribed . . . they are used to determine the shape from the shading, not to detect boundaries" (Churchland and Sejnowski, 1992, pp. 185–6).

References

Bechtel, W. (ed.) 1986: The nature of scientific integration. In *Integrating Scientific Disciplines*, The Hague: Martinus Nijhoff.

Bechtel, W. 1996: What should a connectionist philosophy of science look like? In R. N. McCauley (ed.), *The Churchlands and their Critics*, Oxford: Blackwell.

Bechtel, W., and Abrahamsen, A. A. 1993: Connectionism and the future of folk psychology. In R. G. Burton (ed.), *Natural and Artificial Minds*, Albany: SUNY Press.

Bechtel, W., and Richardson, R. C. 1993: *Discovering Complexity*. Princeton, NJ: Princeton University Press.

Bickle, J. 1992: Mental anomaly and the new mind-brain reductionism. *Philosophy of Science*, 59, 217–30.

Bower, G. H. 1972: Stimulus-sampling theory of encoding variability. In A. W. Melton and E. Martin (eds), *Coding Processes in Human Memory*, Washington: V. H. Winston.

Burton, R. G. 1993: Reduction, elimination, and strategic interdependence. In R. G. Burton (ed.), *Natural and Artificial Minds*, Albany: SUNY Press.

Causey, R. 1972: Uniform microreductions. *Synthese*, 25, 176–218.

Causey, R. 1977: *Unity of Science*. Dordrecht: Reidel.

Christensen, S. M., and Turner, D. R. (eds) 1993: *Folk Psychology and the Philosophy of Mind*. Hillsdale, NJ: Lawrence Erlbaum Associates.

Churchland, P. M. 1979: *Scientific Realism and the Plasticity of Mind*. Cambridge, UK: Cambridge University Press.

Churchland, P. M. 1989: *A Neurocomputational Perspective: The Nature of Mind and the Structure of Science*. Cambridge, MA: MIT Press.

Churchland, P. M., and Churchland, P. S. 1990: Intertheoretic reduction: A neuroscientist's field guide. *Seminars in the Neurosciences*, 2, 249–56.

Churchland, P. S. 1986: *Neurophilosophy*. Cambridge, MA: MIT Press.

Churchland, P. S., Koch, C., and Sejnowski, T. J. 1990: What is computational neuroscience? In E. L. Schwartz (ed.), *Computational Neuroscience*, Cambridge, MA: MIT Press.

Churchland, P. S., and Sejnowski, T. J. 1990: Neural representation and neural computation. In W. Lycan (ed.), *Mind and Cognition: A Reader*, Oxford: Basil Blackwell.

Churchland, P. S., and Sejnowski, T. J. 1992: *The Computational Brain*. Cambridge, MA: MIT Press.

Dennett, D. C. 1987: Three kinds of intentional psychology. In *The Intentional Stance*, Cambridge, MA: MIT Press.

Enc, B. 1983: In defense of the identity theory. *Journal of Philosophy*, 80, 279–98.

Feyerabend, P. K. 1962: Explanation, reduction, and empiricism. In H. Feigl and G. Maxwell (eds), *Minnesota Studies in the Philosophy of Science, Volume III*, Minneapolis: University of Minnesota Press.

Fodor, J. A. 1975: *The Language of Thought*. New York: Thomas Y. Crowell.

Glenberg, A. M. 1976: Monotonic and nonmonotonic lag effects in paired-associate and recognition memory paradigms. *Journal of Verbal Learning and Verbal Behavior*, 15, 1–16.

Greenwood, J. D. (ed.) 1991: *The Future of Folk Psychology*. New York: Cambridge University Press.

Hirst, W., and Gazzaniga, M. 1988: Present and future of memory research and its applications. In M. Gazzaniga (ed.), *Perspectives in Memory Research*, Cambridge, MA: MIT Press.

Hooker, C. 1981: Towards a general theory of reduction. *Dialogue* 20: 38–59, 201–36, 496–529.

Jacoby, L. L. 1978: On interpreting the effects of repetition: solving a problem versus remembering a solution. *Journal of Verbal Learning and Verbal Behavior*, 17, 649–67.

Kuhn, T. 1970: *The Structure of Scientific Revolutions*, 2nd edn. Chicago: University of Chicago Press.

Laudan, L. 1977: *Progress and its Problems*. Berkeley: University of California Press.

Lehky, S. R., and Sejnowski, T. J. 1988: Network model of shape-from-shading: Neural function arises from both receptive and projective fields. *Nature*, 333, 452–4.

Lykken, D. T., McGue, M., Tellegen, A., and Bouchard, T. J. 1992: Emergenesis: Genetic traits that may not run in families. *American Psychologist*, 47, 1565–77.

McCauley, R. N. 1981: Hypothetical identities and ontological economizing: Comments on Causey's program for the unity of science. *Philosophy of Science*, 48, 218–27.

McCauley, R. N. 1986: Intertheoretic relations and the future of psychology. *Philosophy of Science*, 53, 179–99.

McCauley, R. N. 1987: The role of cognitive explanations in psychology. *Behaviorism* (subsequently *Behavior and Philosophy*), 15, 27–40.

McCauley, R. N. 1989: Psychology in mid-stream. *Behaviorism* (subsequently *Behavior and Philosophy*), 17, 75–7.

McCauley, R. N. 1993: Brainwork: A review of Paul Churchland's *A Neurocomputational Perspective*. *Philosophical Psychology*, 6, 81–96.

McCauley, R. N. (forthcoming): Cross-scientific relations: Toward an integrated approach to the study of the emotions.

Nagel, E. 1961: *The Structure of Science*. New York: Harcourt, Brace and World.

Neisser, U. 1967: *Cognitive Psychology*. New York: Appleton-Century-Crofts.

Oppenheim, P., and Putnam, H. 1958: Unity of science as a working hypothesis. In H. Feigl, M. Scriven, and G. Maxwell (eds), *Minnesota Studies in the Philosophy of Science – Volume II*, Minneapolis: University of Minnesota Press.

Richardson, R. 1979: Functionalism and reductionism. *Philosophy of Science*, 46, 533–58.

Schaffner, K. 1967: Approaches to reduction. *Philosophy of Science*, 34, 137–47.

Sejnowski, T. J., and Churchland, P. S. 1989: Brain and cognition. In M. Posner (ed.), *Foundations of Cognitive Science*, Cambridge, MA: MIT Press.

Sejnowski, T. J., and Rosenberg, C. R. 1987: Parallel networks that learn to pronounce English text. *Complex Systems*, 1, 145–68.

Sejnowski, T. J., and Rosenberg, C. 1988: Learning and representation in connectionist Models. In M. Gazzaniga (ed.), *Perspectives in Memory Research*, Cambridge, MA: MIT Press.

Thagard, P. 1992: *Conceptual Revolutions*. Princeton, NJ: Princeton University Press.

Wimsatt, W. C. 1976: Reductionism, levels of organization, and the mind-body problem. In G. Globus, G. Maxwell, and I. Savodnik (eds), *Consciousness and the Brain*, New York: Plenum Press.

24

McCauley's Demand
for a Co-level Competitor

Paul M. Churchland and Patricia S. Churchland

McCauley's is perhaps the most straightforward of the criticisms. He sees some form of *reductive* accommodation as the relation most likely to develop between propositional-attitude psychology, on the one hand, and the underlying neurosciences, on the other. In support of this expectation, he cites the typical co-evolutionary process described by Patricia S. Churchland, wherein theories at adjacent levels gradually knit themselves into some appropriate reductive relation or other. McCauley's crucial move is then to claim that *eliminative* adjustments of theory are never (almost never?) motivated by considerations of cross-level conflict; rather, they are typically or properly motivated only by conflicts of theory at or within the *same* level of organization. In the absence of some compelling and comparably high-level alternative to folk psychology, then, we need not see folk psychology (FP) as facing any real threat of elimination. Accordingly, says McCauley, we should stand back and let the gradual interlevel knitting of theory proceed.

McCauley's portrait of FP's future may be correct. His guess is as good as ours, and a largely retentive reduction remains a live possibility. But the historical pattern he leans on is not so uniform as he suggests, and any probative classification of reality into distinct "levels" is something that is itself hostage to changeable theory. Consider the highly instructive example of astronomy.

For at least two thousand years (roughly from Aristotle to Galileo), the realm of the heavens was regarded as a distinct and wholly different level within the natural order. It was distinguished from the terrestrial, or sublunary, realm in several mutually reinforcing ways. Sheer scale was the first difference, then as now. Thanks to Aristotle, Aristarchos, and Eratosthenes, even the geocentric ancients were aware that the moon was 240,000 miles away, that the sun was at least 5,000,000 miles away, and that the planets and the stars were more distant still. Astronomical phenomena evidently took place on a spatial scale at least four or five orders of magnitude beyond the scale of any human practical experience.

Second, the laws that governed our small-scale sublunary realm had neither place among nor grip upon the obviously special superlunary objects. They moved in their (almost) perfectly circular paths, according to their own laws, in a fashion that had no parallel within the terrestrial domain. Third, the realm of the heavens was immutable and incorruptible, in contrast to our own sorry domain. Centuries may flow by, but the heavens remain unaltered. Fourth and finally, the realm of the heavens was evidently the realm of the divine, the home or doorstep of the gods.

Accordingly, even the most casual of observers could appreciate that the discipline of astronomy was attempting to grasp a level of the natural order far beyond what the Lilliputian mechanics of falling stones, taut ropes, and rolling wagons could ever hope to address. Ptolemy was explicit in rejecting the aspirations of "physics" to explain astronomical phenomena, and his voice reflected an almost universal opinion. Astronomy was an autonomous science attempting to grasp the autonomous laws appropriate to the phenomena at a dramatically distinct level of the natural order.

Further, the ancient astronomical theories actually made good on this conviction. Aristotle's account had a nest of 57 concentric earth-centered spheres, spheres made of the transparent and exclusively superlunary "fifth essence" (Plato's cosmium), each moving at the behest of its own perfectly circular, perfectly uniform *telos*. Ptolemy's different but similarly geocentric account had the familiar nest of perfectly circular deferent circles with eccentrically-placed centers, moving epicycles, and artfully placed "equant" points with which to cheat a bit on the issue of the perfect uniformity of astronomical motions.

We all know what finally happened to these ancient, "high-level" theories. They turned out to be radically false theories, so fundamentally defective that both their principles and their ontologies were eventually displaced, rather than smoothly reduced, by Newton's completed mechanics of motion (cf. the opening sentence of Churchland, 1981). Astronomy as a discipline is still with us, of course, and is more vigorous than ever, but it no longer speaks of crystalline spheres, fifth essences, moving epicycles, and phantom equants. An anisotropic, geocentric, rotating, finite spherical universe was displaced wholesale in favor of an isotropic, earth-indifferent, nonrotating, possibly infinite space. And the laws that govern the heavens turned out to be the very same laws that govern phenomena at the terrestrial level. They are the laws of Newtonian mechanics.

We present this as a presumptive counterexample to McCauley's claim that theories suffer radical displacement only at the hands of co-level competitors, and never at the hands of theories whose primary home is at a different level of scale or organization. Since the Newtonian revolution, modern astronomy has simply *become* the Physics of the Heavens. What remains, then, among the patterns of history, that would preclude modern psychology from simply *becoming* the Neuroscience of very Large and Intricate Brains? Perhaps brains differ from sea-slug ganglia only in the scale of neuronal interactions they involve.

We anticipate the reply, from McCauley, that this historical elimination of an ancient astronomical theory was not a cross-level displacement at all, but rather a displacement by a theory (Newtonian mechanics) that encompassed phenomena at the *same* dynamical level as the old theory. It is just that astronomical phenomena turned out not to be unique or special after all: they are distinguished only by their vast scale.

The reply has a point, and McCauley may succeed in pressing this interpretation upon us. But this reply entails what should have been clear anyway: that science can be profoundly wrong about what counts as a nomically distinct level of phenomena, and profoundly wrong in its estimation of which theories do and do not count as genuinely "co-level" theoretical competitors. And if McCauley accepts this point, as we think it clear he must, then he is in no position to insist that the psychology/neuroscience case must turn out differently from the astronomy/physics case. Psychological phenomena, perhaps, are distinguished only by the unusual scale of the networks that display them.

Our conclusion, then, is as follows. The claim that psychology comprehends a distinct level of phenomena comprehended by a distinct set of laws uniquely appropriate to that level *is not an assumption that our opposition can have for free.* It is part of what is at issue – *empirically* at issue – in this broad debate, and the historical fate of ancient astronomy should caution against any premature convictions in its favor.

Astronomy aside, there are other historical examples that contradict McCauley's generalization about the agents of ontological displacement.[1] Eliminative cross-level impacts on conceptual structure, both upward and downward, seem to us to be historically familiar, not rare or nonexistent. But we need not explore further examples here. Instead, let us explore directly the popular conviction that psychological phenomena really do belong to a more abstract level of analysis. If they do, would that really serve to insulate FP or other propositional-attitude theories from the threat of wholesale displacement?

Not in the least. Even if an abstract or higher-level explanatory framework were somehow essential to grasping psychological phenomena, it would remain an open question whether our current FP is the *correct* framework with which to meet this challenge. Legitimating the office need not legitimate the current office holder. This point is important because a priori there are infinitely many comparably high-level alternatives to FP; and because it is arguable that the conceptual framework of neo-connectionism is one of them.

As we sketched the fate of ancient astronomy a few paragraphs ago, it turned out that astronomical phenomena were not distinctly higher-in-level after all. But we might just as well have expressed the outcome by saying that the assembled laws of Newtonian mechanics turned out to be, when suitably articulated to fit the astronomical context, exactly the high-level theory that was needed to do the relevant high-level job. The analog of this latter stance, within psychology, will now be explored.

The claim on the table is that a psychological-level competitor for FP is already here and is already staring us in the face. It is the framework in which the occurrent *representations* are patterns of activation (or sequences of such patterns) across millions of neurons. It is the framework in which the *computations* are synapse-driven transformations of such patterns (or sequences thereof) into further such patterns across further neuronal populations. It is the framework in which such transformations are dictated by the *learned* patterns of synaptic connection strengths that connect one population of neurons with another. It is the framework, in short, of contemporary connectionist theory.

A frequent judgment about connectionist models of cognition is that they constitute at most an account of how classically conceived cognitive processes might be *implemented* in an underlying neural hardware. A quarter-century from now, we predict, this dismissal will be celebrated as one of the great head-in-the-sand episodes of twentieth-century science. Our confidence here is born not primarily of confidence in the ultimate correctness of connectionist models of cognition. (They must chance their hand to fate along with every other approach.) Rather, it is born of the recognition that the kinematics and dynamics of current connectionism already constitute an account of cognition at a decidedly abstract level. Allow us to explain.

When one sees a standard introduction to the connectionist modeling of cognitive processes, one is typically presented with a diagram of several layers of neuron-like units connected to one another by way of axon-like projections ending in synapse-like contacts (see figure 24.1, for example). One is then told about the variable nature of the weights of such contacts, about the multiplication of each axonal activation level by the synaptic weight it encounters, about the summation of all such products within the contacted neuron, and finally about the great variety of real-world discriminations such networks can be trained to make. We have given such accounts ourselves, and any audience can be forgiven for thinking that they are witness to an account of the underlying wheels and gears that might or might not *realize* the many abstract cognitive faculties that psychology presumes to study.

And so witness they are. But the real story only begins there, and strictly speaking that beginning is inessential. Neuronal details are no more essential to connectionist conceptions of cognition than vacuum-tube or transistor details are essential to the classical conception of cognition embodied in orthodox AI, Fodorean psychology, and FP itself. What is essential is the idea of fleeting high-dimensional patterns being transformed into other such patterns by virtue of their distributed interaction with an even higher-dimensional matrix of relatively stable transforming elements. The fleeting patterns constitute a creature's specific representations of important aspects of its changing environment. And the relatively stable matrix of transforming elements constitutes the creature's background knowledge of the general or chronic features of the world.

The abstract nature of this new conception of cognitive activity is revealed immediately by the fact that such activity can be physically realized in a wide variety of ways: in sundry biological wetwares, in silicon chips etched with parallel architectures, and even in a suitably programmed serial/digital machine, although this third incarnation exacts an absurdly high price in lost speed.

Its abstract or high-level nature is further revealed when we explore its kinematical and dynamical properties. Each *population* of elements (such as the neurons at

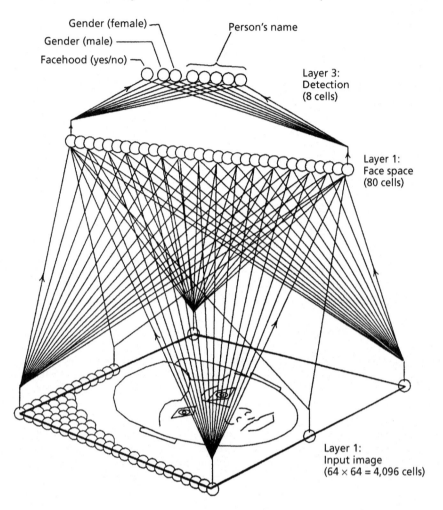

Gender (female)
Gender (male)
Facehood (yes/no)
Person's name

Layer 3:
Detection
(8 cells)

Layer 1:
Face space
(80 cells)

Layer 1:
Input image
(64 × 64 = 4,096 cells)

Figure 24.1   A feedforward network for discriminating facehood, gender, and personal identity as displayed in photographic images (adapted from Cottrell, 1991).

the retina, or at the LGN [lateral geniculate nucleus], or at the primary visual cortex, and so on) defines a high-dimensional *space* of possible activation patterns across that population, patterns that are roughly equiprobable to begin with. But their relative probabilities gradually change over time as the system learns from its ongoing experience. Learning consists in the gradual modification of the many transforming matrices through which each activation pattern must pass as it filters its way through the system's many layers. Each matrix is so modified as to make certain activation patterns at the next layer more likely and other patterns less likely. The space of possible activation patterns at each layer thus acquires an intricate internal structure in the course of training.

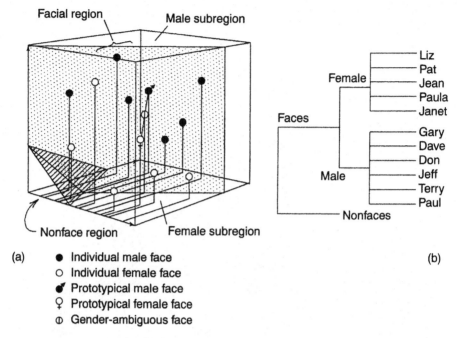

Facial region   Male subregion

Nonface region   Female subregion

(a)   ● Individual male face
       ○ Individual female face
       ⚦ Prototypical male face
       ♀ Prototypical female face
       ⦶ Gender-ambiguous face

(b)

Figure 24.2   (a) An activation state space whose three axes represent three of the 80 units at the middle layer of Cottrell's face-discrimination network. The partitions into subcategories are visible, as are the small volumes that code input images for each of the eleven individuals variously portrayed in the training set. This 3-space is frankly a cartoon, in that the distinctions displayed cannot effectively be drawn within only three of the 80 dimensions available at layer 2. If they could, the network would need only three units at that layer. In fact, only the bulk of those 80 units working together will draw all of those distinctions reliably. However, the 3-space does represent fairly the kind of partitioning that training produces, except that the true partitions are high-dimensional hypersurfaces rather than 2-D planes. (b) A dendrogram representing the same set of hierarchically organized categories. Since they make no attempt to portray partition surfaces and hypersurfaces, dendrograms are indifferent to the dimensionality of the activation space at issue.

Visual models are helpful here, and two standard display types are shown in figure 24.2. Their purpose is to illustrate the background cognitive state of the network of figure 24.1 after it has been trained to discriminate faces from nonfaces and female faces from male faces, and, within each gender, to recognize the specific faces of 11 named individuals displayed in the original set of training images (Cottrell, 1991).

The space in figure 24.2(a) represents the possible activation levels of three of the 80 units that make up layer 2. As you can see, the space has been partitioned into a hierarchy of subspaces. Nonface images (strictly, nonface activation patterns) at the input layer are transformed into activation triplets at layer 2 (strictly, they are transformed into 80-tuples, but we are here ignoring 77 of those dimensions so that we can have a coherent picture to examine), triplets that always fall into the smallish subvolume near the origin of this 3-space. Evidently, most of the dynamic range of the units at layer 2 has been given over to the representation of faces. For all face images at the input layer get transformed into triplets that fall into the much larger subvolume to the right of the small triangular partition.

Within that larger subvolume is a second partition, this time dividing the range of activation triplets that represent female faces from the range of activation triplets for male faces. Activation triplets that fall anywhere on that speckled vertical partition are the network's mature responses to input-layer face images that are highly *ambiguous* as to gender. Activation triplets within each of the 11 small volumes scattered on either side of that partition represent slightly different photographs of the 11 different individuals represented in the training set. The network has thus developed six further subcategories within the male subvolume and five subcategories within the female subvolume. The relevant partitions have been left out of figure 24.2(a) so as to avoid visual clutter, but the 11 prototypical "hot spots" within each final partition are saliently represented.

Figure 24.2(a) indicates how the regularities and variances implicit in the set of training images have come to be represented by an acquired set of structures within the activation space of layer 2. The job of the network's layer 3 is now the relatively easy one of discriminating just where, within this hierarchy of layer 2 subspaces, any fleeting activation pattern happens to fall. This it does well. Overall, Cottrell's network achieved 100 percent reliability on the (roughly 100) images in the training set, in facehood, gender, and individual identity. More importantly, its acquired perceptual skills generalized robustly to images it had never seen before. It remains 100 percent accurate on faces vs. nonfaces; it remains almost 90 percent accurate on arbitrary male and female faces; and to any novel face, it tends to apply the name of the individual among the original eleven to whom that novel face bears the closest resemblance, as judged by relevant proximity in the space of figure 24.2(a).

What we are looking at in this figure is the *conceptual space* of the trained network. (Or, rather, *one* of its conceptual spaces. The fact is, a network with many layers has many distinct conceptual spaces, one for each layer or distinct population of units. These spaces interact with each other in complex ways.) We are looking at the categorial framework with which the network apprehends its perceptual world.

Here it is important to appreciate, once more, that it is the overall activation pattern across all or most of layer 2 that is important for the network's cognitive activities. Because each element of the network contributes such a tiny amount to the overall process, no single unit is crucial and no single synapse is crucial. If any randomly chosen small subset of the units and synapses in the network is made inactive then the quality of the network's responses will be degraded slightly, but its behavioral profile will be little changed. It is the *molar-level* properties of the network – its global activation patterns and its global matrix configurations – that are decisive for reckoning the major features of its ongoing input–output behavior. A single unit is no more crucial than is a single pixel on your TV screen: its failure is unlikely even to be noticed.

Evidently, this "vector/matrix" or "pattern/transformer" conception of cognition comprehends a level of abstraction beyond any of its possible implementation-level counterparts. It is not itself an implementation-level theory. The fact is, we have long been in possession of the relevant implementation-level science: it is neuroscience. Connectionism is something else again. What connectionism brings is a new and revealing way of comprehending the molar-level behavior of cognitive creatures, a way that coheres smoothly with at least two implementational stories: the theory of biological neural networks, and the theory of massively parallel silicon architectures. If McCauley insists upon a suitably high-level competitor for FP, fate has already delivered what he deems necessary. FP is already being tested against a new and quite different conception of cognition.

Note

1   First, the rather feeble conceptual framework of early biology – sporting notions such as *telos*, *animal spirits*, *archeus*, and *essential form* – was eventually displaced by an entirely new framework of biological notions (such as *enzyme*, *vitamin*, *metabolic pathway*, and *genetic code*), notions regularly inspired by the emerging categories of structural and dynamical chemistry, a science that addressed a lower level of natural organization.

Second, the molar-level theory of classical thermodynamics, which identified heat with a macroscopic fluid substance called "caloric," was displaced by the molecular/kinetic account of statistical thermodynamics, a theory that addressed the dynamical behavior of corpuscles at a submicroscopic level.

Third, the well-established conceptual framework of geometrical optics, while a useful tool for understanding many macro-level effects, was shown to be a false model of reality when it turned out that all optical phenomena could be reduced to (i.e. reconstructed in terms of) the propagation of oscillating electromagnetic fields. In particular, it turned out that there is no such thing as a literal *light ray*. Geometrical optics had long been inadequate to diffraction, interference, and polarization effects, anyway, but it took Maxwell's much more general electromagnetic theory to retire it permanently as anything more than an occasionally convenient tool.

Fourth, the old Aristotelian/alchemical conception of physical substance (as consisting of a continuous but otherwise fairly featureless base matter that gets variously informed by sundry insubstantial spirits) was gradually displaced in the nineteenth century by Dalton's atomic/structural conception of matter. Once again, we may count this an intralevel displacement if you wish, but it is clear that most of the details of Dalton's atomism – in particular, the relative atomic weight and the valence of each elemental atom – were inspired by higher-level chemical data concerning the intricate web of constant *weight ratios* experimentally revealed in chemical combinations and dissociations. Bluntly, a maturing chemistry had an enormous and continuing impact on the shape of a still-infantile atomism. In this case, note well, it was a higher-level science that was dictating our theoretical convictions at a lower level of natural organization.

References

Churchland, P. M. 1981: Eliminative materialism and the prepositional attitudes. *Journal of Philosophy*, 78, 67–90.

Cottrell, G. 1991: Extracting features from faces using compression networks: Face, identity, emotions and gender recognition using holons. In D. Touretzky, J. Elman, T. Sejnowski, and G. Hinton, *Connectionist Models: Proceedings of the 1990 Summer School*, San Mateo, CA: Morgan Kaufmann.

Questions for Further Study and Reflection

1 What are the supposed advantages to be had in theoretical reduction generally, and why, in particular, do the Churchlands advocate the reduction of psychology to neuroscience? What are the possible disadvantages or losses that might result from reduction, and does this depend on the sort of reduction attempted?

2 How many different senses of reductionism figure in the exchange between the Churchlands and McCauley, and what are they? Are there others?

3 What are some of McCauley's arguments in favor of explanatory pluralism (co-evolution$_P$), and why does he think it preferable to both co-evolution$_M$ and co-evolution$_S$?

4 Both the Churchlands and McCauley provide historical examples of theoretical reduction from sciences other than psychology and neuroscience. Do you think that these are comparable cases of reduction, with lessons to offer for psychology and neuroscience?

5 The Churchlands suggest that the extent to which psychology is reducible to neuroscience is an empirical question. Do you agree? Might there also be logical, taxonomic, or conceptual issues at stake? If it is an empirical question, what sorts of things might count against it?

6 With respect to the Churchlands' proposed "co-level competitor," do you agree that it is at the same level as psychology, and do you think it is a genuine competitor?

7 Assume, for the moment, that an especially useful and productive element within psychological theory failed to correlate neatly with any corresponding element within neuroscience, and so proved resistant to reduction. How would you handle such a case? Conversely, if a particularly salient entity within neuroscience failed to capture anything distinctive at the psychological level, how would you proceed in this case?

Author Index

Subject Index

CPSIA information can be obtained at www.ICGtesting.com
Printed in the USA
BVOW07s0943090715

407616BV00003B/5/P